Hydrogen Sulfide and Reactive Oxygen Species, Antioxidant Defense, Abiotic Stress Tolerance Mechanisms in Plants

Hydrogen Sulfide and Reactive Oxygen Species, Antioxidant Defense, Abiotic Stress Tolerance Mechanisms in Plants

Editors

Yanjie Xie
Francisco Corpas
Jisheng Li

MDPI • Basel • Beijing • Wuhan • Barcelona • Belgrade • Manchester • Tokyo • Cluj • Tianjin

Editors
Yanjie Xie
Department of Biochemistry and Molecular Biology
Nanjing Agricultural University
Nanjing
China

Francisco Corpas
Department of Biochemistry, Cell and Molecular Biology of Plants
Consejo Superior de Investigaciones Científicas (CSIC),
Granada
Spain

Jisheng Li
College of Life Sciences
Northwest AF University
Yangling
China

Editorial Office
MDPI
St. Alban-Anlage 66
4052 Basel, Switzerland

This is a reprint of articles from the Special Issue published online in the open access journal *International Journal of Molecular Sciences* (ISSN 1422-0067) (available at: www.mdpi.com/journal/ijms/special_issues/897788).

For citation purposes, cite each article independently as indicated on the article page online and as indicated below:

LastName, A.A.; LastName, B.B.; LastName, C.C. Article Title. *Journal Name* **Year**, *Volume Number*, Page Range.

ISBN 978-3-0365-5376-4 (Hbk)
ISBN 978-3-0365-5375-7 (PDF)

Contents

International Journal of *Molecular Sciences*

MDPI

Editorial

Hydrogen Sulfide and Reactive Oxygen Species, Antioxidant Defense, Abiotic Stress Tolerance Mechanisms in Plants

Jing Zhang [1], Francisco J. Corpas [2], Jisheng Li [3] and Yanjie Xie [1,*]

1 Laboratory Center of Life Sciences, College of Life Sciences, Nanjing Agricultural University, Nanjing 210095, China
2 Department of Biochemistry, Cell and Molecular Biology of Plants, Estación Experimental del Zaidín, Consejo Superior de Investigaciones Científicas (CSIC), 18008 Granada, Spain
3 College of Life Sciences, Northwest A&F University, Yangling, Xianyang 712100, China
* Correspondence: yjxie@njau.edu.cn

Citation: Zhang, J.; Corpas, F.J.; Li, J.; Xie, Y. Hydrogen Sulfide and Reactive Oxygen Species, Antioxidant Defense, Abiotic Stress Tolerance Mechanisms in Plants. *Int. J. Mol. Sci.* **2022**, *23*, 9463. https://doi.org/10.3390/ijms23169463

Received: 1 August 2022
Accepted: 18 August 2022
Published: 21 August 2022

Various stress conditions, such as drought, salt, heavy metals, and extreme temperatures, have severe deleterious effects on plant growth and directly lead to a decline in yield and quality. The exposure of plants to such abiotic stresses leads to the uncontrolled overproduction of reactive oxygen species (ROS), which are highly toxic and can impair proteins, lipids, and nucleic acids, ultimately resulting in oxidative stress. Scientists have agreed that ROS are important signaling molecules involved in the regulation of gene expression during plant growth, development, and stress responses [1]. The healthy growth of plants in response to abiotic stress is inseparable from the joint action of many metabolic regulators, among which the behaviors of signaling molecules are universal.

Hydrogen sulfide (H_2S), which was previously considered to be toxic, has now been regarded as a burgeoning endogenous gaseous transmitter [2]. H_2S plays a vital role in the response/adaptation mechanisms to adverse environmental conditions, as well as crosstalk with other signaling molecules, including ROS, by affecting corresponding gene expression and subsequent enzyme activities. H_2S can provoke reversible oxidative posttranslational modification to the cysteine residues of proteins, called persulfidation, which affect the redox status and function of the target proteins. These concepts were the basis of the Special Issue "Hydrogen Sulfide and Reactive Oxygen Species, Antioxidant Defense, Abiotic Stress Tolerance Mechanisms in Plants" which aims to provide the most current findings on the function of signaling molecules, including H_2S and ROS, in higher plants. This Special Issue contains 5 reviews and 9 original research articles. All research articles mainly concentrate on the molecular mechanism of H_2S in plant acclimatization to abiotic stress regarding its crosstalk with other signal molecules, and each of them represents an interesting approach to this topic. The significant participation of several authors and the number of contributions testifies to the considerable interest that the topic is currently receiving in the plant science community. Here, we briefly summarize the contributions included in this Special Issue.

Chilling is the widespread environmental stress caused by drastic and rapid global climate changes, severely restricting plant growth and crop production. Cucumbers (*Cucumis sativus* L.) are typical cold-sensitive plants, and generally suffer chilling injury. Zhang et al. reported on H_2S's important role in improving cucumber chilling tolerance, in which they revealed the crosstalk among H_2S, hydrogen peroxide (H_2O_2), and auxin and its intrinsic molecular mechanism in controlling cucumber chilling tolerance [3,4]. Using a pharmacological method, they established a causal link among the different signaling molecules and suggested that H_2O_2, as a downstream signal of indole-3-acetic acid (IAA), mediates H_2S-induced chilling tolerance in cucumber seedlings [3]. Furthermore, they illustrated the molecular mechanism by which H_2S regulates chilling stress response through the regulation of auxin signaling at the transcriptional level [4]. Transcriptome analyses were able to identify a cucumber auxin response factor (ARF) gene *CsARF5*, which was differentially expressed under H_2S treatment. The overexpression of *CsARF5* enhanced the cold stress tolerance of cucumber. This is because ARF5 can directly activate the expression of the dehydration-responsive element-binding (DREB)/C-repeat binding factor (CBF) gene *CsDREB3*, thus regulating cucumber cold stress tolerance. The regulation of the cold stress response by the CsARF5-CsDREB3 module depended on H_2S, as the application of H_2S scavenger hypotaurine blocked CsARF5-mediated chilling tolerance.

H_2S also plays important role in regulating the abscisic acid (ABA) signaling pathway. ABSCISIC ACID-INSENSITIVE 4 (ABI4) is a versatile transcription factor in ABA signaling pathways. Zhou et al. suggest the molecular link between H_2S and ABI4 through the post-translational modifications of persulfidation during seed germination and early seedling development [5]. They demonstrated that H_2S-mediated persulfidation plays an important role in ABI4-controlled ABA signaling in Arabidopsis. H_2S-mediated persulfidation attenuates the ABI4 degradation during seed germination. As the ABI4 level decreased during seed germination, persulfidation-attenuated ABI4 degradation, in turn, inhibited seed germination and seedling establishment. These results indicated that H_2S has an inhibitory effect on both the germination and postgermination growth of Arabidopsis.

Another study by Zhou et al. provided insights into the molecular mechanisms underlying H_2S-conferred rice drought tolerance and demonstrated that H_2S-regulated rice drought tolerance may be achieved through the persulfidation of NIA2, which is a nitrate reductase (NR) isoform responsible for the main NR activity [6]. They showed that NR activity was decreased under drought stress, along with the increase in H_2S content. The persulfidation of NIA2 led to a decrease in total NR activity in rice. Furthermore, the drought-stress-triggered inhibition of NR activity and persulfidation of NIA2 was intensified in the rice transgenic line with the overproduction of H_2S. In agreement with these observations, mutation of *NIA2* enhanced rice drought tolerance by activating the expression of genes encoding antioxidant enzymes and ABA-responsive genes.

The critical role of H_2S was also suggested for plants counteracting manganese (Mn) stress. Hou et al. investigated the mechanism of H_2S participation and alleviation of Mn stress in Arabidopsis thaliana [7]. H_2S is involved in the alleviation of Mn-induced Arabidopsis seedling growth inhibition by reducing Mn^{2+} content, reducing ROS accumulation, and enhancing antioxidant enzyme activity. They found that the L-cysteine desulfhydrase (AtLCD) is critical for endogenous H_2S production and further regulates Arabidopsis tolerance to Mn stress.

Interestingly, another study reported on H_2S's involvement in plant response to the combined stress of multiple elements [8]. Mercury (Hg) is a toxic metal, even at low levels. Se (selenium) is a beneficial micronutrient for plant growth, which, at proper concentrations, can rescue plants from the toxicity of heavy metals, including Hg. However, Se application with excess concentration shows a synergistic toxic effect with Hg. Using a specific fluorescence probe, Yang et al. found that endogenous H_2S could be triggered as a defensive signal in response to the synergistic toxicity of Hg and Se in *Brassica rapa*. Neither Hg nor Se worked alone. The defensive role of H_2S in response to Hg and Se treatment was evaluated by the manipulation of endogenous H_2S levels, as the increase in endogenous H_2S was associated with the decrease in ROS level, followed by alleviating cell death and recovering root growth. Such findings extend our knowledge of plant H_2S in response to multiple stress conditions.

N-acetyl-5-methoxytryptamine (melatonin) was known to act as a multifunctional molecule to alleviate abiotic and biotic stresses. The review paper by Gu et al. is devoted to the effects of melatonin on the plant response to heavy metal cadmium (Cd) [9]. Melatonin activates several downstream signals, such as nitric oxide (NO), H_2O_2, and salicylic acid (SA), which are required for plant Cd tolerance. The author summarizes the progress in various physiological and molecular mechanisms regulated by melatonin in plants under Cd stress and discussed the complex interactions between melatonin and H_2S in the acquisition of Cd stress tolerance. This will be of considerable interest to researchers working in the field of heavy metals, and also to those investigating the crosstalk among signaling molecules in response to abiotic stress. They also discussed the crosstalk between melatonin and ROS in plant abiotic stress responses [10]. Melatonin often cooperates with other signaling molecules, such as ROS, NO, and H_2S. The interaction between melatonin, NO, H_2S, and ROS orchestrates the responses to abiotic stresses via signaling networks, thus conferring plant tolerance. Gu et al. summarize the roles of melatonin in establishing redox homeostasis through the antioxidant system. They also reviewed the current progress of complex interactions between melatonin, NO, H_2S, and ROS in plant responses to abiotic stresses and highlighted the vital role of respiratory burst oxidase homologs (RBOHs) during these processes.

Plenty of achievements have been announced regarding H_2S working in combination with other signal molecules to adapt to environmental changes. In the review article of Wang et al., the crosstalk and regulatory mechanisms of H_2S and other signal molecules, such as NO, ABA, calcium ion (Ca^{2+}), H_2O_2, SA, ethylene (ETH), jasmonic acid (JA), proline (Pro), and melatonin, have been summarized within the context of plant response to abiotic stresses, including maintaining cellular redox homeostasis, exchanging metal ion transport, regulating stomatal aperture, and altering gene expression and enzyme activities [11].

The review of Khan et al. not only summarizes and discusses the current understanding of the molecular mechanisms of H_2S-induced cellular adjustments and H_2S involvement in various signaling pathways in plants but emphasizes the recent progress in H_2S-mediated protein persulfidation [12]. The authors point out that more fundamental research is required to investigate the fate and regulation of endogenous H_2S production. The direct detection of endogenous H_2S and its potential emission is still a challenge in higher plants. By using an ion-selective microelectrode and a specific gas detector, Muñoz-Vargas et al. measured the endogenous content of H_2S and its emissions among different plant species of agronomical and nutritional interest, including pepper fruits, broccoli, ginger, and different members of the genus Allium, such as garlic, leek, and Welsh and purple onion [13]. These results provide a good example for the accurate quantification of endogenous H_2S production and emissions in a plant.

H_2S plays important roles in prolonging storage life and conserving the quality attributes of horticultural products. Sun et al. reported that H_2S's involvement in calcium deficiency induced the development of a bitter pit on the surface of apple peels [14]. They found that the calcium content, ROS, and H_2S production were the main differences between calcium-deficient and calcium-sufficient apple peels. Four calmodulin-like proteins (CMLs), seven AP2/ERFs, and three bHLHs transcripts were significantly differentially expressed in calcium-deficient apple peels. RT-qPCR and correlation analyses further revealed that *CML5* expression was significantly positively correlated with the expression of *ERF2/17*, *bHLH2*, and H_2S production-related genes. Therefore, the author provides a basis for studying the molecular mechanism of postharvest quality declines in calcium-deficient apples and the potential interaction between Ca^{2+} and endogenous H_2S.

More importantly, H_2S plays an important role in plant development and senescence. Hu et al. demonstrated that the overexpression of tomato *LCD1* increased the endogenous H_2S content, and delayed dark-triggered chlorophyll degradation and ROS accumulation in detached tomato leaves [15]. They found that increases in the expression of chlorophyll degradation genes *NYC1*, *PAO*, *PPH*, *SGR1*, and senescence-associated genes (*SAGs*) during senescence were attenuated by *LCD1* overexpression, whereas *lcd1* mutants showed enhanced senescence-related parameters. In the review of Li et al., the regulation of H_2S on the root system architecture (RSA) was summarized in terms of primary root growth, lateral and adventitious root formation, root hair development, and the formation of nodules [16]. The genes involved in the regulation of the RSA by H_2S, and the relationships with other signal pathways were also discussed. This review provides a comprehensive understanding of the role that H_2S plays in roots during development and under abiotic stress.

In conclusion, the negative consequences for crop growth and food production that are inevitably caused by global climate changes will be a major risk in the coming decades. This explains researchers' unprecedented interest in the fields of plant fitness, productivity, and adaptation to adverse environmental conditions. The studies in this Special Issue add valuable pieces to the puzzle regarding plant response to abiotic stress, and especially the underlying regulatory mechanisms of H_2S and its crosstalk with other signal molecules involved in this process. We are confident that they will inspire further productive research. We wish to thank all the contributors to this Special Issue and hope that it will raise interest in, and further expand our current understanding of, plant H_2S function. Nonetheless, the current Special Issue can only cover a few parts of H_2S's function in plant biology; therefore, we are pleased to release "Hydrogen Sulfide and Reactive Oxygen Species, Antioxidant Defense, Abiotic Stress Tolerance Mechanisms in Plants 2.0", to which we continue to invite our readers to contribute.

Author Contributions: J.Z., writing—original draft preparation; F.J.C., J.L. and Y.X., writing—review and editing. All authors have read and agreed to the published version of the manuscript.

Funding: This work was supported by the Jiangsu Natural Science Foundation for Distinguished Young Scholars (BK20220084 to Y.X.).

Conflicts of Interest: The authors declare no conflict of interest.

References

1. Mittler, R.; Zandalinas, S.I.; Fichman, Y.; Van Breusegem, F. Reactive Oxygen Species Signalling in Plant Stress Responses. *Nat. Rev. Mol. Cell Biol.* **2022**, 1–17. [CrossRef] [PubMed]
2. Zhang, J.; Zhou, M.J.; Zhou, H.; Zhao, D.D.; Gotor, C.; Romero, L.C.; Shen, J.; Ge, Z.L.; Zhang, Z.R.; Shen, W.B.; et al. Hydrogen sulfide (H_2S), a signaling molecule in plant stress responses. *J. Integr. Plant Biol.* **2021**, *63*, 146–160. [CrossRef] [PubMed]
3. Zhang, X.; Zhang, Y.; Xu, C.; Liu, K.; Bi, H.; Ai, X. H_2O_2 Functions as a Downstream Signal of IAA to Mediate H_2S-Induced Chilling Tolerance in Cucumber. *Int. J. Mol. Sci.* **2021**, *22*, 12910. [CrossRef] [PubMed]
4. Zhang, X.; Fu, X.; Liu, F.; Wang, Y.; Bi, H.; Ai, X. Hydrogen Sulfide Improves the Cold Stress Resistance through the CsARF5-CsDREB3 Module in Cucumber. *Int. J. Mol. Sci.* **2021**, *22*, 13229. [CrossRef] [PubMed]
5. Zhou, M.; Zhang, J.; Zhou, H.; Zhao, D.; Duan, T.; Wang, S.; Yuan, X.; Xie, Y. Hydrogen Sulfide-Linked Persulfidation Maintains Protein Stability of ABSCISIC ACID-INSENSITIVE 4 and Delays Seed Germination. *Int. J. Mol. Sci.* **2022**, *23*, 1389. [CrossRef] [PubMed]
6. Zhou, H.; Zhou, Y.; Zhang, F.; Guan, W.; Su, Y.; Yuan, X.; Xie, Y. Persulfidation of Nitrate Reductase 2 Is Involved in L-Cysteine Desulfhydrase-Regulated Rice Drought Tolerance. *Int. J. Mol. Sci.* **2021**, *22*, 12119. [CrossRef] [PubMed]
7. Hou, L.; Wang, Z.; Gong, G.; Zhu, Y.; Ye, Q.; Lu, S.; Liu, X. Hydrogen Sulfide Alleviates Manganese Stress in Arabidopsis. *Int. J. Mol. Sci.* **2022**, *23*, 5046. [CrossRef] [PubMed]
8. Yang, L.; Yang, H.; Bian, Z.; Lu, H.; Zhang, L.; Chen, J. The Defensive Role of Endogenous H_2S in *Brassica rapa* against Mercury-Selenium Combined Stress. *Int. J. Mol. Sci.* **2022**, *23*, 2854. [CrossRef]
9. Gu, Q.; Wang, C.; Xiao, Q.; Chen, Z.; Han, Y. Melatonin Confers Plant Cadmium Tolerance: An Update. *Int. J. Mol. Sci.* **2021**, *22*, 11704. [CrossRef] [PubMed]
10. Gu, Q.; Xiao, Q.; Chen, Z.; Han, Y. Crosstalk between Melatonin and Reactive Oxygen Species in Plant Abiotic Stress Responses: An Update. *Int. J. Mol. Sci.* **2022**, *23*, 5666. [CrossRef] [PubMed]
11. Wang, C.; Deng, Y.; Liu, Z.; Liao, W. Hydrogen Sulfide in Plants: Crosstalk with Other Signal Molecules in Response to Abiotic Stresses. *Int. J. Mol. Sci.* **2021**, *22*, 12068. [CrossRef] [PubMed]
12. Khan, M.S.S.; Islam, F.; Ye, Y.; Ashline, M.; Wang, D.; Zhao, B.; Fu, Z.Q.; Chen, J. The Interplay between Hydrogen Sulfide and Phytohormone Signaling Pathways under Challenging Environments. *Int. J. Mol. Sci.* **2022**, *23*, 4272. [CrossRef]
13. Muñoz-Vargas, M.A.; González-Gordo, S.; Palma, J.M.; Corpas, F.J. H_2S in Horticultural Plants: Endogenous Detection by an Electrochemical Sensor, Emission by a Gas Detector, and Its Correlation with L-Cysteine Desulfhydrase (LCD) Activity. *Int. J. Mol. Sci.* **2022**, *23*, 5648. [CrossRef] [PubMed]
14. Sun, H.-Y.; Zhang, W.-W.; Qu, H.-Y.; Gou, S.-S.; Li, L.-X.; Song, H.-H.; Yang, H.-Q.; Li, W.-J.; Zhang, H.; Hu, K.-D.; et al. Transcriptomics Reveals the ERF2-bHLH2-CML5 Module Responses to H_2S and ROS in Postharvest Calcium Deficiency Apples. *Int. J. Mol. Sci.* **2021**, *22*, 13013. [CrossRef] [PubMed]
15. Hu, K.; Peng, X.; Yao, G.; Zhou, Z.; Yang, F.; Li, W.; Zhao, Y.; Li, Y.; Han, Z.; Chen, X.; et al. Roles of a Cysteine Desulfhydrase LCD1 in Regulating Leaf Senescence in Tomato. *Int. J. Mol. Sci.* **2021**, *22*, 13078. [CrossRef] [PubMed]
16. Li, H.; Chen, H.; Chen, L.; Wang, C. The Role of Hydrogen Sulfide in Plant Roots during Development and in Response to Abiotic Stress. *Int. J. Mol. Sci.* **2022**, *23*, 1024. [CrossRef]

International Journal of
Molecular Sciences

MDPI

Article

H_2O_2 Functions as a Downstream Signal of IAA to Mediate H_2S-Induced Chilling Tolerance in Cucumber

Xiaowei Zhang, Yanyan Zhang, Chenxiao Xu, Kun Liu, Huangai Bi and Xizhen Ai *

State Key Laboratory of Crop Biology, Key Laboratory of Crop Biology and Genetic Improvement of Horticultural Crops in Huanghuai Region, College of Horticulture Science and Engineering, Shandong Agricultural University, Tai'an 271018, China; 2019010077@sdau.edu.cn (X.Z.); zyy725619@163.com (Y.Z.); xcx18853812186@163.com (C.X.); ulking1223@163.com (K.L.); bhg163@163.com (H.B.)
* Correspondence: axz@sdau.edu.cn; Tel.: +86-538-8246218

Abstract: Hydrogen sulfide (H_2S) plays a crucial role in regulating chilling tolerance. However, the role of hydrogen peroxide (H_2O_2) and auxin in H_2S-induced signal transduction in the chilling stress response of plants was unclear. In this study, 1.0 mM exogenous H_2O_2 and 75 μM indole-3-acetic acid (IAA) significantly improved the chilling tolerance of cucumber seedlings, as demonstrated by the mild plant chilling injury symptoms, lower chilling injury index (CI), electrolyte leakage (EL), and malondialdehyde content (MDA) as well as higher levels of photosynthesis and cold-responsive genes under chilling stress. IAA-induced chilling tolerance was weakened by N, N′-dimethylthiourea (DMTU, a scavenger of H_2O_2), but the polar transport inhibitor of IAA (1-naphthylphthalamic acid, NPA) did not affect H_2O_2-induced mitigation of chilling stress. IAA significantly enhanced endogenous H_2O_2 synthesis, but H_2O_2 had minimal effects on endogenous IAA content in cucumber seedlings. In addition, the H_2O_2 scavenger DMTU, inhibitor of H_2O_2 synthesis (diphenyleneiodonium chloride, DPI), and IAA polar transport inhibitor NPA reduced H_2S-induced chilling tolerance. Sodium hydrosulfide (NaHS) increased H_2O_2 and IAA levels, flavin monooxygenase (FMO) activity, and respiratory burst oxidase homolog (*RBOH1*) and FMO-like protein (*YUCCA2*) mRNA levels in cucumber seedlings. DMTU, DPI, and NPA diminished NaHS-induced H_2O_2 production, but DMTU and DPI did not affect IAA levels induced by NaHS during chilling stress. Taken together, the present data indicate that H_2O_2 as a downstream signal of IAA mediates H_2S-induced chilling tolerance in cucumber seedlings.

Keywords: chilling stress; hydrogen sulfide; hydrogen peroxide; indole-3-acetic acid; signaling pathway

Citation: Zhang, X.; Zhang, Y.; Xu, C.; Liu, K.; Bi, H.; Ai, X. H_2O_2 Functions as a Downstream Signal of IAA to Mediate H_2S-Induced Chilling Tolerance in Cucumber. *Int. J. Mol. Sci.* **2021**, *22*, 12910. https://doi.org/10.3390/ijms222312910

Academic Editors: Yanjie Xie, Francisco J. Corpas and Jisheng Li

Received: 12 October 2021
Accepted: 26 November 2021
Published: 29 November 2021

1. Introduction

Cucumbers (*Cucumis sativus* L.) are typical light-loving and cold-sensitive plants, but they are mainly cultivated in solar greenhouses in northern China. When exposed to temperatures below 10 °C, cucumber plants generally suffer chilling injury (Ai et al.) [1]. Therefore, chilling is considered as a crucial limitation to growth and yield in cucumber production. Hydrogen sulfide (H_2S) is a novel gaseous signaling molecule that plays an important role in regulating plant growth and development and defense responses to various abiotic stresses. Previous studies revealed that H_2S upregulated the expression levels of mitogen-activated protein kinase (*MAPK*) and was involved in the upregulation of *MAPK* gene expression caused by cold stress [2]. The exogenous fumigation of H_2S or application of sodium hydrosulfide (NaHS, the H_2S donor), can relieve multiple abiotic stresses, such as chilling, heat, salinity, drought, hypoxia, and heavy metal toxicity [3]. We recently found that NaHS enhances the chilling tolerance of cucumber by scavenging reactive oxygen species (ROS), increasing CO_2 assimilation, and upregulating the expression of cold-responsive genes [4]. Some signaling molecules, such as nitrogen monoxide (NO), Ca^{2+}, abscisic acid (ABA), and indol-3-acetic acid (IAA) are involved in H_2S-induced response to chilling stress in cucumber [4–7]. However, whether any other signaling molecules are involved in the process of H_2S-induced chilling tolerance, the relationship between these signaling molecules remains unclear.

Studies over the last decades have indicated that endogenous hydrogen peroxide (H_2O_2) is induced in plants after exposure to abiotic stress, such as low or high temperature, heavy metals, water stress, etc. [8–10]. H_2O_2 interacts with other plant growth regulators, such as auxins, gibberellins, cytokinins, etc. (as signaling molecules) synergistically or antagonistically, and it mediates plant growth and development and responses to abiotic stresses [10]. Pasternak et al. [11] suggested that the variation of *PINOID* gene expression triggered by H_2O_2 influenced the polar transport of auxin and might alter auxin homeostasis. The application of H_2O_2 induced the formation of adventitious roots in *Linum usitatissimum* by regulating endogenous auxin levels [12]. Zhu et al. [13] demonstrated that ethylene and H_2O_2 play an important role in triggering brassinosteroid-induced salt tolerance in tomato plants. Our recent study suggests that H_2O_2 is involved in H_2S-induced photoprotection in cucumber seedlings after exposure to chilling [14].

Auxin plays an essential role in the regulation of plant growth and development, but information about its role under chilling stress remains limited. Previous studies have revealed that changes in plant growth and development caused by cold stress are closely related to the intracellular auxin gradient, which is regulated by the polar deployment and intracellular trafficking of auxin transporters [15]. Recently, we found that IAA, a main auxin, could increase chilling tolerance by decreasing ROS accumulation, increasing the enzyme activities of photosynthesis and upregulating the expression of cold-responsive genes [4]. IAA also participates in the H_2S-mediated response to chilling stress in cucumber, and it controls the H_2O_2 in the growing part of the root [16]. Therefore, we speculate that crosstalk may exist among H_2O_2, IAA, and H_2S in response to chilling stress. To test this assumption, we investigated the effect of H_2O_2 and IAA on the ROS accumulation, photosynthesis, and relative expression of cold-responsive genes and the role of H_2O_2 and IAA in H_2S-induced chilling tolerance in cucumber seedlings.

2. Results

2.1. H_2O_2 Is Involved in H_2S-Induced Chilling Tolerance in Cucumber

To explore the effect of exogenous H_2O_2 on chilling tolerance in cucumber, we determined the maximum photochemical efficiency of PSII (F_v/F_m), actual photochemical efficiency of PSII (Φ_{PSII}), chilling injury index (CI), electrolyte leakage (EL), and photosynthetic rate (P_n) of cucumber seedlings, which were pretreated with different concentrations of H_2O_2 after exposure to 8/5 °C (day/night) for 24–72 h. As shown in Figure 1, H_2O_2 alleviated chilling injury symptoms in cucumber seedlings, and this alleviation effect was increased at low concentrations of H_2O_2 but was suppressed when the concentration exceeded 1.0 mM. The F_v/F_m, Φ_{PSII}, and P_n of H_2O_2-treated seedlings were much higher, and the CI and EL were much lower than 0 mM H_2O_2 (H_2O) treatments. These results reveal that H_2O_2 improves the chilling tolerance of cucumber seedlings, and its effect is concentration dependent. Thus, we use 1.0 mM H_2O_2 in further experiments.

Figure 1. Effect of H_2O_2 on the chilling tolerance of cucumber seedlings. (**a**) Phenotype characterization of cucumber seedlings pretreated with H_2O_2 or deionized water under chilling stress (8/5 °C) for 48 h. Deionized water-treated seedlings before chilling stress were used as the control. The experiments were repeated three times with similar results. A typical picture is shown here. (**b**) Image of F_v/F_m and Φ_{PSII} in seedlings before (control) and after chilling stress for 24 h. The false color code depicted at top of the image represents the degree of photoinhibition at PSII. (**c**) CI of seedlings before (control) and after chilling stress for 72 h. (**d**) EL of seedlings before (control) and after chilling stress for 48 h. (**e**) Pn of seedlings before (control) and after chilling stress for 24 h. Two-leaf stage cucumber seedlings were foliage sprayed with 0, 0.01, 0.1, 1.0, 2.0 and 3.0 mM H_2O_2 solution for 24 h and subsequently were exposed to 8/5 °C (day/night). The data represent the mean ± SD of three biological replicates. Different letters indicate significant differences ($p < 0.05$), according to Duncan's new multiple range test.

We previously demonstrated that 1.0 mM NaHS markedly increased endogenous H_2O_2 accumulation, and H_2S-induced H_2O_2 plays an important role in CO_2 assimilation and photoprotection in cucumber [14,17]. Consistent with previous results, we found that NaHS induced endogenous H_2O_2 production. However, both N, N′-dimethylthiourea (DMTU, a H_2O_2 scavenger) and diphenyleneiodonium chloride (DPI, a H_2O_2 synthesis inhibitor) markedly inhibited the H_2S-induced increase in H_2O_2 biosynthesis and Respiratory burst oxidase homolog (*RBOH1*) mRNA abundance in seedlings under chilling stress (Figure 2a,b). NaHS obviously decreased the CI and EL and increased F_v/F_m and Φ_{PSII}, but the NaHS-induced decrease in CI and EL or increase in F_v/F_m and Φ_{PSII} in stressed seedlings were weakened by DMTU and DPI (Figure 2c–e). Therefore, we speculate that H_2O_2 is involved in the H_2S-induced response to chilling stress.

2.2. H_2O_2 Participates in IAA-Induced Chilling Tolerance in Cucumber

Our previous study demonstrated that IAA acts as a downstream signaling molecule and is involved in H_2S-induced chilling tolerance in cucumber seedlings [4]. To explore the interactions of H_2S, IAA, and H_2O_2 in response to chilling stress, we studied the interaction between H_2O_2 and IAA in the chilling stress response in cucumber. We first measured the EL, CI, and malondialdehyde (MDA) content in cucumber seedlings pretreated with 75 μM IAA, 1.0 mM H_2O_2, 5.0 mM DMTU + 75 μM IAA, 50 μM 1-naphthylphthalamic acid (NPA, a polar transport inhibitor of IAA) + 1.0 mM H_2O_2, or deionized water, after exposure to 8/5 °C for 48–72 h. Seedlings pretreated with IAA and H_2O_2 showed remarkably lower EL, CI, and MDA content than H_2O-pretreated seedlings during chilling stress (Figure 3a–c). The decrease in EL, CI, and MDA content in IAA treatment was blocked by DMTU, but the values in H_2O_2 pretreated seedlings were not significantly affected by the IAA polar transport inhibitor NPA. The IAA- and H_2O_2 pretreated seedlings exhibited distinctly less damage caused by chilling. The effects of IAA in mitigating in chilling damage in cucumber seedlings was weakened by DMTU, but NPA had minimal effect on H_2O_2-induced remission of chilling damage (Figure 3d).

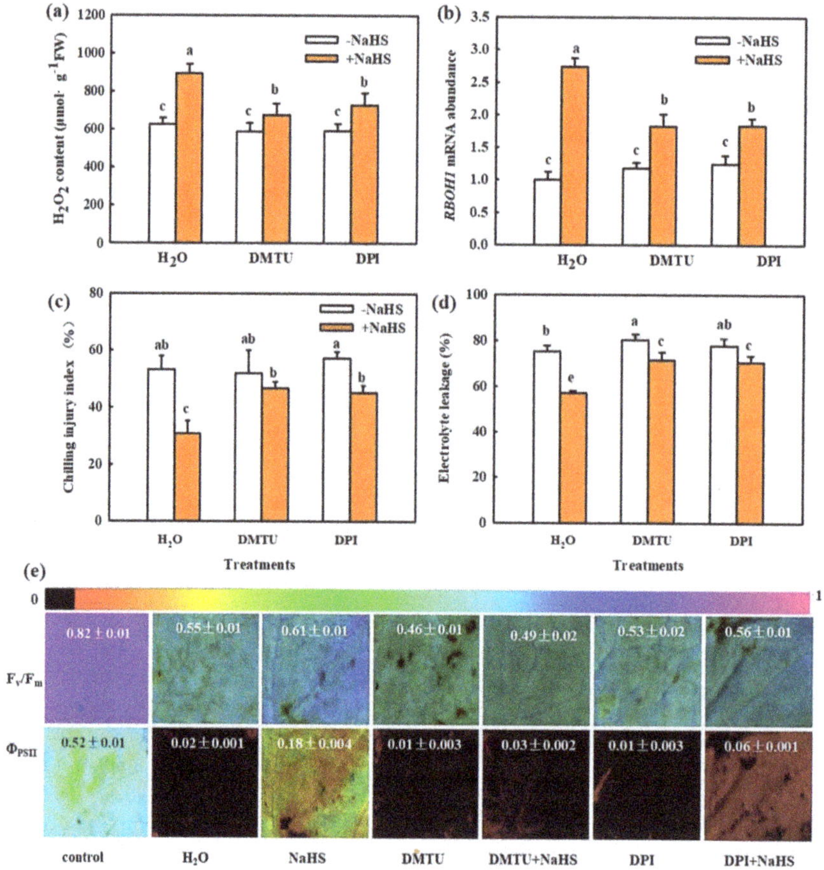

Figure 2. Effects of DMTU and DPI on H_2S-induced H_2O_2 content, *RBOH1* mRNA abundance, and chilling tolerance in cucumber. (**a**) H_2O_2 content in seedlings before (control) and after chilling stress for 9 h; (**b**) mRNA abundance of *RBOH1* in seedlings before (control) and after chilling stress for 9 h. (**c**) CI of seedlings before (control) and after chilling stress for 72 h; (**d**) EL of seedlings before (control) and after chilling stress for 48 h; (**e**) Image of F_v/F_m and Φ_{PSII} of seedlings before (control) and after chilling stress for 24 h. The false color code depicted at top of the image represents the degree of photoinhibition at PSII. Two-leaf stage cucumber seedlings were pretreated with DMTU, DPI, or deionized water and then sprayed with NaHS after 6 h. Twelve hours later, the seedlings were exposed to chilling stress. The data represent mean ± SD of three biological replicates. Different letters indicate significant differences ($p < 0.05$), according to Duncan's new multiple range test.

Figure 3. Interactive effects of IAA and H_2O_2 on the chilling tolerance of cucumber seedlings. Cucumber seedlings were pretreated with 75 μM IAA, 1.0 mM H_2O_2, 5.0 mM DMTU + 75 μM IAA, 50 μM NPA +1.0 mM H_2O_2, or deionized water (control) for 24 h and subsequently were exposed to chilling (8/5 °C, day/night). (**a**) EL of seedlings before (control) and after chilling stress for 48 h. (**b**) CI of seedlings before (control) and after chilling stress for 72 h. (**c**) MDA content of seedlings before (control) and after chilling stress for 48 h; (**d**) Phenotype characterization of different treatments before (control) and after chilling stress for 48 h. Deionized water-treated seedlings before chilling stress were used as the control. The experiments were repeated three times with similar results. A typical picture is shown here. The data represent the mean ± SD of three biological replicates. Different letters indicate significant differences ($p < 0.05$), according to Duncan's new multiple range test.

Then, we detected the interactive effects of IAA and H_2O_2 on the mRNA levels of large and small subunits (*rbcL*, *rbcS*) of ribulose 1, 5-bisphosphate carboxylase/oxygenase (rubisco) and rubisco activase (*RCA*), as well as the P_n, F_v/F_m and Φ_{PSII} under chilling stress. Both IAA and H_2O_2 treatments revealed a marked increase in mRNA levels of *rbcL*, *rbcS*, and *RCA* (Figure 4a–c), and rbcL and RCA protein levels (Figure 4e), compared with H_2O treatment ($p < 0.05$). The application of DMTU distinctly repressed IAA-induced expression of *rbcL*, *rbcS*, and *RCA*, but NPA did not inhibit the effect of H_2O_2 on *rbcL*, *rbcS*, and *RCA* mRNA levels. Chilling stress significantly reduced the P_n of cucumber seedlings. After chilling treatment for 24 h, the decrease in P_n in cucumber seedlings was 77.6%, 53.9%, 58.6%, 74.2%, and 61.8% in the H_2O, IAA, H_2O_2, DMTU + IAA, and NPA+ H_2O_2 treatments respectively, compared to the control (Figure 4d). Figure 4f shows that F_v/F_m and Φ_{PSII} were markedly higher in IAA- and H_2O_2-treated than in H_2O-treated seedlings during chilling stress. The application of DMTU significantly weakened the IAA-induced increase in F_v/F_m and Φ_{PSII}, but NPA showed a minimal influence on the H_2O_2-induced variation of F_v/F_m and Φ_{PSII}. These data suggest that IAA and H_2O_2 mitigate the negative effect of chilling stress on the photosynthetic function by upregulating the mRNA and protein levels of the key photosynthetic enzymes and activating the photoprotective mechanism.

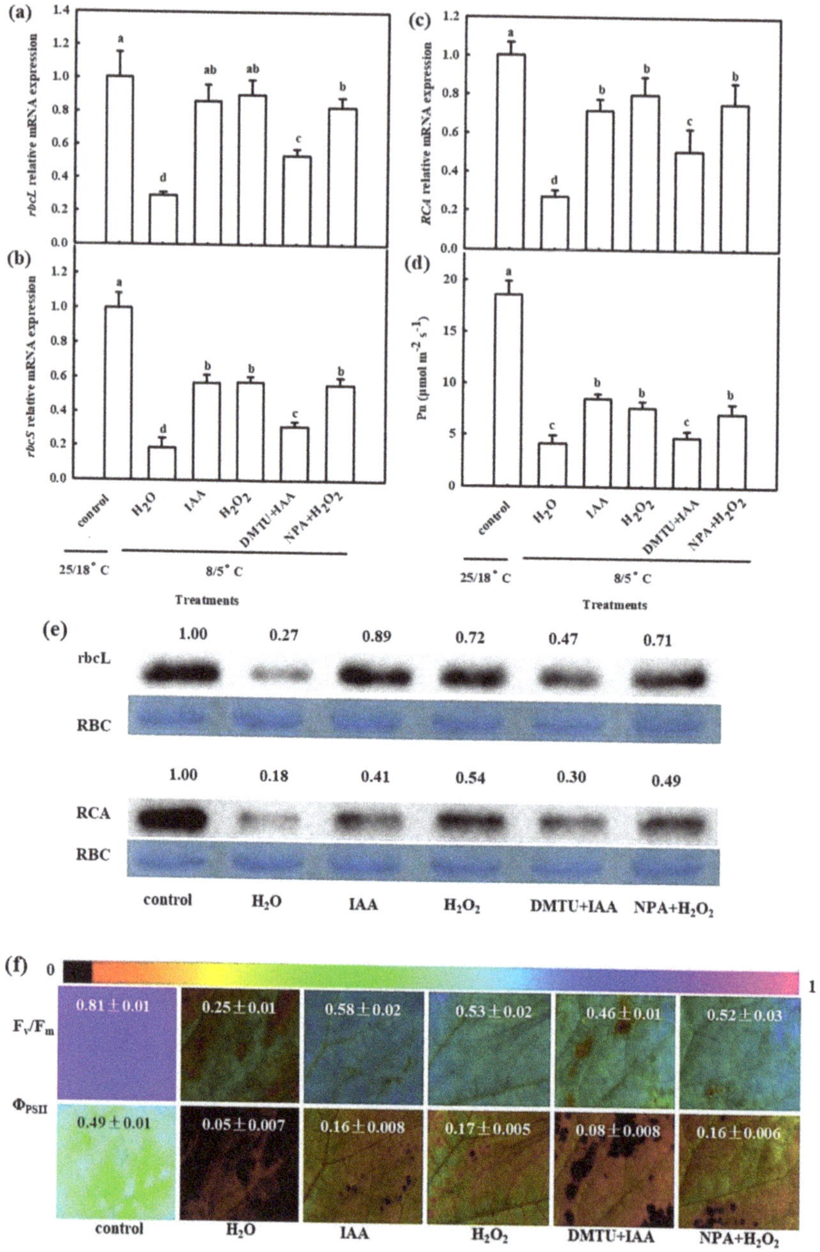

Figure 4. Interactive effects of IAA and H_2O_2 on mRNA abundances of *rbsL*, *rbcS*, and *RCA*, and protein levels of rbsL and RCA in cucumber seedlings under chilling stress. (**a–c**) mRNA abundances of *rbcL*, *rbcS*, and *RCA*; (**d**) Pn; (**e**) Protein levels of rbcL and RCA; (**f**) Image of F_v/F_m and Φ_{PSII}. The false color code depicted at top of the image represents the degree of photoinhibition at PSII. Cucumber seedlings were pretreated with 75 μM IAA, 1.0 mM H_2O_2, 5.0 mM DMTU +75 μM IAA, 50 μM NPA +1.0 mM H_2O_2, or deionized water (control) for 24 h, and subsequently exposed to 5 °C for 24 h. The data represent the mean ± SD of three biological replicates. Different letters indicate significant differences ($p < 0.05$), according to Duncan's new multiple range test.

We also analyzed the effect of IAA and H_2O_2 on the relative expression of the cold responsive genes after seedlings were exposed to chilling stress for 24 h. IAA and H_2O_2 notably increased the mRNA levels of C-repeat-binding factor (*CBF1*), inducer of *CBF* expression (*ICE1*) and cold responsive (*COR47*) genes (Figure 5a–c) as well as CBF1 protein levels (Figure 5d) in cucumber seedlings under chilling stress. The increases in the mRNA and protein levels of the cold responsive genes in IAA-treated seedlings were dramatically weakened by DMTU, whereas those in H_2O_2-treated seedlings were minimally affected by NPA.

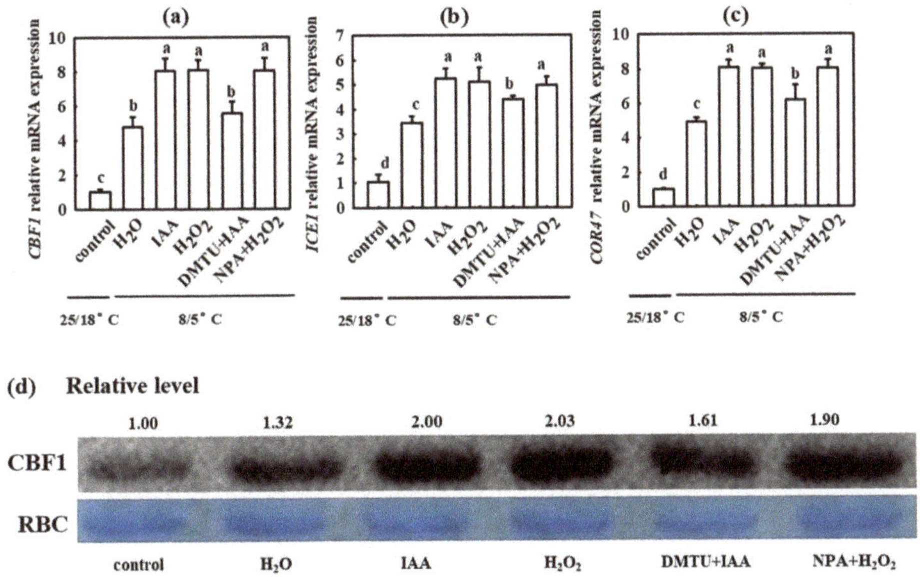

Figure 5. Interactive effects of IAA and H_2O_2 on the level of cold responsive genes in cucumber seedlings under chilling stress. (**a**–**c**) mRNA abundances of *CBF1*, *ICE1*, and *COR47*, respectively; (**d**) CBF1 protein level. Cucumber seedlings were pretreated with 75 µM IAA, 1.0 mM H_2O_2, 5.0 mM DMTU +75 µM IAA, 50 µM NPA +1.0 mM H_2O_2, or deionized water (control) for 24 h and subsequently exposed to 5 °C for 24 h. The data represent the mean ± SD of three biological replicates. Different letters indicate significant differences ($p < 0.05$), according to Duncan's new multiple range test.

We found that 75 µM IAA remarkably increased the RBOH activity (Figure 6a) and H_2O_2 content (Figure 6b) in cucumber seedlings, and the increase was remarkable after treatment for 6 h. However, no remarkable differences in flavin monooxygenase (FMO) activity and IAA content were observed between H_2O_2- and H_2O-treated seedlings (Figure 7). At normal temperature, the H_2O_2-treated seedlings showed similar mRNA expressions of *PIN1* and *AUX2* to the H_2O-treated seedlings. After 9 h or 24 h of chilling stress, *PIN1* and *AUX2* mRNA levels markedly increased in both H_2O_2 and H_2O treatment, but the extent of the increase did not vary and showed no significant differences between H_2O_2 and H_2O-treated seedlings (Supplemental Figure S1). All the above results indicate that IAA affects H_2O_2 signaling in cucumber seedlings under chilling stress. H_2O_2 might play a critical role in the IAA-induced positive response to chilling stress in cucumber seedlings.

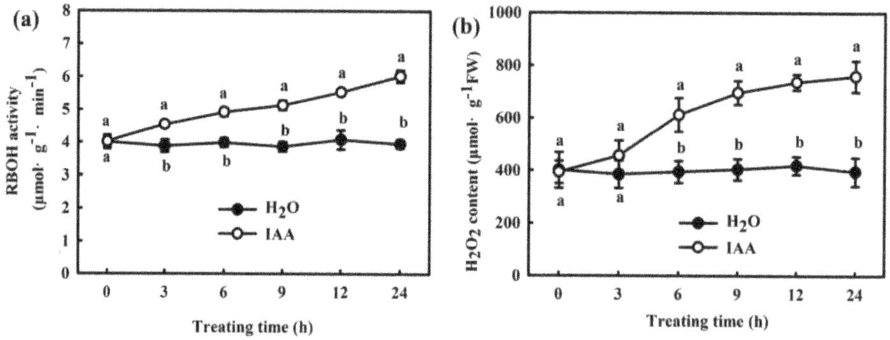

Figure 6. Effect of IAA on the RBOH activity (**a**) and H_2O_2 accumulation (**b**) in cucumber seedlings. Cucumber seedlings were foliar sprayed with 75 μM IAA or deionized water (control), and then, we measured the changes of RBOH activity and H_2O_2 content within 24 h. The data represent the mean ± SD of three biological replicates. Different letters indicate significant differences ($p < 0.05$), according to Duncan's new multiple range test.

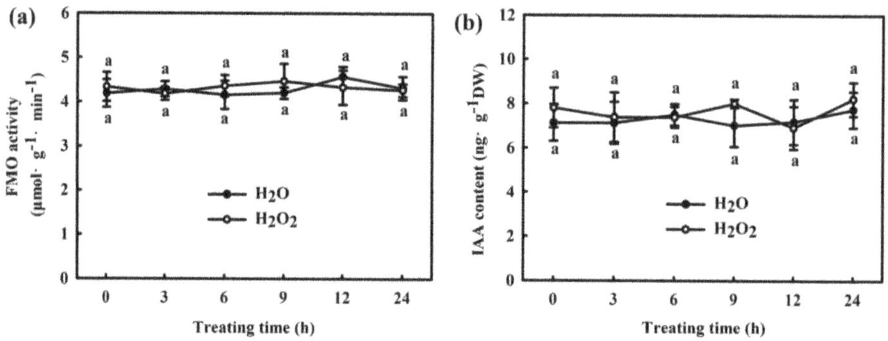

Figure 7. Effects of H_2O_2 on the FMO activity (**a**) and IAA content (**b**) in cucumber seedlings. Cucumber seedlings were foliar sprayed with 1.0 mM H_2O_2 or deionized water (control), and then, we measured the changes of FMO activity and IAA content within 24 h. The data represent the mean ± SD of three biological replicates. Different letters indicate significant differences ($p < 0.05$), according to Duncan's new multiple range test.

2.3. Interaction of IAA and H_2O_2 in H_2S-Induced Chilling Tolerance in Cucumber

To further analyze the upstream and downstream relationship between IAA and H_2O_2 in H_2S-mediated plant stress response, we determined the effect of NPA on H_2S-induced H_2O_2 production and that of the H_2O_2 scavenger DMTU and H_2O_2 synthetic inhibitor DPI on H_2S-induced IAA biosynthesis. As shown in Figure 8, 1.0 mM NaHS markedly increased *RBOH1* mRNA abundance and H_2O_2 accumulation. NPA significantly inhibited the increase in *RBOH1* mRNA abundance and H_2O_2 content induced by NaHS, suggesting that IAA is involved in H_2S-induced H_2O_2 production. NaHS also upregulated FMO-like protein *(YUCCA2)* mRNA abundance, FMO activity, and IAA levels, but DMTU and DPI had minimal effects on H_2S-induced IAA biosynthesis in cucumber leaves (Figure 9). Combining the results of Figure 2, it is further inferred that H_2O_2, a downstream component of IAA, is involved in H_2S-induced chilling tolerance in cucumber seedlings.

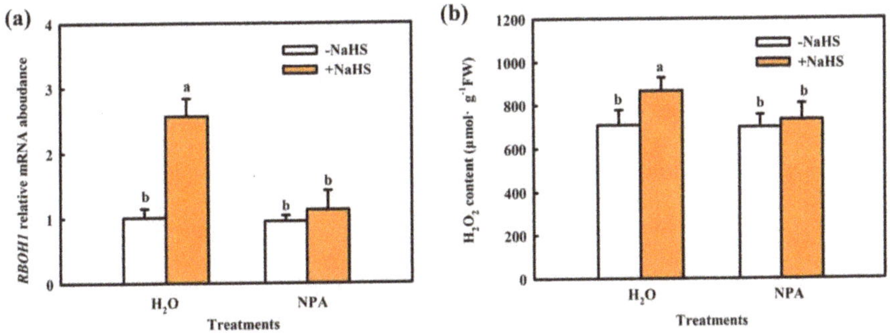

Figure 8. Effect of NPA on H_2S-induced *RBOH1* mRNA abundance (**a**) and H_2O_2 accumulation (**b**) in cucumber seedlings. Cucumber seedlings were pretreated with 50 μM NPA or deionized water and then sprayed with 1.0 mM NaHS after 6 h. Twelve hours later, the seedlings were exposed to 5 °C for 9 h. The data represent the mean ± SD of three biological replicates. Different letters indicate significant differences ($p < 0.05$), according to Duncan's new multiple range test.

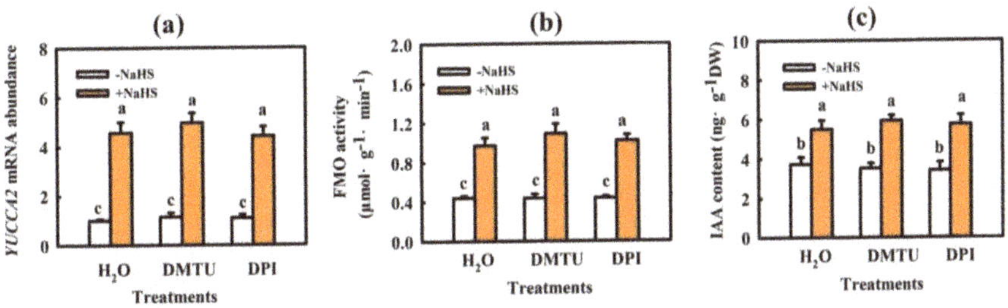

Figure 9. Effect of DMTU and DPI on H_2S-induced IAA production in cucumber seedlings. (**a**) *YUCCA2* mRNA abundance; (**b**) FMO activity; (**c**) IAA accumulation. Cucumber seedlings were pretreated with 5.0 mM DMTU, 100 μM DPI, or deionized water and then sprayed with NaHS after 6 h. Twelve hours later, the seedlings were exposed to 5 °C for 9 h. The data represent the mean ± SD of three biological replicates. Different letters indicate significant differences ($p < 0.05$), according to Duncan's new multiple range test.

3. Discussion

Chilling is a major abiotic stress that affects the growth, development, and geographical distribution of plants [18,19]. Low-temperature stress mainly affects light energy utilization and photosynthetic efficiency by destroying electron transport chains in chloroplasts and mitochondria, leading to ROS accumulation, and eventually inducing cell membrane damage in plants [20]. H_2S, as a major gaseous transmitter, plays a critical role in plant resistance to various stress conditions, such as low temperature, salt, drought, and heavy metals [21,22]. The application of exogenous H_2S can enhance chilling tolerance in *Arabidopsis thaliana* [1], hawthorns [23], and cucumbers [3]. ABA, NO, Ca^{2+}, and SA are involved in H_2S-induced resistance to abiotic stresses in plants [24–26]. Recently, we verified that H_2S interacts with NO, ABA, Ca^{2+}, IAA, and SA to enhance the chilling tolerance in cucumbers [4–7,27]. However, whether H_2O_2 and IAA exhibit synergistic effects on the H_2S-mediated plant stress response remains unclear.

Previous studies have revealed that H_2O_2 is a key molecule of signal transduction and regulates various physiological metabolic processes. For example, H_2O_2 recruited the promoter of the senescence-related transcription factor WRKY53, which in turn activated WRKY53 transcription and led to a senescence of *Arabidopsis* [28]. The H_2O_2 response gene (*HRG1/2*) could quickly respond to exogenous or endogenous H_2O_2 and further regulated *Arabidopsis* seed germination [29]. Islam et al. proved that by inducing production of the reactive carbonyl species (RCS), H_2O_2 could induce stomatal closure of guard cells in *Arabidopsis* [30]. Moreover, H_2O_2 responds to many abiotic stress of plants. Sun et al.

showed that as a signal, on the one hand, Respiratory burst oxidase homologue-dependent H_2O_2 (RBOH-H_2O_2) enhanced the heat tolerance of heat sensitive tomato. On the other hand, RBOH-H_2O_2 regulated the activities of antioxidant enzymes to control the total H_2O_2 at a level conducive to heat stress memory, which in turn maintained a lower level of total H_2O_2 during the future heat stress challenge [8]. H_2O_2 also could interact synergistically with other hormones or regulators, such as IAA, ABA, SA, MeJA, etc., mediating the response to abiotic stress in plants [10,31]. Recently, our results showed that H_2O_2 induced CO_2 assimilation and photoprotection in cucumber seedlings during chilling stress [14]. In this study, we found that H_2O_2 increased the chilling tolerance of cucumber seedlings (Figure 1), suggesting that the response of H_2O_2 to chilling stress is consistent with previous studies. NaHS significantly enhanced H_2O_2 levels, *RBOH1* mRNA abundance, and chilling tolerance in cucumber seedlings. The H_2O_2 scavenger DMTU and inhibitor of H_2O_2 synthesis DPI decreased H_2S-induced H_2O_2 accumulation and chilling tolerance (Figure 2). These results indicate that H_2O_2 may crosstalk with H_2S to improve the chilling stress response in cucumber seedlings.

Auxin is a major phytohormone that controls various aspects of plant growth and development, including cell division and elongation, tissue patterning, and the response to environmental stimuli [32,33], but knowledge about its role and interaction with other signals under chilling stress is limited. Previous investigations have indicated that chilling-induced variation in plant growth and development is closely related to the intracellular auxin gradient. Chilling stress promotes auxin biosynthesis or changes auxin gradient distribution, thus affecting the root gravity response in Arabidopsis, rice, and poplar [15,34–36]. Recently, we learned that NaHS increased endogenous IAA accumulation and improved chilling tolerance. The IAA polar transport inhibitor NPA suppressed H_2S-induced chilling tolerance. IAA reduced the negative effects of chilling stress on growth and photosynthesis, but it showed minimal effects on endogenous H_2S levels. H_2S scavengers did not influence the chilling tolerance induced by IAA [4]. Here, we observed that IAA-induced chilling tolerance was repressed by the H_2O_2 scavenger DMTU, but the IAA inhibitor NPA did not affect H_2O_2–induced tolerance to chilling stress (Figures 3–5). IAA significantly enhanced endogenous H_2O_2 synthesis, but H_2O_2 showed minimal effects on endogenous IAA level in cucumber seedlings (Figures 6 and 7). Thus, we speculate that IAA depends on the H_2O_2 signaling pathway in the regulation to chilling stress response. In addition, NPA significantly decreased H_2S-induced *RBOH1* and H_2O_2 levels (Figure 8), whereas DMTU and DPI showed no marked effect on H_2S-induced *YUCCA2* mRNA abundance, FMO activity, or IAA levels (Figure 9). These results suggest that H_2O_2 lies downstream of IAA in the regulation of H_2S to the chilling stress response.

Based on previous studies and the above results, we proposed a model of H_2O_2 and IAA regulating the H_2S-mediated chilling stress response in cucumber seedlings. Figure 10 shows that endogenous H_2S induced by chilling stress or the application of exogenous NaHS, H_2O_2, and IAA all enhanced chilling tolerance in cucumber seedlings by scavenging excessive ROS, improving photosynthetic capacity, and upregulating the mRNA and protein levels of cold responsive genes. Chilling stress-induced or exogenous H_2S promotes IAA generation, and IAA further triggers H_2O_2 accumulation and subsequently increases chilling tolerance. Thus, H_2O_2 may act as a downstream signal of IAA and play a significant role in H_2S-mediated chilling stress tolerance in cucumber seedlings. Further studies using advanced molecular techniques and mutants are required to better reveal the mechanisms and interactions of H_2S-, IAA-, and H_2O_2-induced chilling tolerance in plants.

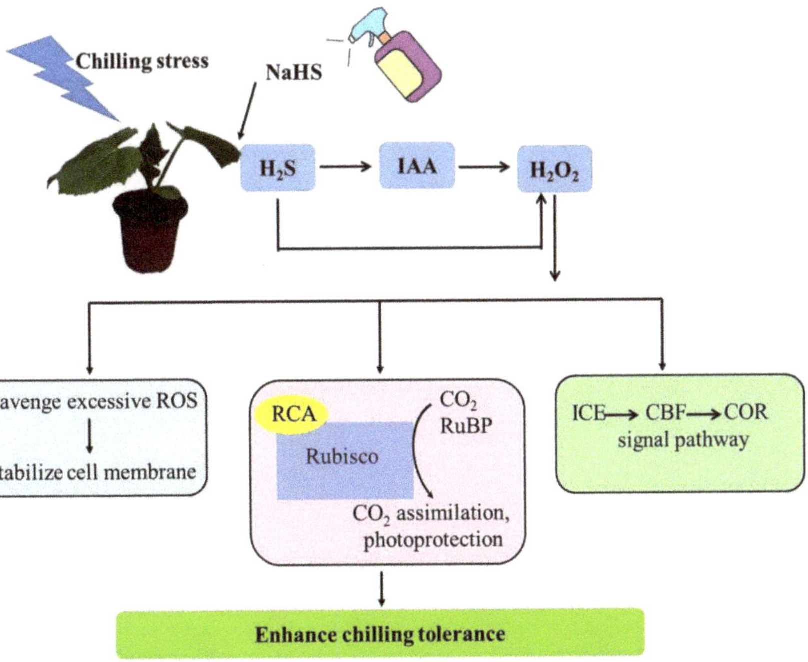

Figure 10. A proposed model for the role of IAA and H_2O_2 in H_2S-induced chilling tolerance in cucumber. Chilling induces the accumulation of H_2S in the plants. H_2S induced by chilling promotes IAA generation, triggers H_2O_2 accumulation, and subsequently increases chilling tolerance by scavenging excessive ROS, improving CO_2 assimilation and photoprotection, and upregulating the levels of cold-responsive genes.

In summary, H_2O_2 and IAA markedly improved the chilling tolerance of cucumber seedlings, as illustrated by the decrease in stress-induced CI and EL, the increase in CO_2 assimilation, and the upregulation in the level of cold-responsive genes. Even more importantly, our results first confirmed that H_2O_2 interacts with IAA signaling and is jointly involved in H_2S-induced chilling tolerance in cucumber. Moreover, H_2O_2 may act as a downstream signaling molecule of IAA, and it plays a critical role in H_2S-mediated chilling stress response in cucumber.

4. Materials and Methods

4.1. Plant Materials and Growth Condition

"Jinyou 35" cucumber (*Cucumis sativus* L.) seedlings were used in the current study. After soaking and germinating, the seeds were sown in nutrition bowls filled with seedling substrate, which consisted of peat, vermiculite, and perlite (5:3:1, v/v), and then transferred to a climate chamber with a photon flux density (PFD) of 600 μmol $m^{-2} \cdot s^{-1}$, a 26/17 °C thermoperiod, an 11 h photoperiod, and 80% relative humidity.

4.2. Experimental Design

4.2.1. Effect of H_2O_2 on the Chilling Tolerance of Cucumber Seedlings

The seedlings with two leaves were foliar sprayed with 0 (control), 0.01, 0.1, 1.0, 2.0, and 3.0 mM H_2O_2, respectively. Twenty-four hours later, the pretreated seedlings were exposed to 8/5 °C to analyze the CI, EL, P_n, F_v/F_m, and Φ_{PSII}.

4.2.2. Effect of H_2O_2 Scavenger or Inhibitor on H_2S-Induced H_2O_2 Biosynthesis and Chilling Tolerance

The seedlings were pretreated with 1.0 mM NaHS, (a H_2S donor), 5.0 mM DMTU (a H_2O_2 scavenger), 100 μM DPI (a H_2O_2 synthesis inhibitor), 5.0 mM DMTU + 1.0 mM NaHS, 100 μM DPI + 1.0 mM NaHS, or deionized water (H_2O). Twenty-four hours later, the pretreated seedlings were subjected to 8/5 °C for 9–72 h to assay the biosynthesis of H_2O_2, EL, CI, F_v/F_m, Φ_{PSII} and relative expression of cold-responsive genes. The H_2O treatment at normal temperature served as the control.

4.2.3. Interaction between IAA and H_2O_2 in Response to Chilling Stress

The seedlings were pretreated with 75 μM IAA, 1.0 mM H_2O_2, 5.0 mM DMTU + 75 μM IAA, 50 μM NPA (a polar transport inhibitor of IAA) +1.0 mM H_2O_2, or deionized water (H_2O). At 24 h after pretreatment, the seedlings were exposed to 8/5 °C to assay the P_n, fluorescence parameters, gene expression and protein expression of key photosynthesis enzymes, and relative expression of cold-responsive genes. H_2O treatment under normal conditions served as the control.

4.2.4. Effect of IAA Inhibitor on H_2S-Induced H_2O_2 Biosynthesis in Cucumber Seedlings

The seedlings were pretreated with 1.0 mM NaHS, 50 μM NPA, 50 μM NPA + 1.0 mM NaHS, or deionized water (H_2O). Twenty-four hours later, the pretreated seedlings were subjected to 5 °C for 9 h to assay the relative expression of *RBOH1* and H_2O_2 content.

4.2.5. Effect of Scavengers and Synthetic Inhibitors of H_2O_2 on H_2S-Induced IAA Biosynthesis in Cucumber Seedlings

The seedlings were pretreated with 1.0 mM NaHS, 5.0 mM DMTU, 100 μM DPI, 5.0 mM DMTU + 1.0 mM NaHS, 100 μM DPI + 1.0 mM NaHS, or deionized water (H_2O). Twenty-four hours later, the pretreated seedlings were subjected to 5 °C for 9 h to assay the FMO activity, *YUCCA2* mRNA abundance, and IAA content.

4.3. CI, EL, and MDA Measurements

The chilling stressed cucumber seedlings were graded based on the Semeniuk et al. [37] standard, and the CI was calculated according to the following formulas: CI = Σ (plants of different grade × grade)/[total plants × 5 (the maximum grade)].

EL was measured as described by Dong et al. [38]. Leaf discs (0.2 g) were incubated at 25 °C in 20 mL deionized water for 3 h, and the electrical conductivity (EC1) was estimated using a conductivity meter (DDB-303A, Shanghai, China). Afterwards, the leaf discs were boiled for 30 min and then cooled to detect EC2. EL was calculated according to the following formula: EL= EC1/EC2 × 100.

MDA content was determined using the thiobarbituric acid (TBA) colorimetric method as described by Heath and Packer [39].

4.4. Detection of P_n and Chlorophyll Fluorescence

The P_n was determined using a portable photosynthetic system (Ciras-3, PP-systems International, Hitchin, Hertfordshire, UK). Constant PFD (600 $\mu mol{\cdot}m^{-2}{\cdot}s^{-1}$), CO_2 concentration (380 $mg{\cdot}L^{-1}$), and leaf temperature (25 ± 1 °C) were maintained during the assessment. F_v/F_m was measured after seedlings were dark-adapted for 45 min, and the Φ_{PSII} was determined after leaves were light-adapted for 30 min using a portable pulse-modulated fluorometer (FMS-2, Hansatech, King's Lynn, Norfolk, UK). The chlorophyll fluorescence parameters were calculated according to Demmig-Adams and Adams [40] and Maxwell [41] as follows: $F_v/F_m = (F_m - F_0)/F_m'$; $\Phi_{PSII} = (F_m' - F_s)/F_m'$. Chlorophyll fluorescence imaging was visualized using a chlorophyll fluorescence imaging system (Imaging PAM, Walz, Wurzburg, Germany) with a computer-operated PAM-control system [42].

4.5. Detection of H_2O_2 Content and RBOH Activity

H_2O_2 content was determined using an H_2O_2 kit (Nanjing Jiancheng Bioengineering Institute, Nanjing, China). RBOH activity was detected with an ELISA kit (Jiangsu Meimian Industrial Co. Ltd., Yancheng, China) according to the instructions.

4.6. IAA Content and FMO Activity Assay

IAA content was determined by the method of Zhang et al. [4]. In brief, 0.3 g sample ground with liquid nitrogen and extracted thrice with 80% methanol (containing 30 $\mu g \cdot mL^{-1}$ sodium diethyldithiocarbamate). Samples were centrifuged (7155 g, 10 min, 4 °C) to obtain the supernatant by rotary evaporation (Shanghai EYELA, N-1210B, Shanghai, China) at 38 °C. The residue was washed with 5 ml of PBS (pH = 8, 0.2 M) and 4 mL of trichloromethane, shaken for 20 min, and allowed to stand for 30 min to remove pigment present in the trichloromethane. The resulting residues were added to 0.15 g of polyvinylpolypyrrolidone (PVPP) to remove phenols. Then, the samples were centrifuged at 7155 g for 10 min, and the resulting supernatant was re-extracted with ethyl acetate thrice and dried with a rotary evaporator in vacuo at 36 °C. The dried material was dissolved in 1.0 mL mobile phases (methanol: 0.04% acetic acid = 45:55, v/v), and the filtrate was used for HPLC–MS (Thermo Fisher Scientific, TSQ Quantum Access, San Jose, CA, USA) analysis followed by the method of Zhang et al. [4].

Flavin monooxygenase (FMO) activity was estimated using an ELISA kit (Jiangsu Meimian Industrial Co. Ltd., Yancheng, China). In brief, the FMO1 antibody was conjugated with standard, sample, and horseradish peroxidase (HRP)-labeled detection antibody and incubated, aspirated, and washed. Then, chromogen solution was added, and the reaction was terminated with sulfuric acid solution. The absorbance was detected at 450 nm with a microplate reader, and FMO activity was calculated using the standard curve [4].

4.7. Quantitative Real-Time PCR Analysis

Total RNA was extracted from cucumber leaves using an RNA extraction kit (TRIzol; Tiangen, Beijing, China). The isolated RNA was reverse transcribed with the PrimeScript® RT Master Mix Perfect Real Time (TaKaRa, Dalian, China). qRT-PCR was performed using the TransStart® TipTop Green q-PCR SuperMix (Cwbio, Beijing, China). The relative expression levels were standardized to those of cucumber β-actin gene (Solyc11g005330). The qRT-PCR primers are shown in Supplemental Table S1.

4.8. SDS-PAGE and Immunoblot Analysis

The extracted total protein of samples was separated using a 10% SDS-PAGE gel, and the resulting proteins were transferred to polyvinylidene difluoride (PVDF) membranes. The PVDF membrane was blocked for 2 h with 5% (w/w) skimmed milk and then incubated with the primary antibody followed by a horseradish peroxidase-conjugated anti-rabbit IgG antibody (ComWin Biotech Co., Ltd., Beijing, China) for 2 h. Finally, the immunoreaction was tested using the eECL Western blot Kit (CW00495, ComWin Biotech Co., Ltd., Beijing, China) and the ChemiDoc™ XRS imaging system (Bio-Rad Laboratories, Inc., Hercules, CA, USA). Primary antibodies against RbcL and RCA (ATCG00490, AT2G39730) were obtained from PhytoAB Co. Ltd. (San Francisco, CA, USA), and the CBF1 antibody was obtained from GenScript Co., Ltd. (Nanjing, China).

4.9. Statistical Analysis

The whole experiment was performed in triplicate, and the results shown are the mean ± standard deviation (SD). Statistical analysis was performed with DPS software, and the comparison of treatments was based on the analysis of variance by Duncan's multiple range test (DMRT) at a significance level of 5% ($p < 0.05$).

Supplementary Materials: The following are available online at https://www.mdpi.com/article/10.3390/ijms222312910/s1.

Author Contributions: X.A. and X.Z. designed the experiment and wrote the paper. X.Z. performed the research and analyzed the data. Y.Z., C.X., K.L. and H.B. worked together with X.Z. to accomplish the experiment. All authors have read and agreed to the published version of the manuscript.

Funding: This research was funded by the National Science Foundation of China (31572170); the Major Science and Technology Innovation of Shandong Province in China (2019JZZY010715); the Special Fund of Modern Agriculture Industrial Technology System of Shandong Province in China (SDAIT-05-10); and the Funds of Shandong "Double Tops" Program (SYL2017YSTD06).

Institutional Review Board Statement: Not applicable.

Informed Consent Statement: Not applicable.

Data Availability Statement: The original contributions presented in the study are included in the article, further inquiries can be directed to the corresponding author.

Conflicts of Interest: The authors agreed with the final manuscript and have no conflicts of interest.

References

1. Goldstein, I.; Chastre, J.; Rouby, J.-J. Novel and Innovative Strategies to Treat Ventilator-Associated Pneumonia: Optimizing the Duration of Therapy and Nebulizing Antimicrobial Agents. *Semin. Respir. Crit. Care Med.* **2006**, *27*, 082–091. [CrossRef]
2. Du, X.; Jin, Z.; Liu, D.; Yang, G.; Pei, Y. Hydrogen sulfide alleviates the cold stress through MPK4 in Arabidopsis thaliana. *Plant Physiol. Biochem.* **2017**, *120*, 112–119. [CrossRef] [PubMed]
3. Banerjee, A.; Tripathi, D.K.; Roychoudhury, A. Hydrogen sulphide trapeze: Environmental stress amelioration and phytohormone crosstalk. *Plant Physiol. Biochem.* **2018**, *132*, 46–53. [CrossRef]
4. Zhang, X.-W.; Liu, F.-J.; Zhai, J.; Li, F.-D.; Bi, H.-G.; Ai, X.-Z. Auxin acts as a downstream signaling molecule involved in hydrogen sulfide-induced chilling tolerance in cucumber. *Planta* **2020**, *251*, 69. [CrossRef] [PubMed]
5. Wu, G.X.; Cai, B.B.; Zhou, C.F.; Li, D.D.; Bi, H.G.; Ai, X.Z. Hydrogen sulfide-induced chilling tolerance of cucumber and involvement of nitric oxide. *J. Plant Biol. Res.* **2016**, *5*, 58–69.
6. Wu, G.X.; Li, D.D.; Sun, C.C.; Sun, S.N.; Liu, F.J.; Bi, H.G.; Ai, X.Z. Hydrogen sulfide interacts with Ca^{2+} to enhance chilling tolerance of cucumber seedlings. *Chin. J. Biochem. Mol. Biol.* **2017**, *33*, 1037–1046. [CrossRef]
7. Li, D.D.; Zhang, X.W.; Liu, F.J.; Pan, D.Y.; Ai, X.Z. Hydrogen sulfide interacting with abscisic acid counteracts oxidative damages against chilling stress in cucumber seedlings. *Acta Hortic. Sin.* **2018**, *45*, 2395–2406. [CrossRef]
8. Sun, M.; Jiang, F.; Zhou, R.; Wen, J.; Cui, S.; Wang, W.; Wu, Z. Respiratory burst oxidase homologue-dependent H2O2 is essential during heat stress memory in heat sensitive tomato. *Sci. Hortic.* **2019**, *258*, 108777. [CrossRef]
9. Nobakht, P.; Ebadi, A.; Parmoon, G.; Bahrami, R.N. Study role H2O2 in Photosynthetic pigments Peppermint (Mentha piperita L) on Water stress conditions. *J. Plant Proc. Func.* **2019**, *7*, 19–30.
10. Nazir, F.; Fariduddin, Q.; Alam Khan, T. Hydrogen peroxide as a signalling molecule in plants and its crosstalk with other plant growth regulators under heavy metal stress. *Chemosphere* **2020**, *252*, 126486. [CrossRef] [PubMed]
11. Pasternak, T.; Potters, G.; Caubergs, R.; Jansen, M. Complementary interactions between oxidative stress and auxins control plant growth responses at plant, organ, and cellular level. *J. Exp. Bot.* **2005**, *56*, 1991–2001. [CrossRef]
12. Takáč, T.; Obert, B.; Rolčík, J.; Šamaj, J. Improvement of adventitious root formation in flax using hydrogen peroxide. *New Biotechnol.* **2016**, *33*, 728–734. [CrossRef] [PubMed]
13. Zhu, T.; Deng, X.; Zhou, X.; Zhu, L.; Zou, L.; Li, P.; Zhang, D.; Lin, H. Ethylene and hydrogen peroxide are involved in brassinosteroid-induced salt tolerance in tomato. *Sci. Rep.* **2016**, *6*, 35392. [CrossRef]
14. Liu, F.; Fu, X.; Wu, G.; Feng, Y.; Li, F.; Bi, H.; Ai, X. Hydrogen peroxide is involved in hydrogen sulfide-induced carbon assimilation and photoprotection in cucumber seedlings. *Environ. Exp. Bot.* **2020**, *175*, 104052. [CrossRef]
15. Rahman, A. Auxin: A regulator of cold stress response. *Physiol. Plant* **2013**, *147*, 28–35. [CrossRef]
16. Ivanchenko, M.G.; Os, D.D.; Monshausen, G.B.; Dubrovsky, J.G.; Bednarova, A.; Krishnan, N. Auxin increases the hydrogen peroxide (H2O2) concentration in tomato (*Solanum lycopersicum*) root tips while inhibiting root growth. *Ann. Bot.* **2013**, *112*, 1107–1116. [CrossRef]
17. Liu, F.; Zhang, X.; Cai, B.; Pan, D.; Fu, X.; Bi, H.; Ai, X. Physiological response and transcription profiling analysis reveal the role of glutathione in H2S-induced chilling stress tolerance of cucumber seedlings. *Plant Sci.* **2020**, *291*, 110363. [CrossRef]
18. Zhou, J.; Wang, J.; Shi, K.; Xia, X.J.; Zhou, Y.H.; Yu, J.Q. Hydrogen peroxide is involved in the cold acclimation-induced chilling tolerance of tomato plants. *Plant Physiol. Biochem.* **2012**, *60*, 141–149. [CrossRef] [PubMed]
19. Ma, X.; Chen, C.; Yang, M.; Dong, X.; Lv, W.; Meng, Q. Cold-regulated protein (SlCOR413IM1) confers chilling stress tolerance in tomato plants. *Plant Physiol. Biochem.* **2018**, *124*, 29–39. [CrossRef]
20. Fan, J.; Hu, Z.; Xie, Y.; Chan, Z.; Chen, K.; Amombo, E.; Chen, L.; Fu, J. Alleviation of cold damage to photosystem II and metabolisms by melatonin in Bermudagrass. *Front. Plant Sci.* **2015**, *6*, 925. [CrossRef]

21. Huang, D.; Huo, J.; Liao, W. Hydrogen sulfide: Roles in plant abiotic stress response and crosstalk with other signals. *Plant Sci.* **2021**, *302*, 110733. [CrossRef]
22. Arif, Y.; Hayat, S.; Yusuf, M.; Bajguz, A. Hydrogen sulfide: A versatile gaseous molecule in plants. *Plant Physiol. Biochem.* **2021**, *158*, 372–384. [CrossRef] [PubMed]
23. Aghdam, M.S.; Mahmoudi, R.; Razavi, F.; Rabiei, V.; Soleimani, A. Hydrogen sulfide treatment confers chilling tolerance in hawthorn fruit during cold storage by triggering endogenous H 2 S accumulation, enhancing antioxidant enzymes activity and promoting phenols accumulation. *Sci. Hortic.* **2018**, *238*, 264–271. [CrossRef]
24. Khan, M.N.; Siddiqui, M.H.; AlSolami, M.A.; Alamri, S.; Hu, Y.; Ali, H.M.; Al-Amri, A.A.; Alsubaie, Q.D.; Al-Munqedhi, B.M.; Al-Ghamdi, A. Crosstalk of hydrogen sulfide and nitric oxide requires calcium to mitigate impaired photosynthesis under cadmium stress by activating defense mechanisms in Vigna radiata. *Plant Physiol. Biochem.* **2020**, *156*, 278–290. [CrossRef] [PubMed]
25. Kour, J.; Khanna, K.; Sharma, P.; Singh, A.D.; Sharma, I.; Arora, P.; Kumar, P.; Devi, K.; Ibrahim, M.; Ohri, P.; et al. Hydrogen sulfide and phytohormones crosstalk in plant defense against abiotic stress. In *Hydrogen Sulfide in Plant Biology*; Singh, S., Singh, V.P., Tripathi, D.K., Prasad, S.M., Chauhan, D.K., Dubey, N.K., Eds.; Elsevier: Amsterdam, The Netherlands, 2021; pp. 267–302. [CrossRef]
26. Chen, J.; Zhou, H.; Xie, Y. SnRK2.6 phosphorylation/persulfidation: Where ABA and H2S signaling meet. *Trends Plant Sci.* **2021**, *26*, 1207–1209. [CrossRef] [PubMed]
27. Pan, D.-Y.; Fu, X.; Zhang, X.-W.; Liu, F.-J.; Bi, H.-G.; Ai, X.-Z. Hydrogen sulfide is required for salicylic acid–induced chilling tolerance of cucumber seedlings. *Protoplasma* **2020**, *257*, 1543–1557. [CrossRef] [PubMed]
28. Lin, W.; Huang, D.; Shi, X.; Deng, B.; Ren, Y.; Lin, W.; Miao, Y. H_2O_2 as a Feedback Signal on Dual-Located WHIRLY1 Associates with Leaf Senescence in Arabidopsis. *Cells* **2019**, *8*, 1585. [CrossRef] [PubMed]
29. Gong, F.; Yao, Z.; Liu, Y.; Sun, M.; Peng, X. H2O2 response gene 1/2 are novel sensors or responders of H2O2 and involve in maintaining embryonic root meristem activity in Arabidopsis thaliana. *Plant Sci.* **2021**, *310*, 110981. [CrossRef]
30. Islam, M.; Ye, W.; Matsushima, D.; Rhaman, M.S.; Munemasa, S.; Okuma, E.; Nakamura, Y.; Biswas, S.; Mano, J.; Murata, Y. Reactive Carbonyl Species Function as Signal Mediators Downstream of H_2O_2 Production and Regulate [Ca^{2+}]cyt Elevation in ABA Signal Pathway in Arabidopsis Guard Cells. *Plant Cell Physiol.* **2019**, *60*, 1146–1159. [CrossRef] [PubMed]
31. Li, H.; Guo, Y.; Lan, Z.; Xu, K.; Chang, J.; Ahammed, G.J.; Ma, J.; Wei, C.; Zhang, X. Methyl jasmonate mediates melatonin-induced cold tolerance of grafted watermelon plants. *Hortic. Res.* **2021**, *8*, 57. [CrossRef]
32. Brumos, J.; Robles, L.M.; Yun, J.; Vu, T.C.; Jackson, S.; Alonso, J.; Stepanova, A.N. Local Auxin Biosynthesis Is a Key Regulator of Plant Development. *Dev. Cell* **2018**, *47*, 306–318.e5. [CrossRef] [PubMed]
33. Lv, B.; Yan, Z.; Tian, H.; Zhang, X.; Ding, Z. Local Auxin Biosynthesis Mediates Plant Growth and Development. *Trends Plant Sci.* **2019**, *24*, 6–9. [CrossRef] [PubMed]
34. Shibasaki, K.; Uemura, M.; Tsurumi, S.; Rahman, A. Auxin Response inArabidopsisunder Cold Stress: Underlying Molecular Mechanisms. *Plant Cell* **2009**, *21*, 3823–3838. [CrossRef] [PubMed]
35. Popko, J.; Hänsch, R.; Mendel, R.-R.; Polle, A.; Teichmann, T. The role of abscisic acid and auxin in the response of poplar to abiotic stress. *Plant Biol.* **2010**, *12*, 242–258. [CrossRef] [PubMed]
36. Du, H.; Liu, H.; Xiong, L. Endogenous auxin and jasmonic acid levels are differentially modulated by abiotic stresses in rice. *Front. Plant Sci.* **2013**, *4*, 397. [CrossRef]
37. Chen, B.; Wu, J. Coupling and Decoupling between Range and Angle. *Synth. Impulse Aperture Radar (SIAR)* **2014**, *111*, 241–257. [CrossRef]
38. Dong, X.; Bi, H.; Wu, G.; Ai, X. Drought-induced chilling tolerance in cucumber involves membrane stabilisation improved by antioxidant system. *Int. J. Plant Prod.* **2013**, *7*, 67–80.
39. Heath, R.L.; Packer, L.J. Photoperoxidation in isolated chloroplasts. i. kinetics and stoichiometry of fatty acid peroxidation. *Arch. Biochem. Biophys.* **1968**, *125*, 189–198. [CrossRef]
40. Demmig-Adams, B.; Adams, W.W. Xanthophyll cycle and light stress in nature: Uniform response to excess direct sunlight among higher plant species. *Planta* **1996**, *198*, 460–470. [CrossRef]
41. Maxwell, K.; Johnson, G.N. Chlorophyll fluorescence—A practical guide. *J. Exp. Bot.* **2000**, *51*, 659–668. [CrossRef]
42. Tian, Y.; Ungerer, P.; Zhang, H.; Ruban, A.V. Direct impact of the sustained decline in the photosystem II efficiency upon plant productivity at different developmental stages. *J. Plant Physiol.* **2017**, *212*, 45–53. [CrossRef] [PubMed]

International Journal of
Molecular Sciences

MDPI

Article

Hydrogen Sulfide Improves the Cold Stress Resistance through the CsARF5-CsDREB3 Module in Cucumber

Xiaowei Zhang, Xin Fu, Fengjiao Liu, Yanan Wang, Huangai Bi and Xizhen Ai *

State Key Laboratory of Crop Biology, Key Laboratory of Crop Biology and Genetic Improvement of Horticultural Crops in Huanghuai Region, College of Horticulture Science and Engineering, Shandong Agricultural University, Tai'an 271018, China; 2019010077@sdau.edu.cn (X.Z.); 15621321275@163.com (X.F.); lfjsdnd@126.com (F.L.); 18864805562@163.com (Y.W.); bhg163@163.com (H.B.)
* Correspondence: axz@sdau.edu.cn; Tel.: +86-538-8246218

Abstract: As an important gas signaling molecule, hydrogen sulfide (H_2S) plays a crucial role in regulating cold tolerance. H_2S cooperates with phytohormones such as abscisic acid, ethylene, and salicylic acid to regulate the plant stress response. However, the synergistic regulation of H_2S and auxin in the plant response to cold stress has not been reported. This study showed that sodium hydrosulfide (NaHS, an H_2S donor) treatment enhanced the cold stress tolerance of cucumber seedlings and increased the level of auxin. *CsARF5*, a cucumber auxin response factor (ARF) gene, was isolated, and its role in regulating H_2S-mediated cold stress tolerance was described. Transgenic cucumber leaves overexpressing *CsARF5* were obtained. Physiological analysis indicated that overexpression of *CsARF5* enhanced the cold stress tolerance of cucumber and the regulation of the cold stress response by *CsARF5* depends on H_2S. In addition, molecular assays showed that *CsARF5* modulated cold stress response by directly activating the expression of the dehydration-responsive element-binding (DREB)/C-repeat binding factor (CBF) gene *CsDREB3*, which was identified as a positive regulator of cold stress. Taken together, the above results suggest that CsARF5 plays an important role in H_2S-mediated cold stress in cucumber. These results shed light on the molecular mechanism by which H_2S regulates cold stress response by mediating auxin signaling; this will provide insights for further studies on the molecular mechanism by which H_2S regulates cold stress. The aim of this study was to explore the molecular mechanism of H_2S regulating cold tolerance of cucumber seedlings and provide a theoretical basis for the further study of cucumber cultivation and environmental adaptability technology in winter.

Keywords: ARF; auxin; cold stress; cucumber; DREB; hydrogen sulfide; module; resistance

Citation: Zhang, X.; Fu, X.; Liu, F.; Wang, Y.; Bi, H.; Ai, X. Hydrogen Sulfide Improves the Cold Stress Resistance through the CsARF5-CsDREB3 Module in Cucumber. *Int. J. Mol. Sci.* **2021**, *22*, 13229. https://doi.org/10.3390/ijms222413229

Academic Editors: Yanjie Xie, Francisco J. Corpas and Jisheng Li

Received: 13 November 2021
Accepted: 5 December 2021
Published: 8 December 2021

1. Introduction

Cucumber (*Cucumis sativus* L.) is one of the most important economic crops worldwide. The cultivation and yield of cucumbers in China have ranked among the top in the world for many years. Cucumbers are typical cold-sensitive plants and are generally grown in solar greenhouses in northern China. Because of the extreme low-temperature conditions, cucumbers in greenhouses are prone to cold injury in winter and early spring. Cucumbers with cold injury showed inhibited growth, wilted and died in severe cases. Therefore, it is of great practical significance to study the effects of low-temperature stress on cucumber growth and development and the response mechanism of cucumber to low-temperature stress.

Hydrogen sulfide (H_2S) is a gaseous compound recognized as the third gas signaling molecule discovered after nitric oxide and carbon monoxide [1,2]. H_2S has been found to be widespread in mammals and has important cellular protective effects [3,4]. However, H_2S is also highly toxic. Low concentrations of H_2S can affect the eyes, respiratory system and central nervous system. Inhaling hydrogen sulfide in small concentrations can be fatal [5,6]. In agricultural production, the application of an appropriate concentration of

H_2S can regulate plant growth and development, such as germination, maturation, root development, senescence and defense [7–9]. Numerous investigations have determined that H_2S plays a key role in the regulation of abiotic stress responses, including cold tolerance [8,10–12]. Low temperatures trigger H_2S biosynthesis [13,14]. H_2S is found to alleviate cold stress tolerance in many plant species, although the mechanisms remain elusive [11,13–15]. Several reports show that H_2S modulates cold stress response, possibly through mitogen-activated protein kinase (MAPK) signaling [11,16,17]. In addition to H_2S, phytohormones, especially auxin, also play a vital positive regulatory role in cold stress response [18,19].

Auxin is involved in plant growth and development in various aspects, including cell division and elongation, tissue patterning and the response to environmental stimuli [20,21]. Since auxin was identified for the first time in the 1930s as indole-3-acetic acid, there has been a major breakthrough in the molecular mechanisms of auxin perception and signal transduction. Many genetic and biochemical approaches have elucidated that the TRANSPORT INHIBITOR RESPONSE1 (TIR1) protein functions as the receptor to perceive auxin signaling based on the reduced auxin response of *tir1* mutations [22,23]. Additionally, the SCF^{TIR1} ubiquitin-ligase complex is regarded as a central regulator of auxin signaling, and it-mediated proteolysis of auxin/indole acetic acid (Aux/IAA) proteins is responsible for auxin signaling transduction [24,25]. In this signaling pathway, Aux/IAA proteins are direct targets of TIR1. Aux/IAA proteins directly interact with auxin response factors (ARFs) to repress their activities [26]. Upon exposure to auxin, the F-box protein TIR1 recruits the Aux/IAA proteins for degradation, which leads to the release of various auxin response factors, including *Small Auxin-up RNAs* (*SAURs*), *GH3s* and *Aux/IAAs*, and consequently regulates diverse auxin-mediated plant growth [27,28].

ARFs are vital transcription factors (TFs) that regulate the expression of auxin response genes [27,29,30]. To date, 23 and 25 *ARF* genes have been isolated in *Arabidopsis* and rice, respectively [27,31,32]. Most ARF members consist of a DNA-binding domain (DBD), a variable middle region and a carboxy-terminal dimerization domain (CTD) [27,33]. The DBD is classified as a plant-specific B3-type and functions to bind to TGTCTC/GAGACA sites (AuxREs) in vitro [33,34]. The middle region includes two types: activation domain-type (AD) and repression domain-type (RD), which are used as the basis of classification between transcription activators and transcription repressors [30,33]. Additionally, the CTD domain is responsible for protein-protein interactions by dimerizing with Aux/IAA proteins as well as other ARFs [35,36]. Extensive studies have suggested that ARF proteins are involved in distinct developmental processes. In *Arabidopsis*, numerous ARF genes have been implicated in embryogenesis (ARF5 and ARF17) [37], root growth (ARF7, ARF10, ARF16, and ARF19) [38–41], hypocotyl growth (ARF6, ARF7, ARF8, and ARF19) [42–44], shoot regeneration (ARF4 and ARF5) [45], flower development (ARF2, ARF3, ARF6, and ARF8) [46,47] and senescence (ARF1 and ARF2) [48]. In the case of rice, genetic studies show that the functions of ARFs are different from the functions of *Arabidopsis*. OsARF1 is involved in root initiation and seed development [49,50]. OsARF12 regulates root elongation, iron accumulation and phosphate homeostasis [51,52]. OsARF16 regulates phosphate transport, phosphate starvation and iron deficiency responses [53–55]. OsARF19 controls leaf angles [56]. OsARF11, OsARF12, OsARF16 and OsARF17 are involved in antiviral defences [57,58]. A recent study showed that OsARF6 regulates rice yields [59]. Although a number of ARF members have been functionally characterized in *Arabidopsis* and rice, as mentioned earlier, little is known about the functions of *ARF* genes in cucumber.

In this study, the molecular mechanisms by which H_2S regulates cold stress response in cucumber were explored. The study suggested that H_2S treatment could improve cold resistance and auxin content of cucumber. *CsARF5*, a transcriptional regulator in auxin signaling, was responsive to cold stress and H_2S treatments, and overexpression of *CsARF5* improved the cold stress tolerance of cucumber. Further studies indicated that *CsARF5* modulated cold stress response by directly activating the expression of the dehydration-responsive element-binding (DREB)/C-repeat binding factor (CBF) gene *CsDREB3*.

2. Results

2.1. Sodium Hydrosulfide (NaHS) Improves Cold Tolerance in Cucumber Seedlings

A previous study demonstrated that NaHS could improve the cold tolerance of cucumber seedlings in a concentration-dependent manner, and 1.0 mM NaHS treatment showed a very significant difference compared with the control [60]. Here, Figure 1 shows that 1.0 mM NaHS significantly reduced cold stress injury, accumulation of H_2O_2 and superoxide anion ($O_2^{\cdot -}$), as well as electrolyte leakage (EL) in cucumber seedlings after exposure to 5 °C for 48 h. However, 0.15 mM H_2S scavenger hypotaurine (HT) increased cold stress injury, H_2O_2, $O_2^{\cdot -}$ and EL, compared with the deionized water (H_2O, as a comparison)-treated seedlings (Figure 1A–F). The mRNA abundance of *CsCBF1* and *CsCOR* in NaHS-treated seedlings also increased by 1.38-fold and 3.14-fold, respectively, under cold stress, but no obvious differences were observed between H_2O and HT treatments (Figure 1G,H). Therefore, the results further confirmed that H_2S improves cold tolerance in cucumber.

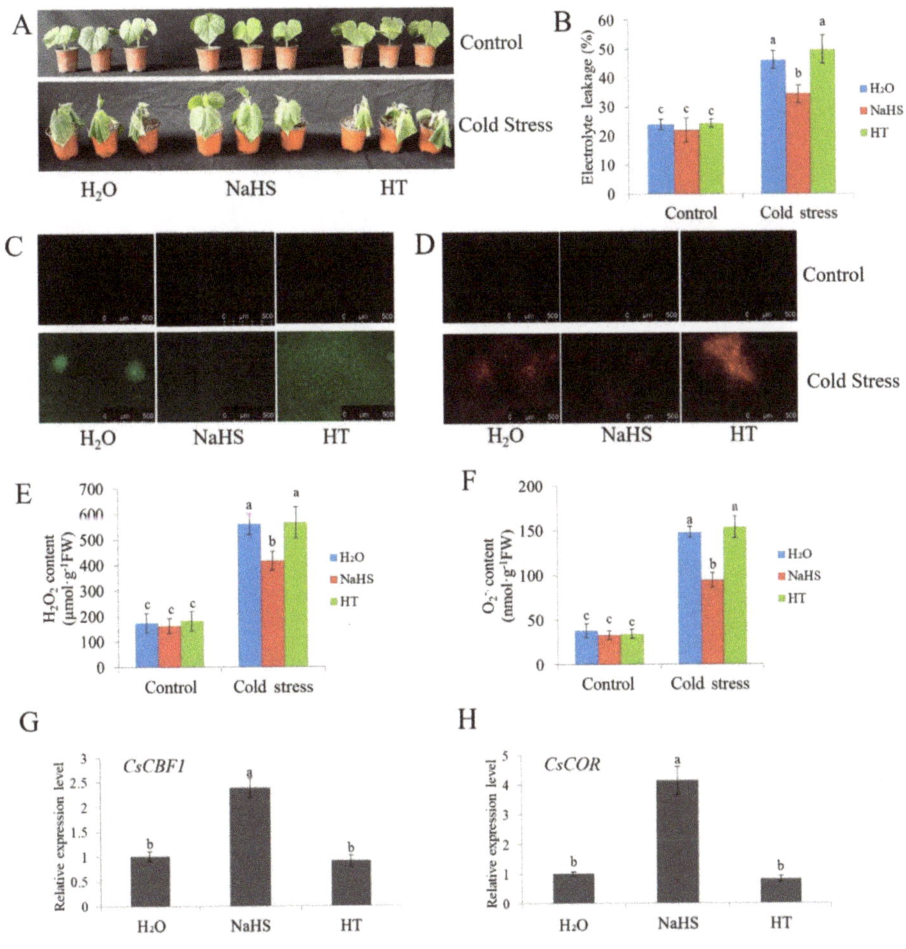

Figure 1. Effects of NaHS and HT on the cold tolerance of cucumber seedlings. Cucumber seedlings were treated with NaHS, HT or deionized water for 24 h and subsequently were exposed to cold stress. (**A**) Phenotypic characterization of cucumber seedlings before (control) and after cold stress (5 °C for 48 h). Each treatment contained 5–10 cucumber seedlings.

The experiments were repeated three times with similar results. A typical picture is shown here. (**B**) The EL results of cucumber seedlings before (control) and after cold stress for 48 h. (**C**,**D**) Inverted fluorescence microscopy imaging of H_2O_2 and $O_2^{\cdot -}$ levels in cucumber seedling leaves before (control) and after cold stress treatment for 48 h. (**E**,**F**) Detection of H_2O_2 and $O_2^{\cdot -}$ content of cucumber seedlings before (control) and after cold stress treatment for 48 h. (**G**,**H**) Expression of *CsCBF1* and *CsCOR* genes in cucumber seedlings under cold stress treatment for 6 h. qRT-PCR was performed simultaneously with three biological replicates and three technical replicates. The value of the water treatment was used as the reference and was set to 1. Error bars denote standard deviations. Different letters indicate significant differences ($p < 0.05$) based on Duncan's multiple range tests.

2.2. NaHS Treatment Affects Auxin Signaling

To explore the molecular mechanism by which H_2S improves cold resistance in cucumber, transcriptome analyses of cucumber seedlings treated with NaHS and H_2O were performed. A total of 1952 cucumber genes were analyzed from transcriptome data (Figure 2A). Among these cucumber genes, 118 genes were downregulated, and 54 genes were upregulated (Figure 2A). The upregulated genes were further analyzed (Figure 2B). One of the genes that caught our attention was an auxin response gene (accession number: CsaV3_3G045690, Figure 2B). The NCBI database comparison found that it was the *CsARF5* gene.

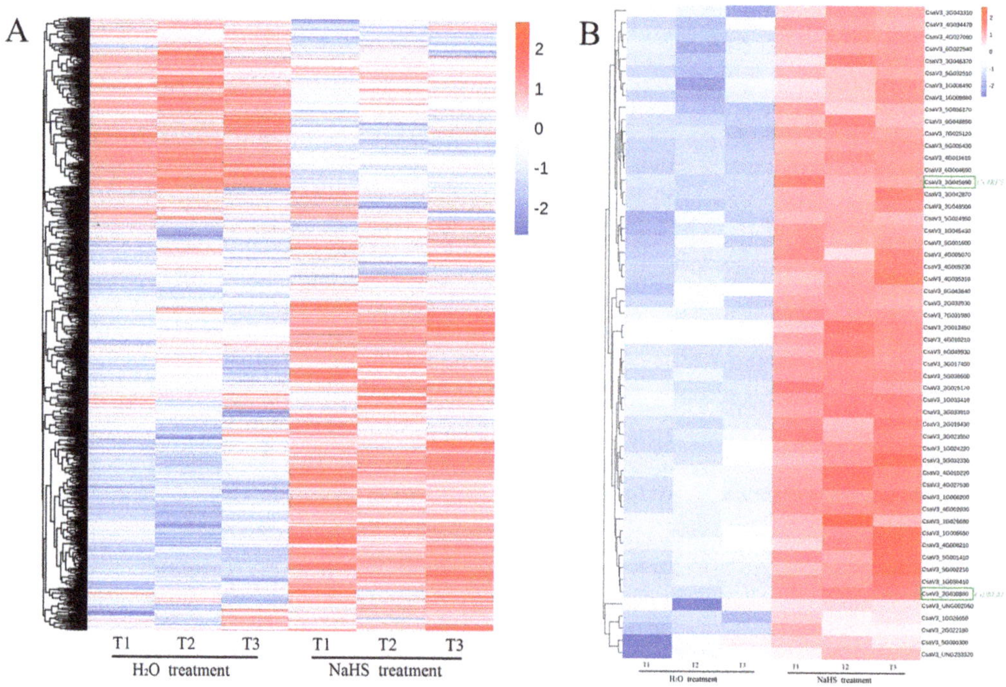

Figure 2. RNA-seq analysis of cucumber seedlings with or without NaHS treatment. (**A**) Normalized heat map showing the changes in gene expression after NaHS treatment at 5 °C for 6 h. All experiments were performed in triplicate. (**B**) Normalized heat map of upregulated gene expression after NaHS treatment at 5 °C for 6 h. All experiments were performed in triplicate.

To study the effect of NaHS on auxin signaling, the change in auxin content in cucumber seedlings treated with NaHS or HT was estimated. NaHS (1.0 mM) markedly increased endogenous indole-3-acetic acid (IAA) accumulation and flavin monooxygenase (FMO, a key enzyme in auxin synthesis) activity. However, HT treatment revealed lower or similar IAA content and FMO activity compared with H_2O treatment (Figure 3A,B). In addition, NaHS treatment also significantly upregulated the mRNA levels of *CsARF5* and *CsDREB3*, while HT-treated seedlings showed no marked differences in the relative mRNA expression of *CsARF5* and *CsDREB3* relative to the H_2O treatment (Figure 3C,D). These data indicate that H_2S affects auxin signaling in cucumber seedlings under cold stress.

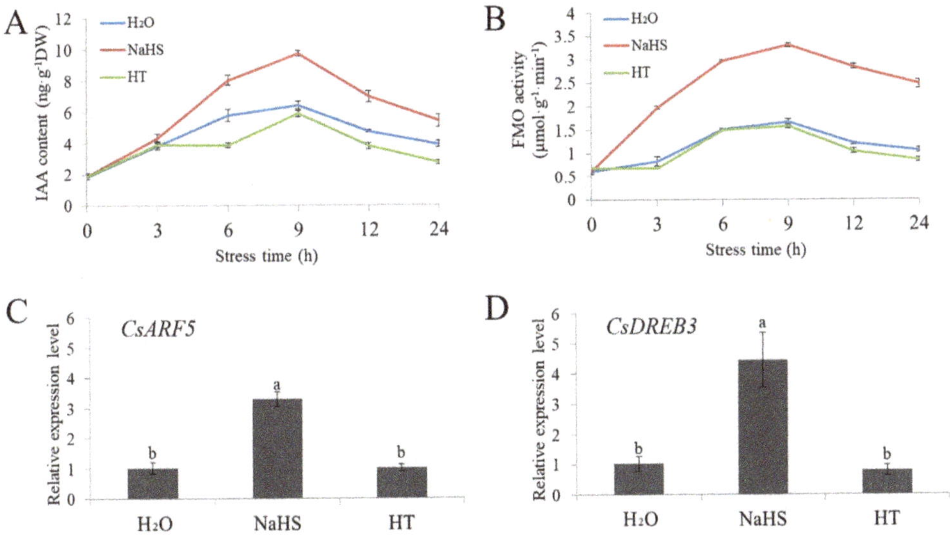

Figure 3. NaHS treatment affects IAA levels in cucumber. (**A**) The IAA content of cucumber seedlings treated with NaHS or HT under cold stress treatment for 24 h. (**B**) The FMO activity of cucumber seedlings treated with NaHS or HT under cold stress treatment for 24 h. (**C**) Expression of the *CsARF5* gene in cucumber seedlings treated with NaHS or HT under cold stress treatment for 6 h. (**D**) Expression of the *CsDREB3* gene in cucumber seedlings treated with NaHS or HT under cold stress treatment for 6 h. qRT- PCR was performed simultaneously with three biological replicates and three technical replicates. The value of 0 h was used as the reference and was set to 1. Error bars denote standard deviations. Different letters indicate significant differences ($p < 0.05$) based on Duncan's multiple range tests.

2.3. *IAA Treatment Improves Cold Resistance of Cucumber Seedlings*

A previous study showed that 75 μM IAA enhances the cold tolerance of cucumber seedlings [60]. Here, Figure 4 shows that 75 μM IAA significantly reduced the EL, and accumulation of H_2O_2 and $O_2^{\cdot -}$ caused by cold stress, while 50 μM 1-naphthylphthalamic acid (NPA, a polar transport inhibitor) treatment showed no remarkable difference relative to H_2O treatment under cold stress (Figure 4A–F). As an auxin response factor, the relative mRNA expression of *CsARF5* was upregulated in IAA-treated seedlings under cold stress. However, no remarkable difference was found in the mRNA expression of *CsARF5* between the NPA and H_2O treatments (Figure 4G). The mRNA expression levels of *CsCBF1* and *CsCOR* were significantly upregulated in IAA-treated seedlings but downregulated or not influenced in HT-treated seedlings compared with H_2O-treated seedlings when exposed to cold stress (Figure 4H,I). The latest results are in keeping with the earlier findings [60], so the results further confirm that IAA enhances cold tolerance in cucumber.

2.4. *CsARF5 Positively Regulates Cold Stress Tolerance of Cucumber*

qRT-PCR results showed that cold stress increased the mRNA abundance of the *CsARF5* gene, and the expression reached a peak after seedlings were exposed to cold for 3 h and then decreased gradually (Figure 5A). Compared with the control and HT treatments, NaHS treatment further increased the expression of *CsARF5* (Figure 5B). These results demonstrate that *CsARF5* is responsive to cold stress and H_2S treatment.

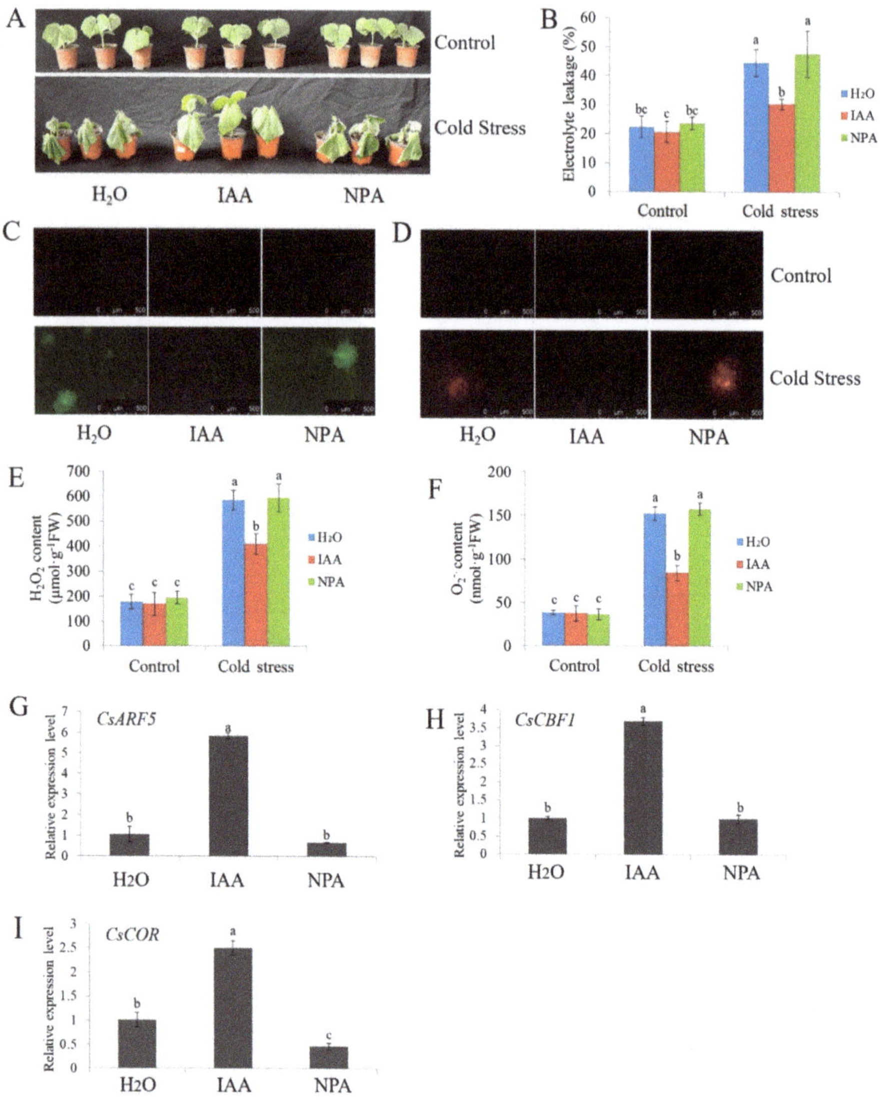

Figure 4. Effects of IAA and NPA on the cold resistance of cucumber seedlings. Cucumber seedlings were treated with IAA, NPA or deionized water for 24 h and subsequently were exposed to cold stress. (**A**) Phenotypic characterization of cucumber seedlings before (control) and after cold stress (5 °C for 48 h). Each treatment contained 5–10 cucumber seedlings. The experiments were repeated three times with similar results. A typical picture is shown here. (**B**) The EL results of cucumber seedlings before (control) and after cold stress treatment for 48 h. (**C**,**D**) Inverted fluorescence microscopy imaging of H_2O_2 and $O_2^{\cdot-}$ levels in cucumber seedling leaves before (control) and after cold stress treatment for 48 h. (**E**,**F**) Detection of H_2O_2 and $O_2^{\cdot-}$ accumulation of cucumber seedlings before (control) and after cold stress treatment for 48 h. (**G**–**I**) Expression of *CsARF5*, *CsCBF1* and *CsCOR* genes in cucumber seedlings under cold stress for 6 h. qRT-PCR was performed simultaneously with three biological replicates and three technical replicates. The value of the water treatment was used as the reference and was set to 1. Error bars denote standard deviations. Different letters indicate significant differences ($p < 0.05$) based on Duncan's multiple range tests.

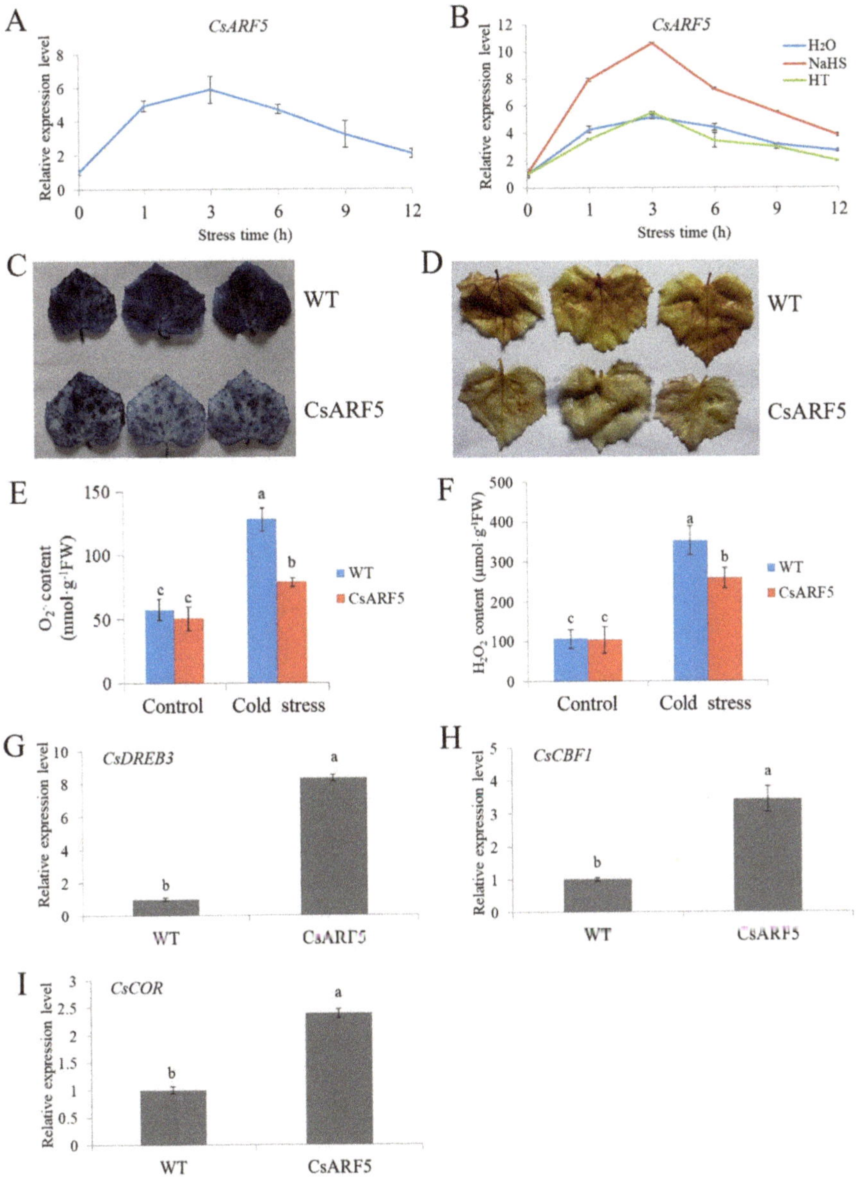

Figure 5. Overexpression of *CsARF5* improves cold tolerance of cucumber. (**A**) Expression of the *CsARF5* gene in cucumber seedlings under cold stress treatment for 12 h. (**B**) Expression of the *CsARF5* gene in cucumber seedlings treated with NaHS or HT under cold stress treatment for 12 h. (**C**,**D**) NBT and DAB staining of empty vector control (WT) and *CsARF5* transient transgenic cucumber leaves treated with cold stress for 12 h. Each genotype contained 5–10 cucumber leaves. The experiments were repeated three times with similar results. A typical picture is shown here. (**E**,**F**) Detection of $O_2^{\cdot-}$ and H_2O_2 contents of transgenic cucumber leaves before (control) and after cold stress treatment for 12 h. (**G**–**I**) Expression of *CsDREB3*, *CsCBF1* and *CsCOR* genes in transgenic cucumber leaves under cold stress for 3 h. qRT-PCR was performed simultaneously with three biological replicates and three technical replicates. The value of WT was used as the reference and was set to 1. Error bars denote standard deviations. Different letters indicate significant differences ($p < 0.05$) based on Duncan's multiple range tests.

To further explore the role of *CsARF5* in response to cold stress in cucumber, we obtained *CsARF5* transgenic cucumber leaves through *Agrobacterium*-mediated transient genetic transformation (Supplemental Figure S1A). Then, the accumulation of reactive oxygen species (ROS) in transgenic leaves after exposure to cold stress for 12 h were observed, using nitroblue tetrazolium (NBT) and 3, 3-diaminobenzidine (DAB) staining. The results showed that the accumulation of ROS in cucumber leaves overexpressing *CsARF5* (CsARF5) was lower than that of the empty vector control (WT) under cold stress (Figure 5C,D). In addition, the contents of $O_2^{\cdot-}$ and H_2O_2 were measured with biochemical analysis, and the results were in agreement with the NBT and DAB staining images (Figure 5E,F). qRT-PCR results revealed that *CsARF5* overexpression upregulated the mRNA level of cold stress-responsive genes *CsDREB3*, *CsCBF1* and *CsCOR* (Figure 5G–I). These data suggest that *CsARF5* is a positive regulator of cold stress response.

2.5. HT Treatment Affects CsARF5-Mediated Cold Stress Tolerance

Considering that H_2S induces the expression of *CsARF5*, and that *CsARF5* positively regulates cold stress resistance, the role of *CsARF5* in H_2S-mediated cold stress was further explored. The H_2S scavenger HT was applied to *CsARF5*-overexpressing cucumber leaves to observe ROS accumulation. The NBT and DAB staining results showed that the application of HT alleviated the CsARF5-decreased ROS accumulation in detached cucumber leaves (Figure 6A,B). The biochemical analysis for $O_2^{\cdot-}$ and H_2O_2 was consistent with the NBT and DAB staining results (Figure 6C,D). qRT-PCR results showed that the application of HT inhibited the promotion of *CsDREB3*, *CsCBF1* and *CsCOR* mRNA expression levels caused by CsARF5 (Figure 6E–G). These results suggest that the regulation of the cold stress response by *CsARF5* depends on H_2S.

2.6. Overexpression of CsDREB3 Enhances Cold Stress Tolerance of Cucumber

DREB/CBF TFs play essential roles in the regulation of the plant cold stress response [61,62]. *CsDREB3* was induced by NaHS (Figure 2B), which prompted us to explore whether CsDREB3 was involved in the cold stress response in cucumber. As shown in Figure 7A, cold stress induced the expression of the *CsDREB3* gene, and the expression reached a peak at 3 h, and then decreased gradually. Compared with the control and HT treatments, NaHS treatment further increased the expression of *CsDREB3* (Figure 7B).

To investigate the role of *CsDREB3* in response to cold stress in cucumber, *CsDREB3* transient transgenic cucumber leaves were obtained (Supplemental Figure S1B). NBT and DAB staining results showed that overexpression of *CsDREB3* decreased ROS accumulation after cold stress treatment (Figure 7C,D). In addition, $O_2^{\cdot-}$ and H_2O_2 detection results also revealed that the accumulation of $O_2^{\cdot-}$ and H_2O_2 was significantly lower in leaves of overexpressing *CsDREB3* than in WT leaves (Figure 7E,F). qRT-PCR results suggested that overexpression of *CsDREB3* increased the expression of *CsCBF1* and *CsCOR* (Figure 7G,H). These data demonstrate that *CsDREB3* positively regulates the cold tolerance of cucumber.

2.7. CsARF5 Directly Activates the Expression of CsDREB3

CsARF5 acts as an auxin response factor and can bind to the AuxRE motif in the promoters of target genes [27]. Considering the similar expression patterns of *CsARF5* and *CsDREB3* in cold stress and H_2S treatment, as well as the key role of DREB/CBF TFs in the regulation of the cold stress response, we hypothesized that CsARF5 might be involved in the cold stress response by mediating the expression of *CsDREB3*. Then, the sequence of the *CsDREB3* gene promoter region was analyzed, and a putative AuxRE motif was found (Figure 8A). Fortunately, the direct binding between the CsARF5 protein and the promoter of *CsDREB3* was detected by electromobility shift assay (EMSA) (Figure 8B). To test how CsARF5 regulated the expression of *CsDREB3*, dual luciferase assays in tobacco leaves were performed. The CsARF5 effector construct was expressed under the 35S promoter, and the promoter of *CsDREB3* was fused to the Luc gene as a reporter (Figure 8C). The results showed that co-expression of 35Spro:CsARF5 with CsDREB3pro: Luc led to an obvious increase in luminescence intensity (Figure 8D,E), while the binding site was mutated, and the activation was abolished (Figure 8D,E). These results suggest that CsARF5 transactivates the expression of *CsDREB3* in cucumber.

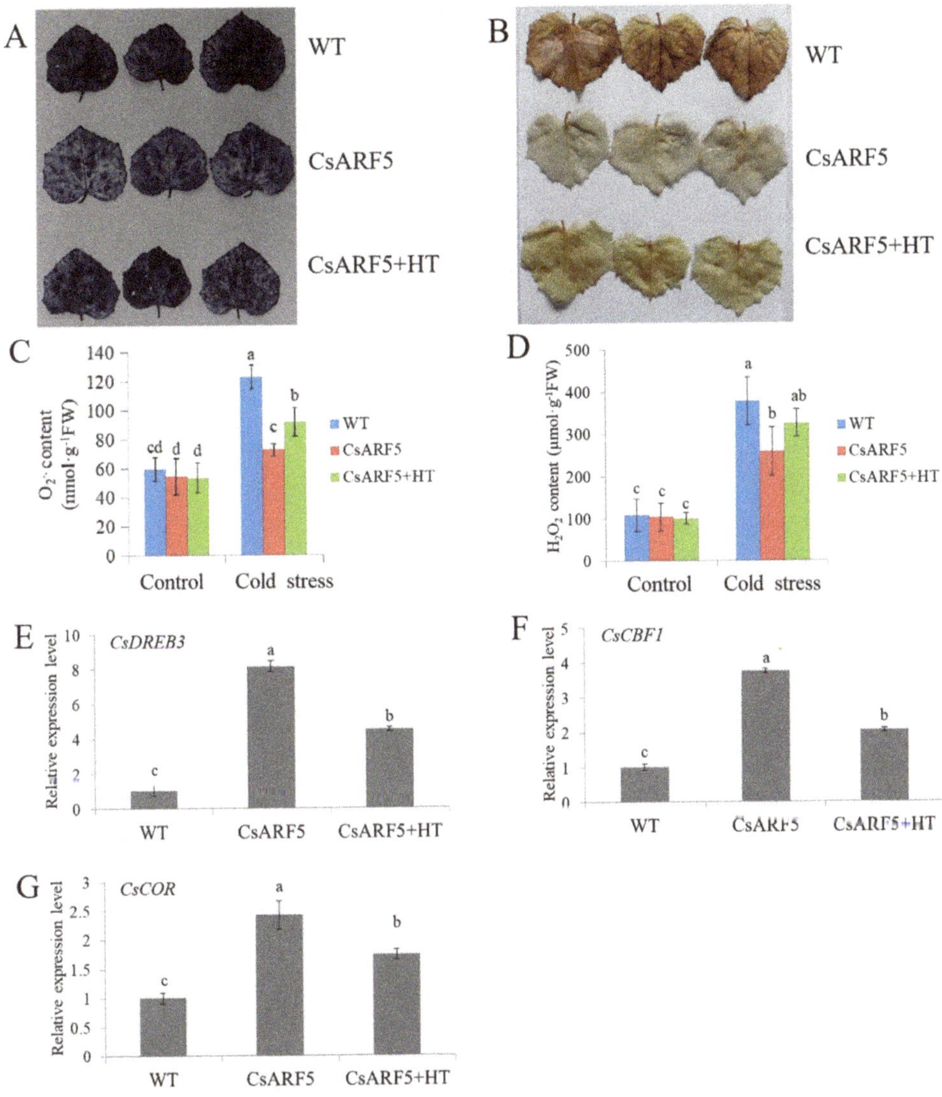

Figure 6. Effect of HT on the cold resistance of *CsARF5* transgenic cucumber. (**A**,**B**) NBT and DAB staining of empty vector control (WT), *CsARF5* transient transgenic cucumber leaves (CsARF5) and CsARF5 sprayed with HT (CsARF5+HT) treated with cold stress for 12 h. Each genotype contained 5–10 cucumber leaves. The experiments were repeated three times with similar results. A typical picture is shown here. (**C**,**D**) Detection of $O_2^{\cdot -}$ and H_2O_2 contents of transgenic cucumber leaves before (control) and after cold stress or HT treatment for 12 h. (**E**–**G**) Expression of *CsDREB3*, *CsCBF1* and *CsCOR* genes in transgenic cucumber leaves under cold stress for 3 h. qRT- PCR was performed simultaneously with three biological replicates and three technical replicates. The value of WT was used as the reference and was set to 1. Error bars denote standard deviations. Different letters indicate significant differences ($p < 0.05$) based on Duncan's multiple range tests.

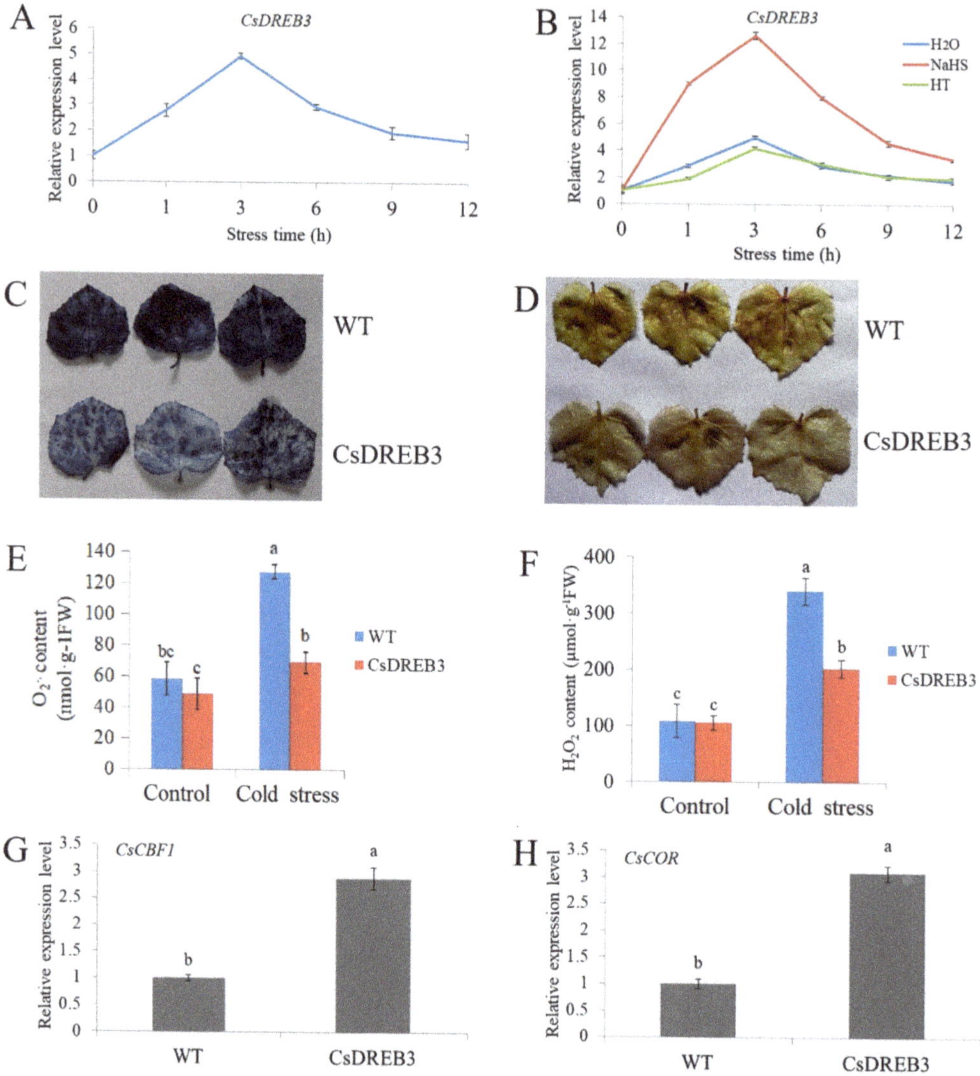

Figure 7. Overexpression of *CsDREB3* improves the cold tolerance of cucumber. (**A**) Expression of the *CsDREB3* gene in cucumber seedlings under cold stress treatment for 12 h. (**B**) Expression of the *CsDREB3* gene in cucumber seedlings treated with NaHS or HT under cold stress treatment for 12 h. (**C**,**D**) NBT and DAB staining of empty vector control (WT) and *CsDREB3* transient transgenic cucumber leaves treated with cold stress for 12 h. Each genotype contained 5–10 cucumber leaves. The experiments were repeated three times with similar results. A typical picture is shown here. (**E**,**F**) Detection of $O_2^{\cdot-}$ and H_2O_2 contents of transgenic cucumber leaves before (control) and after cold stress treatment for 12 h. (**G**,**H**) Expression of *CsCBF1* and *CsCOR* genes in transgenic cucumber leaves under cold stress for 3 h. qRT-PCR was performed simultaneously with three biological replicates and three technical replicates. The value of WT was used as the reference and was set to 1. Error bars denote standard deviations. Different letters indicate significant differences ($p < 0.05$) based on Duncan's multiple range tests.

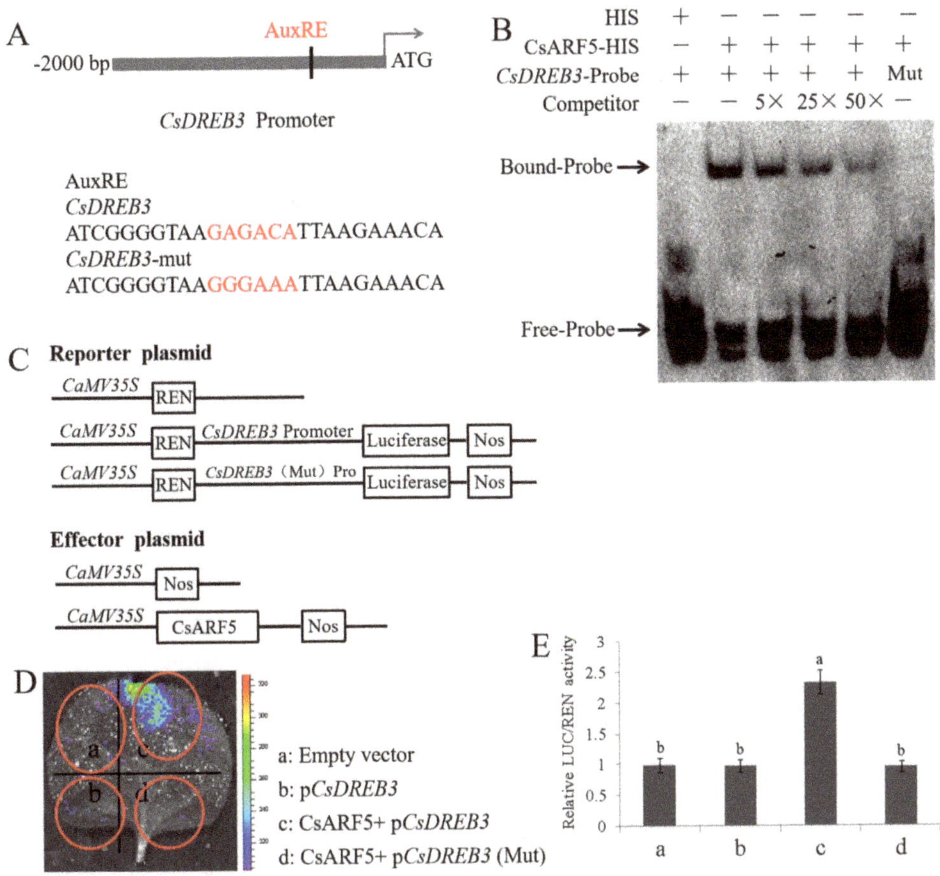

Figure 8. CsARF5 directly modulates the expression of CsDREB3. (**A**) Schematic diagram showing the *CsDREB3* promoter probe used for EMSAs. Mutated probe (*CsDREB3*-mut) in which the 5′-GAGACA-3′ motif was replaced by 5′-GGGAAA-3′. (**B**) EMSAs show that CsARF5 binds to the *MdDREB3* promoter. The experiments were repeated three times with similar results. A typical picture is shown here. (**C**) Schematic representation of the LUC reporter vector containing the *CsDREB3* promoter and the effector vectors expressing *CsARF5* under the control of the 35S promoter. (**D**) Dual luciferase tests in tobacco leaves showing that CsARF5 activates *CsDREB3* transcription. Mutated promoter sequence (pCsDREB3-Mut) in which the 5′-GAGACA-3′ motif was replaced by 5′-GGGAAA-3′. The experiments were repeated three times with similar results. A typical picture is shown here. (**E**) LUC/REN activity detection to verify that CsARF5 activates the transcription of *CsDREB3*. Empty vector was used as the reference and set to 1. Error bars denote standard deviations. Different letters indicate significant differences ($p < 0.05$) based on Duncan's multiple range tests. All experiments were performed three times with similar results, and representative data from one repetition are shown.

3. Discussion

H_2S, as a gaseous signaling molecule, plays a crucial role in plant relevance to various stress conditions, such as low temperature, salt, drought and heavy metals [12,63]. A report showed that the exogenous application of the H_2S donor NaHS could effectively improve plant growth and stress response [64]. In this study, NaHS treatment reduced the accumulation of ROS and activated the expression of cold stress-responsive genes, thus improving the cold-tolerance of cucumber seedlings (Figure 1), suggesting that the response of H_2S to cold stress is consistent in different species [64]. Previous reports indicate that H_2S and phytohormones have synergistic effects on the H_2S-mediated plant stress response [65]. For example, H_2S interacts with abscisic acid (ABA) and ethylene and is involved in the plant response to drought stress and in the regulation of stomatal closure [66,67]. Salicylic acid (SA) may play a key role in H_2S-alleviated heavy metal stress and low-temperature stress [68,69]. Jasmonic acid (JA) stimulation increases the H_2S level of protective cells and induces stomatal closure in *Vicia faba* [70]. Here, the study found that NaHS treatment promoted IAA synthesis and upregulated the expression of auxin-responsive genes (Figures 2 and 3), indicating that H_2S may crosstalk with auxin to regulate the cold stress response of cucumber.

Several plant hormones, such as JA, SA and ethylene, have been shown to play key roles in the plant's response to cold stress [66–69,71–73]. However, the role of auxin under cold stress is limited. Auxin is a crucial phytohormone that is involved in a variety of plant physiological and developmental processes, including the regulation of the cold stress response [74,75]. Previous investigations have determined that cold stress promotes auxin biosynthesis or changes the auxin gradient distribution, thus affecting the root gravity response in *Arabidopsis*, rice and poplar [18,19,76,77]. A recent study showed that the auxin signaling repressor IAA14 plays an important role in integrating microRNAs with auxin and cold reactions in *Arabidopsis* [78]. However, studies on the effects of auxin on plant cold stress response are not sufficient. This study showed that the application of auxin improved cold resistance, whereas the application of the polar transport inhibitor NPA slightly reduced the cold resistance of cucumber seedlings (Figure 4), demonstrating that auxin is a positive regulator of the cold stress response.

As a master regulator of auxin signaling, ARF TFs have been functionally characterized in *Arabidopsis*, rice and *populus trichocarpa* [27,31,32,79]. An increasing number of studies have indicated that ARFs regulate multiple plant developmental processes such as root growth [40,41], flower development [46,47], senescence [48] and stress response [19,80]. Among these processes, the auxin-regulated cold stress response has become a major focus of biotechnology in cucumber, with the task of identifying the cold resistance genes and improving the yield of cucumber. In this study, an ARF TF, CsARF5, was isolated from cucumber, and its expression was induced by NaHS and cold stress treatments, suggesting that CsARF5 may be involved in H_2S-mediated cold tolerance (Figure 5A, B).

To investigate the functions of CsARF5, transgenic cucumber leaves overexpressing *CsARF5* were generated (Supplemental Figure S1A). As hypothesized, the physiology and genetic analysis indicated that the overexpression of *CsARF5* decreased the accumulation of ROS and increased the expression level of cold stress-responsive genes, thus improving the cold stress resistance of cucumber (Figure 5). These results suggest that CsARF5 positively regulates the cold stress tolerance of cucumber. Meanwhile, the H_2S scavenger HT was applied to *CsARF5*-overexpressing cucumber leaves to study the role of *CsARF5* in H_2S-mediated cold stress. The results showed that HT inhibited the cold stress resistance increased by CsARF5 (Figure 6), indicating that the regulation of the cold stress response by CsARF5 depends on H_2S.

DREB/CBF proteins play important roles in the regulation of the plant cold stress response [64,65]. In *Arabidopsis*, DREB1A, DREB2C, CBF1, CBF2 and CBF3 are involved in the cold stress response [62,81–83]. In rice, OsDREB1A, OsDREB1B, OsDREB1C, OsDREB1F, OsDREB2B and OsDREBL are responsive to cold stress treatment [84–87]. Here, cucumber *DREB* genes, namely *CsDREB3*, were induced by NaHS and cold stress treatments, and overexpression of *CsDREB3* significantly enhanced the cold stress tolerance of cucumber (Figure 7, Supplemental Figure S1B), revealing that CsDREB3 is a positive regulator of cold stress.

ARF5 is known to function as a transcriptional activator [27]. The remarkable expression of *CsDREB3* in CsARF5 transgenic leaves prompted us to consider whether CsARF5 can directly regulate CsDREB3 expression (Figure 5G). The putative AuxRE elements that were recognized by CsARF5 were searched in the promoter region of *CsDREB3*. Fortunately, one AuxRE site was found, and the results of EMSAs and dual luciferase assays provided evidence to show that CsARF5 could bind to the promoter of *CsDREB3* and activate its expression (Figure 8).

Based on previous studies in model plant species and our results in this study, a working model of CsARF5 regulating the H_2S-mediated cold stress response was proposed (Figure 9). CsIAA proteins interact with CsARF5 and interfere with the transcriptional regulation of *CsDREB3* by CsARF5. When H_2S signaling was detected, auxin levels in cucumber were increased, and CsARF5 was freed from the CsIAA protein complex. Then, CsARF5 specifically activated the expression of *CsDREB3* to improve the cold tolerance of cucumber. Our study elucidates the molecular mechanism by which H_2S regulates the cold stress response in cucumber by mediating auxin signaling, which will provide insights for further studies on the molecular mechanisms by which H_2S regulates cold stress. A better understanding of the function and signal transduction of CsARF5 in cucumber is helpful to regulate the resistance of cucumber to low-temperature stress to obtain high-quality fruit.

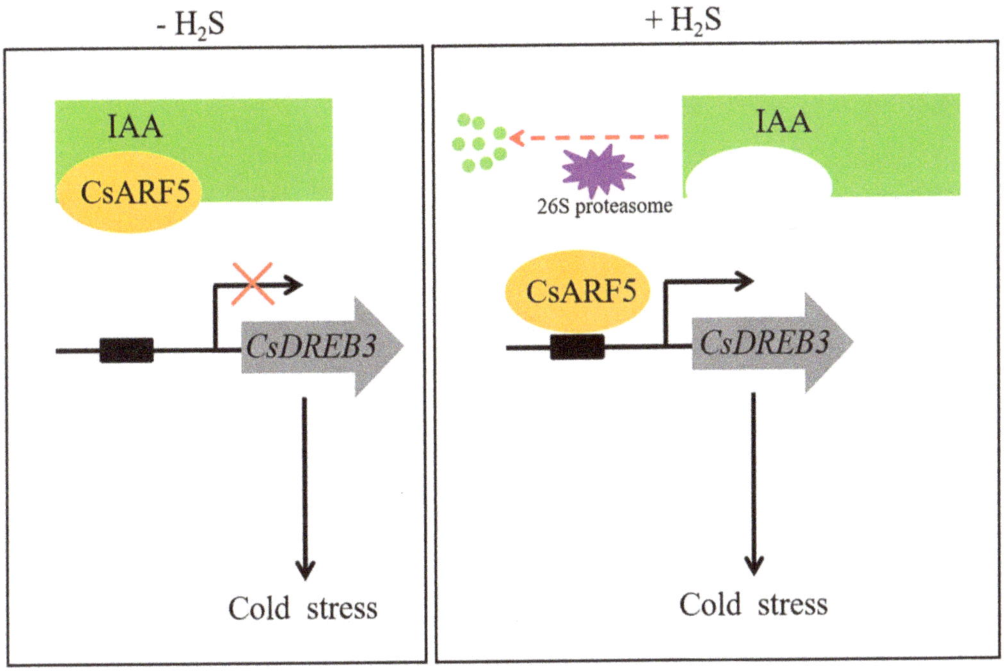

Figure 9. A regulatory model elucidates that H_2S mediates the cold stress response in cucumber through auxin signaling. In the absence of H_2S, IAA repressor proteins inhibit the expression of CsARF5, which in turn inhibits the cold stress response mediated by the CsARF5-*CsDREB3* module. In the presence of H_2S, IAA repressor proteins release CsARF5, which promotes the cold stress response by activating *CsDREB3* expression.

4. Materials and Methods

4.1. Plant Material and Growth Conditions

'Jinyou 35' cucumber seedlings were used for cold stress treatment and genetic transformation. After soaking and germinating, the cucumber seeds were sown in nutrition bowls and transferred to a climate chamber with a PFD of 600 μmol $m^{-2}{\cdot}s^{-1}$, a 25 °C/16 °C thermo-period, an 11-h photoperiod and 80% relative humidity.

4.2. Vector Construction and Transient Transformation

To generate *CsARF5* and *CsDREB3* overexpression vectors, full-length *CsARF5* and *CsDREB3* were inserted into the pCAMBIA1300 plasmid.

For the transient transformation of detached cucumber leaves, cucumber leaves with the same growth conditions were taken, and *Agrobacterium tumefaciens* LBA4404 (Weidi, Shanghai, China) with overexpression vector was injected into cucumber leaves through a medical syringe.

4.3. Cold Stress Treatments

To evaluate the effect of H_2S and IAA on the cold resistance of cucumber, cucumber seedlings with two leaves were foliar sprayed with 1.0 mM NaHS (an H_2S donor; Shanghai Macklin Biochemical Co., Ltd., Shanghai, China), 0.15 mM HT (a specific scavenger of H_2S; Sigma-Aldrich, Shanghai, China) and deionized water (H_2O), respectively, or pretreated with 75 μM IAA (Solarbio, Beijing, China), 50 μM NPA (a polar transport inhibitor of IAA; Shanghai Aladdin Biochemical Technology Co., Ltd., Shanghai, China) or deionized water (H_2O), respectively. Twenty-four hours later, half of the treatments were exposed to low temperatures (5 °C), and the other half of the seedlings were placed at normal temperatures as the control. The EL and accumulation of ROS were determined at 48 h after exposure of seedlings to 5 °C. Leaf samples of seedlings pretreated with H_2O and 1.0 mM NaHS were collected from 3 plants ($n = 3$) for transcriptome analysis after 6 h of cold stress.

To compare the difference in cold tolerance between WT and transgenic cucumber leaves, the pCAMBIA1300 empty vectors and overexpressed *CsARF5* and *CsDREB3* vectors were injected into the first leaf, which was just flat of cucumber. Twelve hours later, WT leaves, *CsDREB3* overexpressing leaves, and some *CsARF5* overexpressing leaves were exposed to 5 °C. Other *CsARF5* overexpressing leaves were treated with HT and then exposed to 5 °C after the water droplets on the leaves were absorbed and dried. The gene expression of *CsDREB3*, *CsCBF1* and *CsCOR* in transgenic cucumber leaves was measured at 3 h after exposure to cold stress. NBT staining, DAB staining and ROS content were detected after cold treatment for 12 h.

4.4. qRT-PCR Analysis

The transcription levels of *CsARF5*, *CsDREB3*, *CsCBF1* and *CsCOR* were examined using specific primers CsARF5 (qRT)-F/R, CsDREB3 (qRT)-F/R, CsCBF1 (qRT)-F/R and CsCOR (qRT)-F/R, respectively. β-Actin was used as an internal reference. All of the primers used are shown in Supplemental Table S1. qRT-PCR was carried out simultaneously with three biological replicates and three technical replicates.

4.5. Detection of EL

EL was detected according to methods described by Dong et al. (2013) [88]. Leaf discs (0.2 g) were immersed in 20 mL deionized water and incubated at 25 °C for 3 h. The electrical conductivity (EC1) was estimated using a conductivity meter (DDB-303A, Shanghai, China). The leaf discs were boiled for 10 min and then cooled to detect EC2. EL was calculated according to the following formula: EL= EC1/EC2 × 100.

4.6. IAA Content and FMO Activity Assay

IAA content was measured using high-performance liquid chromatography-triple quadrupole mass spectrometry (Thermo Fisher Scientific, TSQ Quantum Access, San Jose, CA, USA) according to the method of Li et al. (2014) [89], with minor modifications by Zhang et al. (2020) [60]. In brief, leaf samples were extracted with 80% methanol (containing 30 $\mu g \cdot mL^{-1}$ sodium diethyldithiocarbamate), and the supernatant was retained by rotary evaporation (Shanghai EYELA, N-1210B, Shanghai, China). Pigment and phenolic impurities of samples were removed using trichloromethane and polyvinylpolypyrrolidone (PVPP), respectively. Auxin was further extracted with ethyl acetate, and then the ester phase was collected. Finally, rotation drying at 36 °C and drying were dissolved in 1.0 mL mobile phases (methanol: 0.04% acetic acid = 45:55, v/v). The filtrate could then be used directly for HPLC–MS analysis.

The activity of FMO was detected using an enzyme-linked immunosorbent assay (ELISA) kit (Jiangsu Meimian Industrial Co. Ltd., Yancheng, China) as described by Zhang et al. (2020) [60].

4.7. Determination of H_2O_2 and $O_2^{\cdot-}$ Contents

H_2O_2 content was estimated with the H_2O_2 kit (Nanjing Jiancheng Bioengineering Institute, Nanjing, China) according to the instructions. The $O_2^{\cdot-}$ content was detected according to the method of Wang and Luo (1990) [90]. Cellular H_2O_2 and $O_2^{\cdot-}$ were fluorescently stained with 2′,7′-dichlorodihydrofluorescein diacetate (H_2DCFDA, the fluorescent probe of H_2O_2) (MCE, Cat. No. HY-D0940, Shanghai, China) and dihydroethidium (DHE, $O_2^{\cdot-}$ fluorescent probe) (Fluorescence Biotechnology Co. Ltd., Cat. No. 15200, Beijing, China), respectively, as described by Galluzzi and Kroemer (2014) [91] and modified by Zhang et al. (2020) [60]. In brief, the samples were infiltrated in a 20 μM H_2O_2 fluorescent probe at 25 °C under dark conditions for 30 min. Then the tissues were rinsed with HEPES-NaOH buffer (pH 7.5). Under excitation at 488 nm and emission at 522 nm of an inverted microscope (Leica DMi8), cellular H_2O_2 showed obvious green fluorescent coloration. For cellular $O_2^{\cdot-}$ measurements, the samples were infiltrated in 10 μM DHE at 37 °C under dark conditions for 30 min. After fixation, the tissues were rinsed with Tris-HCl buffer (pH 7.5). $O_2^{\cdot-}$ showed strong red fluorescence under excitation at 490 nm and emission at 520 nm under an inverted microscope (Leica DMi8).

4.8. NBT and DAB Staining

NBT staining of $O_2^{\cdot-}$ was performed according to the method of Jabs et al. (1996) [92] with minor modifications. The fresh leaves were washed with distilled water, immersed in 0.5 mM NBT in a vacuum and stained at 28 °C for 1 h. Then, the leaves were boiled in ethanol:lactic acid:glycerol (3:1:1) mixed solution to remove pigments, and $O_2^{\cdot-}$ was visualized in blue-purple coloration. DAB staining of H_2O_2 was carried out as described by Thordal-Christensen et al. (1997) [93]. The cleaned fresh leaves were soaked in 1 mM DAB staining solution (pH 3.8) in a vacuum and stained at 28 °C for 8 h. Then, the leaves were boiled in ethanol:lactic acid:glycerol (3:1:1) mixed solution to remove pigments and H_2O_2 was visualized in reddish-brown coloration.

4.9. EMSAs

The CsARF5-HIS fusion protein and biotin labeled probes were prepared for EMSAs. The CsARF5-HIS fusion protein was obtained by inducing *Escherichia coli* BL21 (TransGen Biotech, Beijing, China) with isopropyl β-D-thiogalactoside (IPTG). The biotin-labeled probes were synthesized by Sangon Biotech (Shanghai, China) Co., Ltd. To perform the EMSAs, the fusion protein was mixed with the probe and incubated at 24 °C for 30 min. The protein-probe mixture was separated by nondenatured acrylamide gel electrophoresis.

4.10. Dual Luciferase Assay

The promoter sequence of *CsDREB3* was amplified and cloned into pGreenII 0800-LUC to generate the reporter construct pCsDREB3-LUC. The effector plasmid was constructed by inserting full-length *CsARF5* into pGreenII 62-SK. Different plasmid combinations were injected into tobacco (*Nicotiana benthamiana*) leaves by *Agrobacterium tumefaciens* LBA4404. The leaves were sprayed with 100 mM luciferin, and luminescence was detected after being placed in darkness for 3 min. Fluorescence images were obtained with a live imaging system (Xenogen, Alameda, California, USA). The fluorescence activity was determined using a fluorescence activity detection kit (Promega, Madison, WI, USA).

4.11. Statistical Analysis

All experiments were performed at least in triplicate, and the results are expressed as the mean ± standard deviation (SD) of three replicates. The data were analyzed statistically with DPS software. Duncan's multiple range test was used to compare differences among treatments, and $p < 0.05$ was considered statistically significant.

4.12. Accession Numbers

CsARF5 (CsaV3_3G045690), CsDREB3 (CsaV3_2G030880), CsCBF1 (XM_004140746), CsCOR (XM_011659051).

5. Conclusions

H_2S treatment increases cold tolerance and auxin content of cucumber. The auxin response factor CsARF5 directly activates the expression of *CsDREB3* to improve the cold tolerance of cucumber in response to H_2S treatment. This study elucidates the molecular mechanism by which H_2S regulates the cold stress response in cucumber by mediating auxin signaling, which will provide insights for further studies on the molecular mechanisms by which H_2S regulates cold stress.

Supplementary Materials: The following are available online at https://www.mdpi.com/article/10.3390/ijms222413229/s1.

Author Contributions: X.A. and X.Z. designed the experiment; X.Z. performed the research and analyzed the data; X.F., F.L., Y.W. and H.B. worked together with X.Z. to accomplish the experiment X.A. and X.Z. wrote the paper. All authors have read and agreed to the published version of the manuscript.

Funding: This research was funded by The National Science Foundation of China (31572170), The National Key Research and Development Program of China (2018YFD1000800), The Major Science and Technology Innovation of Shandong Province in China (2019JZZY010715) and The Special Fund of Vegetable Industrial Technology System of Shandong Province (SDAIT-05-10).

Institutional Review Board Statement: Not applicable.

Informed Consent Statement: Not applicable.

Data Availability Statement: The data discussed in this publication have been deposited in NCBI's Gene Expression Omnibus along with our lab's previous publication and are accessible through SRA accession: PRJNA579777 (https://www.ncbi.nlm.nih.gov/sra/PRJNA579777, accessed on 26 October 2019).

Conflicts of Interest: All authors agreed with the final manuscript and have no conflicts of interest.

References

1. Li, L.; Rose, P.; Moore, P.K. Hydrogen sulfide and cell signaling. *Annu. Rev. Pharmacol.* **2011**, *51*, 169–187. [CrossRef] [PubMed]
2. Olas, B. Hydrogen sulfide in signaling pathways. *Clin. Chim. Acta* **2015**, *439*, 212–218. [CrossRef] [PubMed]
3. Aboubakr, E.M.; Taye, A.; El-Moselhy, M.A.; Hassan, M.K. Protective effect of hydrogen sulfide against cold restraint stress-induced gastric mucosal injury in rats. *Arch. Pharm. Res.* **2013**, *36*, 1507–1515. [CrossRef] [PubMed]
4. Zaorska, E.; Tomasova, L.; Koszelewski, D.; Ostaszewski, R.; Ufnal, M. Hydrogen Sulfide in Pharmacotherapy, Beyond the Hydrogen Sulfide-Donors. *Biomolecules* **2020**, *10*, 323. [CrossRef]
5. Truong, D.H.; Eghbal, M.A.; Hindmarsh, W.; Roth, S.H.; O'Brien, P.J. Molecular mechanisms of hydrogen sulfide toxicity. *Drug Metab. Rev.* **2006**, *38*, 733–744. [CrossRef]
6. Jiang, J.; Chan, A.; Ali, S.; Saha, A.; Haushalter, K.J.; Lam, W.L.M.; Glasheen, M.; Parker, J.; Brenner, M.; Mahon, S.B.; et al. Hydrogen sulfide—mechanisms of toxicity and development of an antidote. *Sci. Rep.* **2016**, *6*, 20831. [CrossRef]
7. Li, Z.G.; Min, X.; Zhou, Z.H. Hydrogen sulfide: A signal molecule in plant cross-adaptation. *Front. Plant Sci.* **2016**, *7*, 1621. [CrossRef]
8. Hancock, J.T. Hydrogen sulfide and environmental stresses. *Environ. Exp. Bot.* **2019**, *161*, 50–56. [CrossRef]
9. Xuan, L.; Li, J.; Wang, X.; Wang, C. Crosstalk between hydrogen sulfide and other signal molecules regulate plant growth and development. *Int. J. Mol. Sci.* **2020**, *21*, 4593. [CrossRef]
10. Shi, H.; Ye, T.; Han, N.; Bian, H.; Liu, X.; Chan, Z. Hydrogen sulfide regulates abiotic stress tolerance and biotic stress resistance in *Arabidopsis*. *J. Integr. Plant Biol.* **2015**, *57*, 628–640. [CrossRef] [PubMed]
11. Du, X.; Jin, Z.; Liu, D.; Yang, G.; Pei, Y. Hydrogen sulfide alleviates the cold stress through MPK4 in *Arabidopsis thaliana*. *Plant Physiol. Bioch.* **2017**, *120*, 112–119. [CrossRef]

12. Zhang, J.; Zhou, M.; Zhou, H.; Zhao, D.; Gotor, C.; Romero, L.C.; Shen, J.; Ge, Z.; Zhang, Z.; Shen, W.; et al. Hydrogen sulfide, a signaling molecule in plant stress responses. *J. Integr. Plant Biol.* **2021**, *63*, 146–160. [CrossRef]
13. Fu, P.; Wang, W.; Hou, L.; Liu, X. Hydrogen sulfide is involved in the chilling stress response in *Vitis vinifera* L. *Acta Soc. Bot. Pol.* **2013**, *82*. [CrossRef]
14. Shi, H.; Ye, T.; Chan, Z. Exogenous application of hydrogen sulfide donor sodium hydrosulfide enhanced multiple abiotic stress tolerance in bermudagrass (*Cynodon dactylon* (L). Pers.). *Plant Physiol. Bioch.* **2013**, *71*, 226–234. [CrossRef]
15. Nasibi, F.; Kalantari, K.M.; Tavakoli, Z.M. Effects of Hydrogen Sulfide on Cold-Induced Oxidative Damage in *Cucumis sativus* L. *Int. J. Hortic. Sci. Technol.* **2020**, *7*, 199–211. [CrossRef]
16. Ba, Y.; Zhai, J.; Yan, J.; Li, K.; Xu, H. H_2S improves growth of tomato seedlings involving the MAPK signaling. *Sci. Hortic.* **2021**, *288*, 110366. [CrossRef]
17. Du, X.; Jin, Z.; Liu, Z.; Liu, D.; Zhang, L.; Ma, X.; Yang, G.; Liu, S.; Guo, Y.; Pei, Y. H_2S Persulfidated and Increased Kinase Activity of MPK4 to Response Cold Stress in *Arabidopsis*. *Front. Mol. Biosci.* **2021**, *8*, 81. [CrossRef]
18. Shibasaki, K.; Uemura, M.; Tsurumi, S.; Rahman, A. Auxin response in *Arabidopsis* under cold stress: Underlying molecular mechanisms. *Plant Cell* **2009**, *21*, 3823–3838. [CrossRef]
19. Rahman, A. Auxin: A regulator of cold stress response. *Physiol. Plant.* **2013**, *147*, 28–35. [CrossRef]
20. Zhao, Y. Auxin biosynthesis and its role in plant development. *Annu. Rev. Plant Biol.* **2010**, *61*, 49–64. [CrossRef]
21. Lv, B.; Yan, Z.; Tian, H.; Zhang, X.; Ding, Z. Local Auxin Biosynthesis Mediates Plant Growth and Development. *Trends Plant Sci.* **2019**, *24*, 6–9. [CrossRef] [PubMed]
22. Dharmasiri, N.; Dharmasiri, S.; Estelle, M. The F-box protein TIR1 is an auxin receptor. *Nature* **2005**, *435*, 441–445. [CrossRef]
23. Kepinski, S.; Leyser, O. The *Arabidopsis* F-box protein TIR1 is an auxin receptor. *Nature* **2005**, *435*, 446–451. [CrossRef] [PubMed]
24. Gray, W.M.; Kepinski, S.; Rouse, D.; Leyser, O.; Estelle, M. Auxin regulates SCF TIR1-dependent degradation of AUX/IAA proteins. *Nature* **2001**, *414*, 271–276. [CrossRef] [PubMed]
25. Kepinski, S.; Leyser, O. Auxin-induced SCF^{TIR1}–Aux/IAA interaction involves stable modification of the SCFTIR1 complex. *Proc. Natl. Acad. Sci. USA* **2004**, *101*, 12381–12386. [CrossRef] [PubMed]
26. Liscum, E.; Reed, J.W. Genetics of Aux/IAA and ARF action in plant growth and development. *Plant Mol. Biol.* **2002**, *49*, 387–400. [CrossRef]
27. Guilfoyle, T.J.; Hagen, G. Auxin response factors. *Curr. Opin. Plant Biol.* **2007**, *10*, 453–460. [CrossRef] [PubMed]
28. Weijers, D.; Friml, J. SnapShot: Auxin signaling and transport. *Cell* **2009**, *136*, 1172. [CrossRef] [PubMed]
29. Chandler, J.W. Auxin response factors. *Plant Cell Environ.* **2016**, *39*, 1014–1028. [CrossRef]
30. Li, S.B.; Xie, Z.Z.; Hu, C.G.; Zhang, J.Z. A review of auxin response factors (ARFs) in plants. *Front. Plant Sci.* **2016**, *7*, 47. [CrossRef]
31. Wang, D.; Pei, K.; Fu, Y.; Sun, Z.; Li, S.; Liu, H.; Tang, K.; Han, B.; Tao, Y. Genome-wide analysis of the auxin response factors (ARF) gene family in rice (*Oryza sativa* L). *Gene* **2007**, *394*, 13–24. [CrossRef] [PubMed]
32. Shen, C.; Wang, S.; Bai, Y.; Wu, Y.; Zhang, S.; Chen, M.; Guilfoyle, T.J.; Wu, P.; Qi, Y. Functional analysis of the structural domain of ARF proteins in rice (*Oryza sativa* L.). *J. Exp. Bot.* **2010**, *61*, 3971–3981. [CrossRef]
33. Tiwari, S.B.; Hagen, G.; Guilfoyle, T. The roles of auxin response factor domains in auxin-responsive transcription. *Plant Cell* **2003**, *15*, 533–543. [CrossRef] [PubMed]
34. Ulmasov, T.; Hagen, G.; Guilfoyle, T.J. Dimerization and DNA binding of auxin response factors. *Plant J.* **1999**, *19*, 309–319. [CrossRef] [PubMed]
35. Kim, J.; Harter, K.; Theologis, A. Protein-protein interactions among the Aux/IAA proteins. *Proc. Natl. Acad. Sci. USA* **1997**, *94*, 11786–11791. [CrossRef]
36. Piya, S.; Shrestha, S.K.; Binder, B.; Stewart, C.N.; Hewezi, T. Protein-protein interaction and gene co-expression maps of ARFs and Aux/IAAs in *Arabidopsis*. *Front. Plant Sci.* **2014**, *5*, 744. [CrossRef]
37. Mallory, A.C.; Bartel, D.P.; Bartel, B. MicroRNA-directed regulation of *Arabidopsis* auxin response factor17 is essential for proper development and modulates expression of early auxin response genes. *Plant Cell* **2005**, *17*, 1360–1375. [CrossRef]
38. Okushima, Y.; Overvoorde, P.J.; Arima, K.; Alonso, J.M.; Chan, A.; Chang, C.; Ecker, J.R.; Hughes, B.; Lui, A.; Nguyen, D.; et al. Function genomic analysis of the auxin response factor gene family members in *Arabidopsis thaliana*: Unique and overlapping functions of ARF7 and ARF19. *Plant Cell* **2005**, *17*, 444–463. [CrossRef] [PubMed]
39. Fukaki, H.; Tasaka, M. Hormone interactions during lateral root formation. *Plant Mol. Biol.* **2009**, *69*, 437–449. [CrossRef]
40. Orosa-Puente, B.; Leftley, N.; Wangenheim, D.V.; Banda, J.; Srivastava, A.K.; Hill, K.; Truskina, J.; Bhosale, R.; Morris, E.; Srivastava, M.; et al. Root branching toward water involves posttranslational modification of transcription factor ARF7. *Science* **2018**, *362*, 1407–1410. [CrossRef]
41. Lee, H.W.; Cho, C.; Pandey, S.K.; Park, Y.; Kim, M.J.; Kim, J. LBD16 and LBD18 acting downstream of ARF7 and ARF19 are involved in adventitious root formation in *Arabidopsis*. *BMC Plant Biol.* **2019**, *19*, 46. [CrossRef]
42. Liu, K.; Li, Y.; Chen, X.; Li, L.; Liu, K.; Zhao, H.; Wang, Y.; Han, S. ERF72 interacts with ARF6 and BZR1 to regulate hypocotyl elongation in *Arabidopsis*. *J. Exp. Bot.* **2018**, *69*, 3933–3947. [CrossRef] [PubMed]
43. Reed, J.W.; Wu, M.F.; Reeves, P.H.; Hodgens, C.; Yadav, V.; Hayes, S.; Pierik, R. Three Auxin Response Factors Promote Hypocotyl Elongation. *Plant Physiol.* **2018**, *178*, 864–875. [CrossRef]

44. Wang, X.; Yu, R.; Wang, J.; Lin, Z.; Han, X.; Deng, Z.; Fan, L.; He, H.; Deng, X.W.; Chen, H. The Asymmetric Expression of SAUR Genes Mediated by ARF7/19 Promotes the Gravitropism and Phototropism of Plant Hypocotyls. *Cell Rep.* **2020**, *31*, 107529. [CrossRef]
45. Zhang, M.M.; Zhang, H.K.; Zhai, J.F.; Zhang, X.S.; Sang, Y.L.; Cheng, Z.J. ARF4 regulates shoot regeneration through coordination with ARF5 and IAA12. *Plant Cell Rep.* **2021**, *40*, 315–325. [CrossRef]
46. Nagpal, P.; Ellis, C.M.; Weber, H.; Ploense, S.E.; Barkawi, L.S.; Guilfoyle, T.J.; Hagen, G.; Alonso, J.M.; Cohen, J.D.; Farmer, E.E.; et al. Auxin response factors ARF6 and ARF8 promote jasmonic acid production and flower maturation. *Development* **2005**, *132*, 4107–4118. [CrossRef]
47. Finet, C.; Fourquin, C.; Vinauger, M.; Berne-Dedieu, A.; Chambrier, P.; Paindavoine, S.; Scutt, C.P. Parallel structural evolution of auxin response factors in the angiosperms. *Plant J.* **2010**, *63*, 952–959. [CrossRef]
48. Ellis, C.M.; Nagpal, P.; Young, J.C.; Hage, G.; Guilfoyle, T.J.; Reed, J.W. AUXIN RESPONSE FACTOR1 and AUXIN RESPONSE FACTOR2 regulate senescence and floral organ abscission in *Arabidopsis thaliana*. *Development* **2005**, *132*, 4563–4574. [CrossRef] [PubMed]
49. Inukai, Y.; Sakamoto, T.; Ueguchi-Tanaka, M.; Shibata, Y.; Gomi, K.; Umemura, I.; Matsuoka, M. Crown rootless1, which is essential for crown root formation in rice, is a target of an AUXIN RESPONSE FACTOR in auxin signaling. *Plant Cell* **2005**, *17*, 1387–1396. [CrossRef] [PubMed]
50. Attia, K.A.; Abdelkhalik, A.F.; Ammar, M.H.; Wei, C.; Yang, J.; Lightfoot, D.A.; El-Shemy, H.A. Antisense phenotypes reveal a functional expression of OsARF1, an auxin response factor, in transgenic rice. *Curr. Issues Mol. Biol.* **2009**, *11*, I29–I34. [CrossRef]
51. Qi, Y.; Wang, S.; Shen, C.; Zhang, S.; Chen, Y.; Xu, Y.; Jiang, D. OsARF12, a transcription activator on auxin response gene, regulates root elongation and affects iron accumulation in rice (*Oryza sativa*). *New Phytol.* **2012**, *193*, 109–120. [CrossRef] [PubMed]
52. Wang, S.; Zhang, S.; Sun, C.; Xu, Y.; Chen, Y.; Yu, C.; Qian, Q.; Jiang, D.; Qi, Y. Auxin response factor (OsARF12), a novel regulator for phosphate homeostasis in rice (*Oryza sativa* L.). *New Phytol.* **2014**, *201*, 91–103. [CrossRef]
53. Shen, C.; Wang, S.; Zhang, S.; Xu, Y.; Qian, Q.; Qi, Y.; Jiang, D.A. OsARF16, a transcription factor, is required for auxin and phosphate starvation response in rice (*Oryza sativa* L.). *Plant Cell Environ.* **2013**, *36*, 607–620. [CrossRef]
54. Shen, C.; Yue, R.; Yang, Y.; Zhang, L.; Sun, T.; Tie, S.; Wang, H. OsARF16 is involved in cytokinin-mediated inhibition of phosphate transport and phosphate signaling in rice (*Oryza sativa* L.). *PLoS ONE* **2014**, *9*, e112906. [CrossRef]
55. Shen, C.; Yue, R.; Sun, T.; Zhang, L.; Yang, Y.; Wang, H. OsARF16, a transcription factor regulating auxin redistribution, is required for iron deficiency response in rice (*Oryza sativa* L.). *Plant Sci.* **2015**, *231*, 148–158. [CrossRef]
56. Zhang, S.; Wang, S.; Xu, Y.; Yu, C.; Shen, C.; Qian, Q.; Qi, Y. The auxin response factor, OsARF19, controls rice leaf angles through positively regulating OsGH3-5 and OsBRI1. *Plant Cell Environ.* **2015**, *38*, 638–654. [CrossRef]
57. Qin, Q.; Li, G.; Jin, L.; Huang, Y.; Wang, Y.; Wei, C.; Xu, Z.; Yang, Z.; Wang, H.; Li, Y. Auxin response factors (ARFs) differentially regulate rice antiviral immune response against rice dwarf virus. *PLoS Pathog.* **2020**, *16*, e1009118. [CrossRef]
58. Zhang, H.; Li, L.; He, Y.; Qin, Q.; Chen, C.; Wei, Z.; Tan, X.; Xie, K.; Zhang, R.; Hong, G.; et al. Distinct modes of manipulation of rice auxin response factor OsARF17 by different plant RNA viruses for infection. *Proc. Natl. Acad. Sci. USA* **2020**, *117*, 9112–9121. [CrossRef]
59. Qiao, J.; Jiang, H.; Lin, Y.; Shang, L.; Wang, M.; Li, D.; Fu, X.; Geisler, M.; Qi, Y.; Gao, Z.; et al. A Novel miR167a-OsARF6-OsAUX3 Module Regulates Grain Length and Weight in Rice. *Mol. Plant* **2021**, *14*, 1683–1698. [CrossRef]
60. Zhang, X.W.; Liu, F.J.; Zhai, J.; Li, F.D.; Bi, H.G.; Ai, X.Z. Auxin acts as a downstream signaling molecule involved in hydrogen sulfide-induced chilling tolerance in cucumber. *Planta* **2020**, *251*, 1–19. [CrossRef] [PubMed]
61. Zhou, M.L.; Ma, J.T.; Pang, J.F.; Zhang, Z.L.; Tang, Y.X.; Wu, Y.M. Regulation of plant stress response by dehydration responsive element binding (DREB) transcription factors. *Afr. J. Biotechnol.* **2010**, *9*, 9255–9269. [CrossRef]
62. Lata, C.; Prasad, M. Role of DREBs in regulation of abiotic stress responses in plants. *J. Exp. Bot.* **2011**, *62*, 4731–4748. [CrossRef] [PubMed]
63. Huang, D.; Huo, J.; Liao, W. Hydrogen sulfide: Roles in plant abiotic stress response and crosstalk with other signals. *Plant Sci.* **2021**, *302*, 110733. [CrossRef]
64. Guo, H.M.; Xiao, T.Y.; Zhou, H.; Xie, Y.J.; Shen, W.B. Hydrogen sulfide: A versatile regulator of environmental stress in plants. *Acta Physiol. Plant.* **2016**, *38*, 16. [CrossRef]
65. Li, Z.G.; Xiang, R.H.; Wang, J.Q. Hydrogen sulfide–phytohormone interaction in plants under physiological and stress conditions. *J. Plant Growth Regul.* **2021**, *40*, 2476–2484. [CrossRef]
66. Jin, Z.P.; Xue, S.W.; Luo, Y.A.; Tian, B.H.; Fang, H.H.; Li, H.; Pei, Y.X. Hydrogen sulfide interacting with abscisic acid in stomatal regulation responses to drought stress in *Arabidopsis*. *Plant Physiol. Biochem.* **2013**, *62*, 41–46. [CrossRef]
67. Hou, Z.H.; Wang, L.X.; Liu, J.; Hou, L.X.; Liu, X. Hydrogen sulfide regulates ethylene-induced stomatal closure in *Arabidopsis thaliana*. *J. Integr. Plant Biol.* **2013**, *55*, 277–289. [CrossRef] [PubMed]
68. Li, Z.G.; Xie, L.R.; Li, X.J. Hydrogen sulfide acts as a downstream signal molecule in salicylic acid-induced heat tolerance in maize (*Zea mays* L.) seedlings. *J. Plant Physiol.* **2015**, *177*, 121–127. [CrossRef]
69. Qiao, Z.J.; Jing, T.; Liu, Z.Q.; Zhang, L.P.; Jin, Z.P.; Liu, D.M.; Pei, Y.X. H_2S acting as a downstream signaling molecule of SA regulates Cd tolerance in *Arabidopsis*. *Plant Soil* **2015**, *393*, 137–146. [CrossRef]
70. Hou, Z.; Liu, J.; Hou, L.; Li, X.; Liu, X. H_2S may function downstream of H_2O_2 in jasmonic acid-induced stomatal closure in *Vicia faba*. *Chin. Bull. Bot.* **2011**, *46*, 396–406. [CrossRef]

71. Miura, K.; Tada, Y. Regulation of water, salinity, and cold stress responses by salicylic acid. *Front. Plant Sci.* **2014**, *5*, 4. [CrossRef]
72. Kazan, K. Diverse roles of jasmonates and ethylene in abiotic stress tolerance. *Trends Plant Sci.* **2015**, *20*, 219–229. [CrossRef]
73. Hu, Y.; Jiang, Y.; Han, X.; Wang, H.; Pan, J.; Yu, D. Jasmonate regulates leaf senescence and tolerance to cold stress: Crosstalk with other phytohormones. *J. Exp. Bot.* **2017**, *68*, 1361–1369. [CrossRef]
74. Zhao, Y. Essential Roles of Local Auxin Biosynthesis in Plant Development and in Adaptation to Environmental Changes. *Annu. Rev. Plant Biol.* **2018**, *69*, 417–435. [CrossRef]
75. Blakeslee, J.J.; Rossi, T.S.; Kriechbaumer, V. Auxin biosynthesis: Spatial regulation and adaptation to stress. *J. Exp. Bot.* **2019**, *70*, 5041–5049. [CrossRef]
76. Popko, J.; Hänsch, R.; Mendel, R.R.; Polle, A.; Teichmann, T. The role of abscisic acid and auxin in the response of poplar to abiotic stress. *Plant Biol.* **2010**, *12*, 242–258. [CrossRef] [PubMed]
77. Du, H.; Liu, H.; Xiong, L. Endogenous auxin and jasmonic acid levels are differentially modulated by abiotic stresses in rice. *Front. Plant Sci.* **2013**, *4*, 397. [CrossRef]
78. Aslam, M.; Sugita, K.; Qin, Y.; Rahman, A. Aux/IAA14 Regulates microRNA-Mediated Cold Stress Response in *Arabidopsis* Roots. *Int. J. Mol. Sci.* **2020**, *21*, 8441. [CrossRef] [PubMed]
79. Kalluri, U.C.; DiFazio, S.P.; Brunner, A.M.; Tuskan, G.A. Genome-wide analysis of *Aux/IAA* and *ARF* gene families in *Populus trichocarpa*. *BMC Plant Biol.* **2007**, *7*, 1–14. [CrossRef]
80. He, X.J.; Mu, R.L.; Cao, W.H.; Zhang, Z.G.; Zhang, J.S.; Chen, S.Y. AtNAC2, a transcription factor downstream of ethylene and auxin signaling pathways, is involved in salt stress response and lateral root development. *Plant J.* **2005**, *44*, 903–916. [CrossRef]
81. Liu, Q.; Kasuga, M.; Sakuma, Y.; Abe, H.; Miura, S.; Yamaguchi-Shinozaki, K.; Shinozaki, K. Two transcription factors, DREB1 and DREB2, with an EREBP/AP2 DNA binding domain separate two cellular signal transduction pathways in drought-and low-temperature-responsive gene expression, respectively, in *Arabidopsis*. *Plant Cell* **1998**, *10*, 1391–1406. [CrossRef] [PubMed]
82. Gilmour, S.J.; Zarka, D.G.; Stockinger, E.J.; Salazar, M.P.; Houghton, J.M.; Thomashow, M.F. Low temperature regulation of the *Arabidopsis* CBF family of AP2 transcriptional activators as an early step in cold-induced *COR* gene expression. *Plant J.* **1998**, *16*, 433–442. [CrossRef]
83. Lee, S.J.; Kang, J.Y.; Park, H.J.; Kim, M.D.; Bae, M.S.; Choi, H.I.; Kim, S.Y. DREB2C interacts with ABF2, a bZIP protein regulating abscisic acid-responsive gene expression, and its overexpression affects abscisic acid sensitivity. *Plant Physiol.* **2010**, *153*, 716–727. [CrossRef]
84. Chen, J.Q.; Dong, Y.; Wang, Y.J.; Liu, Q.; Zhang, J.S.; Chen, S.Y. An AP2/EREBP-type transcription-factor gene from rice is cold-inducible and encodes a nuclear-localized protein. *Theor. Appl. Genet.* **2003**, *107*, 972–979. [CrossRef] [PubMed]
85. Dubouzet, J.G.; Sakuma, Y.; Ito, Y.; Kasuga, M.; Dubouzet, E.G.; Miura, S.; Seki, M.; Shinozaki, K.; Yamaguchi-Shinozaki, K. *OsDREB* genes in rice, *Oryza sativa* L., encode transcription activators that function in drought-, high-salt-and cold-responsive gene expression. *Plant J.* **2003**, *33*, 751–763. [CrossRef]
86. Wang, Q.; Guan, Y.; Wu, Y.; Chen, H.; Chen, F.; Chu, C. Overexpression of a rice OsDREB1F gene increases salt, drought, and low temperature tolerance in both *Arabidopsis* and rice. *Plant Mol. Biol.* **2008**, *67*, 589–602. [CrossRef]
87. Matsukura, S.; Mizoi, J.; Yoshida, T.; Todaka, D.; Ito, Y.; Maruyama, K.; Shinozaki, K.; Yamaguchi-Shinozaki, K. Comprehensive analysis of rice DREB2-type genes that encode transcription factors involved in the expression of abiotic stress-responsive genes. *Mol. Genet. Genomics* **2010**, *283*, 185–196. [CrossRef]
88. Dong, X.B.; Bi, H.G.; Wu, G.X.; Ai, X.Z. Drought-induced chilling tolerance in cucumber involves membrane stabilisation improved by antioxidant system. *Int. J. Plant. Prod.* **2013**, *7*, 67–80.
89. Li, Y.; Xu, J.; Zheng, L.; Li, M.; Yan, X.; Luo, Q. Simultaneous determination of ten phytohormones in five parts of *Sargasumfusiforme* (Hary.) Seichell by high performance liquid chromatography-triple quadrupole mass spectrometry. *Chin. J. Chromatogr.* **2014**, 32, 861–866. [CrossRef]
90. Wang, A.G.; Luo, G.H. Quantitative relation between the reaction of hydroxylamine and superoxide anion radicals in plants. *Plant Physiol. Commun.* **1990**, *26*, 55–57. [CrossRef]
91. Galluzzi, L.; Kroemer, G. *Conceptual Background and Bioenergetic/Mitochondrial Aspects of Oncometabolism*; Elsevier: Amsterdam, The Netherlands, 2014; p. 542.
92. Jabs, T.; Dietrich, R.A.; Dangl, J.L. Initiation of runaway cell death in an *Arabidopsis* mutant by extracellular superoxide. *Science* **1996**, *27*, 1853–1856. [CrossRef]
93. Thordal-Christensen, H.; Zhang, Z.; Wei, Y.; Collinge, D.B. Subcellular localization of H_2O_2 in plants: H_2O_2 accumulation in papillae and hypersensitive response during the barley-powdery mildew interaction. *Plant J.* **1997**, *11*, 1187–1194. [CrossRef]

International Journal of
Molecular Sciences

MDPI

Article

Hydrogen Sulfide-Linked Persulfidation Maintains Protein Stability of ABSCISIC ACID-INSENSITIVE 4 and Delays Seed Germination

Mingjian Zhou [1], Jing Zhang [1], Heng Zhou [1], Didi Zhao [1], Tianqi Duan [1], Shuhan Wang [1], Xingxing Yuan [2] and Yanjie Xie [1,*]

[1] Laboratory Center of Life Sciences, College of Life Sciences, Nanjing Agricultural University, Nanjing 210095, China; 2020216038@stu.njau.edu.cn (M.Z.); 2018216035@njau.edu.cn (J.Z.); hengzhou@njau.edu.cn (H.Z.); 2019116100@njau.edu.cn (D.Z.); 2021116100@stu.njau.edu.cn (T.D.); 2021116101@stu.njau.edu.cn (S.W.)

[2] Institute of Industrial Crops, Jiangsu Academy of Agricultural Sciences, Nanjing 210014, China; yxx@jaas.ac.cn

* Correspondence: yjxie@njau.edu.cn

Abstract: Hydrogen sulfide (H_2S) is an endogenous gaseous molecule that plays an important role in the plant life cycle. The multiple transcription factor ABSCISIC ACID INSENSITIVE 4 (ABI4) was precisely regulated to participate in the abscisic acid (ABA) mediated signaling cascade. However, the molecular mechanisms of how H_2S regulates ABI4 protein level to control seed germination and seedling growth have remained elusive. In this study, we demonstrated that ABI4 controls the expression of L-CYSTEINE DESULFHYDRASE1 (DES1), a critical endogenous H_2S-producing enzyme, and both ABI4 and DES1-produced H_2S have inhibitory effects on seed germination. Furthermore, the ABI4 level decreased during seed germination while H_2S triggered the enhancement of the persulfidation level of ABI4 and alleviated its degradation rate, which in turn inhibited seed germination and seedling establishment. Conversely, the mutation of ABI4 at Cys250 decreased ABI4 protein stability and facilitated seed germination. Moreover, ABI4 degradation is also regulated via the 26S proteasome pathway. Taken together, these findings suggest a molecular link between DES1 and ABI4 through the post-translational modifications of persulfidation during early seedling development.

Keywords: hydrogen sulfide; persulfidation; DES1; ABI4; protein stability

Citation: Zhou, M.; Zhang, J.; Zhou, H.; Zhao, D.; Duan, T.; Wang, S.; Yuan, X.; Xie, Y. Hydrogen Sulfide-Linked Persulfidation Maintains Protein Stability of ABSCISIC ACID-INSENSITIVE 4 and Delays Seed Germination. *Int. J. Mol. Sci.* **2022**, *23*, 1389. https://doi.org/10.3390/ijms23031389

Academic Editor: Karen Skriver

Received: 4 October 2021
Accepted: 14 January 2022
Published: 26 January 2022

1. Introduction

Growing bodies of reports have demonstrated that H_2S, a gaseous signaling molecule, is involved in a variety of physiological processes during plant growth and development, such as autophagy, flowering, and stomatal closure [1–4]. The multiple H_2S generation pathways have been correspondingly identified, including L/D-cysteine desulfhydrase, β-cyanoalanine synthase, and *O*-acetylserine(thiol)lyase [5,6]. Particularly, cytoplasmic L-Cys desulfhydrase DES1 was considered as one of the most critical H_2S-producing enzymes in *Arabidopsis*. DES1 exhibited a higher affinity to L-cysteine and a strong catalytic capacity to degrade L-cysteine to sulfide, ammonia, and pyruvate [7]. In parallel, endogenous H_2S levels reduced by 30% in the *Arabidopsis des1* mutant [7], which has brought great convenience for us to investigate the physiological effects of H_2S in plants [8,9]. A greater understanding of the mechanism for H_2S biological functions in plants has been achieved, accompanied by the key feature of H_2S at protein persulfidation, one type of post-translational modification of oxidized cysteine residues to form a persulfide group (RSSH) in target proteins [10,11]. To date, persulfidated proteins in plants have been widely reported and the function of persulfation has exhibited great complexity and diversity in different tissues and organs of plants [12,13]. Persulfidation significantly impacts protein biological functions, such as enzymatic activity, subcellular localization, and protein interaction [14–16]. At least 5214 proteins, 13% of the entire annotated proteome in *Arabidopsis* roots, were susceptible to persulfidation [17]. However, our current understanding of persulfidation remains relatively limited compared to other protein modifications such as phosphorylation and acetylation.

Seed germination is an essential stage in the whole life cycle of plants, which is crucial for plant propagation and requires the precise coordination of multiple external and internal signals [18]. A sequence of mechanisms has been adopted to fine-tune intercellular homeostasis during seed germination in plants due to its high vulnerability to biotic and abiotic stresses [19,20]. ABSCISE ACID-INSENSITIVE 4 (ABI4), an APETALA2 (AP2)/ ETHYLENE-RESPONSIVE ELEMENT BINDING PROTEIN (EREBP) domain-containing transcription factor (TF), acts as a positive modulator in the ABA signaling pathway [21]. During seed germination and post-germination process, ABI4 is involved in regulating a wide range of important biological events, including stress response, chloroplast and mitochondria retrograde, and lipid metabolism [22–25]. ABI4 integrates with various phytohormones signaling during seed germination and flowering [23,26,27]. Interestingly, ABI4 acts as either activator or repressor to control the transcription of downstream responsive genes by recognizing coupling 1 elements (CE1) (CACCG and CCAC motif) within their promoters [21,28]. For example, ABI4 could directly control the MAPK cascade by specifically binding to the CE1 motif of the *MAPKKK18* promoter [16]. However, ABI4 inhibits the transcript abundance of the mitochondrial retrograde signal gene *ALTERNATIVE OXIDASE1a* (*AOX1a*) by targeting the CGTGAT element in the promoter [29].

Our current understandings of the mechanism of H_2S-regulated signal transduction have been achieved in genetic and molecular studies. Our recent study has illustrated a DES1-ABI4 loop molecular regulatory mechanism to control plant responses to ABA. ABA-triggered accumulation of H_2S leads to persulfidation of ABI4 at Cys250 [16]. Meanwhile, ABI4 binds specifically to the CE1 motif of the *DES1* promoter to control *DES1* expression and further H_2S production. However, ABI4 transcript and protein levels are fairly low and even undetectable during vegetative growth in mature overexpressing transgenic plants [30], implying that ABI4 is precisely regulated by a complex network while exercising its function. The general mechanism behind the protein stability of ABI4 controlled by protein persulfidation and the underlying signaling pathways remain to be characterized.

In this study, we show that ABI4 controls the expression of *DES1*, and both ABI4 and DES1-produced H_2S have inhibitory effects on seed germination. H_2S-linked persulfidation on ABI4 Cys250 enhances ABI4 stability, which in turn inhibits seed germination and post-germination development in *Arabidopsis*. We establish a molecular mechanism for DES1 and ABI4 synergistically in the regulation of seed germination and seedling establishment through ABI4 protein stability.

2. Results

2.1. Inhibition of Germination and Seedling Growth in Arabidopsis by Exogenous H_2S

Our previous studies have revealed that persulfidation plays an indispensable role in regulating plant ABA responses, including the induction of stomatal closure [14] and the inhibition of primary root growth [16]. To determine whether H_2S is involved in the *Arabidopsis* germination process, we measured the relative expression of *DES1* during the WT seed germination period. Quantitative PCR (qPCR) results showed that the relative expression of *DES1* decreased in a time-dependent manner from dormant seeds to 6-day-old seedlings (Figure 1A). Meanwhile, the enzyme activity of L-cysteine desulfhydrase (L-CDes) was also detected. Mutation of *DES1* resulted in a marked decrease of L-CDes activity in germinating seeds (Figure 1B). Moreover, we observed that the enzyme activity of total L-CDes decreased significantly during the progress of germination in WT but not in *des1* mutant (Figure 1B). These results further proved that L-CDes activity decrease was closely related to DES1, which may function as a negative regulator during seed germination.

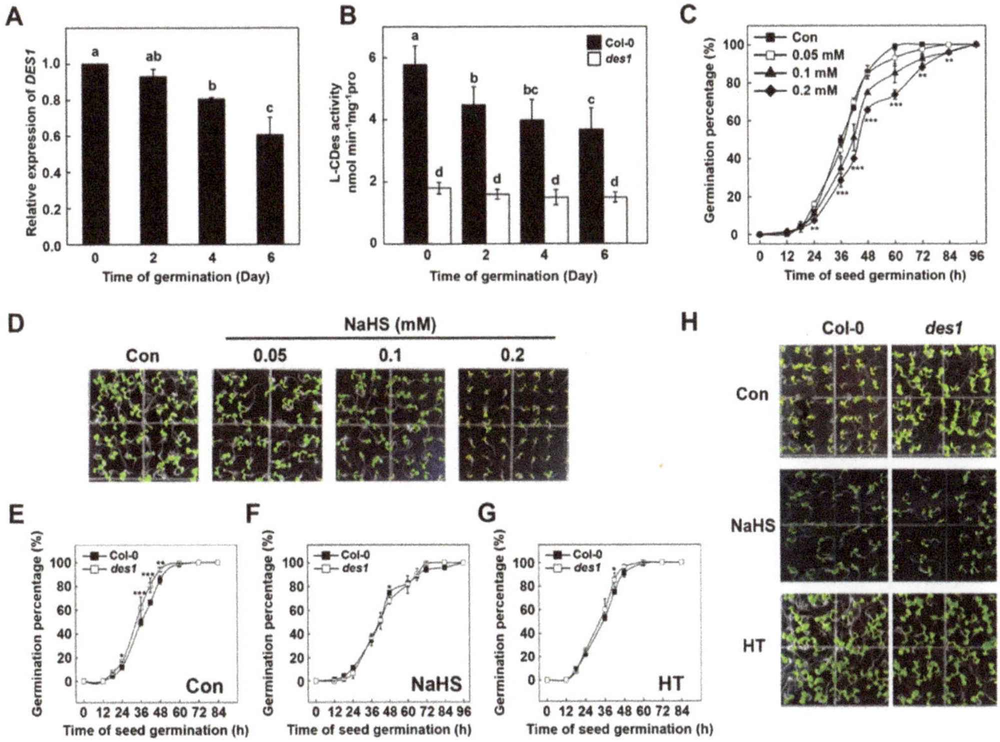

Figure 1. The inhibitory effect of H_2S on the seed germination and post-germination of *Arabidopsis thaliana*. (**A**) Time-course changes of relative expression of *DES1* in wild-type (Col-0) *Arabidopsis* seedlings. *Arabidopsis* seeds were incubated on $^1/_2$ MS medium. Afterward, the relative expression of *DES1* was determined at the indicated time points after germination. Expression levels are presented relative to that of *UBQ10*, with *DES1* expression in the 0-day set to 100%. (**B**) Time-course changes of L-CDes activity in wild-type (Col-0) and *des1* germinating seeds. L-CDes activity was determined at the indicated time points after germination. (**C**) Time-course changes of germination percentage in Col-0 *Arabidopsis* seedlings upon NaHS treatment. *Arabidopsis* seeds were incubated on $^1/_2$ MS medium containing 0 (control), 0.05, 0.1, or 0.2 mM NaHS, respectively. The germination percentage (%) of seeds was counted at the indicated time points. Con: control. (**D**) Photographs of wild-type *Arabidopsis* seeds germinated seeds five days after the initiation of germination on $^1/_2$ MS medium containing NaHS (0, 0.05, 0.1, 0.2 mM). (**E–G**) Time-course changes of germination percentage in *Arabidopsis* seedlings upon NaHS or HT treatment. (**H**) Photographs of germinated seeds of wild-type and *des1* five days after the initiation of germination on $^1/_2$ MS medium containing NaHS (0.1 mM) or HT (1 mM). HT: hypotaurine. Data are mean ± SE of three independent experiments with three biological replicates for each individual experiment. Bars with different letters indicate significant differences at $p < 0.05$ according to one-way ANOVA (post-hoc Tukey's HSD test). Asterisks represent significant differences between treatments according to Student's *t*-test (* $p < 0.05$, ** $p < 0.01$, *** $p < 0.001$).

To further analyze the effect of H_2S on *Arabidopsis* seed germination, NaHS, a well-known H_2S donor, was applied into an MS medium with different concentrations. As expected, the inhibitory effects of NaHS on *Arabidopsis* seed germination were observed over a wide range of NaHS concentrations from 0.05 to 0.2 mM (Figure 1C). These results indicated that NaHS treatment could delay the beginning of germination in *Arabidopsis* and prolong the full germination period in a concentration-dependent manner. However,

NaHS-treated seeds can fully germinate at 96 h after treatment (Figure 1C). Subsequently, we examined NaHS affect the post-germination process. We observed that *Arabidopsis* cultured in different NaHS concentrations displayed growth inhibition as shown by a lower greening degree and smaller leaves of the seedlings. NaHS strengthened this inhibitory effect in a concentration-dependent manner (Figure 1D). Primary root length was measured to evaluate the effect of NaHS on post-germination growth. Accordingly, NaHS treatment inhibited primary root growth of 5-day-old *Arabidopsis* seedlings, suggesting H_2S inhibits seedling growth (Supplementary Figure S1). Taken together, these results revealed that exogenous H_2S delayed the initiation of seed germination and inhibited post-germination growth.

2.2. DES1-Produced H_2S Inhibits Arabidopsis Seed Germination and Post-Germination Growth

To test whether DES1-produced H_2S impacted seed germination, the germination performance of the *des1* mutant was compared with WT. Under normal conditions, *des1* mutant germinates faster compared to WT, especially 36 to 48 h after germination. However, both WT and *des1* mutant completely germinated after 72 h (Figure 1E). Subsequently, the *des1* mutant was treated with NaHS. As shown, the germination rate of both WT and *des1* displayed a similar, overall downward trend with the treatment of NaHS (Figure 1F). Long treatment with NaHS can lead to sulfide oxidation and thus form polysulfides. In this study, the treatment of chemical H_2S donor GYY4137 resulted in a similar inhibitory effect on both WT and *des1* as NaHS, suggesting that the effects on seed germination observed are related to H_2S directly (Supplementary Figure S2). Meanwhile, both WT and *des1* mutant seed germination was significantly accelerated by hypotaurine (HT), an H_2S scavenger (Figure 1G).

The post-germination phenotype was also investigated. *des1* mutant displayed a faster growth trend than WT during the post-germination process (Figure 1H). The greening degree of *des1* was significantly higher than that of WT, and the leaves were larger after five days of germination under control conditions. Meanwhile, the application of NaHS inhibited the development of the leaf and delayed the greening of both *des1* and WT. By contrast, treatment of HT accelerated seedling growth and abolished the phenotypic differences between WT and *des1* (Figure 1H). Therefore, these experimental data revealed that DES1-produced H_2S has an inhibitory effect on both germination and post-germination growth of *Arabidopsis*.

2.3. Functional of DES1 and ABI4 on Seed Germination and Post-Germination Growth

Our previous studies demonstrated that ABI4 can activate *DES1* transcription by binding to its promoter, and Cys250 from ABI4 is critical for its binding [16]. In this study, the qPCR analysis showed that the expression of *DES1* was significantly blocked in the 5-day-old *abi4* mutant seedling (Figure 2A). Correspondingly, *DES1* expression was higher in transgenic plants overexpressing *ABI4* but not *the* $ABI4^{Cys250Ala}$ variant (Figure 2A), implying that Cys250 is crucial for ABI4-activated *DES1* expression in *Arabidopsis* seedlings. Subsequently, the total L-CDes activity was also measured in *Arabidopsis* seedlings of WT, *abi4*, *35S:ABI4 abi4*, and $35S:ABI4^{Cys250Ala}$ *abi4*. *abi4* and $35S:ABI4^{Cys250Ala}$ *abi4* exhibited decreased L-CDes activities compared with WT, whereas L-CDes activity was increased in *35S:ABI4 abi4* (Figure 2B). These results demonstrated that *DES1* expression was controlled by functional ABI4 in *Arabidopsis* seedlings.

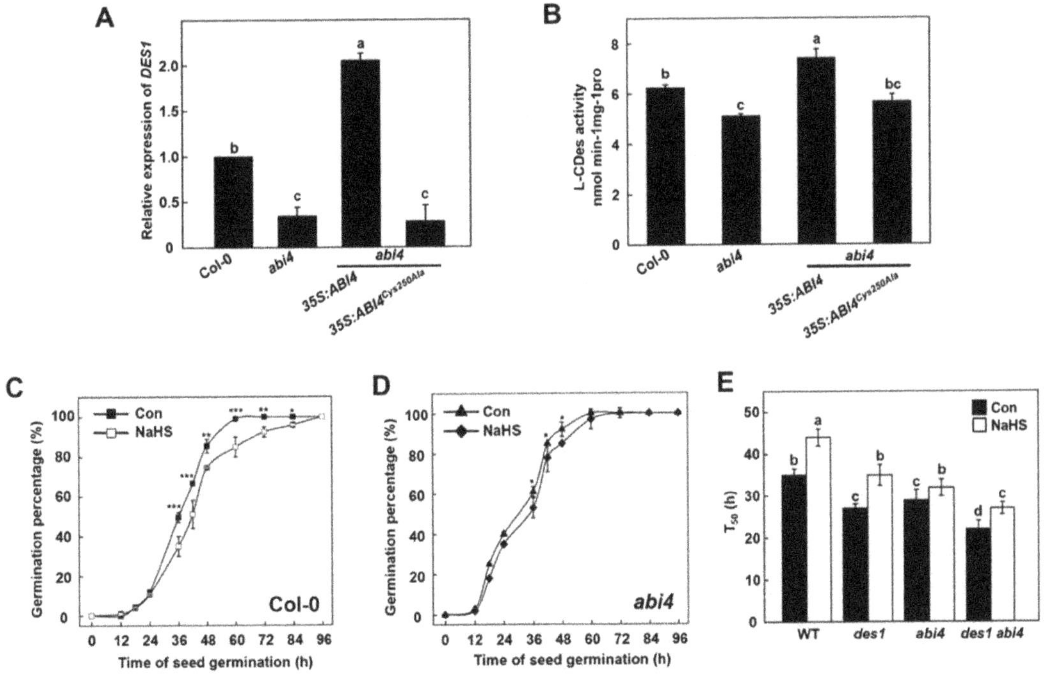

Figure 2. Functional analysis of DES1 and ABI4 on seed germination and post-germination growth. (**A**,**B**) Relative expression of *DES1* (**A**) and L-CDes activity (**B**) in Col-0, *abi4*, *35S:ABI4 abi4*, and *35S:ABI4*^{Cys250Ala} *abi4 Arabidopsis* seedlings. Five-day-old seedlings were collected for qRT-PCR analysis and measurement of L-CDes enzyme activity, respectively. Expression levels are presented relative to that of *UBQ10*, with *DES1* expression in the Col-0 set to 100%. (**C**,**D**) Time-course changes of germination percentage in Col-0 and *abi4 Arabidopsis* seedlings upon NaHS treatment. *Arabidopsis* seeds were incubated on $^1/_2$ MS medium containing 0.1 mM NaHS, respectively. The germination percentage (%) of seeds was counted at the indicated time points. (**E**) The time to obtain 50% of germinated seeds in WT, *des1*, *abi4*, and *des1 abi4* seeds with or without 0.1 mM NaHS. Data are mean ± SE of three independent experiments with three biological replicates for each individual experiment. Bars with different letters indicate significant differences at $p < 0.05$ according to one-way ANOVA (post-hoc Tukey's HSD test). Con: control. Asterisks represent significant differences between treatments according to Student's *t*-test (* $p < 0.05$, ** $p < 0.01$, *** $p < 0.001$).

Time-course changes of germination percentage were examined in WT and *abi4 Arabidopsis* seedlings upon NaHS treatment. While the germination speed of *abi4* is faster than wild-type, NaHS-triggered inhibitory response in germination is weakened by the mutation of ABI4 (Figure 2C,D). The germination rates of *des1* and *abi4* mutant lines were further compared. Mutation of either DES1 or ABI4 reduced the time to have 50% germinated seeds (T50; Figure 2E). Meanwhile, the T50 index is significantly shortened in *des1 abi4* double mutant than its parental lines, all of which were significantly increased by NaHS treatment. Furthermore, DES1 and ABI4 also showed additive effects with regard to seedling primary root growth (Supplementary Figure S3). Taken together, these results demonstrated that H_2S and ABI4 have additive inhibitory effects during seed germination and the post-germination stages.

2.4. ABI4 Protein and Its Persulfidation Level Decreases during Germination and Post-Germination Stages

ABI4 is a positive regulator of ABA regulation, and ABA has an inhibitory effect on the germination of *Arabidopsis* [21]. However, the expression of ABI4 was very low in vegetative growth [30]. In this study, *35S:ABI4-GFP* transgenic seedlings were used. Meanwhile, the effect of NaHS on ABI4 protein level during seed germination (2 days) and post-germination (6 days) stages were compared. The results showed that the level of ABI4 protein in the post-germination stage decreased by half compared with that of the germination stage (Figure 3A,B). However, compared to the control sample, treatment with NaHS caused a 40% increase in ABI4 protein level after two days of sampling. Meanwhile, the decreased ABI4 protein level at 6-days after germination increased by 100% after the addition of NaHS treatment. In conclusion, the ABI4 protein level decreased markedly during seed germination and post-germination growth, which can be future attenuated by the addition of H_2S.

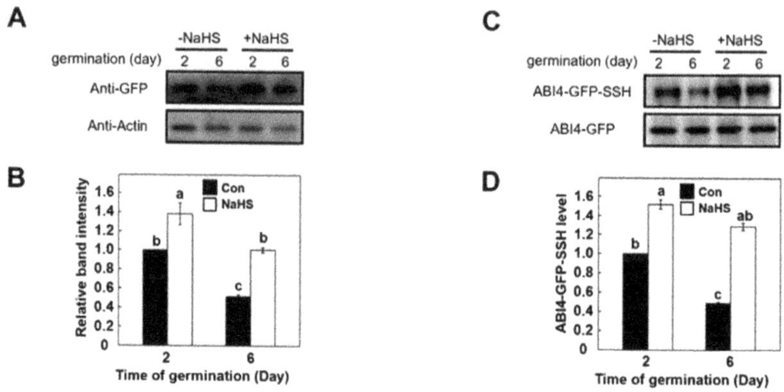

Figure 3. The stability and persulfidation level of ABI4 during germination and post-germination growth. (**A**) NaHS affects the ABI4 protein level In vivo. *35S:ABI4-GFP* transgenic seedlings grown on $^1/_2$ MS medium with or without NaHS (0.1 mM) were harvested at the indicated times. Total proteins were extracted and then checked by immunoblot analysis. ABI4 proteins were detected using an anti-GFP antibody. Relative amounts of proteins were determined by densitometry normalized to actin. (**B**) Quantification of the relative band intensity shown in (**A**). (**C**) Effect of the NaHS on the persulfidation of ABI4 In vivo. Sample treatments were as described for (**A**). Total proteins were extracted and then subjected to the biotin-switch assay to analyze persulfidation levels (ABI4-GFP-SSH). Persulfidated ABI4-GFP protein was detected with anti-GFP antibody after tag-switch labeling and streptavidin purification. The bottom panels show the total ABI4-GFP used for the tag-switch assay as a loading control. (**D**) Quantification of the ABI4-GFP-SSH level is shown in (**C**). For (**B**,**D**), the bars indicate the relative abundance of the corresponding protein (**B**) or persulfidated protein (**D**) compared with that of the control un-treated sample (set to 1.0). Signals from two independent experiments were quantified. Different letters indicate significant differences at $p < 0.05$ according to one-way ANOVA (post-hoc Tukey's HSD test).

To verify whether the above NaHS-driven responses were related to ABI4 persulfidation, we further checked the persulfidation level of ABI4 by a biotin-switch method in the following experiment [15]. As shown in Figure 3C, the ABI4 persulfidation level of seedlings decreased by approximately 50% after six days of growth, suggesting the ABI4 presulfidation level decreased during seed germination and seedling establishment. However, the declined level of ABI4 persulfidation was reversed to a higher level in the presence of NaHS, compared with that of the 2-day sample (Figure 3D). Besides, the level of ABI4 persulfidation increased dramatically in both 2-day and 6-day samples after exposure to NaHS. These results suggested that both protein and corresponding persulfidation level of ABI4 was regulated by H_2S.

2.5. *Persulfidation at Cys250 of ABI4 Inhibits Its Degradation*

We hypothesized that persulfidation of ABI4 may regulate its protein level during germination. To test whether persulfidation affects ABI4 stability, we set up a cell-free assay using ABI4-His recombinant protein purified from *Escherichia coli*, treating with either NaHS or dithiothreitol (DTT) followed by dialysis. The level of ABI4-His recombinant protein was detected after 3–18 h incubation with total protein extract from WT, followed by using immunoblotting with anti-His antibody. With an increased incubation period, the ABI4-His protein level was gradually decreased, which was further attenuated or exacerbated by NaHS or DTT treatment, respectively (Figure 4A,B). For example, the ABI4-His protein level decreased by 60% after 9 h of incubation, while this value is 20% or 85% after treatment of NaHS or DTT, respectively. Moreover, the decreased tendency of ABI4-His protein level became faster when its Cys250 was mutated to Ala (Figure 4C,D), which showed a maximum decrease of 40% at 6 h of incubation. Meanwhile, treatment of NaHS or DTT enhance or decrease the persulfidation level of recombinant ABI4 protein, whereas Cys250Ala mutation almost fully abolished the persulfidation of recombinant ABI4 protein (Supplementary Figure S4). These results further indicated persulfidation on Cys250 is critical for ABI4 stability.

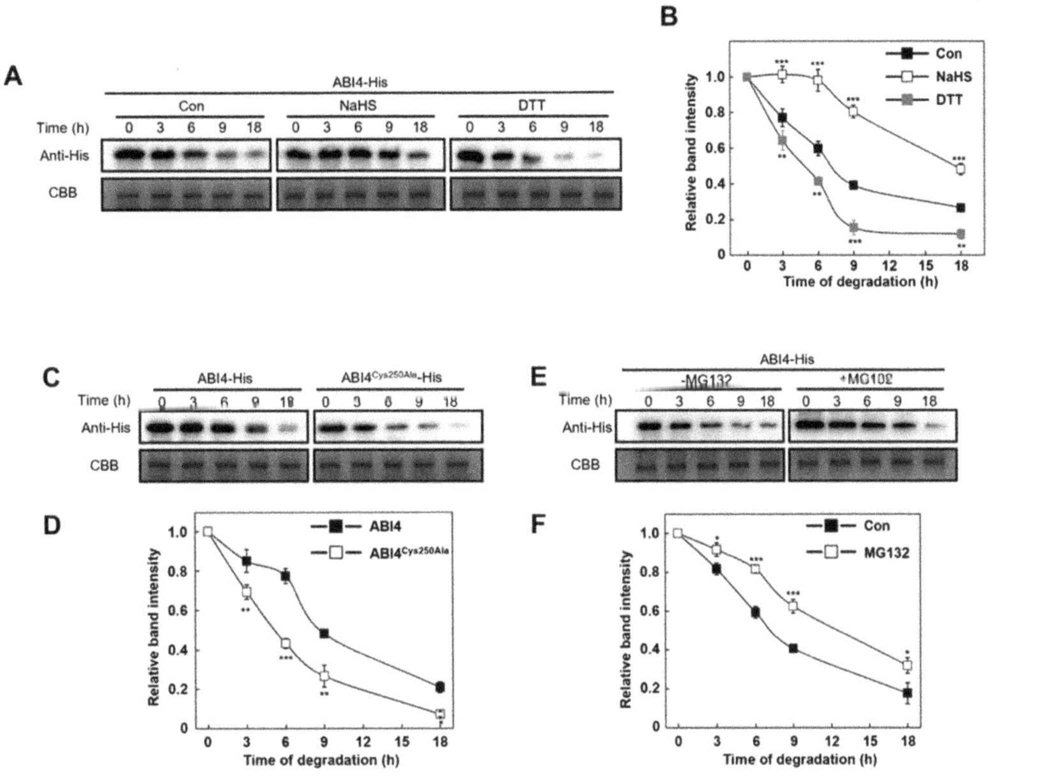

Figure 4. Persulfidation at Cys250 of ABI4 inhibits its degradation In vitro. (**A**) Cell-free degradation of ABI4-His recombinant protein in protein extracts from Col-0. Purified ABI4-His recombinant protein was treated with NaHS (0.1 mM) or DTT (1 mM) for 1 h, dialyzed, and then incubated with extracts of 10-day-old *Arabidopsis* seedlings at 25 °C. Protein without chemical treatment was set as

control (Con). During incubation, samples were collected at indicated time points for immunoblot analysis using anti-His antibody. Samples collected before the addition of protein extracts were set as internal control, as normalized to load equal amounts determined by Coomassie brilliant blue staining (CCB). (**B**) Quantification of the relative band intensity shown in (**A**). (**C**) Cell-free degradation analysis of ABI4 mutant recombinant proteins. Purified ABI4-His and ABI4^{Cys250Ala}-His recombinant protein were dialyzed and then incubated with extracts of 10-day-old *Arabidopsis* seedlings. Sample treatments and measurements were as described for (**A**). (**D**) Quantification of the relative band intensity shown in (**C**). (**E**) The effect of MG132 on the stability of ABI4 in vitro. Purified ABI4-His recombinant proteins were dialyzed and then incubated with extracts containing 50 μm MG132 of 10-day-old *Arabidopsis* seedlings. Sample treatments and measurements were as described for (**A**). (**F**) Quantification of the relative band intensity shown in (**E**). For the quantification of relative band intensity, the data indicate the relative abundance of the corresponding protein compared with that of the control sample (set to 1.0). Signals from two independent experiments were quantified. Asterisks represent significant differences between treatments according to Student's *t*-test (* $p < 0.05$, ** $p < 0.01$, *** $p < 0.001$).

To further investigate whether the decrease of ABI4 level is also attributed to the ubiquitin-26S proteasome pathway, MG132, a specific 26S proteasome inhibitor of ubiquitin-mediated protein degradation, was used. In this study, ABI1 protein, an important component of ABA signaling, where its degradation was regulated by ubiquitination [31,32], was used as a positive control. In vitro and In vivo experiments demonstrated that MG132 sufficiently blocks ABI1 degradation (Supplementary Figure S5). As expected, the decrease level of ABI4-His recombinant protein was relieved by the application MG132 to some extent, with maximum mitigation effect at 6 h by 25% (Figure 4E,F). These results indicated that the ABI4 Cys250 persulfidation regulates its degradation In vitro, and this process is partially related to 26S proteasome activity.

2.6. The Stability of ABI4 Is Regulated by Its Persulfidation

To investigate the effect of persulfidation on protein stability of ABI4 In vivo, a protoplast-based time course degradation assay was performed. The protoplasts from WT plants were transfected with equal amounts of plasmids expressing either ABI4-GFP or ABI4^{Cys250Ala}-GFP, respectively. The protoplasts expressing ABI4-GFP were treated with or without NaHS before lysed. Cycloheximide (CHX) was used to inhibit protein synthesis during protein extraction. As shown in Figure 5A,B, NaHS treatment attenuated the decrease of ABI4-GFP level during the whole detection period compared with the control sample, indicating that NaHS enhanced the stability of ABI4-GFP In vivo. However, the possibility that NaHS might inhibit proteasome activity and thus indirectly enhance the stability of ABI4 cannot be ruled out.

Cys250 mutation almost completely abolished ABI4 persulfidation (Supplementary Figure S4). Next, the degradation rate between ABI4-GFP and ABI4^{Cys250Ala}-GFP were compared. Notably, our results exhibited that the ABI4^{Cys250Ala}-GFP mutant version caused a decrease rate which increased to approximately 80% at 9 h, whereas ABI4-GFP only decreased by 50% (Figure 5C,D). These results demonstrated that Cys250 residue is necessary for maintaining the ABI4-GFP level. Subsequently, the following experiment demonstrated that MG132 attenuated the decreased tendency of the ABI4-GFP level in a time-dependent manner (Figure 5E,F), indicating the involvement of the 26S proteasome-mediated degradation pathway. The persulifdation level of ABI4-GFP was increased by NaHS, regardless of the presence of MG132 (Supplementary Figure S6). Furthermore, the degradation rate of ABI4-GFP was further inhibited by the application of NaHS in the presence of MG132, especially during 0–3 h of treatment. After 9 h of incubation, the degradation rate was inhibited by 40% with the treatment of both MG132 and NaHS (Figure 5E,F). Taken together, these results indicated that persulfidation regulates ABI4 protein stability, in which the pathway of 26S proteasome may involve.

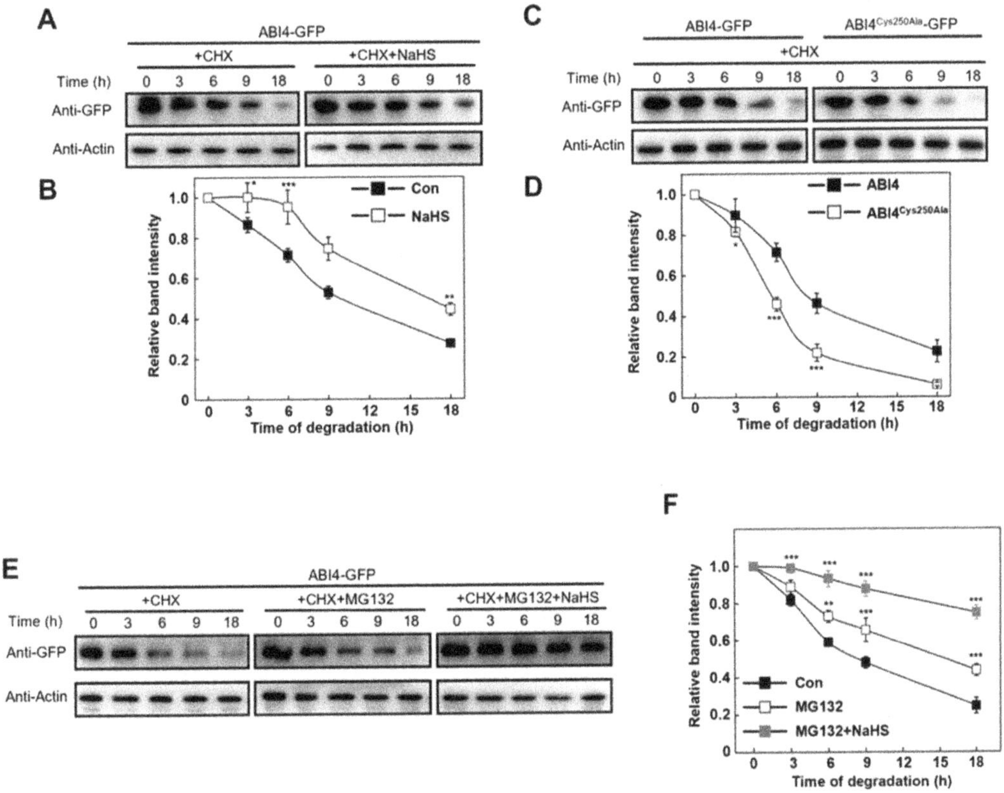

Figure 5. The stability of ABI4 is regulated by its persulfidation In vivo. (**A**) The effect of NaHS on the stability of ABI4 In vivo. *Arabidopsis* protoplasts isolated from the Col-0 lines were transfected with ABI4-GFP expressing plasmid constructs. After incubation in low light for 12 h, protoplasts were treated with 0.1 mM NaHS (or distilled water, as a control) and 150 μM CHX (to block protein translation) for 1 h. Then protoplasts were lysed and incubated at 30 °C after being divided into five tubes. Samples were stopped at indicated time points and checked by immunoblot analysis. Proteins were detected using an anti-GFP antibody, and relative amounts of proteins were determined by densitometry normalized to actin. (**B**) Quantification of the relative band intensity shown in (**A**). (**C**) The stability of wild-type ABI4 or its Cys250Ala mutated version In vivo. *Arabidopsis* protoplasts isolated from the wild-type lines were transfected with ABI4-GFP or $ABI4^{Cys250Ala}$-GFP expressing plasmid constructs, respectively. Sample harvest and measurements were as described for (**A**). (**D**) Quantification of the relative band intensity shown in (**C**). (**E**) The effect of MG132 and NaHS on the stability of ABI4 In vivo. Protoplasts were treated with 50 μm MG132 or 50 μm MG132 and 0.1 mM NaHS, respectively (or distilled water, as a control) for 1 h, and other sample treatments and measurements were as described for (**A**). (**F**) Quantification of the relative band intensity shown in (**E**). For the quantification of relative band intensity, the data indicate the relative abundance of the corresponding protein compared with that of the control sample (set to 1.0). Signals from two independent experiments were quantified. Asterisks represent significant differences between treatments according to Student's *t*-test (* $p < 0.05$, ** $p < 0.01$, *** $p < 0.001$).

3. Discussion

In this study, we focused on the effects of H_2S and ABI4 on seed germination and post-germination growth under normal physiological conditions. While ABA levels decreased gradually during seed germination, we found that ABI4 and DES1-produced H_2S

have an inhibitory effect on seed germination synergistically. We established a molecular framework for H_2S-linked persulfidation maintaining protein stability of ABI4 in the regulation of *Arabidopsis* seed germination and post-germination process. During this process, ABI4 protein was degraded concurrently and its persulfidation level decreased, which was closely related to H_2S/DES1. ABI4 controls the expression of *DES1*, and both ABI4 and DES1-produced H_2S have inhibitory effects during seed germination and post-germination growth. We further discovered persulfidation regulates ABI4 stability during seed germination and post-germination growth in *Arabidopsis*.

Multiple lines of evidence obtained from genetic and physiological studies demonstrate that DES1-produced H_2S has an inhibitory effect on both germination and post-germination growth of *Arabidopsis*. This idea is supported by the observation that seed germination and post-germination development was inhibited by NaHS in a concentration-dependent manner (Figure 1D). Consistently, the *des1* mutant showed faster germination compared with WT. Our data further suggested that ABI4 regulates *DES1* expression, as *DES1* expression and L-CDes activities in *abi4* mutants was severely reduced from that in the WT, respectively. These results may be supported by the fact that ABI4 activates *DES1* transcription by binding to its promoter [16]. Interestingly, the germination of WT, *des1*, *abi4*, and *des1 abi4* seeds were all significantly inhibited by the treatment of NaHS (Figure 2E). Based on these findings, this study provides new evidence for the notion that ABI4 integrates with DES1 as a negative regulator to participate in seed germination coordinately.

ABI4 plays an essential role as one of the positive regulation transcription factors mediating ABA-dependent stress response [25]. The ABI4 protein level is precisely regulated by a complex network under a vegetative state [30]. In this study, protein degradation analysis revealed that MG132, a specific 26S proteasome, relieved the decrease level of ABI4 protein, indicating that ABI4 degradation was also regulated via the 26S proteasome pathway, which was also found in ABI1 and ABI5 [32,33]. However, a recent study has found that phytoplasma SAP05 (secreted AY-WB proteins) mediates the concurrent degradation of SPL and GATA developmental regulators via hijacking the plant ubiquitin receptor RPN10 independent of substrate ubiquitination [34]. Thus, the degradation of ABI4 may also not be dependent on ubiquitination entirely. Interestingly, the accumulation of ABI4 in seeds was observed at the early stage of germination, and its steady-state mRNA levels dropped sharply a few days after germination [35]. Importantly, the ABI4 level in the NaHS treated seedlings was evidently higher than in the control. These results further implied that H_2S might also be involved in the mechanism of regulation of ABI4 stability to control seed germination.

H_2S is emerging as a potential messenger molecule involved in the modulation of various aspects of physiological processes in plants [36,37]. Signaling by H_2S is proposed to occur via persulfidation, a post-translational modification of protein Cys residues (R-SHs) to form persulfides (R-SSH). Our results revealed that the ABI4 presulfidation level decreased during seed germination (Figure 3C), and the ABI4 protein stability could be attenuated or accelerated by NaHS or DTT (Figure 4A). Furthermore, mutation of Cys250 to Ala accelerated ABI4 decrease tendency (Figures 4C and 5C). Collectively, these results illustrated that the persulfidation of ABI4 was also linked to ABI4 stability.

In conclusion, we discovered that H_2S could greatly trigger the enhancement of the persulfidation level of ABI4 and alleviate its degradation rate, which in turn inhibits *Arabidopsis* germination. The discovery of ABI4 protein stability regulated by its persulfidation expands our understanding of H_2S role in plant signal transduction networks and establishes a molecular framework for the crosstalk between different post-translational modifications during seed germination.

4. Materials and Methods

4.1. Plant Materials and Growth Conditions

Arabidopsis mutants of *des1* (SALK_103855; Col-0) and *abi4* (SALK_080095; Col-0) mutants were obtained from the Arabidopsis Biological Resource Center (http://www.

arabidopsis.org/abrc, accessed on 23 March 2020). Cross *abi4* with *des1* to obtain the double mutant *des1 abi4*. *35S:ABI4-GFP* transgenic materials were obtained from Dr. Wei Chi (Institute of Botany, The Chinese Academy of Sciences, Beijing, China). Seeds were disinfected with sodium hypochlorite for 20 min and then washed three times with sterile water. They were cultured in Petri dishes on semi-solid Murashige and Skoog ($^1/_2$ MS) medium (pH 5.8). The plants were grown in a 16 h/8 h (23 °C/18 °C) growth chamber using a bulb-type fluorescent lamp with a light intensity of 100 mmol photons m^{-2} s^{-1} irradiation.

4.2. *Molecular Cloning*

4.2.1. For Expression in *Escherichia coli*

The fragment of ABI4 amplified by PCR was introduced into the PET28a vector (for His fusion) using a homologous recombination technique (Vazyme, Nanjing, China) with the enzyme digestion sites *NdeI* and *XhoI*. Site-directed mutagenesis was performed using a Mut Express II Fast Mutagenesis Kit (Vazyme). All procedures followed the manufacturer's manual. The specific primers used for *ABI4* are listed in Supplementary Data 1.

4.2.2. For Transient Expression in *Arabidopsis* Protoplasts

Homologous recombination technology (Vazyme) was used to transfer PCR amplified fragments into the PAN580 vector at *XbaI* and *BamHI*. *Arabidopsis* protoplasts were extracted according to the method previously described by Yoo et al. [38] with some modifications. The constructed transient expression vector PAN580 was transformed into *Arabidopsis* protoplasts combined with PEG calcium-mediated method and cultured in the dark at 22 °C for 16 h. The protoplasts expressing target GFP were identified by fluorescence microscopy and used for transient expression analysis.

4.2.3. For Expression in Planta

ABI4 fragments were cloned into the pCAMBIA 1302 vector. The constructed plasmid was transferred into a competent cell of *Agrobacterium tumefaciens* and then transformed into *Arabidopsis* using the inflorescence infection method. $^1/_2$ MS medium with 50 mg/mL hygromycin B was used to select transgenic plants. PCR, fluorescence observation, and western blot analysis were combined to identify transgenic plants.

4.3. *Real-Time RT-PCR Analysis*

Mature *Arabidopsis* seedlings were collected for RNA extraction. According to the manufacturer's instructions, seedlings were ground using a mortar and pestle in liquid nitrogen until a fine powder appeared, and then separated total RNA using RNA-easy Isolation Reagent (Vazyme, Nanjing, China). 1000 ng RNA from seedlings was used to synthesize the first-strand cDNA in a 20 μL reaction volume (Vazyme, Nanjing, China) using 1 μM of primers. According to the manufacturer's instructions, the AceQ qPCR SYBR green master mix (Vazyme) was used for real-time RT-PCR in the Mastercycler®ep realplex real-time PCR system (Eppendorf, Germany). The specific primers used are listed in Supplementary Data 1.

4.4. *Determination of Activity of L-Cysteine Desulfhydrase*

The method for determining L-cysteine desulfhydrase activity was described by Riemenschneider et al. [39] with some modifications. 0.2 g of seedlings was collected and ground with liquid nitrogen and the soluble protein was extracted using 1 mL 20 mM pH 8.0 Tris-HCl. The protein concentration was determined using a BCA kit (Takara, Dalian, China) and calibrated to be consistent. The release of H_2S was determined to evaluate L-cysteine desulfhydrase activity. 100 mM Tris HCl (pH 9.0), 2.5 mM DTT, 0.8 mM L-cysteine and 10 μg protein solutions were mixed to 1 mL. The reaction was initiated by the addition of L-cysteine and terminated by 100 μL 30 mM $FeCl_3$ and 100 μL 20 mM N, N-dimethyl-p-phenylenediamine dihydrochloride after incubation at 37 °C for 15 min in the dark. The content of H_2S was measured by colorimetric at 670 nm. Taking the known

Na_2S concentration as the standard curve, the activity of L-cysteine desulfhydrase was expressed as nmol g^{-1} FW min^{-1}.

4.5. Seed Germination and Green Open Cotyledon Assays

The seeds were germinated and grown on $^1/_2$ MS medium with or without NaHS, if there were no other instructions in the text. The percentage of germinated seeds and green cotyledons were recorded, and the seedlings were photographed at the designated time points.

4.6. Expression and Purification of Recombinant Protein

The recombinant protein was expressed and purified in BL21 competent cells (Vazyme). 0.1 mM IPTG was added, and the bacteria were grown to OD600 = 0.4 to 0.6 at 16 °C for 12 h. After enriching the bacterial solution, suspend the pellet in PBS buffer and use an ultrasonic breaker. The protein was broken and centrifuged at 12,000× *g* for 30 min, and then the extract was collected for purification. NI-NTA pre-packaged gravity column (Sangon Biotech, Shanghai, China) was used to purify the His-labeled protein, and the protein purification procedure was performed in accordance with the column specifications.

4.7. SDS-PAGE and Immunoblotting

Protein extracts were separated by 12.5% SDS-PAGE. The electrophoresis was ended when the bromophenol blue was moved to 5 mm below the gel, and the gel was transferred to a polyvinylidene fluoride membrane (Roche, Shanghai, China) for 60 min at 100 V on ice using a wet transfer method. The membrane was rinsed with deionized water, then immersed in a blocking buffer (5% skim milk), placed on a decolorizing shaker, and incubated slowly for one hour or overnight at 4 °C. Place the membrane in TBST with a sufficient amount of primary antibody and incubate at room temperature for 2 h with gentle shaking. After the primary antibody incubation, the membrane was washed three times with TBST for 7 min each time. The appropriate HRP-labeled secondary antibody was labeled according to the source of the primary antibody, it was diluted in the corresponding proportion, and shaken gently at room temperature for 1 h. The horseradish peroxidase HRP-ECL luminescence method was used to perform the immunoblot analysis. The software ImageJ (https://imagej.nih.gov/ij/, accessed on 31 December 2021) was used to quantify protein abundance, and signals from two independent experiments were quantified. Full-sized membrane scans are presented in Supplemental Figure S7.

4.8. Immunochemical Detection of S-Persulfidated Proteins

The protein persulfidation level was detected with a tag-switch method described by Aroca et al. [15] with modifications. *35S:ABI4-GFP* transgenic seedlings were grown in $^1/_2$ MS medium. Seedling extracts were dissolved with protein extraction buffer (25 mM pH8.0 Tris-HCl, 100 mM NaCl, and 0.2% Triton X-100) containing protease inhibitor (Yeasen, Shanghai, China). The extract was centrifuged at 4 °C for 10 min and centrifuged at 12,000× *g* for 10 min. Take 80 μL supernatant and add 320 μL blocking buffer (50 mM methylsulfonylbenzothiazole was dissolved in tetrahydrofuran), incubated at 37 °C for 1 h to block free sulfhydryl groups. The protein was precipitated twice with acetone, the pellet was resuspended with buffer (50 mM pH 8.0 Tris-HCl, 2.5% SDS, and 20 mM HPDP-biotin), and then incubated at 37 °C in the dark for 3 to 4 h. 33 μL of protein loading buffer was added without β-mercaptoethanol. It was then incubated at 95 °C for 5 min, the protein was separated with 12.5% SDS-PAGE, transferred to the PVDF membrane, and the level of persulfidation was detected with an anti-GFP antibody (Beyotime, Shanghai, China).

4.9. Cell-Free Protein Degradation Assay

The cell-free protein degradation assay was performed as described with some modifications [32]. 10-day-old WT seedlings were harvested, homogenized in liquid nitrogen, and suspended in degradation buffer (25 mM pH 7.4 Tris-HCl, 10 mM $MgCl_2$, 50 mM NaCl,

1 mM DTT, 0.2% Triton X-100, 5 mM ATP, and 1 mM PMSF). The lysis was centrifuged at 12,000× *g* at 4 °C for 10 min. 100 ng purified recombinant ABI4-His protein was incubated with 500 µg total proteins in a total volume of 100 µL for each reaction. Samples were kept at 25 °C. The reactions were stopped at indicated times by adding a 5× SDS loading buffer. The samples were incubated at 95 °C for 5 min and subjected to western blot analysis with anti-His antibody (Beyotime, Shanghai, China).

4.10. Protoplast-Based Protein Degradation Assay

Protoplasts were lysed in 200 µL degradation buffer (25 mM pH 7.4 Tris-HCl, 50 mM NaCl, 1 mM DTT, 1 mM PMSF, 5 mM ATP and 0.2% Triton X-100) with 150 µM CHX. The lysis was divided into 5 tubes (80 µL each) and incubated at 30 °C. The reaction was terminated by adding 20 µL 5 × SDS loading buffer at the indicated time points. Samples were kept on ice until all the reactions were completed, then incubated at 95 °C for 5 min and subjected to Western blot analysis with anti-GFP antibody (Beyotime, Shanghai, China).

4.11. Statistical Analysis

The statistical analysis and graph construction were performed using SPSS v16.0 (https://www.ibm.com/products/spss-statistics, accessed on 30 December 2021). Differences were considered significant at $p < 0.001$, 0.01, or 0.05.

Supplementary Materials: The following are available online at https://www.mdpi.com/article/10.3390/ijms23031389/s1.

Author Contributions: M.Z., J.Z., H.Z., D.Z., T.D., S.W., X.Y. and Y.X., designed the study. M.Z., J.Z., H.Z., T.D. and S.W., performed the experiments. M.Z., J.Z., H.Z., D.Z., T.D., S.W., X.Y. and Y.X. analyzed the data. M.Z., J.Z., H.Z. and Y.X wrote the paper. All authors have read and agreed to the published version of the manuscript.

Funding: This work was supported by grants from the National Natural Science Foundation of China (32101671, 31670255), the National Natural Science Foundation of China of Jiangsu Province (BK20200561), and the China Postdoctoral Science Foundation (2019M661860).

Institutional Review Board Statement: Not applicable.

Informed Consent Statement: Not applicable.

Data Availability Statement: Not applicable.

Conflicts of Interest: The authors declare no conflict of interest.

References

1. Frederick, D.D.; Nair, S.P.; Ward, P.D. Increased growth and germination success in plants following hydrogen sulfide administration. *PLoS ONE* **2013**, *8*, e62048.
2. Laureano-Marín, A.M.; Aroca, Á.; Pérez-Pérez, M.E.; Yruela, I.; Jurado-Flores, A.; Moreno, I.; Gotor, C. Abscisic acid-triggered persulfidation of the cysteine protease ATG4 mediates regulation of autophagy by sulfide. *Plant Cell* **2020**, *32*, 3902–3920. [CrossRef] [PubMed]
3. Zhang, H.; Hu, S.L.; Zhang, Z.J.; Hu, L.Y.; Jiang, C.X.; Wei, Z.J.; Liu, J.; Wang, H.L.; Jiang, S.T. Hydrogen sulfide acts as a regulator of flower senescence in plants. *Postharvest Biol. Technol.* **2011**, *60*, 251–257. [CrossRef]
4. Scuffi, D.; Álvarez, C.; Laspina, N.; Gotor, C.; Lamattina, L.; García-Mata, C. Hydrogen sulfide generated by L-cysteine desulfhydrase acts upstream of nitric oxide to modulate ABA-dependent stomatal closure. *Plant Physiol.* **2014**, *166*, 2065–2076. [CrossRef]
5. Arenas-Alfonseca, L.; Gotor, C.; Romero, L.C.; Garcıá, I. ß-cyanoalanine synthase action in root hair elongationis exerted at early steps of the root hair elongation pathway and is independent of direct cyanide inactivation of NADPH oxidase. *Plant Cell Physiol.* **2018**, *59*, 1072–1083. [CrossRef] [PubMed]
6. Zhang, J.; Zhou, M.J.; Zhou, H.; Zhao, D.D.; Gotor, C.; Romero, L.C.; Shen, J.; Ge, Z.L.; Zhang, Z.R.; Shen, W.B.; et al. Hydrogen sulfide (H_2S), a signaling molecule in plant stress responses. *J. Integr. Plant Biol.* **2021**, *63*, 146–160. [CrossRef] [PubMed]
7. Álvarez, C.; Calo, L.; Romero, L.C.; Garcia, I.; Gotor, C. An O-acetylserine(thiol)lyase homolog with L-cysteine desulfhydrase activity regulates cysteine homeostasis in *Arabidopsis*. *Plant Physiol.* **2010**, *152*, 656–669. [CrossRef] [PubMed]
8. Jin, Z.P.; Xue, S.W.; Luo, Y.N.; Tian, B.H.; Fang, H.H.; Li, H.; Pei, Y.X. Hydrogen sulfide interacting with abscisic acid in stomatal regulation responses to drought stress in *Arabidopsis*. *Plant Physiol. Biochem.* **2013**, *62*, 41–46. [CrossRef]

9. Zhang, J.; Zhou, M.J.; Ge, Z.L.; Shen, J.; Zhou, C.; Gotor, C.; Romero, L.C.; Duan, X.L.; Liu, X.; Wu, D.L.; et al. ABA-triggered guard cell L-cysteine desulfhydrase function and in situ H_2S production contributes to heme oxygenase-modulated stomatal closure. *Plant Cell Environ.* **2020**, *43*, 624–636. [CrossRef]
10. Aroca, A.; Gotor, C.; Romero, L.C. Hydrogen sulfide signaling in plants: Emerging roles of protein persulfidation. *Front. Plant Sci.* **2018**, *9*, 1369. [CrossRef]
11. Moseler, A.; Dhalleine, T.; Rouhier, N.; Couturier, J. *Arabidopsis thaliana* 3-mercaptopyruvate sulfurtransferases interact with and are protected by reducing systems. *J. Biol. Chem.* **2021**, *296*, 100429. [CrossRef]
12. Arif, Y.; Hayat, S.; Yusuf, M.; Bajguz, A. Hydrogen sulfide: A versatile gaseous molecule in plants. *Plant Physiol. Biochem.* **2021**, *158*, 372–384. [CrossRef] [PubMed]
13. Filipovic, M.R.; Jovanović, V.M. More than just an intermediate: Hydrogen sulfide signalling in plants. *J. Exp. Bot.* **2017**, *68*, 4733–4736. [CrossRef] [PubMed]
14. Shen, J.; Zhang, J.; Zhou, M.J.; Zhou, H.; Cui, B.M.; Gotor, C.; Romero, L.C.; Fu, L.; Yang, J.; Foyer, C.H.; et al. Persulfidation-based modification of cysteine desulfhydrase and the NADPH oxidase RBOHD controls guard cell abscise acid signaling. *Plant Cell* **2020**, *32*, 1000–1017. [CrossRef] [PubMed]
15. Aroca, A.; Benito, J.M.; Gotor, C.; Romero, L.C. Persulfidation proteome reveals the regulation of protein function by hydrogen sulfide in diverse biological processes in *Arabidopsis*. *J. Exp. Bot.* **2017**, *68*, 4915–4927. [CrossRef]
16. Zhou, M.J.; Zhang, J.; Shen, J.; Zhou, H.; Zhao, D.D.; Gotor, C.; Romero, L.C.; Fu, L.; Li, Z.M.; Yang, J.; et al. Hydrogen sulfide-linked persulfidation of ABSCISIC INSENSITIVE 4 controls *Arabidopsis* ABA responses through the transactivation of mitogen-activated protein kinase kinase kinase 18. *Mol. Plant* **2021**, *14*, 1–16. [CrossRef] [PubMed]
17. Jurado-Flores, A.; Romero, L.C.; Gotor, C. Label-free quantitative proteomic analysis of nitrogen starvation in *Arabidopsis* root reveals new aspects of H_2S signaling by protein persulfidation. *Antioxidants* **2021**, *10*, 508. [CrossRef] [PubMed]
18. Weitbrecht, K.; Müller, K.; Leubner-Metzger, G. First off the mark: Early seed germination. *J. Exp. Bot.* **2011**, *62*, 3289–3309. [CrossRef] [PubMed]
19. Nambara, E.; Okamoto, M.; Tatematsu, K.; Yano, R.; Seo, M.; Kamiya, Y. Abscisic acid and the control of seed dormancy and germination. *Seed Sci. Res.* **2010**, *20*, 55–67. [CrossRef]
20. Zhou, M.J.; Zhou, H.; Shen, J.; Zhang, Z.R.; Gotor, C.; Romero, L.C.; Yuan, X.X.; Xie, Y.J. H_2S action in plant life cycle. *Plant Growth Regul.* **2021**, *94*, 1–9. [CrossRef]
21. Chandrasekaran, U.; Luo, X.F.; Zhou, W.G.; Shu, K. Multifaceted signaling networks mediated by Abscisic Acid Insensitive 4. *Plant Commun.* **2020**, *1*, 10040. [CrossRef]
22. Shu, K.; Zhang, H.W.; Wang, S.F.; Chen, M.L.; Wu, Y.R.; Tang, S.Y.; Liu, C.Y.; Feng, Y.Q.; Cao, X.F.; Xie, Q. ABI4 regulates primary seed dormancy by regulating the biogenesis of abscisic acid and gibberellins in *Arabidopsis*. *PLoS Genet.* **2013**, *9*, e1003577. [CrossRef]
23. Shu, K.; Chen, Q.; Wu, Y.R.; Liu, R.J.; Zhang, H.W.; Wang, P.F.; Li, Y.L.; Wang, S.F.; Tang, S.Y.; Liu, C.Y.; et al. ABI4 mediates antagonistic effects of abscisic acid and gibberellins at transcript and protein levels. *Plant J.* **2016**, *85*, 348–361. [CrossRef]
24. Nott, A.; Jung, H.S.; Koussevitzky, S.; Chory, J. Plastid-to-nucleus retrograde signaling. *Annu. Rev. Plant Biol.* **2006**, *57*, 739–759. [CrossRef]
25. Wind, J.J.; Peviani, A.; Snel, B.; Hanson, J.; Smeekens, S.C. ABI4: Versatile activator and repressor. *Trends Plant Sci.* **2012**, *18*, 125–132. [CrossRef] [PubMed]
26. Huang, X.; Zhang, X.; Gong, Z.; Yang, S.; Shi, Y. ABI4 represses the expression of type-A ARRs to inhibit seed germination in *Arabidopsis*. *Plant J.* **2017**, *89*, 354–365. [CrossRef] [PubMed]
27. Aroca, A.; Gotor, C.; Bassham, D.C.; Romero, L.C. Hydrogen sulfide: From a toxic molecule to a key molecule of cell life. *Antioxidants* **2020**, *9*, 621. [CrossRef] [PubMed]
28. Luo, X.F.; Dai, Y.J.; Zheng, C.; Yang, Y.Z.; Chen, W.; Wang, Q.C.; Chandrasekaran, U.; Du, J.B.; Liu, W.G.; Shu, K. The ABI4-RbohD/VTC2 regulatory module promotes reactive oxygen species (ROS) accumulation to decrease seed germination under salinity stress. *New Phytol.* **2021**, *229*, 950–962. [CrossRef]
29. Giraud, E.; Van Aken, O.; Ho, L.H.M.; Whelan, J. The transcription factor ABI4 is a regulator of mitochondrial retrograde expression of ALTERNATIVE OXIDASE1a. *Plant Physiol.* **2009**, *150*, 1286–1296. [CrossRef] [PubMed]
30. Shkolnik-Inbar, D.; Bar-Zvi, D. Expression of ABSCISIC ACID INSENSITIVE 4 (ABI4) in developing *Arabidopsis* seedlings. *Plant Signal. Behav.* **2011**, *6*, 694–696. [CrossRef]
31. Pan, W.B.; Lin, B.Y.; Yang, X.Y.; Liu, L.J.; Xia, R.; Li, J.G.; Wu, Y.R.; Xie, Q. The UBC27-AIRP3 ubiquitination complex modulates ABA signaling by promoting the degradation of ABI1 in *Arabidopsis*. *Proc. Natl. Acad. Sci. USA* **2020**, *117*, 27694–27702. [CrossRef]
32. Kong, L.Y.; Cheng, J.K.; Zhu, Y.J.; Ding, Y.L.; Meng, J.J.; Chen, Z.Z.; Xie, Q.; Guo, Y.; Li, J.G.; Yang, S.H.; et al. Degradation of the ABA co-receptor ABI1 by PUB12/13 U-box E3 ligases. *Nat. Commun.* **2015**, *6*, 8630. [CrossRef] [PubMed]
33. Albertos, P.; Romero-Puertas, M.C.; Tatematsu, K.; Mateos, I.; Sánchez-Vicente, I.; Nambara, E.; Lorenzo, O. S-nitrosylation triggers ABI5 degradation to promote seed germination and seedling growth. *Nat. Commun.* **2015**, *6*, 8669. [CrossRef]
34. Huang, W.J.; MacLean, A.M.; Sugio, A.; Kuo, C.H.; Kuo, R.G.H.; Hogenhout, S.A. Parasitic modulation of host development by ubiquitin-independent protein degradation. *Cell* **2021**, *184*, 1–14. [CrossRef] [PubMed]
35. Söderman, E.M.; Brocard, I.M.; Lynch, T.J.; Finkelstein, R.R. Regulation and function of the *Arabidopsis* ABA-insensitive 4 gene in seed and abscisic acid response signaling networks. *Plant Physiol.* **2000**, *124*, 1752–1765. [CrossRef]

36. Chen, S.S.; Jia, H.L.; Wang, X.F.; Shi, C.; Wang, X.; Ma, P.Q.; Wang, J.; Ren, M.J.; Li, J.S. Hydrogen sulfide positively regulates abscisic acid signaling through persulfidation of SnRK2.6 in Guard Cells. *Mol. Plant* **2020**, *13*, 732–744. [CrossRef] [PubMed]
37. Chen, S.S.; Wang, X.F.; Jia, H.L.; Li, F.L.; Ma, Y.; Liesche, J.; Liao, M.Z.; Ding, X.T.; Liu, C.X.; Chen, Y.; et al. Persulfidation-induced structural change in SnRK2.6 establishes intramolecular interaction between phosphorylation and persulfidation. *Mol. Plant* **2021**, *14*, 1814–1830. [CrossRef]
38. Yoo, S.D.; Cho, Y.H.; Sheen, J. *Arabidopsis* mesophyll protoplasts: A versatile cell system for transient gene expression analysis. *Nat. Protoc.* **2007**, *2*, 1565–1572. [CrossRef]
39. Riemenschneider, A.; Nikiforova, V.; Hoefgen, R.; De Kok, L.J.; Papenbrock, J. Impact of elevated H_2S on metabolite levels, activity of enzymes and expression of genes involved in cysteine metabolism. *Plant Physiol. Biochem.* **2005**, *43*, 473–483. [CrossRef] [PubMed]

International Journal of *Molecular Sciences*

MDPI

Article

Persulfidation of Nitrate Reductase 2 Is Involved in L-Cysteine Desulfhydrase-Regulated Rice Drought Tolerance

Heng Zhou [1], Ying Zhou [1], Feng Zhang [1], Wenxue Guan [1], Ye Su [1], Xingxing Yuan [2] and Yanjie Xie [1,*]

[1] Laboratory Center of Life Sciences, College of Life Sciences, Nanjing Agricultural University, Nanjing 210095, China; hengzhou@njau.edu.cn (H.Z.); 2020116101@stu.njau.edu.cn (Y.Z.); 2018116103@njau.edu.cn (F.Z.); 2016116114@njau.edu.cn (W.G.); 2017116117@njau.edu.cn (Y.S.)

[2] Institute of Industrial Crops, Jiangsu Academy of Agricultural Sciences, Nanjing 210014, China; yxx@jaas.ac.cn

* Correspondence: yjxie@njau.edu.cn

Abstract: Hydrogen sulfide (H_2S) is an important signaling molecule that regulates diverse cellular signaling pathways through persulfidation. Our previous study revealed that H_2S is involved in the improvement of rice drought tolerance. However, the corresponding enzymatic sources of H_2S and its regulatory mechanism in response to drought stress are not clear. Here, we cloned and characterized a putative L-*cysteine desulfhydrase* (*LCD*) gene in rice, which encodes a protein possessing H_2S-producing activity and was named *OsLCD1*. Overexpression of *OsLCD1* results in enhanced H_2S production, persulfidation of total soluble protein, and confers rice drought tolerance. Further, we found that nitrate reductase (NR) activity was decreased under drought stress, and the inhibition of NR activity was controlled by endogenous H_2S production. Persulfidation of NIA2, an NR isoform responsible for the main NR activity, led to a decrease in total NR activity in rice. Furthermore, drought stress-triggered inhibition of NR activity and persulfidation of NIA2 was intensified in the *OsLCD1* overexpression line. Phenotypical and molecular analysis revealed that mutation of *NIA2* enhanced rice drought tolerance by activating the expression of genes encoding antioxidant enzymes and ABA-responsive genes. Taken together, our results showed the role of *OsLCD1* in modulating H_2S production and provided insight into H_2S-regulated persulfidation of NIA2 in the control of rice drought stress.

Keywords: hydrogen sulfide; persulfidation; drought stress; nitrate reductase; L-cysteine desulfhydrase

Citation: Zhou, H.; Zhou, Y.; Zhang, F.; Guan, W.; Su, Y.; Yuan, X.; Xie, Y. Persulfidation of Nitrate Reductase 2 Is Involved in L-Cysteine Desulfhydrase-Regulated Rice Drought Tolerance. *Int. J. Mol. Sci.* **2021**, *22*, 12119. https://doi.org/10.3390/ijms222212119

Academic Editor: Kenji Miura

Received: 22 September 2021
Accepted: 5 November 2021
Published: 9 November 2021

1. Introduction

Drought is the most widespread and damaging of all environmental stresses, restricting global crop production and food security [1]. Plants can mitigate the effects of drought through the collaboration of complex signal networks. It is well documented that maintaining redox homeostasis and activating ABA signaling could improve plant drought stress tolerance [2,3]. Hydrogen sulfide (H_2S) has been recognized as a newly gaseous signaling molecule in both animals and plants [4,5]. During the past decades, numerous studies have suggested that H_2S is involved in various developmental and stress response processes during the whole lifespan in plants [6–9]. For example, H_2S is involved in the improvement of drought tolerance by interacting with abscisic acid (ABA) and ion fluxes, thus regulating stomal movement and downstream genes expression in *Arabidopsis* [10,11]. Pretreatment with exogenous NaHS (a H_2S donor) alleviates drought stress responses by increasing ABA levels through the expression of ABA synthesis genes in wheat and rice [12,13]. Although those studies have demonstrated that H_2S is involved in regulating many metabolic processes or improving plant tolerance to abiotic stresses, they mainly rely on exogenous application of H_2S donors, scavengers, and inhibitors to manipulate endogenous H_2S content [13,14].

In plants, cysteine desulfhydrases (CDes) are one of the most important clusters of H_2S-producing enzymes catalyzing the degradation of cysteine into H_2S, pyruvate, and ammonium [15]. There are two types of CDes in plants: L-cysteine desulfhydrase (LCD) and D-cysteine desulfhydrase (DCD) with L-cysteine (L-Cys) or D-cysteine (D-Cys) as substrate, respectively [16]. H_2S can also be generated as a side reaction of cysteine biosynthesis catalyzed by serine acetyltransferase (SAT) and O-acetyl-serine(thiol)lyase (OAS-TL) [17,18]. Interestingly, an OAS-TL isoform CYSTEINE SYNTHASE (CS)-LIKE protein (CS-LIKE) has been reported that actually catalyzes the desulfuration of L-Cys to H_2S plus ammonia and pyruvate [18]. Thus, CS-LIKE is a novel L-cysteine desulfhydrase and has been designated as *DES1*. In *Arabidopsis*, *LCD* and *DES1*-mediated endogenous H_2S production has been widely reported as an important role in facilitating tolerance to various environmental stimuli, including heavy metal and drought stress [11,18–20]. However, to date, little information is available about the LCD in rice. A recent study revealed that a putative rice L-cysteine desulfhydrase *LCD* actually encodes a true L-cysteine synthetase [21], suggesting the enzymatic sources of endogenous H_2S production still need to be further explored.

Signaling by H_2S is proposed to occur via persulfidation, the oxidative post-translational modification of protein Cys residues (R-SHs) by covalent addition of thiol groups to form persulfides (R-SSHs) [9,22]. Persulfidation modulates protein functions by affecting its biochemical activity and subcellular distribution, thus providing a robust and flexible mechanism for biological regulation in response to metabolic stimuli and environmental cues [23,24]. Recently, by using a comparative and label-free quantitative proteomic analysis approach, almost 13% of the entire annotated proteome proteins were identified as being persulfidated in *Arabidopsis* [23,25]. These proteins are involved in a wide range of biological functions, regulating important processes such as primary metabolism, plant responses to stresses, growth and development, RNA translation, and protein degradation. In guard cells, a complex interaction of H_2S-mediated persulfidation and ABA signaling has also been described. In the presence of ABA, L-cysteine desulfhydrase1 (DES1) is activated by H_2S through persulfidation resulting in a burst of H_2S in guard cells [26]. The increase in H_2S, in turn, facilitates the over-accumulation of ROS via persulfidation of the NADPH oxidase RESPIRATORY BURST OXIDASE HOMOLOG D(RBOHD), thereby inducing stomatal closure [26]. Besides that, H_2S-induced persulfidation of ABSCISIC INSENSITIVE 4 (ABI4) is involved in the ABA signaling pathway [27]. These results clearly indicated that H_2S exerts its biological function through precisely persulfidation of target protein in plants. Previously, we found that exogenous application of NaHS could significantly improve rice drought tolerance by reestablishing redox homeostasis and activation of ABA biosynthesis and signaling [13]. However, the underlying regulatory mechanisms of endogenous H_2S are not clear.

Nitrate reductase (NR) is a key enzyme in plant nitrogen assimilation, which catalyzes the reduction in nitrate to nitrite in plants [28]. NR plays an important role in plant response to a variety of biotic and abiotic stresses [29]. A study in *Arabidopsis* showed the rate of water loss due to water transpiration was significantly slower in *nia1/nia2* double mutant than in wild-type plants, with *nia1/nia2* double mutant showing the higher expression of ABA-responsive genes and drought tolerance [30], demonstrating plant drought tolerance is negatively regulated by *NR* abundance.

The aim of this study is to explore and characterize the enzymatic sources of endogenous H_2S production and elucidate the underlying mechanism of how H_2S confers rice drought tolerance. We cloned and characterized the function of a true *LCD* (*OsLCD1*) from rice. The corresponding biochemical characteristics of purified LCD1 proteins showed that this enzyme predominantly processes H_2S producing activity. We found that overexpression of *OsLCD1* enhanced rice drought tolerance by activating the expression of related genes encoding antioxidant enzymes and ABA-responsive gene. Further, we demonstrated that persulfidation of NIA2, an NR isoform responsible for the main NR activity, led to a decrease in total NR activity, thus controlling the above genes expression. By combining genetic and molecular analysis, we provide evidence here that H_2S might through, at least partially, persulfidation-mediated inhibition of NR activity to improve rice drought tolerance.

2. Results

2.1. Cloning and Functional Characterization of the OsLCD

In order to characterize the putative LCD protein in rice plants, the *Arabidopsis LCD* (*At3g62130*) was used as a query sequence to search the homologous gene in *Oryza sativa* by using uniprot-BLAST (https://www.ncbi.nlm.nih.gov/, accessed on 10 July 2019). We found a putative L-cysteine desulfhydrase (*OsLCD1*, *LOC_Os01g18640*) sharing the highest similarity (56%) with *AtLCD*, which encode an OsLCD protein with 482 amino acids residues (Figure 1). The molecular mass of OsLCD1 is 55 kDa, and the theoretical isoelectric point is 5.836 (http://isoelectric.ovh.org, accessed on 20 July 2019). Subsequently, the sequences alignment of the OsLCD1 and CDes homology and OAS-TL family proteins from other species were performed. The results showed that OsLCD1 shares a higher sequence identity with CDes homology from *Panicum miliaceum*, *Dichanthelium oligosanthes*, *Zea mays*, and *Arabidopsis thaliana* in comparison with that of OAS-TL family members from *Arabidopsis thaliana*, including AtDES1. Furthermore, the phylogenetic tree and homology tree were created with MAGE and DNAMAN software with default parameters, respectively. Among those proteins, OsLCD1 is more closely related to the LCD homology proteins from plants (Figure S1).

To validate the biochemical properties of OsLCD1, the corresponding full-length cDNA was cloned, and the recombinant OsLCD1 protein was expressed in *E. coli* as a 6× His N-terminally tagged fusion protein using *pET-28a*(+) vector. The *OsLCD1* fusion protein was purified by nickel affinity chromatography using nickel-nitrilotriacetic acid agarose (Ni-NAT) under non-denaturing conditions to preserve the enzymatic activity. A band appeared in the SDS-PAGE gel at the position corresponding to that of the His-tagged OsLCD1 protein (55 kDa, Figure 2A). The band size and specificity of the OsLCD1 protein were further verified by Western blot analysis using His antibody (Figure 2B). We were able to recover 0.14 mg purified protein per 150 mL of *E. coli* culture with a yield of 36.33% (Table 1). To confirm OsLCD1 functioned as a true LCD, the LCD and OAS-TL activities of both bacterial extracts and purified recombinant OsLCD1 protein were detected, respectively. As shown in Table 1, after purification, the specific LCD or OAS-TL activity (nmol/min/mg pro) of OsLCD1 protein changed from 8.02 or 1900 to 23.93 or 720, with a corresponding purification factor of 2.98 and 0.38. These results suggested that purified OsLCD1 protein might predominately catalyze the degradation of L-cysteine, and the OAS-TL reaction might be a side reaction. This proposition was also reinforced by the results of Km value, showing that the Km for OAS or Na_2S in the OAS-TL reaction is 25- or 54-fold higher than that for L-cysteine in LCD-catalytic reaction (Table 2), further suggesting a much higher affinity of OsLCD1 for L-cysteine as a substrate. Subsequently, biochemical analysis showed that the optimum temperature range of purified OsLCD1 protein was 50 to 80 °C (Figure 2C). The rate of the LCD reaction increased to its maximum value at a temperature of 60 °C and declined thereafter. Meanwhile, the LCD activity of OsLCD1 under different pH was determined at 60 °C, and results showed that the optimal pH of OsDCD1 was 9.5 (Figure 2D).

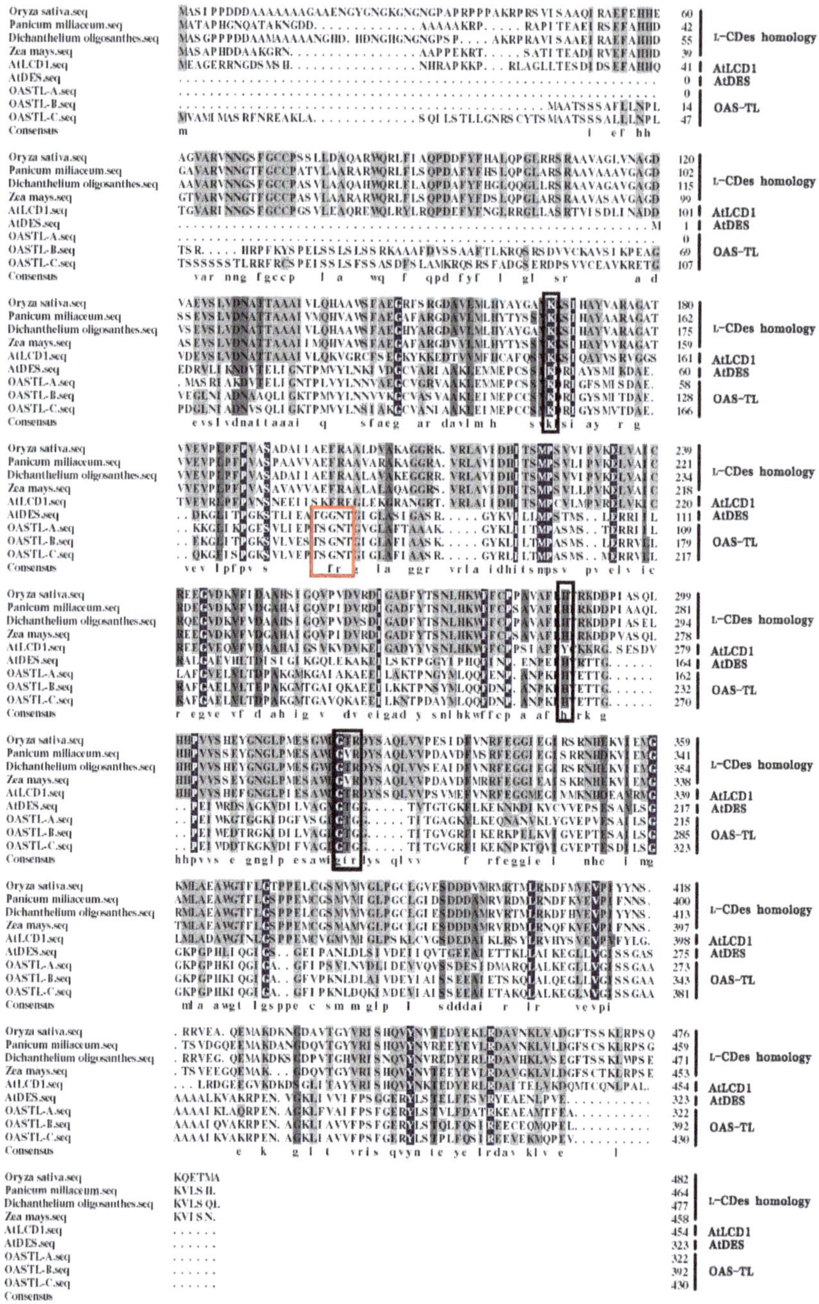

Figure 1. Sequence analysis of CDes, CDes homology, and OAS-TL-A, B, C proteins from the plant. The sequence alignment of L-CDes homology from *Oryza sativa* (XP_015613237), *Panicum miliaceum* (RLN24808), *Dichanthelium oligosanthes* (OEL32418), and *Zea mays* (PWZ10688) and AtLCD1 (NP_001327694), AtDES (OAO92103), OAS-TL-A (AEE83514), OAS-TL-B (AEC10318), and OAS-TL-C (AEE79963) from *Arabidopsis thaliana* was created by DNAMAN with default parameter. Amino acids with blue, black, and gray backgrounds indicate completely or highly conserved residues. The PLP binding sites are shown with a black box, the substrates binding sites of OAS-TL are shown with a red box.

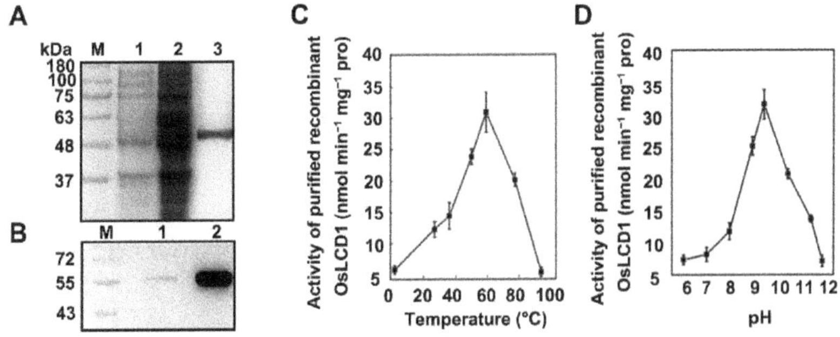

Figure 2. Biochemical characterization of purified recombinant *His*-tagged *OsLCD1* expressed in *E. coli.* (**A**) Expression and purification of OsLCD1 recombinant protein. *OsLCD1* expression was induced with 0.2 mM IPTG for 12 h and then purified by Ni-affinity chromatography. M: Molecular marker; Lane1: Total protein without IPTG induction of *BL21(DE3)//pET28α-OsLCD1;* Lane2 total protein after IPTG induction of *BL21(DE3)/pET28α-OsLCD1;* Lane3: Expressed protein purified by Ni-affinity chromatography. (**B**) Western blot of purified OsLCD1 recombinant protein. M: Molecular marker; Lane1: immunoblot of purified protein developed with the polyclonal antiserum against the putative OsLCD1 eluted by NTA-50 buffer; Lane2: immunoblot of purified protein developed with the polyclonal antiserum against the putative OsLCD1 eluted by NTA-50 buffer NTA-100. (**C**,**D**) Temperature and pH dependence of the OsLCD1 reaction. Data shown are means ± SD from three independent measurements.

Table 1. The purification and catalytic activity of *OsLCD1* expressed in *E. coli*. The recombinant *His*-tagged *OsLCD1* protein was purified using the Ni-NTA Purification System. CDes and OAS-TL activities were measured as described.

Purification Step	Protein (mg)	Specific Activity (nmol min^{-1} mg^{-1} pro) CDes OAS-TL	Total Activity (nmol min^{-1}) CDes OAS-TL	Purification Factor CDes OAS-TL	Yield (%) CDes OAS-TL
Crude extract	1.15	8.02 1.90 × 10^3	9.22 2.77 × 10^3	——	——
Ni-NTA chromatography	0.14	23.93 0.72 × 10^3	3.35 0.13 × 10^3	2.98 0.38	36.33 4.69

Table 2. Catalytic properties of the recombinant *His*-tagged *OsLCD1* for the CDes and OAS-TL enzymatic reactions. L-Cys was used as a substrate for the DES reaction, whereas OAS and Na_2S were used as co-substrates for the OAS-TL reaction. Lineweaver–Burk plot was performed to calculate the kinetic constants.

Km (mM Cys)	Vmax (μmol H_2S min^{-1} mg^{-1} pro)	Km (mM OAS)	Km (mM Na_2S)	Vmax (μmol L-Cys min^{-1} mg^{-1} pro)
0.15 ± 0.02	0.04 ± 0.01	3.76 ± 0.41	8.13 ± 0.72	1.76 ± 0.32

2.2. Overexpression of OsLCD1 Enhance Endogenous H_2S Production and Drought Tolerance in Rice

To investigate the physiological role of *OsLCD1* in rice, two independent *35S:OsLCD1-GFP* overexpression lines (*OX1* and *OX2*) were generated by introducing the *pCAMBIA1305-OsLCD1-GFP* expression construct into Wuyunjing 7. Firstly, the overexpression of *OsLCD1* was confirmed by immunoblot analysis. The results showed that the band signal of *OsLCD1-GFP* was detected in protein extracts from both two transgenic lines, but not wild-type (cv. Wuyunjing 7) (Figure 3A). Furthermore, biochemical characterization results revealed that the total LCD activity in *OX1* and *OX2* was increased by 43.7% and 71.8% compared to the wild-type plants (Figure 3B). This result further confirms that the LCD1 protein is a true LCD enzyme. Accordingly, the endogenous H_2S content in *OX1* and *OX2* was about 47.5% and 102.1% higher than those of wild-type plants (Figure 3C).

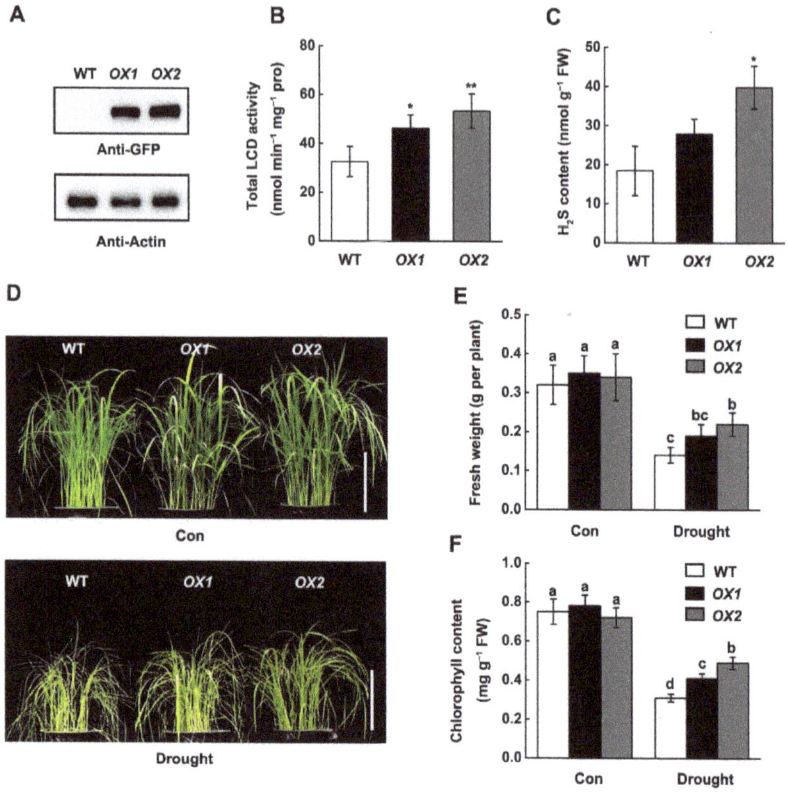

Figure 3. Overexpression of *OsLCD1* enhances rice drought tolerance. (**A**) Identification of *35S:OsLCD1-GFP* overexpression lines. The total protein from two independent overexpression lines (*OX1* and *OX2*) and wild-type (cv. Wuyunjing 7, WT) was extracted and was analyzed by immunoblotting with an anti-GFP antibody. (**B**,**C**) Overexpression of *OsLCD1* increases LCD activity and H_2S production in 14-day-old rice plants. (**D**) Drought stress tolerance assay. Well-irrigated 14-day-old wild-type, *OX1*, and *OX2* rice seedlings were exposed to drought stress by withholding water for 6 days. Pictures were then taken. Scale bar = 10 cm. (**E**,**F**) The related fresh weight and chlorophyll content were determined. Data are means ± SD (n = 3). Statistical comparisons were performed by independent samples t-test (two-tailed) between leaves from wild-type and *OsLCD1* overexpression lines (* $p < 0.05$, ** $p < 0.01$). Different lower case letters indicate significant differences at $p < 0.05$ (one-way ANOVA, Duncan's multiple range tests).

Our previous study has shown that exogenous H_2S could alleviate rice drought stress [13]. We then wonder whether the overexpression of *OsLCD1* would affect rice drought tolerance. Thus, two-week-old rice seedlings (WT, *OX1*, and *OX2*) were subjected to drought stress for 10 days. We observed that overexpression of *OsLCD1* significantly improved the growth performance of rice seedlings under drought stress (Figure 3D). Compared with the wild type, the fresh weight was increased by 29% and 36% (Figure 3E), and the chlorophyll content was increased by 28% and 39% in *OX1* and *OX2* plants (Figure 3F), respectively. These results indicated that overexpression of *OsLCD1* improves rice drought tolerance.

To investigate the molecular mechanism of *OsLCD1* in response to drought stress, the expression profiles of genes involved in drought stress response were determined. RT-qPCR results showed that the transcript of genes encoding antioxidant enzymes, including *ascorbate peroxidase 2* (*APX2*) and *catalase* (*CATA*), and *a basic leucine zipper* (*bZIP*) *transcription factor 23* (*bZIP23*) and *a dehydration responsive element-binding protein* (*DREB*) were increased by drought stress in wild-type plants, while this induction was further enhanced in *OX2* plant (Figure S2). These results indicated that overexpression of *OsLCD1* may improve rice drought tolerance via modulating the expression of genes involved in drought stress response.

H_2S-mediated persulfidation has been reported that regulate diverse cellular signaling pathways [9]. To investigate whether overexpression of *OsLCD1* affects the protein persulfidation level in rice seedlings, we determined the persulfidation level of total protein from wild-type *OX1* and *OX2* plants under normal and drought stress conditions. A tag-switch assay in which persulfidated Cys was labeled with cyan-biotin and could specifically be detected by anti-biotin immunoblot analysis was used [23,24]. The immunoblotting results showed that drought stress significantly enhanced the protein persulfidation level in all rice plants, while the protein persulfidation level was higher in *OX1* and *OX2* compared with wild-type (Figure 4). These results indicated that *OsLCD1*-mediated persulfidation may involve in rice drought tolerance.

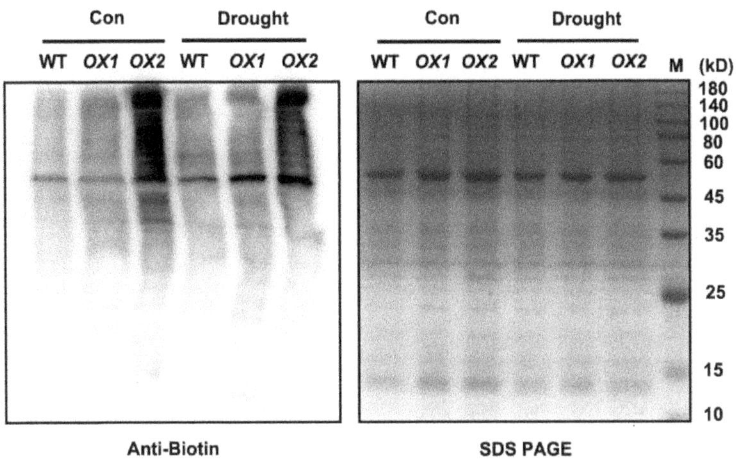

Figure 4. *OsLCD1*-mediated persulfidation of rice total protein. Fourteen-day-old rice seedlings of wild-type (cv. Wuyunjing 7, WT), *OX1*, and *OX2* were isolated from cultivated culture and placed in filter paper for dehydration for 3 h. Afterward, proteins were extracted from 0.2 g of leaves and subjected to the modified biotin switch method, and the labeled proteins were detected using protein blot analysis with antibodies against biotin. Coomassie brilliant blue-stained gels were presented to show that equal amounts of proteins were loaded. Numbers on the right of the panels indicate the position of the protein markers in kDa.

2.3. *Dehydration-Triggered Inhibition of NR Activity Was Correlated with Endogenous H_2S Content*

Previous studies revealed that NR plays an important role in plant stress response [31]. To investigate whether NR is involved in the endogenous H_2S-enhanced rice drought tolerance, we detected the changes of NR activity in rice leaves in response to drought stress. In comparison with the control plants, NR activity was decreased in rice seedling leaves after dehydration (Figure 5A). For example, NR activity was rapidly decreased by 33.3% within 1 h after dehydration and unchanged until 3 h, and then further decreased by 67.2% at 6 h. The changes in NR activity showed the opposite tendency with endogenous H_2S production in response to dehydration stress [13]. Meanwhile, the pretreatment of NaHS reinforced the decrease in NR activity after dehydration, indicating that dehydration-triggered inhibition of NR activity may regulate by endogenous H_2S (Figure 5B). To further verify this, hypotaurine (HT, a H_2S scavenger) [32] or DL-propargylglycine (PAG, an L-DES inhibitor) [33] was used. With respect to the rapidly decreased NR activity in response to dehydration stress by exogenous application of NaHS, pretreated with HT or PAG significantly alleviated dehydration-induced inhibition of NR activity (Figure 5C,D). Thus, these results clearly indicated that dehydration-triggered inhibition of NR activity was correlated with endogenous H_2S content.

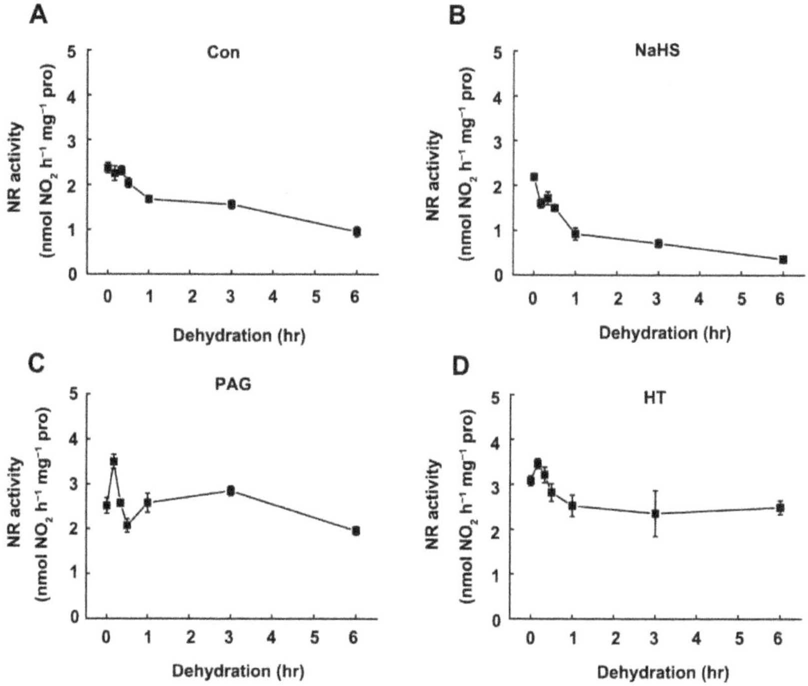

Figure 5. Regulation of NR activity by H_2S. The 14-day-old rice seedlings cultured under normal conditions (**A**) or retreated with NaHS (100 uM, **B**), PAG (1 mM, **C**), or HT (1 mM **D**). For the dehydration time-course experiment, Leaf blade branches were isolated from cultivated soil and placed in filter paper. The leaves samples were harvested, and the total NR activity was measured at indicated time points. Data are means ± SD (n = 3).

To determine whether the decreased NR activity was caused by transcriptional level regulation or post-translational modification, the expression profiles of genes encoding NR were verified. In rice, there are two NR encoding genes, *NIA1* and *NIA2*. The RT-qPCR result showed that the transcripts level of both *NIA1* and *NIA2* was gradually decreased in rice seedling leaves after dehydration (Figure 6A,B). However, pretreatment of NaHS has no significant effect on the dehydration-inhibited gene expression of *NIA1* and *NIA2*, indicating H_2S may regulate NR activity at the post-translational level.

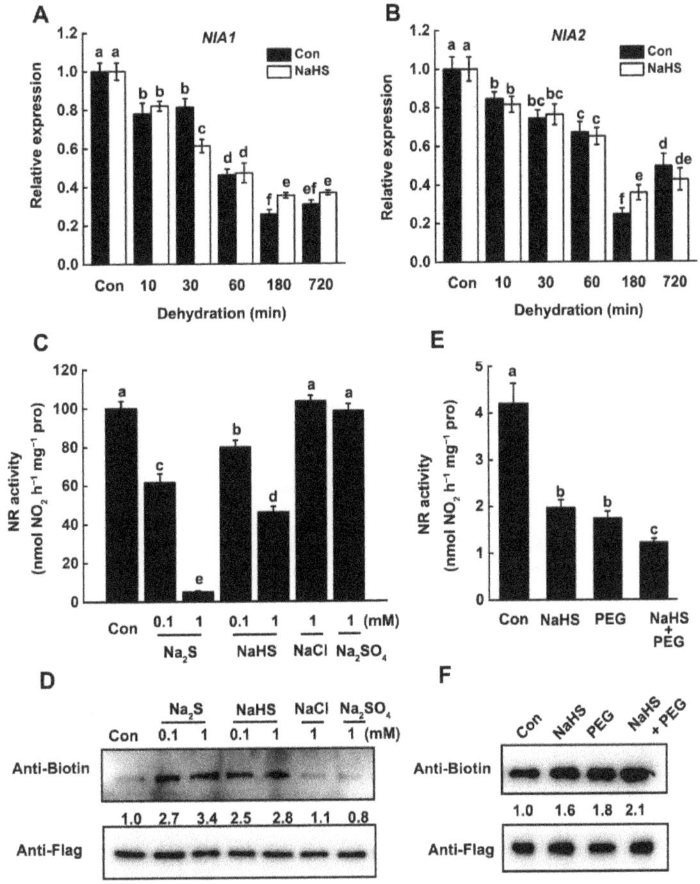

Figure 6. H_2S-mediated persulfidation inhibits NR activity in response to drought stress. (**A**,**B**) Time-course expression profiles of *NIA1* and *NIA2* in rice seedlings pretreated with or without NaHS followed by dehydration. Fourteen-day-old wild-type (cv. Wuyunjing 7, WT) rice seedlings were pretreated with or without NaHS (100 μM) for 2 h and then placed in filter paper for dehydration treatment to mimic drought stress. Leaves were harvested for RT-qPCR analysis at the indicated time point. Expression levels are relative to corresponding untreated wild-type samples (control) after normalization to reference genes of *OsActin1* and *OsActin2*. (**C**) Persulfidation inhibits NIA2-related NR activity in tobacco. *N. benthamiana* leaves were transiently overexpressed *35S:NIA2-Flag* construct. Total proteins were extracted, and NIA2-Flag protein was harvested by immunoprecipitation with anti-Flag agarose beads. NIA2-Flag protein-related NR activity was detected after being treated with or without different sulfur-containing chemicals for 30 min. (**D**) The persulfidation level of NIA2-Flag protein from (**C**) was analyzed by immunoblotting with an anti-Biotin and anti-Flag antibody. The persulfidation levels are relative to corresponding untreated control samples after normalization to the anti-Flag signal abundance. (**E**) Persulfidation-inhibited NIA2-related NR activity upon osmotic stress in rice. Rice protoplasts of the *nia2* mutant (cv. Dongjin) were transfected with *35S:NIA2-Flag*. After 12 h incubation, protoplasts were treated with or without NaHS (100 μM) in the absence or presence of PEG6000 (10%) for 1 h. Total proteins were extracted for the determination of NR activity. (**F**) The persulfidation level of protein from (**E**) was analyzed by immunoblotting with an anti-Biotin and anti-Flag antibody. The related persulfidation level is relative to corresponding untreated control samples after normalization to the anti-Flag signal abundance. Data are means ± SD ($n = 3$). Lower case letters indicate significant differences at $p < 0.05$ (Duncan's multiple range tests).

2.4. H_2S-*Mediated Persulfidation-Inhibited NR Activity*

It is plausible that NR activity was regulated by H_2S-mediated persulfidation. Subsequently, we determine the effects of the exogenous application of H_2S donors on NR activity. In rice, the transcriptional level of *NIA2* is markedly higher than that of *NIA1*, and *NIA2* mutation causes more than 90% loss in NR activity [34,35]. Thus, we first clone the rice *NIA2* gene and transiently overexpressed *35S:NIA2-Flag* construct in *N. benthamiana* leaves. After 12 h incubation, the NIA2 protein was immunoprecipitated by using an anti-Flag antibody and treated with different H_2S donors. The results showed that treatment of both well-known H_2S donors, NaHS and Na_2S, significantly decreased NR activity (Figure 6C). When 1 mM NaHS was applied, the NR activity was decreased by 55%, while 1 mM Na_2S caused a 93% loss in NR activity. It should be mentioned that treatment with NaCl or Na_2SO_4 fails to reduce the NR activity. These results suggested that H_2S or HS^-, rather than other compounds regulates the NR activity. These changes in NR activity were consistent with the corresponding persulfidation level of NIA2, which was enhanced by both H_2S donors rather than NaCl or Na_2SO_4 (Figure 6D).

To determine whether NIA2 could be persulfidated in vivo, the protoplasts from *nia2* rice mutant (cv. Dongjin) with transiently overexpressed *35S:NIA2-Flag* were treated with or without NaHS in the presence or absence of polyethylene glycol (PEG), which further mimic drought stress. As expected, immunoblotting results showed that NIA2 protein was persulfidated in rice protoplasts, and NaHS pretreatment enhanced its persulfidation level (Figure 6E). Importantly, the persulfidation of NIA2 protein was significantly enhanced by PEG treatment, while this could be further strengthened by NaHS. These results on the persulfidation level of NIA2 protein were consistent with the changes in their enzymatic activity (Figure 6F), indicating that dehydration-triggered inhibition of NR activity was controlled by H_2S-mediated persulfidation.

In order to validate the contribution of *OsLCD1* in PEG-induced NIA2 persulfidation, we examined the persulfidation level of NIA2 protein in wild-type (cv. Wuyunjing 7, WT) and *35S:OsLCD1-GFP* overexpression rice plants (*OX2*). With this aim in mind, the *35S:NIA2-Flag* construct was separately transiently expressed into protoplasts of wild type and *35S:OsLCD1-GFP* overexpression line (*OX2*). The immunoblotting result showed that PEG treatment induced the persulfidation of NIA2 protein in protoplasts of wild type, while this was further intensified in *the OX2* line (Figure 7A). Moreover, we found that the NR activity was decreased faster in the *OX2* line as compared to wild-type plants upon the dehydration stress (Figure 7B). These results demonstrated that persulfidation-mediated inhibition of NR activity may confer rice drought tolerance.

2.5. *Knock down of NIA2 Enhances Rice Drought Tolerance*

To investigate the biological role of NR inhibition in rice drought stress response, drought stress tolerance of wild-type and *nia2* mutant was compared. We observed that mutation of *NIA2* significantly improved the growth performance of rice seedlings under drought stress (Figure 8A). Compared with the wild-type (cv. Dongjin, WT), the fresh weight was increased by 36% and 29% (Figure 8B), as the chlorophyll content was increased by 39% in the *nia2* plant (Figure 8C), respectively. These results indicated that mutation of *NIA2* improves rice drought tolerance.

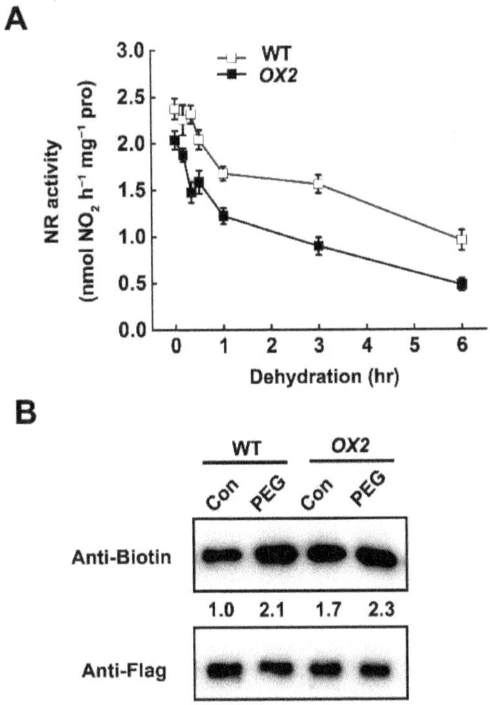

Figure 7. Overexpression of *OsLCD1* enhances persulfidation and the activity decrease in NR. (**A**) Dehydration-inhibited NR activity. The 14-day-old wild-type (cv. Wuyunjing 7, WT) and *OX2* rice seedlings were cultured under normal conditions. For the dehydration time-course experiment, leaf blade branches were isolated from cultivated soil and placed in filter paper. The leaves samples were harvested, and the total NR activity was measured at indicated time points. Data are means ± SD ($n = 3$). (**B**) Wild-type and *OX2* rice protoplasts with transiently expressed *NIA2-Flag* were treated with or without 100 μM NaHS in the absence or presence of 10% PEG6000 1 h, and then the total proteins were extracted and analyzed by immunoblotting with an anti-Biotin and anti-Flag antibody.

To further investigate the molecular mechanism of *NIA2* in response to drought stress, the expression profiles of genes involved in drought stress response in the *nia2* mutant were determined. RT-qPCR results showed that the transcript of genes encoding antioxidant enzymes, including *APX2* and *CATA*, were significantly higher in *nia2* mutant as compared to wild-type plants upon drought stress (Figure 8D,E). Meanwhile, after drought stress, the induction of *bZIP23* and *DREB* genes in *nia2* mutants was further enhanced compared with the wild type (Figure 8F,G). Based on these findings, we concluded that *NIA2* negatively regulates rice drought tolerance.

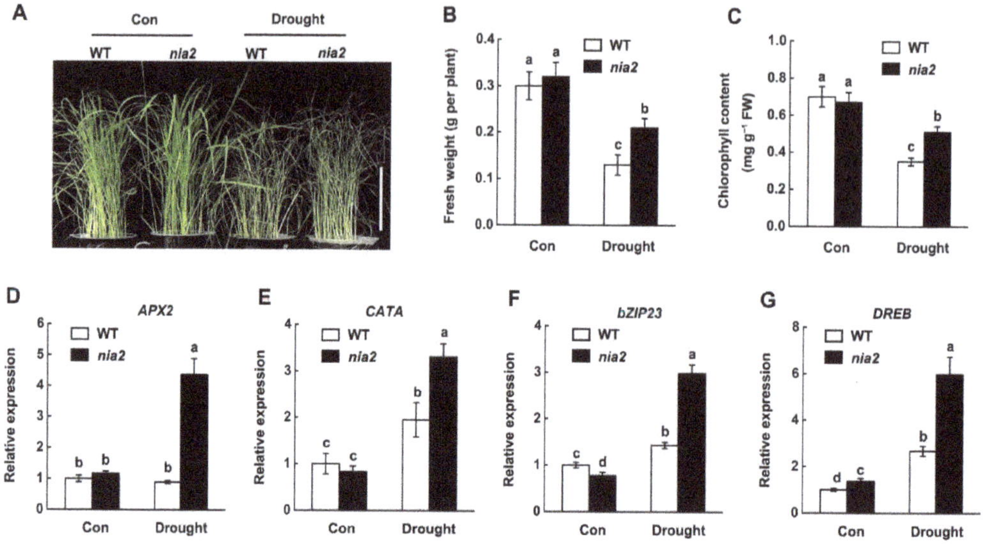

Figure 8. Knockdown of *NIA2* enhances rice drought tolerance. (**A**) Photographs of 14-day-old well-irrigated wild-type (cv. Dongjin, WT) and *nia2* mutant rice seedlings were withdrawn from irrigation for 6 days. Scale bar = 10 cm. (**B**,**C**) The related fresh weight and chlorophyll content were determined. (**D**–**G**) Relative transcript levels of genes involved in drought stress response in rice seedling leaves were quantified after 4 days of drought stress by qRT-PCR. Expression levels are relative to corresponding untreated wild-type samples (control) after normalization to *OsActin1* and *OsActin2*. Data are means ± SD ($n = 3$). Lower case letters indicate significant differences at $p < 0.05$ (Duncan's multiple range tests).

3. Discussion

3.1. A True L-Cysteine Desulfhydrase Confers Rice Drought Tolerance

The importance of cysteine (Cys) in plants is defined not only by its role as an amino acid in primary and secondary metabolisms but also by its function as a metabolic precursor of essential biomolecules [36,37]. In plant cells, H_2S is generated through enzymatic pathways that are closely related to Cys metabolism [16]. L-cysteine DESULFHYDRASE (LCD) and D-cysteine DESULFHYDRASE (DCD) degrade L/D-Cys to H_2S, pyruvate, and ammonia and contribute to the production and biological function of H_2S in the cell [38]. Here, we discovered and characterized a rice LCD encoding gene that shares the highest similarity (56%) with the AtLCD [10,39]. The phylogenetic and homology analysis showed that OsLCD1 is more closely related to the LCD homology proteins from plants rather than the OAS-TL family (Figure 1 and Figure S1).

Up to now, progress has been made in the characterization of the CDes, which usually possess bi-functional activities [9]. For example, the AtDCD1 (*At1G48420*) was identified as a D-CDes and also possessed ACCD activity [40,41]. AtDCD2 (*At3g26115*) catalyzes the release of H_2S from D-cysteine as well as L-cysteine [16]. Moreover, due to the reversibility of catalytic reactions, those enzymes sometimes exhibit opposite activities. For example, *Arabidopsis* DES1, a member of the OAS-TL family, is involved in L-cysteine degradation rather than biosynthesis [18]. By contrast, LCD2, a rice AtLCD homolog, predominantly exhibits cysteine biosynthesis activity and is, therefore, a true cysteine synthetase [21]. Our results showed that the Km of recombinant OsLCD1 protein for OAS or Na_2S in the OAS-TL reaction is 25- or 54-fold higher than that for L-cysteine in LCD-catalytic reaction, indicating OsLCD1 predominantly catalyzes the degradation of L-cysteine and thus is a true LCD (Table 2).

The deduction that OsLCD1 is a true LCD was also supported by the analysis of its overexpression rice plants. The total LCD activity in *OX1* and *OX2* was increased by 43.7% and 71.8% compared to the wild-type plants (Figure 3B). Accordingly, the endogenous

H_2S content in *OX1* and *OX2* was about 47.5% and 102.1% higher than those of wild-type plants (Figure 3C). Previous studies reported the involvement of CDes in drought resistance [10,11]. Our result also demonstrated that overexpression of *OsLCD1* improves rice drought tolerance (Figure 3D). As the main source of H_2S production in plant cells, the biological function of CDes largely relies on H_2S [19,42,43]. *OsLCD1*-improved rice drought tolerance may derive from the increase in endogenous H_2S content.

3.2. Molecular Mechanisms Underlying the Effects of H_2S on Drought Tolerance

Numerous biochemical and genetic results have undoubtedly established that the signaling action of H_2S in cells through persulfidation has important consequences for many physiological processes in plants [22,43,44]. Here, we found that the persulfidation widely exists in rice proteome under normal conditions and was differentially changed by drought stress (Figure 4). This further indicates that persulfidation may involve in rice drought stress response.

Nitrate reductase (NR) is a key enzyme for nitrogen assimilation and acquisition and plays a central role in plant biology and signaling transduction [28,45]. Previous results showed that the NR activity declined rapidly in response to drought stress, indicating NR may act as a negative regulator in plant drought stress response [46]. Most recently, it was reported that suppression of nitrate assimilation by regulating the expression of *NR* under drought stress could contribute to drought tolerance [47]. Similarly, our results showed that the expression of *NR* genes (Figure 6A,B) and related NR activity (Figure 5A) were gradually decreased under drought stress, indicating drought stress-induced inhibition of NR activity may attribute to the transcription regulation. These results were consistent with a previous study in maize leaves, which shows the decrease in maximal extractable NR activity was accompanied by a decrease in NR transcripts [46]. However, we observed that along with the decrease in NR activity, the endogenous H_2S content was gradually decreased after dehydration stress [13], while pretreatment of NaHS could promote the dehydration stress-induced inhibition of NR activity (Figure 5A,B). It indicated that dehydration-triggered inhibition of NR activity may correlate with endogenous H_2S content. This deduction was further confirmed by the application of H_2S scavenger, HT and LCD inhibitor, PAG, which delay or attenuate the inhibition of NR activity under dehydration stress (Figure 5C,D). Moreover, pretreatment of NaHS has no significant effect on the abundance of *NIA1* and *NIA2* (Figure 6B), illustrating that H_2S-promoted inhibition of NR activity may occur at the post-translational level.

NR is a highly regulated enzyme that is regulated at a variety of levels, including transcriptional-level regulation and post-translational modification in response to various environmental stimuli. For instance, the activity of NR in plants changes rapidly in response to various environmental stimuli, such as nitrate, light, plant hormones, low temperature, and drought stress [46,48–50]. Recent study on the interplay of persulfidation and phosphorylation of SnRK2.6 in Arabidopsis stomata regulation and drought tolerance [51] provide a good example for understanding the regulatory mechanism of NR in response to environmental stimuli. Previous studies demonstrated that NR activity was controlled by phosphorylation/dephosphorylation in plant cells. This regulatory model allows the NR transformation between high activity and low activity state [52,53]. Interestingly, both *NIA1* and *NIA2* protein was found in the *Arabidopsis* persulfidation proteome [23], revealing a new regulatory mechanism for NR functions. Here, our study showed that the persulfidation modification was detected in the OsNIA2 protein (Figure 6D,F), which is responsible for more than 90% NR activity in rice [35]. Drought stress significantly induced persulfidation of NIA2 protein, while this could be further enhanced by NaHS pretreatment or overexpression of *OsLCD1* (Figures 6F and 7B). The drought stress or NaHS treatment triggers the persulfidation of NIA2 protein and thus inhibits its activity. These results demonstrated that NR activity was also controlled by H_2S-mediated persulfidation in response to drought stress.

Rice seedlings grown in nitrate-deficient conditions are more tolerant to drought stress than of nitrate-sufficient conditions, indicating that decreased nitrogen assimilation contributes to the drought tolerance of rice [47]. This could be a strategy for plants balancing growth and defense responses under stress conditions. Consistently, the loss-of-function mutants of *OsNR1.2* [47] and *nia2* mutant, which both impaired nitrogen assimilation, are more tolerant to drought stress (Figure 8). A zinc finger transcription factor DROUGHT AND SALT TOLERANCE (DST) was specifically responsible for the suppression of *OsNR1.2* expression, but not *OsNIA2* in response to drought stress. As a consequence, *osnr1.2* mutant plants exhibited similar enhanced stomatal closure and drought tolerance as *dst* mutant plants. As the side reaction during NR-catalyzed nitrogen assimilation, the production of NO, an important signaling molecule, also contributed to the biological function of NR [31]. However, Since NO-deficient plants are markedly resistant to water deficit, the reduced water losses in NO-deficient plants may be due to hypersensitivity to ABA, thus leading to NO-independent inhibition of stomata opening and enhanced closure by ABA. In *Arabidopsis*, ABA-mediated regulation of stomata closure may not be necessarily dependent on de novo biosynthesis of NO through any of the proposed NR-mediated pathways [30]. In our study, the enhanced expression of ABA-responsive genes in the nia2 mutant was observed (Figure 8F,G), further confirming the importance of the NO-independent pathway in plants' response to drought stress.

Thus, the effects of H_2S-mediated persulfidation on NIA2 suggest a new mechanism for the modification of the NR protein itself in response to drought stress. More importantly, our results indicated that H_2S regulates signaling pathways in response to drought stress through persulfidation of NR protein, which led to the faster and more efficient inhibition of NR activity than through transcription regulation. These data provide new information that will benefit future studies on NR functional regulation in plants and expand the biological function of gasotransmitter H_2S.

4. Conclusions

In summary, we cloned and characterized a gene encoding an H_2S-producing enzyme in rice and named *OsLCD1*. Overexpression of OsLCD1 results in enhanced endogenous H_2S production, persulfidation of total soluble protein, and confers rice drought stress. We further elucidated a key mechanism of OsLCD1/H_2S-improved rice drought stress. Upon drought stress, H_2S induces persulfidation of NIA2, an NR isoform responsible for the main NR activity, thus decreasing total NR activity in rice. The inhibition of NIA2 activity improved the drought-responsive genes expression and further led to enhancement of drought tolerance in rice, as proved by the *nia2* mutant analysis. Combined with our previous knowledge of H_2S beneficial role on plant growth performance under various environmental stresses, our results contribute to the effective use of H_2S in agriculture, not only by exogenous administration of H_2S donors but also by genetic manipulation regarding H_2S metabolic pathways. Moreover, our results shed new light on the understanding of crop genetic improvement strategies through exploring and manipulating the other components that effectively regulate NR activity to balance crop growth/nitrogen assimilation and adaptation to stress.

5. Materials and Methods

5.1. Plant Materials, Growth Condition, and Treatment

Rice (*Oryza sativa* L., Wuyunjing 7 [54], and Dongjin [35]) was used in this study. Seeds were surface-sterilized and germinated in distilled water for 2 days at 28 °C. For drought stress experiments, germinated seeds were sowed into a 550 mL black opaque plastic beaker with soil in the glasshouse. The soil was taken from a field experiment site in Nanjing Agricultural University in Nanjing, Jiangsu. After two weeks, seedlings were withdrawn for irrigation for 8 days. After treatments, the corresponding phenotypes, including fresh weight and chlorophyll content, were measured.

5.2. Sequence Alignment and Phylogenetic Analysis

The alignment and phylogenetic tree of L-CDes homology from *Oryza sativa* (XP_015613237), *Panicum miliaceum* (RLN24808), *Dichanthelium oligosanthes* (OEL32418), and *Zea mays* (PWZ10688) and AtLCD1 (NP_001327694), AtDES (OAO92103), OAS-TL-A (AEE83514), OAS-TL-B (AEC10318), and OAS-TL-C (AEE79963) from *Arabidopsis thaliana* was performed and constructed according to the method described previously [8].

5.3. Cloning, Expression, and Purification of Recombinant OsLCD1

Total RNA was extracted from leaves of 14-day-old rice plants using Trizol reagent (Invitrogen, Gaithersburg, MD, USA) according to the manufacturer's instructions. The reverse transcription reaction was carried out to obtain cDNA by using the Super Script First-Strand Synthesis System for RT-PCR (Transgene, Beijing, China). To obtain the putative *L-CDes1* cDNA from Oryza sativa, the forward primer (5′-ATGGCGTCGATCCCGCCGGAT-3′) and the reverse primer (5′-TCAGGCCATCGTTTCCTGCTTC-3′) were used. The full length of *OsLCD1* was introduced into the *pET-28a(+)* vector at the sides of X*hol* and B*amHI* using a homologous recombination technique (Vazyme). After that, the recombinant vector was transferred into *E. coli* strain *Rosetta* (*DE3*) for protein expression. Briefly, the freshly inoculated Rosetta strain was grown at 37 °C with vigorous shaking for 4 h, at which point the OD600 of the culture was 0.5~0.6. Then 0.2 mM isopropyl-β-D-thiogalactopyranoside (IPTG) was added and cultivated for 12 h at 16 °C. The purification was performed under non-denaturing conditions by affinity to nickel resin using the Ni-NTA Purification System (Invitrogen) according to the manufacturer's instructions.

5.4. SDS-PAGE of Recombinant OsLCD1 and Western Blotting

Recombinant OsLCD1 protein was purified and then subjected to 12.5% SDS-PAGE. After electrophoresis, the protein was transferred from gel to the polyvinylidene difluoride (PVDF) membrane. The membrane was incubated in phosphate-buffered saline (PBS) with 5% bovine serum albumin (BSA) for 1 h at room temperature. After being washed with PBS/Tween buffer three times, immunoblot analysis was performed with relevant antibodies. The anti-His antibody was used at 1:5000 dilution. A secondary antibody was also used at 1:5000 dilution. The bands were visualized using enhanced chemiluminescence (ECL) reagents (Vazyme).

5.5. Enzyme Activity Measurements

The OsLCD1 activity was measured by the release of H_2S from L-cysteine. The assay contained a total of 3 mL 100 mM Tris/HCl pH 8.5, various amounts of different protein extracts, and 2.5 mM DTT. The reaction was started by the addition of 1 mM L-cysteine, incubated for 30 min at 37 °C, and terminated by adding 300 µL of 30 mM $FeCl_3$ dissolved in 1.2 N HCl and 300 µL 20 mM N,N-dimethyl-pphenylenediamine dihydrochloride dissolved in 7.2 N HCl [55,56]. The formation of methylene blue was determined at OD670 nm by using a spectrophotometer. Solutions with different concentrations of Na_2S were prepared used for the quantification of the enzymatically formed H_2S. OAS-TL activity was measured using the method described previously [57] in soluble bacterial or purified protein extracts. Nitrate activity was indicated by active nitrate reductase (NRAact). Briefly, the leaf samples or protoplasts were harvested and ground in the extraction buffer containing 25 mM potassium phosphate buffer (pH 8.8) and 10 mM cysteine. The protein extracted in the presence of excess Mg^{2+} is considered to be the NRAact in situ in leaf tissues, while NRAmax is measured in the presence and preincubation of EDTA for 30 min. The reaction mixture contained 0.4 mL of the extracted aliquots, 1.2 mL of a 0.1 mM potassium phosphate buffer (pH 7.5), 0.1 mM KNO_3, and 0.4 mL of 0.25 mM nicotinamide adenine dinucleotide (NADH). NRA was expressed as µmol $NO^{2-} g^{-1}$ FW h^{-1}.

5.6. Construction and Characterization of OsLCD1 Overexpression Lines

Transgenic lines (*OsLCD1* overexpression lines) were generated by Biorun Biotechnology. To obtain the transgenic plants overexpressing *OsLCD1*, the full-length coding DNA sequence of *OsLCD1* was inserted into the plant binary vector *pCAMBIA1305-GFP*. Then, the *OsLCD1* gene under the control of *CaMV 35S* promoter was transformed into rice (cv. Wuyunjing 7) by the Agrobacterium-mediated transformation method [58]. The progeny was selected by hygromycin and Western blotting with anti-GFP antibodies. Homozygous T3 seeds of the transgenic plants were used for further analysis.

5.7. Protoplast Preparation and Transiently Expression of OsNIA2

Stem and sheath tissues from 100 10-day-old rice seedlings were cut into approximately 0.5 mm strips and were used for protoplast isolation [59]. The method and details of *OsNIA2* gene clone and transient expression in rice protoplast were according to a previous study [60]. Briefly, the *1300221-OsNIA2-Flag* plasmid was transfected into 1 mL rice protoplast from WT, *OX2*, or *nia2* plants using a PEG-calcium-mediated method. After 12 h incubation, the protoplasts were harvested by centrifugation.

5.8. Immunochemical Detection of S-Persulfidated Proteins

S-persulfidated proteins were detected using a modified tag-switch method [23]. The total protein was extracted from rice seedlings with buffer (25 mM of Tris, 100 mM of NaCl, 0.2% Triton X-100, pH 8.0). Blocking buffer consisting of 50 mM of methylsulfonylbenzothiazole that was dissolved in tetrahydrofuran was added to an equal amount of extracted protein solutions and was incubated at 37 °C for 1 h to block free sulfhydryl groups. Proteins were precipitated by acetone to remove the excess and were resuspended in buffer (50 mM of Tris, 2.5% (*w*/*v*) SDS, 20 mM of CN-biotin, pH 8.0) and incubated 3 h at 37 °C. After that, the excess was removed by acetone. The final pellet was resuspended in buffer (50 mM of Tris, 0.5% (*w*/*v*) SDS, pH 8.0). The samples were run on SDS-PAGE and then transferred to a polyvinylidene fluoride membrane. The Western blot was performed with 1:10,000 dilution anti-biotin-HRP (Abcam, Cambridge, MA, USA). Coomassie brilliant blue-stained gels are present to show that equal amounts of proteins were loaded.

5.9. Real-Time RT-PCR Analysis

Total RNA was isolated from rice leaves using the Trizol reagent (Invitrogen) according to the manufacturer's instructions. Real-time quantitative reverse-transcription PCR was performed on a Mastercycler ep® realplex real-time PCR system (Eppendorf, Hamburg, Germany) in a 20 μL PCR amplification using SYBR® Premix Ex Taq™ (TaKaRa, San Jose, CA, USA) according to the manufacturer's instructions. Related primers and locus numbers of those genes are shown in Supplementary Table S1. The expression level of target genes was presented as x-fold changes relative to the appropriate control experiment after normalized against that of *OsActin1* (*LOC_Os03g50890*) and *OsActin2* (*LOC_Os10g36650*). Each experiment was performed with three replicates (each biological replicate was measured three times).

5.10. Statistical Analysis

Statistical analysis was performed using the software of SPSS 17.0. Statistical comparisons were performed by independent samples *t*-test (two-tailed, * $p < 0.05$, ** $p < 0.01$). Multiple comparisons were performed using a one-way ANOVA. Differences were considered significant at $p < 0.05$. All experimental data are presented as mean ± SD.

Supplementary Materials: The Supplementary Materials are available online at https://www.mdpi.com/article/10.3390/ijms222212119/s1.

Author Contributions: H.Z. and Y.X. conceived and designed the experiments. H.Z. constructed transgenic plants and performed Western blotting experiments. Y.Z. and F.Z. performed phenotypical and molecular experiments. W.G. cloned, expressed, and characterized OsLCD1 protein. Y.S.

prepared reagents and data analysis. H.Z., Y.X. and X.Y. wrote the manuscript. All authors have read and agreed to the published version of the manuscript.

Funding: This work was supported by grants from the National Natural Science Foundation of China (32101671, 31670255), the National Natural Science Foundation of China of Jiangsu Province (BK20200561, BK20200282), and the China Postdoctoral Science Foundation (2019M661860).

Institutional Review Board Statement: Not applicable.

Informed Consent Statement: Not applicable.

Acknowledgments: We gratefully acknowledge the assistance of Yali Zhang (State Key Laboratory of Crop Genetics and Germplasm Enhancement, and Key Laboratory of Plant Nutrition and Fertilization in Low-Middle Reaches of the Yangtze River, Ministry of Agriculture, Nanjing Agricultural University, China) in providing the *nia2* rice mutant seeds.

Conflicts of Interest: The authors declare that they have no conflict of interest with the contents of this article.

References

1. Seki, M.; Kamei, A.; Yamaguchi-Shinozaki, K.; Shinozaki, K. Molecular responses to drought, salinity and frost: Common and different paths for plant protection. *Curr. Opin. Biotechnol.* **2003**, *14*, 194–199. [CrossRef]
2. Miller, G.; Suzuki, N.; Ciftci-Yilmaz, S.; Mittler, R. Reactive oxygen species homeostasis and signalling during drought and salinity stresses. *Plant Cell Environ.* **2010**, *33*, 453–467. [CrossRef]
3. Zhu, J.K. Abiotic stress signaling and responses in plants. *Cell* **2016**, *167*, 313–324. [CrossRef]
4. Wang, R. Physiological implications of hydrogen sulfide: A whiff exploration that blossomed. *Physiol. Rev.* **2012**, *92*, 791–896. [CrossRef] [PubMed]
5. Corpas, F.J.; González-Gordo, S.; Cañas, A.; Palma, J.M. Nitric oxide and hydrogen sulfide in plants: Which comes first? *J. Exp. Bot.* **2019**, *70*, 4391–4404. [CrossRef]
6. Xie, Y.; Zhang, C.; Lai, D.; Sun, Y.; Samma, M.K.; Zhang, J.; Shen, W. Hydrogen sulfide delays GA-triggered programmed cell death in wheat aleurone layers by the modulation of glutathione homeostasis and heme oxygenase-1 expression. *J. Plant Physiol.* **2014**, *171*, 53–62. [CrossRef] [PubMed]
7. Guo, H.M.; Xiao, T.Y.; Zhou, H.; Xie, Y.J.; Shen, W.B. Hydrogen sulfide, a versatile regulator of environmental stress in plants. *Acta Physiol. Plant* **2016**, *38*, 1–13. [CrossRef]
8. Zhou, H.; Guan, W.; Zhou, M.; Shen, J.; Liu, X.; Wu, D.; Yin, X.; Xie, Y. Cloning and characterization of a gene encoding true D-cysteine desulfhydrase from *Oryza Sativa*. *Plant Mol. Biol. Rep.* **2020**, *38*, 95–113. [CrossRef]
9. Zhang, J.; Zhou, M.; Zhou, H.; Zhao, D.; Gotor, C.; Romero, L.C.; Shen, J.; Ge, Z.; Zhang, Z.; Shen, W.; et al. Hydrogen sulfide, a signaling molecule in plant stress responses. *J. Integr. Plant Biol.* **2021**, *63*, 146–160. [CrossRef]
10. Jin, Z.; Xue, S.; Luo, Y.; Tian, B.; Fang, H.; Li, H.; Pei, Y. Hydrogen sulfide interacting with abscisic acid in stomatal regulation responses to drought stress in *Arabidopsis*. *Plant Physiol. Biochem.* **2013**, *62*, 41–46. [CrossRef] [PubMed]
11. Jin, Z.; Wang, Z.; Ma, Q.; Sun, L.; Zhang, L.; Liu, Z.; Liu, D.; Hao, X.; Pei, Y. Hydrogen sulfide mediates ion fluxes inducing stomatal closure in response to drought stress in *Arabidopsis thaliana*. *Plant Soil* **2017**, *419*, 141–152. [CrossRef]
12. Ma, D.; Ding, H.; Wang, C.; Qin, H.; Han, Q.; Hou, J.; Lu, H.; Xie, Y.; Guo, T. Alleviation of drought stress by hydrogen sulfde is partially related to the abscisic acid signaling pathway in wheat. *PLoS ONE* **2016**, *11*, 1–16. [CrossRef] [PubMed]
13. Zhou, H.; Chen, Y.; Zhai, F.; Zhang, J.; Zhang, F.; Yuan, X.; Xie, Y. Hydrogen sulfide promotes rice drought tolerance via reestablishing redox homeostasis and activation of ABA biosynthesis and signaling. *Plant Physiol. Biochem.* **2020**, *155*, 213–220. [CrossRef] [PubMed]
14. Guo, H.; Zhou, H.; Zhang, J.; Guan, W.; Xu, S.; Shen, W.; Xu, G.; Xie, Y.; Foyer, C.H. L-cysteine desulfhydrase-related H_2S production is involved in *OsSE5*-promoted ammonium tolerance in roots of *Oryza sativa*. *Plant Cell Environ.* **2017**, *40*, 1777–1790. [CrossRef] [PubMed]
15. Rausch, T.; Wachter, A. Sulfur metabolism: A versatile platform for launching defence operations. *Trends Plant Sci.* **2005**, *10*, 503–509. [CrossRef]
16. Papenbrock, J.; Riemenschneider, A.; Kamp, A.; Schulz-Vogt, H.N.; Schmidt, A. Characterization of cysteine-degrading and H_2S-releasing enzymes of higher plants-from the field to the test tube and tack. *Plant Biol.* **2007**, *9*, 582–588. [CrossRef] [PubMed]
17. Burandt, P.; Schmidt, A.; Papenbrock, J. Cysteine synthesis and cysteine desulfuration in Arabidopsis plants at different developmental stages and light conditions. *Plant Physiol. Biochem.* **2001**, *9*, 861–870. [CrossRef]
18. Álvarez, C.; Calo, L.; Romero, L.C.; García, I.; Gotor, C. An O-acetylserine(thiol)lyase homolog with L-cysteine desulfhydrase activity regulates cysteine homeostasis in *Arabidopsis*. *Plant Physiol.* **2010**, *152*, 656–669. [CrossRef]
19. Zhang, J.; Zhou, M.; Ge, Z.; Shen, J.; Zhou, C.; Gotor, C.; Romero, L.C.; Duan, X.; Liu, X.; Wu, D.; et al. Abscisic acid-triggered guard cell L-cysteine desulfhydrase function and in situ hydrogen sulfide production contributes to heme oxygenase-modulated stomatal closure. *Plant Cell Environ.* **2020**, *43*, 624–636. [CrossRef]

20. Zhang, J.; Zhou, H.; Zhou, M.; Ge, Z.; Zhang, F.; Foyer, C.H.; Yuan, X.; Xie, Y. The coordination of guard-cell autonomous ABA synthesis and DES1 function *in situ* regulates plant water deficit responses. *J. Adv. Res.* **2020**, *27*, 191–197. [CrossRef]
21. Shen, J.; Su, Y.; Zhou, C.; Zhang, F.; Zhou, H.; Liu, X.; Wu, D.; Yin, X.; Xie, Y.; Yuan, X. A putative rice L-*cysteine desulfhydrase* encodes a true L-cysteine synthase that regulates plant cadmium tolerance. *Plant Growth Regul.* **2019**, *89*, 217–226. [CrossRef]
22. Filipovic, M.R. Persulfidation (S-sulfhydration) and H_2S. *Handb. Exp. Pharmacol.* **2015**, *230*, 29–59. [PubMed]
23. Aroca, A.; Benito, J.M.; Gotor, C.; Romero, L.C. Persulfidation proteome reveals the regulation of protein function by hydrogen sulfide in diverse biological processes in *Arabidopsis*. *J. Exp. Bot.* **2017**, *68*, 4915–4927. [CrossRef] [PubMed]
24. Filipovic, M.R.; Zivanovic, J.; Alvarez, B.; Banerjee, R. Chemical biology of H_2S signaling through persulfidation. *Chem. Rev.* **2018**, *118*, 1253–1337. [CrossRef]
25. Jurado-Flores, A.; Romero, L.C.; Gotor, C. Label-free quantitative proteomic analysis of nitrogen starvation in *Arabidopsis* root reveals new aspects of H_2S signaling by protein persulfidation. *Antioxidants* **2021**, *10*, 508. [CrossRef] [PubMed]
26. Shen, J.; Zhang, J.; Zhou, M.; Zhou, H.; Cui, B.; Gotor, C.; Romero, L.C.; Fu, L.; Yang, J.; Foyer, C.H.; et al. Persulfdation-based modifcation of cysteine desulfhydrase and the NADPH oxidase RBOHD controls guard cell abscisic acid signaling. *Plant Cell* **2020**, *32*, 1000–1017. [CrossRef] [PubMed]
27. Zhou, M.; Zhang, J.; Shen, J.; Zhou, H.; Zhao, D.; Gotor, C.; Romero, L.C.; Fu, L.; Li, Z.; Yang, J.; et al. Hydrogen sulfide-linked persulfidation of ABI4 controls ABA responses through the transactivation of MAPKKK18 in *Arabidopsis*. *Mol. Plant* **2021**, *14*, 921–936. [CrossRef] [PubMed]
28. Chamizo-Ampudia, A.; Sanz-Luque, E.; Llamas, A.; Galvan, A.; Fernandez, E. Nitrate reductase regulates plant nitric oxide homeostasis. *Trends Plant Sci.* **2017**, *22*, 163–174. [CrossRef]
29. Fu, Y.F.; Zhang, Z.W.; Yuan, S. Putative connections between nitrate reductase *S*-nitrosylation and NO synthesis under pathogen attacks and abiotic stresses. *Front Plant Sci.* **2018**, *9*, 474. [CrossRef]
30. Lozano-Juste, J.; León, J. Enhanced abscisic acid-mediated responses in nia1nia2noa1-2 triple mutant impaired in NIA/NR- and AtNOA1-dependent nitric oxide biosynthesis in *Arabidopsis*. *Plant Physiol.* **2010**, *152*, 891–903. [CrossRef]
31. Fancy, N.N.; Bahlmann, A.K.; Loake, G.J. Nitric oxide function in plant abiotic stress. *Plant Cell Environ.* **2017**, *40*, 462–472. [CrossRef] [PubMed]
32. Ortega, J.A.; Ortega, J.M.; Julian, D. Hypotaurine and sulfhydryl containing antioxidants reduce H_2S toxicity in erythrocytes from a marine invertebrate. *J. Exp. Biol.* **2008**, *211*, 3816–3825. [CrossRef]
33. Lisjak, M.; Teklic, T.; Wilson, I.D.; Whiteman, M.; Hancock, J.T. Hydrogen sulfide: Environmental factor or signalling molecule. *Plant Cell Environ.* **2013**, *36*, 1607–1616. [CrossRef] [PubMed]
34. Cao, Y.; Fan, X.; Sun, S.; Xu, G.; Jiang, H.; Shen, Q. Effect of nitrate on activities and transcript levels of nitrate reductase and glutamine synthetase in rice. *Pedosphere* **2008**, *18*, 664–673. [CrossRef]
35. Sun, H.; Bi, Y.; Tao, J.; Huang, S.; Hou, M.; Xue, R.; Liang, Z.; Gu, P.; Yoneyama, K.; Xie, X.; et al. Strigolactones are required for nitric oxide to induce root elongation in response to nitrogen and phosphate deficiencies in rice. *Plant Cell Environ.* **2016**, *39*, 1473–1484. [CrossRef] [PubMed]
36. Gruhlke, M.C.; Slusarenko, A.J. The biology of reactive sulfur species (RSS). *Plant Physiol. Biochem.* **2012**, *59*, 98–107. [CrossRef] [PubMed]
37. Romero, L.C.; Aroca, M.Á.; Laureano-Marín, A.M.; Moreno, I.; García, I.; Gotor, C. Cysteine and cysteine-related signaling pathways in *Arabidopsis thaliana*. *Mol. Plant* **2014**, *7*, 264–276. [CrossRef]
38. Kopriva, S. Regulation of sulfate assimilation in *Arabidopsis* and beyond. *Ann. Bot.* **2006**, *97*, 479–495. [CrossRef]
39. Shen, J.J.; Qiao, Z.J.; Xing, T.J.; Zhang, L.P.; Liang, Y.L.; Jin, Z.P.; Yang, G.D.; Wang, R.; Pei, Y.X. Cadmium toxicity is alleviated by AtLCD and AtDCD in *Escherichia coli*. *J. Appl. Microbiol.* **2012**, *113*, 130–1138. [CrossRef] [PubMed]
40. Riemenschneider, A.; Wegele, R.; Schmidt, A.; Papenbrock, J. Isolation and characterization of a D-cysteine desulfhydrase protein from *Arabidopsis thaliana*. *FEBS J.* **2005**, *272*, 1291–1304. [CrossRef] [PubMed]
41. McDonnell, L.; Plett, J.M.; Andersson-Gunnerås, S.; Kozela, C.; Dugardeyn, J.; Straeten, D.V.D.; Glick, B.; Undberg, B.; Regan, S. Ethylene levels are regulated by plant encoded 1-inocyclopropane-1-arboxylic acid deaminase. *Physiol. Plant* **2009**, *136*, 94–109. [CrossRef] [PubMed]
42. Mei, Y.; Zhao, Y.; Jin, X.; Wang, R.; Xu, N.; Hu, J.; Huang, L.; Guan, R.; Shen, W. L-Cysteine desulfhydrase-dependent hydrogen sulfide is required for methane-induced lateral root formation. *Plant Mol. Biol.* **2019**, *99*, 283–298. [CrossRef] [PubMed]
43. Gotor, C.; García, I.; Aroca, Á.; Laureano-Marín, A.M.; Arenas-Alfonseca, L.; Jurado-Flores, A.; Moreno, I.; Romero, L.C. Signaling by hydrogen sulfide and cyanide through post-translational modification. *J. Exp. Bot.* **2019**, *70*, 4251–4265. [CrossRef] [PubMed]
44. Aroca, A.; Zhang, J.; Xie, Y.; Romero, L.C.; Gotor, C. Hydrogen sulfide signaling in plant adaptations to adverse conditions: Molecular mechanisms. *J. Exp. Bot.* **2021**, *72*, 5893–5904. [CrossRef] [PubMed]
45. Costa-Broseta, Á.; Castillo, M.; León, J. Nitrite Reductase 1 is a target of nitric oxide-mediated post-translational modifications and controls nitrogen flux and growth in Arabidopsis. *Int. J. Mol. Sci.* **2020**, *21*, 7270. [CrossRef]
46. Foyer, C.H.; Valadier, M.H.; Migge, A.; Becker, T.W. Drought-induced effects on nitrate reductase activity and mRNA and on the coordination of nitrogen and carbon metabolism in maize leaves. *Plant Physiol.* **1998**, *117*, 283–292. [CrossRef] [PubMed]
47. Han, M.L.; Lv, Q.Y.; Zhang, J.; Wang, T.; Zhang, C.X.; Tan, R.J.; Wang, Y.L.; Zhong, L.Y.; Gao, Y.Q.; Chao, Z.F.; et al. Decreasing nitrogen assimilation under drought stress by suppressing DST-mediated activation of *Nitrate Reductase 1.2* in rice. *Mol. Plant* **2021**, in press. [CrossRef]

48. Srivastava, H.S. Regulation of nitrate reductase activity in higher plants. *Phytochemistry* **1980**, *19*, 725–733. [CrossRef]
49. Kaiser, W.M.; Kandlbinder, A.; Stoimenova, M.; Glaab, J. Discrepancy between nitrate reduction rates in intact leaves and nitrate reductase activity in leaf extracts: What limits nitrate reduction in situ? *Planta* **2000**, *210*, 801–807. [CrossRef]
50. Creighton, M.T.; Sanmartín, M.; Kataya, A.R.A.; Averkina, I.O.; Heidari, B.; Nemie-Feyissa, D.; Sánchez-Serrano, J.J.; Lillo, C. Light regulation of nitrate reductase by catalytic subunits of protein phosphatase 2A. *Planta* **2017**, *246*, 701–710. [CrossRef]
51. Chen, S.; Wang, X.; Jia, H.; Li, F.; Ma, Y.; Liesche, J.; Liao, M.; Ding, X.; Liu, C.; Chen, Y.; et al. Persulfidation-induced structural change in SnRK2.6 establishes intramolecular interaction between phosphorylation and persulfidation. *Mol. Plant* **2021**, *14*, 1814–1830. [CrossRef] [PubMed]
52. Kaiser, W.M.; Huber, S.C. Post-translational regulation of nitrate reductase: Mechanism, physiological relevance and environmental triggers. *J. Exp. Bot.* **2001**, *52*, 1981–1989. [CrossRef] [PubMed]
53. Lillo, C.; Meyer, C.; Lea, U.S.; Provan, F.; Oltedal, S. Mechanism and importance of post-translational regulation of nitrate reductase. *J. Exp. Bot.* **2004**, *55*, 1275–1282. [CrossRef] [PubMed]
54. Chen, J.; Liu, X.; Liu, S.; Fan, X.; Zhao, L.; Song, M.; Fan, X.; Xu, G. Co-Overexpression of *OsNAR2.1* and *OsNRT2.3a* increased agronomic nitrogen use efficiency in transgenic rice plants. *Front. Plant Sci.* **2020**, *11*, 1245. [CrossRef] [PubMed]
55. Siegel, M. A direct microdetermination for sulfide. *Anal. Biochem.* **1965**, *11*, 126–132. [CrossRef]
56. Xie, Y.J.; Lai, D.W.; Mao, Y.; Zhang, W.; Shen, W.B.; Guan, R.Z. Molecular cloning, characterization, and expression analysis of a novel gene encoding L-cysteine desulfhydrase from *Brassica napus*. *Mol. Biotechnol.* **2013**, *54*, 737–746. [CrossRef] [PubMed]
57. Barroso, C.; Vega, J.M.; Gotor, C. A new member of the cytosolic O-acetylserine(thiol)lyase gene family in *Arabidopsis thaliana*. *FEBS Lett.* **1995**, *363*, 1–5. [CrossRef]
58. Hiei, Y.; Ohta, S.; Komari, T.; Kumashiro, T. Efficient transformation of rice (*Oryza sativa* L.) mediated by Agrobacterium and sequence analysis of the boundaries of the T-DNA. *Plant J.* **1994**, *6*, 271–282. [CrossRef]
59. Zhang, Y.; Su, J.; Duan, S.; Ao, Y.; Dai, J.; Liu, J.; Wang, P.; Li, Y.; Liu, B.; Feng, D.; et al. A highly efficient rice green tissue protoplast system for transient gene expression and studying light/chloroplast-related processes. *Plant Methods* **2011**, *7*, 30. [CrossRef]
60. Zhou, H.; Wu, H.H.; Zhang, F.; Su, Y.; Guan, W.X.; Xie, Y.J.; Giroldo, J.P.; Shen, W.B. Molecular basis of cerium oxide nanoparticle enhancement of rice salt tolerance and yield. *Environ. Sci. Nano* **2021**. [CrossRef]

International Journal of *Molecular Sciences*

MDPI

Article

Hydrogen Sulfide Alleviates Manganese Stress in *Arabidopsis*

Lixia Hou †, Zhaoxia Wang †, Guangxia Gong, Ying Zhu, Qing Ye, Songchong Lu and Xin Liu *

Key Lab of Plant Biotechnology in University of Shandong Province, College of Life Science, Qingdao Agricultural University, Qingdao 266109, China; houlixia78@163.com (L.H.); 15771397276@163.com (Z.W.); gong286681@163.com (G.G.); zhuying0505@163.com (Y.Z.); 18500058569@126.com (Q.Y.); fudanlsc@126.com (S.L.)

* Correspondence: liuxin6080@126.com

† These authors contributed equally to this work.

Abstract: Hydrogen sulfide (H_2S) has been shown to participate in various stress responses in plants, including drought, salinity, extreme temperatures, osmotic stress, and heavy metal stress. Manganese (Mn), as a necessary nutrient for plant growth, plays an important role in photosynthesis, growth, development, and enzymatic activation of plants. However, excessive Mn^{2+} in the soil can critically affect plant growth, particularly in acidic soil. In this study, the model plant *Arabidopsis thaliana* was used to explore the mechanism of H_2S participation and alleviation of Mn stress. First, using wild-type *Arabidopsis* with excessive Mn^{2+} treatment, the following factors were increased: H_2S content, the main H_2S synthetase L-cysteine desulfhydrase enzyme (AtLCD) activity, and the expression level of the *AtLCD* gene. Further, using the wild-type, *AtLCD* deletion mutant (*lcd*) and overexpression lines (*OE5* and *OE32*) as materials, the phenotype of *Arabidopsis* seedlings was observed by exogenous application of hydrogen sulfide donor sodium hydrosulfide (NaHS) and scavenger hypotaurine (HT) under excessive Mn^{2+} treatment. The results showed that NaHS can significantly alleviate the stress caused by Mn^{2+}, whereas HT aggravates this stress. The *lcd* mutant is more sensitive to Mn stress than the wild type, and the overexpression lines are more resistant. Moreover, the mechanism of H_2S alleviating Mn stress was determined. The Mn^{2+} content and the expression of the Mn transporter gene in the mutant were significantly higher than those of the wild-type and overexpression lines. The accumulation of reactive oxygen species was significantly reduced in NaHS-treated *Arabidopsis* seedlings and *AtLCD* overexpression lines, and the activities of various antioxidant enzymes (SOD, POD, CAT, APX) also significantly increased. In summary, H_2S is involved in the response of *Arabidopsis* to Mn stress and may alleviate the inhibition of Mn stress on *Arabidopsis* seedling growth by reducing Mn^{2+} content, reducing reactive oxygen species content, and enhancing antioxidant enzyme activity. This study provides an important basis for further study of plant resistance to heavy metal stress.

Keywords: *Arabidopsis*; hydrogen sulfide; manganese stress; L-cysteine desulfhydrase; antioxidant enzyme

Citation: Hou, L.; Wang, Z.; Gong, G.; Zhu, Y.; Ye, Q.; Lu, S.; Liu, X. Hydrogen Sulfide Alleviates Manganese Stress in *Arabidopsis*. *Int. J. Mol. Sci.* **2022**, *23*, 5046. https://doi.org/10.3390/ijms23095046

Academic Editors: Yanjie Xie, Francisco J. Corpas and Jisheng Li

Received: 29 January 2022
Accepted: 28 April 2022
Published: 2 May 2022

1. Introduction

As a necessary trace element for plant growth, manganese (Mn) is mainly absorbed by plants in the form of divalent Mn ion (Mn^{2+}), which plays an important role in plant growth, development, and metabolism [1]. However, when the concentration of Mn^{2+} in the soil exceeds a certain threshold, it will be toxic to the plant, thus affecting normal growth. Mn stress usually occurs in acidic soil. When the pH value of the soil is lower than 5.5, a large amount of soluble Mn^{2+} is released. In these conditions, the concentration of Mn^{2+} in the soil increases sharply, which leads to Mn^{2+} accumulation in the plant [2]. With rapid industrialization and changes in tillage methods, the area of acidic soils in the world has expanded. Mn stress has become the second-largest plant-growth-limiting factor, after aluminum toxicity [3]. Therefore, it is of great significance to explore the mechanisms of Mn stress affecting plant growth and how to alleviate it.

When subjected to Mn stress, plants will behave differently at different growth stages [4]. Overall, Mn stress has a greater effect on the above-ground part than on the root system, and the leaf is the main target of Mn. Nearly 90% of the Mn absorbed by the plant is transferred to the above-ground tissue. Excess Mn^{2+} can inhibit leaf photosynthesis [5]. Mn stress decreases the activity of many important enzymes in plants [6], in addition to affecting the absorption, transport, and distribution of other nutrients that destroy the root structure [7]. In plants, Mn^{2+} is absorbed and transported through Mn^{2+} transporters. Most of these transporters are transmembrane proteins, which can transport and store Mn^{2+} in the inner membrane organelles [8,9]. However, the mechanism of how this process is initiated by plants has not fully been revealed.

H_2S is an important gas signal molecule. Studies on endogenous H_2S in plants date back decades. Wilson et al. (1978) observed that the leaves of cucumber, corn, and soybeans can release H_2S [10]. Rennenberg et al. (1987) found that in *Arabidopsis*, L-cysteine desulfhydrase and D-cysteine desulfhydrase used L/D-cysteine as a substrate to produce H_2S. L/D-cysteine desulfhydrase is a key enzyme for H_2S synthesis in plants, and it is also the most studied H_2S synthetase [11]. In recent years, it has been found that H_2S can participate in plant growth, development, and metabolism, such as enhancing plant photosynthesis, delaying flowering and senescence, and promoting seed germination [12–14]. Additionally, H_2S can increase plant resistance to a variety of environmental stresses, including drought, high salt, extreme temperatures, and various heavy metal stresses such as chromium [15,16], cadmium [17], and aluminum [18]. It has been reported that H_2S can alleviate aluminum toxicity by reducing the absorption of Al^{3+} and increasing the antioxidant capacity in barley [19]. H_2S can regulate the AsA–GSH cycle and alleviate As toxicity to peas [20]. It can also alleviate the inhibition of Cr^{6+} on the roots of *Arabidopsis* by upregulating the heavy metal (HM) chelator synthase-encoding genes, such as *PCS1*, *PCS2*, and *MT2A. Increased content of metallothionein and phytochelatins increases Arabidopsis* tolerance to Cr stress [21]. However, no reports have been made on H_2S response to Mn stress.

Using wild-type *Arabidopsis*, the AtLCD defective mutant *lcd*, and the AtLCD over-expression lines *OELCD* (*OE5* and *OE32*) as materials, we performed physiological and biochemical methods to explore the function and mechanism of H_2S in response to Mn stress in *Arabidopsis*.

2. Results

2.1. Effects of Mn Stress on H_2S Content, AtLCD Enzyme Activity, and Gene Expression in Arabidopsis Seedlings

It can be seen from Figure 1 that, under hydroponic conditions, H_2S content in roots treated using 4 mM Mn^{2+} for 3 h increased significantly, reached a high level, and then gradually decreased (Figure 1A). Both the activity of the H_2S synthase AtLCD and the expression of *AtLCD* reached maximum levels at 3 h and then decreased (Figure 1C,E). The H_2S-related indicators of the shoots were tested. It was also found that Mn^{2+} caused the H_2S content to increase, reaching the highest level at 9 h (Figure 1B). The main synthase activity and *AtLCD* expression had the same trend (Figure 1D,F). Thus, it is speculated that H_2S may participate in the response to Mn stress.

2.2. Effects of H_2S on Phenotype and Growth of Arabidopsis under Mn Stress

To investigate the role of H_2S in Mn stress, using wild-type *Arabidopsis* as the material, the seedling phenotype and growth indicators were observed by exogenous application of H_2S donor NaHS and scavenger HT under 4 mM Mn^{2+} treatment after 5 d. Compared with the control, it can be seen from Figure 2A that the growth of *Arabidopsis* seedlings was inhibited by Mn^{2+} treatment, the phenotype of Mn stress was significantly alleviated after NaHS treatment, and the phenotype of Mn stress was aggravated after HT treatment. Furthermore, the main root length, chlorophyll content, fresh weight, and dry weight were calculated, and the results were consistent with the phenotype results (Figure 2B,E). We also measured whether NaHS and HT affected H_2S content under Mn stress. The results showed that H_2S content increased after 5 days of Mn stress; moreover, NaHS could further enhance the content of H_2S, while HT decreased H_2S concentration (Figure S2). This further suggests the participation of H_2S in response to Mn stress, i.e., alleviating the stress caused by Mn^{2+} to *Arabidopsis* seedlings.

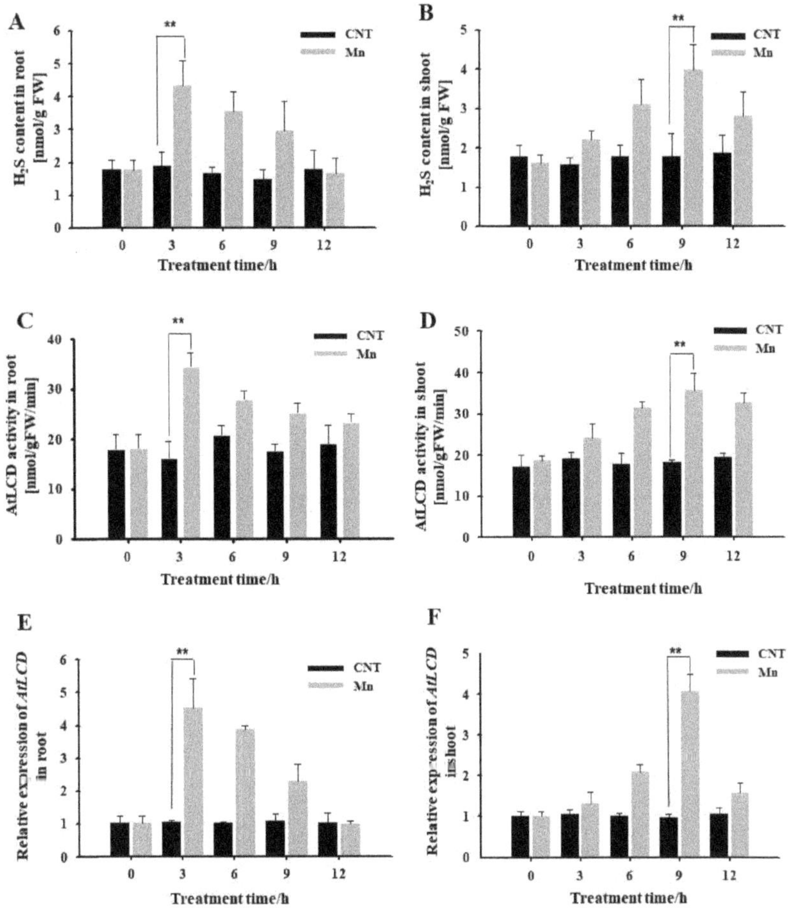

Figure 1. Effects of Mn stress on H_2S content, AtLCD enzyme activity, and gene expression in *Arabidopsis* seedlings. Effects of Mn stress on H_2S content in *Arabidopsis* roots (**A**) and shoot (**B**); AtLCD enzyme activity in *Arabidopsis* roots (**C**) and shoot (**D**); the relative expression of *AtLCD* in *Arabidopsis* roots (**E**) and shoot (**F**). Three independent experimental replications were conducted. Values are the means ± SE of three independent experiments (** $p < 0.01$).

2.3. *Effects on the Phenotype of lcd and OELCD Lines under Mn Stress*

To provide genetic evidence of H_2S participation in Mn stress, *lcd* mutant and two lines overexpressing *AtLCD* (*OE5, OE32*) were obtained and identified, and the results are shown in Supplementary Data (Figure S1). The phenotype and survival of each line after Mn stress were observed. The results are shown in Figure 3A. Overall, compared with *Arabidopsis* seedlings after 2 mM Mn^{2+} application, the phenotypes of all lines of Mn stress were alleviated after NaHS treatment, whereas the growth state of *Arabidopsis* seedlings was poor, and the leaves turned yellow after HT treatment. From a single treatment, after the Mn^{2+}, Mn^{2+} and NaHS treatment, and Mn^{2+} and HT treatment, the phenotypes of both overexpression lines were significantly better than those of the wild type, whereas mutants showed poor growth, and the leaves turned yellow. The survival statistics were consistent with those of the phenotype (Figure 3B).

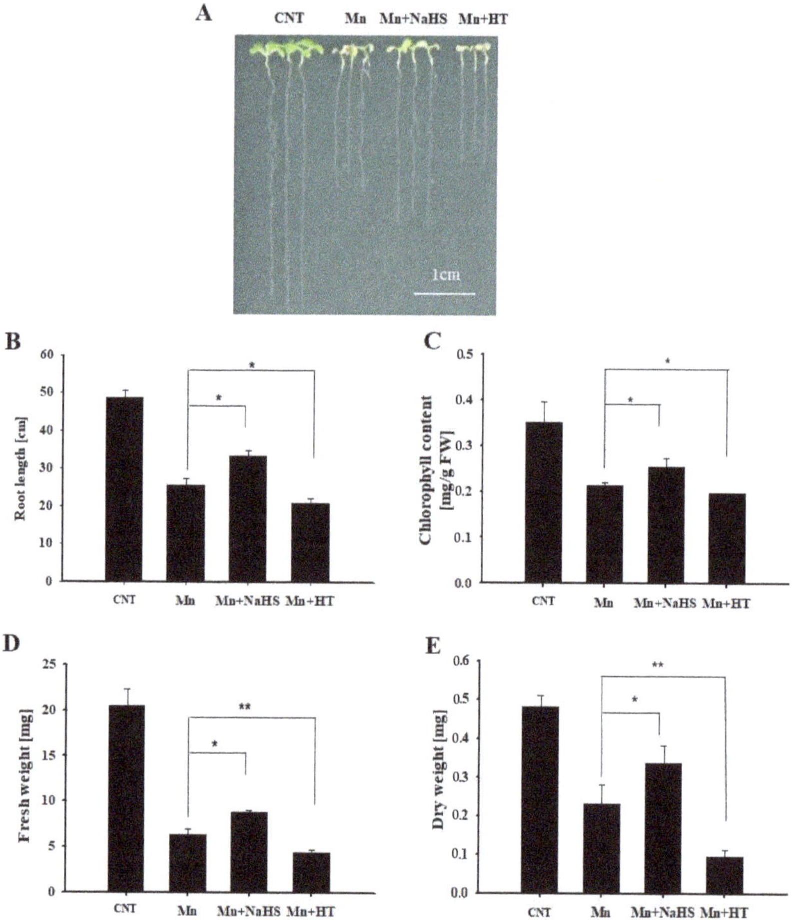

Figure 2. Effects of NaHS and HT on the growth of wild-type *Arabidopsis* seedlings under Mn stress. Effects of NaHS and HT on the phenotype (**A**), root length (**B**), chlorophyll content (**C**), fresh weight (**D**), and dry weight (**E**) of wild-type *Arabidopsis* under Mn stress. Three independent experimental replications were conducted. Values are the means ± SE of three independent experiments (* $p < 0.05$, ** $p < 0.01$). Scale bar = 1 cm.

2.4. Effects on Mn Transporter-Related Gene Expression in Roots of *lcd* and *OELCD* under Mn Stress

Hematoxylin is often used to observe the distribution of metal ions in plant roots [22]. Under Mn stress, the combination of hematoxylin with Mn^{2+} in root cells shows a purple color; the darker the purple color, the more Mn^{2+} in root cells. After 4 mM Mn^{2+} treatment of WT, *lcd*, and *OELCD*, hematoxylin staining was performed, and the results are shown in Figure 4A. The roots of *lcd* were stained deeper than those of the wild type, and the roots of *OE5* and *OE32* were stained shallowly. As hematoxylin staining is not specific to Mn, Mn content in roots was analyzed further by inductively coupled plasma atomic emission spectrophotometry (ICP–MS). The differences between wild type and other lines were compared. The content of Mn in *OE5* and *OE32* lines was significantly lower than that in the wild type under Mn stress, whereas, in *lcd* mutant, Mn content was significantly higher than that of the wild type (Figure 4B). It is, therefore, speculated that H_2S may reduce the root tissue Mn content under Mn stress.

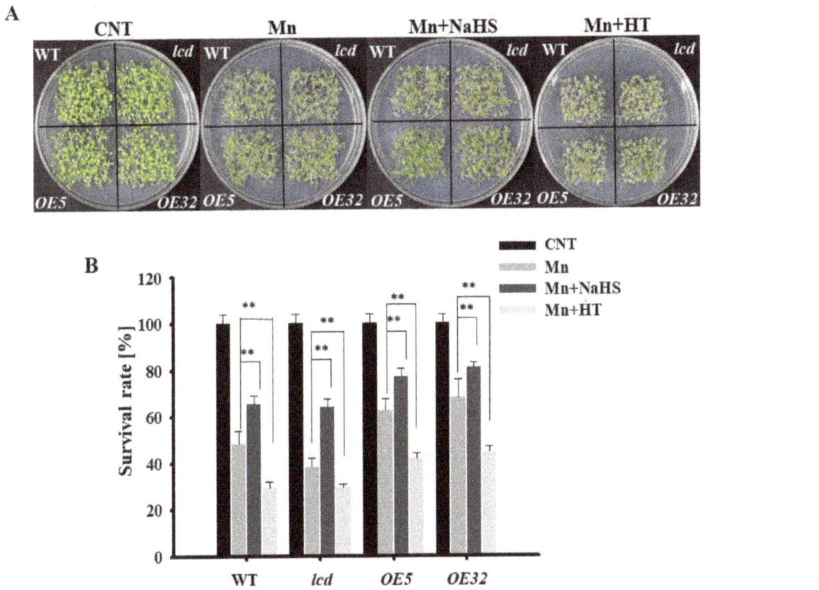

Figure 3. Effects of NaHS and HT on the growth of *lcd* and *OELCD* under Mn stress. Effects of NaHS and HT on the phenotype (**A**) and survival rates (**B**) of wild type, *lcd*, and two independent overexpression *Arabidopsis* seedlings after Mn stress. Three independent experimental replications were conducted. Values are the means ± SE of three independent experiments (** $p < 0.01$).

Furthermore, we analyzed whether a decrease in Mn content in root tissue by H_2S was related to the Mn transporter. There are numerous Mn transport-related proteins in plant roots, in which AtNramp1 is located in the cell membrane and is responsible for absorbing Mn^{2+} from the external environment; AtCAX2, AtMTP11, AtECA1 are located in the endomembrane system, such as in vacuole membranes, endoplasmic omentum, and Golgi membranes, and they are responsible for transporting excess Mn^{2+} to the cell organelles when the concentration of cytoplasmic Mn^{2+} is too high, thereby alleviating Mn stress [23]. Does H_2S reduce Mn^{2+} content in roots by regulating the expression of Mn transporter genes? The expression of Mn transporter-related genes in root tissues of WT, *lcd*, and *OELCD* was detected. The results reveal that 4 mM Mn treatments strongly induced the expression levels of four genes in all lines. The differences between wild type and other lines were further compared. The expression of the Mn^{2+}-uptake-related gene *AtNramp1 in OE5* and *OE32* lines was significantly lower than that in the wild type under Mn stress, whereas, in *lcd* mutant, the gene expression level increased but did not reach a significant level (Figure 4C). At the same time, compared with wild type, the expression levels of *AtCAX2*, *AtMTP11*, and *AtECA1* in *lcd* lines exhibited reduction but not significantly, while those of *OE5* and *OE32* lines were significantly higher than that of wild type (Figure 4D,F). The expression levels of four manganese-transport-related genes changed slightly in the mutant. Therefore, it is speculated that other H_2S synthesis genes and *AtLCD* have functional redundancy. Furthermore, it is inferred that H_2S may have partially prevented the root intake of Mn^{2+} by reducing the gene expression of

Mn^{2+}-uptake-related proteins, thereby promoting the transport of Mn^{2+} to organelles by increasing partial transporter gene expression.

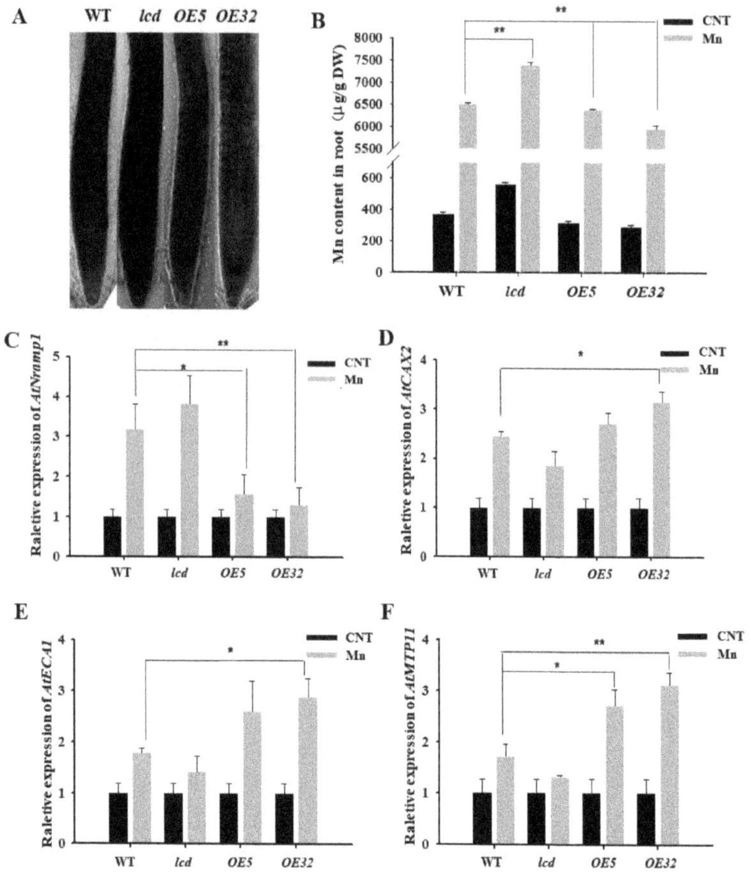

Figure 4. Effects on Mn transporter-related gene expression in roots of *lcd* and *OELCD* under Mn stress. Hematoxylin staining (**A**), Mn content (**B**), and the expression of *AtNramp1* (**C**), *AtCAX2* (**D**), *AtECA1* (**E**), and *AtMTP11* (**F**) in the roots of wild type, *lcd*, and two independent overexpression lines under Mn stress. Three independent experimental replications were conducted. Values are the means ± SE of three independent experiments (* $p < 0.05$; ** $p < 0.01$).

2.5. Effects on Reactive Oxygen Species Content of Arabidopsis Seedlings under Mn Stress

The content of reactive oxygen species in plants increases under abiotic stresses, such as heavy metal stress. Using wild-type *Arabidopsis* as material, the content of superoxide anion and hydrogen peroxide was detected by exogenous application of NaHS and HT in 4 mM Mn^{2+} treatments. Figure 5A,B show that the concentrations of O_2^- and H_2O_2 in *Arabidopsis* seedlings after Mn^{2+} treatment were significantly higher than that of control. The content of reactive oxygen speeies (ROS) was reduced with NaHS, compared with the Mn treatment, whereas the exogenous application of HT increased its content. Thus, it is speculated that exogenous H_2S can reduce the reactive oxygen species in *Arabidopsis* seedlings under Mn stress.

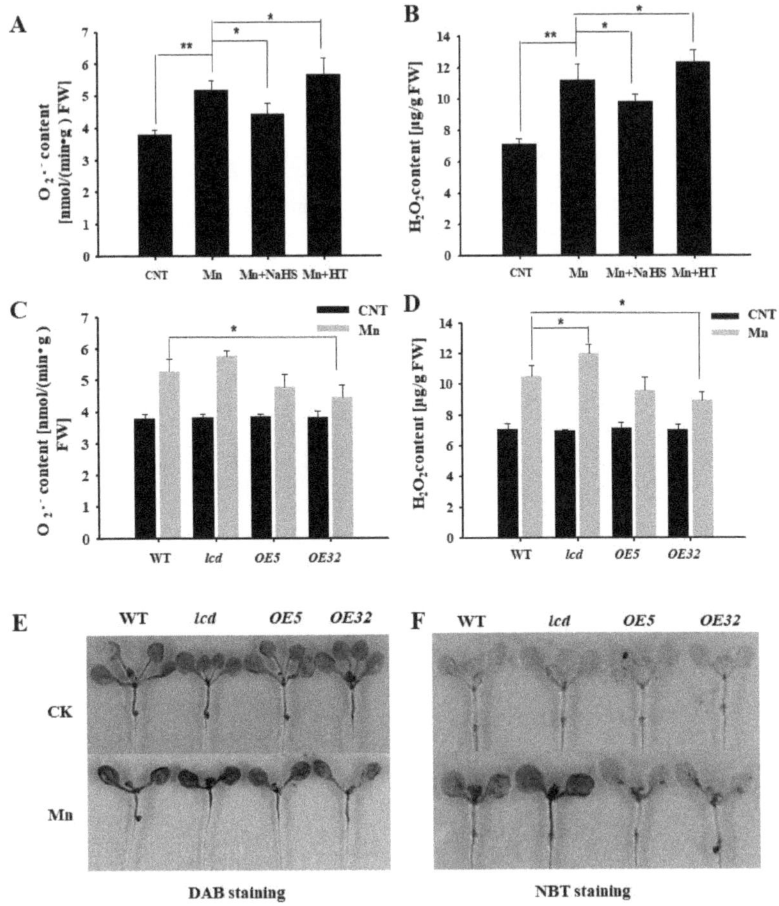

Figure 5. Effects of H_2S on H_2O_2 and O_2^- contents in *Arabidopsis* seedlings under Mn stress. Effects of NaHS and HT on the quantitative measurement of O_2^- (**A**) and H_2O_2 (**B**) concentrations in wild-type *Arabidopsis* seedlings under Mn stress. Quantitative measurement of O_2^- (**C**) and H_2O_2 (**D**) concentrations in wild type, *lcd*, and two independent overexpression lines seedlings treated with and without manganese. In situ accumulations of H_2O_2 (**E**) and O_2^- (**F**) before and after Mn treatment were revealed by DAB and NBT staining, respectively. Three independent experimental replications were conducted. Values are the means ± SE of three independent experiments (* $p < 0.05$; ** $p < 0.01$).

Further evidence from genetics was provided. The concentrations of O_2^- and H_2O_2 in *AtLCD* deficient mutants and overexpression lines were quantified under 4 mM Mn stress. The results showed that the concentrations of O_2^- and H_2O_2 in *OELCD* lines were significantly lower than that in wild type, whereas *lcd* lines were higher than that in wild type (Figure 5C,D), and the results of DAB and NBT were consistent with the above-described results (Figure 5E,F). Thus, it is speculated that H_2S may alleviate Mn stress in *Arabidopsis* seedlings by reducing reactive oxygen species content.

2.6. Effects on Antioxidant Enzyme Activity of lcd and OELCD under Mn Stress

Figure 5 shows that Mn stress increased the content of ROS in *Arabidopsis*, so the antioxidant enzyme activity was further examined. From Figure 6A–D, it can be deduced that the SOD, POD, CAT, and APX activities of *Arabidopsis* under 4 mM Mn^{2+} treatment were significantly higher than those of control. Additionally, the antioxidant enzyme activity of *Arabidopsis* with Mn treatment was upregulated significantly after NaHS treatment, whereas the activity of the antioxidant enzyme decreased after HT application. Additional results found that antioxidant enzyme activity of overexpression lines was significantly higher than that in wild type, and deletion mutants were significantly lower than that in the wild type under Mn^{2+} treatment (Figure 6E–H). It is, therefore, suggested that H_2S may alleviate Mn stress in *Arabidopsis* seedlings by increasing antioxidant enzyme activity.

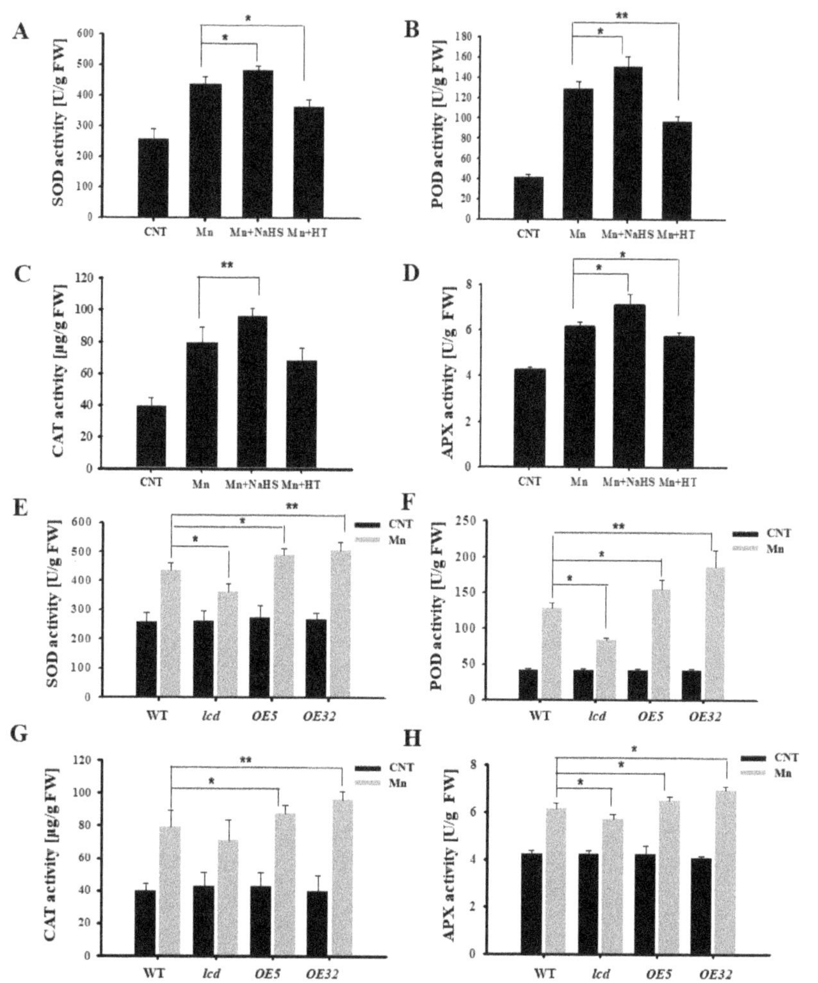

Figure 6. Effects of H_2S on antioxidant enzyme activities in *Arabidopsis* seedlings under Mn stress. Effects of NaHS and HT on the activity of SOD (**A**), POD (**B**), CAT (**C**), and APX (**D**) in wild-type *Arabidopsis* seedlings under Mn stress. The activity of SOD (**E**), POD (**F**), CAT (**G**), and APX (**H**) in wild type, *lcd*, and two independent overexpression line seedlings treated with and without manganese. Three independent experimental replications were conducted. Values are the means ± SE of three independent experiments (* $p < 0.05$; ** $p < 0.01$).

3. Discussion

It has been reported that H_2S participates in a variety of growth and development processes, as well as stress responses, in plants [24]. In this study, we found that Mn stress can induce increased H_2S content, and we also examined the enzyme activity and gene expression changes of the key synthase AtLCD in the H_2S synthesis pathway. We speculate that H_2S may participate in the Mn stress response (Figure 1). Through pharmacological experiments, wild-type *Arabidopsis* were used to observe the phenotype under Mn stress by external application of NaHS and HT, and the results showed that H_2S was involved in the Mn stress activities and could alleviate the phenotype of Mn stress (Figure 2). Based on the results shown in Figure 1, the change in H_2S content induced by Mn stress is from the AtLCD pathway, so further genetic evidence was provided. In Figure 3, we used 4 mM Mn at first, but the difference in phenotypes was not obvious, and the seedlings showed poor growth. It is speculated that the seeds were sowed directly in the culture medium to observe phenotype. And the time of treatment was long, the concentration was too high (4 mM), resulting in considerable damage to them. therefore, the concentration treatment was reduced (2 mM). Using *lcd* and *OELCD* as materials with Mn^{2+}, Mn^{2+} and NaHS, and Mn^{2+} and HT treatments, phenotypic observations demonstrated that H_2S alleviated Mn stress, and *lcd* was more pronounced than the wild-type Mn stress phenotype (Figure 3), which provided more evidence for H_2S participation in Mn stress response. However, current studies have found that there are many types of synthetic H_2S pathways in plants [25], and different synthetic pathways may be involved in different biological processes. Therefore, a question arises: Are there other sources of H_2S in *Arabidopsis* besides the AtLCD pathway under Mn stress? Further studies are needed.

During plant response to heavy metal stress, the cell membrane can prevent or reduce the entry of metal ions into the cell. Additionally, plants can leave Mn in subcellular compartments, such as vacuoles, endoplasmic reticulum, Golgi, and cell walls to resist the toxic effects [23,26]. In *Arabidopsis*, AtNramp1 of the Nramp family is the main Mn transporter that participates in Mn uptake. AtNramp1 is localized to the plasma membrane of the epidermal cells of the root tips, and the expression of *AtNramp1* is upregulated when Mn is deficient [27]. The transport protein of *Arabidopsis* vacuole cation exchanger (CAX) is mainly involved in the transport of Mn^{2+} to the vacuole. The T-DNA knockout mutant of *AtCAX2* has a lower content of Mn^{2+} in the vacuole than that of the wild type. Overexpression of *AtCAX2 in tobacco increases resistance to Mn toxicity by mediating Mn chelation into the vacuole [28]. Endoplasmic reticulum-localized* Ca^{2+}*-ATPase (ECA1) is another transporter that is intended to reduce the concentration of* Mn^{2+} *in the cytoplasm, as ECA1 can pump* Mn^{2+} *from the cytoplasm into the endoplasmic reticulum. Under high Mn conditions, the Arabidopsis* mutant *eca1* exhibited severe Mn stress symptoms, and overexpression of *ECA1* could restore the growth of *eca1* mutants to normal [29]. Metal tolerance protein (MTP), which is a Mn transporter in the CDF family [30], is responsible for transporting Mn into the vacuole and Golgi bodies. AtMTP11 may be in the irregular compartment of the trans-Golgi body [9]. For this group, *AtMTP11* has the highest expression and is more resistant to Mn in plant overexpression of *AtMTP11* [31]. Thus, we detected the expression of Mn-transport-related genes in the above transporter. The results showed that the expression of Mn-uptake-related gene *AtNramp1 in the mutant lcd* line was higher than that in the wild type under Mn stress, while both the expression levels of *OE5* and *OE32* were significantly lower than that in wild type (Figure 4C). This suggests that H_2S may have prevented the uptake of Mn^{2+} by root cells, in part, by inhibiting the expression of the *AtNramp1* gene. This result is consistent with those related to hematoxylin staining (Figure 4A) and Mn content (Figure 4B). However, the expression levels of *AtCAX2*, *AtMTP11*, and *AtECA1 in lcd* lines were lower than that of wild type, while those of *OE5* and *OE32* were significantly higher than that of wild type (Figure 4D,F), which suggests that H_2S may alleviate Mn stress by promoting Mn^{2+} transport to organelles.

Heavy metal stress can lead to excessive buildup of ROS, which may cause oxidative damage to biomolecules in plants [32]. H_2S is a reductive substance and can directly scavenge ROS [33]. Therefore, we examined the content of ROS and the activity of antioxidant enzymes after Mn^{2+} treatment, and the results showed that Mn stress induced excessive ROS in plants (Figure 5). Both endogenous and exogenous H_2S alleviated Mn stress in *Arabidopsis* seedlings by reducing ROS and significantly increasing antioxidant enzyme activity (Figure 6). It has previously been found that exogenous H_2S can alleviate the degree of peroxidation caused to rice by mercury, thus alleviating the stress on rice and improving rice resistance [34]. Some studies have found that H_2S alleviates oxidative stress and ionic toxicity in the cadmium-induced *Arabidopsis* roots through the hydrogen sulfide–cysteine circulatory system, thereby increasing the tolerance to cadmium [35]. Previous studies have found that plants respond to heavy metal copper ions, and H_2S alleviates the *Arabidopsis* copper oxide stress process through a circulatory system with cysteine [36]. More research is warranted on whether cysteine participates in H_2S involvement with Mn stress.

As a signaling molecule, how does H_2S signaling occur when plants are subjected to stress? In recent years, the molecular mechanism by which H_2S mediates the protein cysteine residue process in plants and animals, i.e., S-sulfhydration in post-translational modification, has been found [8,37]. Ethylene-induced hydrogen sulfide negatively regulates ethylene biosynthesis by the S-sulfhydration of ACO in tomatoes under osmotic stress [38]. Recently, it has been found that the persulfidation of SnRK2.6/OST1, a key regulatory protein of stomatal closure by H_2S, promotes the activity of SnRK2.6 and its interaction with transcription factors downstream of the ABA signal, to promote stomatal closure and inhibit stomatal opening and improve the drought resistance of plants [39]. This characteristic of H_2S provides an effective theoretical basis for finding its downstream regulatory proteins. Is the effect of H_2S on the activity of plant antioxidant enzymes due to the S-sulfhydration modification or through other action modes? These issues need to be further studied to understand the diverse mechanisms of plant responses to Mn stress.

4. Materials and Methods

4.1. Experimental Materials

For *Arabidopsis*, the Columbia (Col-0) ecotype was taken as the genetic background, and the T-DNA insertion mutant (SALK_082099, named *lcd*) of *AtLCD* was purchased from the American *Arabidopsis* Biological Resource Center (ABRC). *AtLCD* overexpressing *Arabidopsis* was named *OELCD* and included 2 lines (*OE5* and *OE32*).

4.2. Material Constructs, Cultivation, and Treatment

Full-length *AtLCD* (At3g62130) was obtained from the *Arabidopsis* Information Resource (TAIR). The cDNA fragment was amplified by PCR with the primers as follows: forward primer, 5′- CCCAAGCTTATGGAGGCGGGAGAGCG-3′ with restriction site *Xba* I (TaKaRa, Maebashi, Japan), reverse primer, 5′-GGGGTACCCTACAATGCAGGAAGGTTTT GAC-3′ with restriction site *Kpn* I (TaKaRa, Maebashi, Japan). Then, the cDNA fragment was inserted into the *p-Super1300* vector (containing 35S promoter and *GFP* reporter gene) between restriction sites *Xba* I and *Kpn* I (TaKaRa, Maebashi, Japan). The construct was confirmed by restriction digestion and sequence analysis and then was named p-Super1300-*AtLCD*. Subsequently, the construct was transformed into *Agrobacterium tumefaciens* strain GV3101. The flower dip method was used to transform the *Arabidopsis* [40]. Two independent lines (*OE5* and *OE32*) from the T3 generation were used in this research. T-DNA insertion mutant (*lcd*) was identified by the three-primer method. The primers sequence were as follows: LP, 5′-CACTTTTGCAAGCTTGGTTTC-3′; RP, 5′-TCAATCCAGTTGAATAAGCGC-3′; LBb1: 5′-GCCTTTTCAGAAATGGATAAATAGCCT TGCTTCCT-3′. The mutant *lcd* was homozygous.

The full-fledged seeds were placed in a dark treatment at 4 °C for 2 to 4 days to break the dormancy; then, the seeds were cultivated in 2% MGRL culture solution (pH = 5.8), and the incubator was put in a light environment, which was adjusted to 22 °C with a light–dark cycle of 16 h/8 h. After 7 days, the seedlings were treated with 4 mM $MnCl_2$, which was added to the MGRL culture solution; then, the material was collected for 0, 3, 6, 9, and 12 h to detect H_2S content, AtLCD enzymatic activity, and *AtLCD* gene expression.

For *Arabidopsis* culture in a solid medium, *Arabidopsis* seeds were treated with 10% NaClO for 10–15 min, after which the seeds were washed with aseptic water until there was no NaClO residue. The seeds were then placed at 4 °C for dark treatment for 2 to 4 days to break dormancy, sowed to a solid medium (pH = 5.8), and placed in a light incubator (22 °C, light–dark cycle 16 h/8 h).

Growth of wild-type *Arabidopsis* was allowed for 4–5 days in the solid medium (pH = 5.8); the seedlings were then moved to 1/2 MS solid medium, 1/2 MS solid medium with 4 mM $MnCl_2$, 1/2 MS solid medium with 4 mM $MnCl_2$ and 0.1 mM NaHS (Sigma, St. Louis, MO, USA), and 1/2 MS solid medium with 4 mM $MnCl_2$ and 0.02 mM HT (Sigma, St. Louis, MO, USA). All seedlings continued to grow for 5 days and were sampled for phenotype observation, detection of the growth indicators, and H_2S content.

Growth of *Arabidopsis* was allowed for 4–5 days in the solid medium (pH = 5.8). Wild-type, *lcd*, *OE5*, and *OE32 Arabidopsis* seeds were moved to 1/2 MS solid medium and 1/2 MS solid medium with 4 mM $MnCl_2$. All seedlings continued to grow for 5 days and were sampled for the physiological index.

Wild-type, *lcd*, *OE5*, and *OE32 Arabidopsis* seeds (100 seedlings of each kind of *Arabidopsis* in each treatment) breaking dormancy were sowed on 1/2 MS solid medium, 1/2 MS solid medium containing 2 mM $MnCl_2$, 1/2 MS solid medium containing 2 mM $MnCl_2$ and 0.1 mM NaHS, and 1/2 MS solid medium containing 2 mM $MnCl_2$ and 0.02 mM HT, and then the phenotypes were observed and photographed after two weeks of growth.

Wild-type, *lcd*, *OE5*, and *OE32 Arabidopsis* seeds were allowed for 4–5 days in the solid medium (pH = 5.8); the seedlings were moved to 1/2 MS solid medium with 4 mM $MnCl_2$ for 5 days, the roots of *Arabidopsis* seedlings were sampled for Mn content and Mn-transporter-related gene expression detection.

4.3. Detection of the Growth Indicators

The main root length, which was the distance from the base of the main root to the root tip was measured. The whole seedling from the medium was dried using filter paper, and the fresh weight was measured. Then, the whole seedling was dried at 80 °C for 16 h, and the weight was measured. The determination of chlorophyll content was performed according to the method of Lichtenthaler [41].

4.4. Determination of H_2S-Related Indicators

H_2S content detection was performed according to the methylene blue method described in Li et al. [42]; the determination of AtLCD enzyme activity followed the method of Riemenschneider et al. [43].

4.5. Determination of Physiological Indicators

Determination of hydrogen peroxide and superoxide anion content was achieved using the method of Zhao et al. [44], and the determination of SOD, POD, and CAT activities were determined following He et al. [45]; lastly, nitrogen-blue tetrazolium (NBT) (Macklin, China) and 3, 3′-diaminobenzidine (DAB) (Macklin, Shanghai, China) staining followed Jiang et al. [46].

4.6. Hematoxylin Staining

Hematoxylin staining followed the method of Ownby (1993), with minor modifications [22]. *Arabidopsis* seedlings were treated with 4 mM $MnCl_2$ for 24 h, and then the residual treatment solution was washed with deionized water. The roots of *Arabidopsis* seedlings were dyed in hematoxylin dye (0.2 g of hematoxylin and 0.02 g of potassium iodide, a constant volume of 0.1 L, and stored away from light) for 2 h. Then, the dye solution was washed with deionized water and observed by a microscope.

4.7. Detection of Mn Content

Mn content detection was performed using the method of inductively coupled plasma atomic emission spectrophotometry (ICP–MS), as described in Delhaize et al. [31].

4.8. RNA Extraction and qRT-PCR

Total RNA was extracted using a TRIzol reagent (Invitrogen, Waltham, MA, USA), following the manufacturer's instructions, and the cDNA was obtained by reverse transcription using the M-MLV RT Kit (Promega, Madison, WI, USA). With β-actin as an internal reference, qRT-PCR was performed in the presence of SYBR green I (BioWhittaker Molecular Applications, Walkersville, MD, USA) in the amplification mixture, and the data were analyzed using a MyiQ Detection System. The qRT-PCR procedure included 95 °C for 5 min, 95 °C for 30 s, 58 °C for 30 s, and 72 °C for 30 s, for 40 cycles. Three replicates were run for each sample. qRT-PCR primers are shown in Table 1.

Table 1. Primer sequences of qRT-PCR.

Gene Name	Primers' Sequences (5′-3′)
AtActin	FP: GGTAACATTGTGCTCAGTGG RP: CACGACCTTAATCTTCATGC
AtLCD	FP: TGTATGTGAGGAGGAGGC RP: GTTTCATACTGATGCTGCTC
AtNramp1	FP: GCTGGACAATATGTAATGCAGG RP: CACCGATGAGAGCAACAATTAG
AtCAX2	FP: GCCTCTTAAATGCTACATTCGG RP: TCCTTTGTCAAAGACTTGGTCT
AtECA1	FP: GTACACACACAGTAGCTTCATG RP: GTTTGAGTCGAACGAGAAAGTC
AtMTP11	FP: CAATACGGACATGGTCAATGAC RP: AATGAGAGCCAAATGTGTATGC

4.9. Statistical Methodology

Statistical analysis for all experiments was carried out using SPSS. Data were analyzed with independent *t*-tests ($p < 0.05$). All the values presented are means of replicates ± SE of three independent experiments.

5. Conclusions

In summary, H_2S is involved in the response of *Arabidopsis* to Mn stress and may alleviate the inhibition of Mn stress on *Arabidopsis* seedling growth by reducing Mn^{2+} content, reducing reactive oxygen species content, and enhancing antioxidant enzyme activity. This study provides an important basis for further study of plant resistance to heavy metal stress.

Supplementary Materials: The following supporting information can be downloaded at: https://www.mdpi.com/article/10.3390/ijms23095046/s1.

Author Contributions: All authors contributed to the study's conception and design. Material preparation was performed by Z.W., G.G. and Y.Z. Data collection and analysis were performed by L.H., Q.Y. and S.L. The first draft of the manuscript was written by L.H. and Z.W. The manuscript was revised by X.L. All authors have read and agreed to the published version of the manuscript.

Funding: This research was funded by the National Natural Science Foundation of China (Grant Nos. 31770275, 31900235, 31701063, and 31872082).

Institutional Review Board Statement: Not applicable.

Informed Consent Statement: Not applicable.

Data Availability Statement: Data are contained within the article or Supplementary Materials.

Conflicts of Interest: The authors declare no conflict of interest.

Abbreviation

APX	Ascorbate peroxidase
CAT	Catalase
DAB	3, 3′-diaminobenzidine
H2S	Hydrogen sulfide
L-cysteine desulfhydrase	LCD
NaHS	Sodium hydrosulfide
Mn	Manganese
NBT	Nitrogen-blue tetrazolium
POD	Peroxides
ROS	Reactive oxygen species
SOD	Super oxide dismutase

References

1. You, X.; Yang, L.T.; Lu, Y.B.; Li, H.; Zhang, S.Q.; Chen, L.S. Proteomic changes of *Citrus* roots in response to long-term manganese toxicity. *Trees-Struct. Funct.* **2014**, *28*, 1383–1399. [CrossRef]
2. Pittman, J.K. Managing the manganese: Molecular mechanisms of manganese transport and homeostasis. *New Phytol.* **2005**, *167*, 733–742. [CrossRef] [PubMed]
3. Niu, L.; Yang, F.; Xu, C.; Yang, H.; Liu, W. Status of metal accumulation in farmland soils across China: From distribution to risk assessment. *Environ. Pollut.* **2013**, *176*, 55–62. [CrossRef] [PubMed]
4. Xue, S.; Zhu, F.; Kong, X.; Wu, C.; Huang, L.; Huang, N.; Hartley, W. A review of the characterization and revegetation of bauxite residues (Red mud). *Environ. Sci. Pollut. R.* **2016**, *23*, 1120–1132. [CrossRef]
5. Weng, X.Y.; Zhao, L.L.; Zheng, C.J.; Zhu, J.W. Characteristics of the hyperaccumulator plant *Phytolacca acinosa* (*Phytolaccaceae*) in response to excess manganese. *J. Plant Nutr.* **2013**, *36*, 1355–1365. [CrossRef]
6. Shi, Q.; Zhu, Z.; Xu, M.; Qian, Q.; Yu, J. Effect of excess manganese on the antioxidant system in *Cucumis sativus* L. under two light intensities. *Environ. Exp. Bot.* **2006**, *58*, 197–205. [CrossRef]
7. Millaleo, R.; Reyes-Díaz, M.; Ivanov, A.G.; Mora, M.L.; Alberdi, M. Manganese as essential and toxic element for plants: Transport, accumulation and resistance mechanisms. *J. Soil Sci. Plant Nutr.* **2010**, *10*, 476–494. [CrossRef]
8. Aroca, A.; Benito, J.M.; Gotor, C.; Romero, L.C. Persulfidation proteome reveals the regulation of protein function by hydrogen sulfide in diverse biological processes in *Arabidopsis*. *J. Exp. Bot.* **2017**, *68*, 4915–4927. [CrossRef] [PubMed]
9. Li, J.; Jia, Y.; Dong, R.; Huang, R.; Liu, P.; Li, X.; Wang, Z.; Liu, G.; Chen, Z. Advances in the mechanisms of plant tolerance to manganese toxicity. *Int. J. Mol. Sci.* **2019**, *20*, 5096. [CrossRef] [PubMed]
10. Wilson, L.G.; Bressan, R.A.; Filner, P. Light-dependent emission of hydrogen sulfide from plants. *Plant Physiol.* **1978**, *61*, 184–189. [CrossRef] [PubMed]
11. Rennenberg, H.; Arabatzis, N.; Grundel, I. Cysteine desulphydrase activity in higher plants: Evidence for the action of L- and D-cysteine specific enzymes. *Phytochemistry* **1987**, *26*, 1583–1589. [CrossRef]
12. Jin, Z.; Pei, Y. Physiological implications of hydrogen sulfide in plants: Pleasant exploration behind its unpleasant odour. *Oxid. Med. Cell Longev.* **2015**, *2015*, 397502. [CrossRef] [PubMed]
13. Chen, J.; Wu, F.H.; Wang, W.H.; Zheng, C.J.; Lin, G.H.; Dong, X.J.; He, J.X.; Pei, Z.M.; Zheng, H.L. Hydrogen sulphide enhances photosynthesis through promoting chloroplast biogenesis, photosynthetic enzyme expression, and thiol redox modification in *Spinacia oleracea* seedlings. *J. Exp. Bot.* **2011**, *62*, 4481–4493. [CrossRef] [PubMed]
14. Zhang, H.; Hu, S.L.; Zhang, Z.J.; Hu, L.Y.; Jiang, C.X.; Wei, Z.J.; Liu, J.; Wang, H.L.; Jiang, S.T. Hydrogen sulfide acts as a regulator of flower senescence in plants. *Postharvest Biol. Tec.* **2011**, *60*, 251–257. [CrossRef]
15. Fang, H.; Jing, T.; Liu, Z.; Zhang, L.; Jin, Z.; Pei, Y. Hydrogen sulfide interacts with calcium signaling to enhance the chromium tolerance in *Setaria italica*. *Cell Calcium* **2014**, *56*, 472–481. [CrossRef]
16. Fang, H.; Liu, Z.; Long, Y.; Liang, Y.; Jin, Z.; Zhang, L.; Liu, D.; Li, H.; Zhai, J.; Pei, Y. The Ca^{2+}/calmodulin2-binding transcription factor TGA3 elevates *LCD* expression and H_2S production to bolster Cr^{6+} tolerance in *Arabidopsis*. *Plant J.* **2017**, *91*, 1038–1050. [CrossRef]
17. Liu, Z.; Fang, H.; Pei, Y.; Jin, Z.; Zhang, L.; Liu, D. WRKY transcription factors down-regulate the expression of H_2S-generating genes, *LCD* and *DES* in *Arabidopsis thaliana*. *Sci. Bull.* **2015**, *60*, 995–1001. [CrossRef]
18. Zhang, H.; Tan, Z.Q.; Hu, L.Y.; Wang, S.H.; Luo, J.P.; Jones, R.L. Hydrogen sulfide alleviates aluminum toxicity in germinating wheat seedlings. *J. Integr. Plant Biol.* **2010**, *52*, 556–567. [CrossRef]
19. Dawood, M.; Cao, F.; Jahangir, M.M.; Zhang, G.; Wu, F. Alleviation of aluminum toxicity by hydrogen sulfide is related to elevated ATPase, and suppressed aluminum uptake and oxidative stress in barley. *J. Hazard Mater.* **2012**, *209–210*, 121–128. [CrossRef]
20. Singh, V.P.; Singh, S.; Kumar, J.; Prasad, S.M. Hydrogen sulfide alleviates toxic effects of arsenate in pea seedlings through up-regulation of the ascorbate-glutathione cycle: Possible involvement of nitric oxide. *J. Plant Physiol.* **2015**, *181*, 20–29. [CrossRef]

21. Fang, H.; Liu, Z.; Jin, Z.; Zhang, L.; Liu, D.; Pei, Y. An emphasis of hydrogen sulfide-cysteine cycle on enhancing the tolerance to chromium stress in *Arabidopsis*. *Environ. Pollut.* **2016**, *213*, 870–877. [CrossRef] [PubMed]
22. Ownby, J.D. Mechanisms of reaction of hematoxylin with aluminium-treated wheat roots. *Physiol. Plantarum.* **1993**, *87*, 371–380. [CrossRef]
23. Baldisserotto, C.; Ferroni, L.; Anfuso, E.; Pagnoni, A.; Fasulo, M.P.; Pancaldi, S. Responses of *Trapa natans* L. floating laminae to high concentrations of manganese. *Protoplasma* **2007**, *231*, 65–82. [CrossRef]
24. Fang, T.; Cao, Z.; Li, J.; Shen, W.; Huang, L. Auxin-induced hydrogen sulfide generation is involved in lateral root formation in tomato. *Plant Physiol. Biochem.* **2014**, *76*, 44–51. [CrossRef] [PubMed]
25. Li, Z. Analysis of some enzymes activities of hydrogen sulfide metabolism in plants. *Methods Enzymol.* **2015**, *555*, 253–269.
26. Yang, S.; Yi, K.; Chang, M.M.; Ling, G.Z.; Zhao, Z.K.; Li, X.F. Sequestration of Mn into the cell wall contributes to Mn tolerance in sugarcane (*Saccharum officinarum* L.). *Plant Soil* **2019**, *436*, 475–487. [CrossRef]
27. Shao, J.F.; Yamaji, N.; Shen, R.F.; Ma, J.F. The key to Mn homeostasis in plants: Regulation of Mn transporters. *Trends Plant Sci.* **2017**, *22*, 215–224. [CrossRef]
28. Hirschi, K.D.; Korenkov, V.D.; Wilganowski, N.L.; Wagner, G.J. Expression of *Arabidopsis* CAX2 in tobacco. Altered metal accumulation and increased manganese tolerance. *Plant Physiol.* **2000**, *124*, 125–133. [CrossRef]
29. Wu, Z.; Liang, F.; Hong, B.; Young, J.C.; Sussman, M.R.; Harper, J.F.; Sze, H. An endoplasmic reticulum-bound Ca^{2+}/Mn^{2+} pump, ECA1, supports plant growth and confers tolerance to Mn^{2+} Stress. *Plant Physiol.* **2002**, *130*, 128–137. [CrossRef]
30. Montanini, B.; Blaudez, D.; Jeandroz, S.; Sanders, D.; Chalot, M. Phylogenetic and functional analysis of the Cation Diffusion Facilitator (CDF) family: Improved signature and prediction of substrate specificity. *BMC Genom.* **2007**, *8*, 107–116. [CrossRef]
31. Delhaize, E.; Gruber, B.D.; Pittman, J.K.; White, R.G.; Leung, H.; Miao, Y.; Jiang, L.; Ryan, P.R.; Richardson, A.E. A role for the AtMTP11 gene of *Arabidopsis* in manganese transport and tolerance. *Plant J* **2007**, *51*, 198–210. [CrossRef] [PubMed]
32. Emamverdian, A.; Ding, Y.; Mokhberdoran, F.; Xie, Y. Heavy metal stress and some mechanisms of plant defense response. *Scientific World J.* **2015**, *2015*, 756120. [CrossRef] [PubMed]
33. He, H.; Li, Y.; He, L. The central role of hydrogen sulfide in plant responses to toxic metal stress. *Ecotoxicol Environ. Saf.* **2018**, *157*, 403–408. [CrossRef] [PubMed]
34. Chen, Z.; Chen, M.; Jiang, M. Hydrogen sulfide alleviates mercury toxicity by sequestering it in roots or regulating reactive oxygen species productions in rice seedlings. *Plant Physiol. Biochem.* **2017**, *111*, 179–192. [CrossRef] [PubMed]
35. Jia, H.; Wang, X.; Dou, Y.; Liu, D.; Si, W.; Fang, H.; Zhao, C.; Chen, S.; Xi, J.; Li, J. Hydrogen sulfide-cysteine cycle system enhances cadmium tolerance through alleviating cadmium-induced oxidative stress and ion toxicity in *Arabidopsis* roots. *Sci. Rep.* **2016**, *6*, 39702. [CrossRef]
36. Jia, H.; Yang, J.; Liu, H.; Liu, K.; Ma, P.; Chen, S.; Shi, W.; Wei, T.; Ren, X.; Guo, J.; et al. Hydrogen sulfide—Cysteine cycle plays a positive role in *Arabidopsis* responses to copper oxide nanoparticles stress. *Environ. Exp. Bot.* **2018**, *155*, 195–205. [CrossRef]
37. Ge, S.N.; Zhao, M.M.; Wu, D.D.; Chen, Y.; Wang, Y.; Zhu, J.H.; Cai, W.J.; Zhu, Y.Z.; Zhu, Y.C. Hydrogen sulfide targets EGFR Cys797/Cys798 residues to induce Na^+/K^+-ATPase endocytosis and inhibition in renal tubular epithelial cells and increase sodium excretion in chronic salt-loaded rats. *Antioxid Redox Signal* **2014**, *21*, 2061–2082. [CrossRef]
38. Jia, H.; Chen, S.; Liu, D.; Liesche, J.; Shi, C.; Wang, J.; Ren, M.; Wang, X.; Yang, J.; Shi, W.; et al. Ethylene-induced hydrogen sulfide negatively regulates ethylene biosynthesis by persulfidation of ACO in tomato under osmotic stress. *Front. Plant Sci.* **2018**, *871*, 1517. [CrossRef]
39. Chen, S.; Jia, H.; Wang, X.; Shi, C.; Wang, X.; Ma, P.; Wang, J.; Ren, M.; Li, J. Hydrogen sulfide positively regulates abscisic acid signaling through persulfidation of SnRK2.6 in guard cells. *Mol. Plant* **2020**, *13*, 732–744. [CrossRef]
40. Clough, S.J.; Bent, A.F. Floral dip: A simplified method for *Agrobacterium*-mediated transformation of *Arabidopsis thaliana*. *Plant J.* **1998**, *16*, 735–743. [CrossRef]
41. Lichtenthaler, H.K. Chlorophylls and carotenoids: Pigments of photosynthetic Biomembranes. *Methods Enzymol.* **1987**, *148*, 350–382.
42. Li, Z.G. Quantification of hydrogen sulfide concentration using methylene blue and 5,5′-dithiobis(2-nitrobenzoic acid) methods in plants. *Methods Enzymol.* **2015**, *554*, 101–110. [PubMed]
43. Riemenschneider, A.; Nikiforova, V.; Hoefgen, R.; De Kok, L.J.; Papenbrock, J. Impact of elevated H_2S on metabolite levels, activity of enzymes and expression of genes involved in cysteine metabolism. *Plant Physiol. Biochem.* **2005**, *43*, 473–483. [CrossRef] [PubMed]
44. Zhao, Q.; Zhong, M.; He, L.; Wang, B.; Liu, Q.; Pan, Y.; Jiang, B.; Zhang, L. Overexpression of a *chrysanthemum* transcription factor gene *DgNAC1* improves drought tolerance in chrysanthemum. *Plant Cell Tiss. Org.* **2018**, *135*, 119–132. [CrossRef]
45. He, F.; Sheng, M.; Tang, M. Effects of rhizophagus irregularis on photosynthesis and antioxidative enzymatic system in *robinia pseudoacacia* L. under drought stress. *Front. Plant Sci.* **2017**, *8*, 183. [CrossRef]
46. Jiang, Y.; Qiu, Y.; Hu, Y.; Yu, D. Heterologous expression of at WRKY57 confers drought tolerance in *oryza sativa*. *Front. Plant Sci.* **2016**, *7*, 145. [CrossRef]

International Journal of *Molecular Sciences*

MDPI

Communication

The Defensive Role of Endogenous H_2S in *Brassica rapa* against Mercury-Selenium Combined Stress

Lifei Yang [1,2], Huimin Yang [1], Zhiwei Bian [1], Haiyan Lu [3], Li Zhang [4] and Jian Chen [3,*]

1 Department of Horticulture, College of Horticulture, Nanjing Agricultural University, Nanjing 210095, China; lfy@njau.edu.cn (L.Y.); 2019104074@njau.edu.cn (H.Y.); 201410408@njau.edu.cn (Z.B.)
2 Hexian New Countryside Development Research Institute, Nanjing Agricultural University, Hexian 238200, China
3 Laboratory for Food Quality and Safety-State Key Laboratory Cultivation Base of Ministry of Science and Technology, Institute of Food Safety and Nutrition, Jiangsu Academy of Agricultural Sciences, Nanjing 210014, China; luhaiyan8282@163.com
4 Department of Tobacco, College of Plant Protection, Shandong Agricultural University, Taian 271018, China; lilizhang324@163.com
* Correspondence: chenjian@jaas.ac.cn

Abstract: Plants are always exposed to the environment, polluted by multiple trace elements. Hydrogen sulfide (H_2S), an endogenous gaseous transmitter in plant cells, can help plant combat single elements with excess concentration. Until now, little has been known about the regulatory role of H_2S in response to combined stress of multiple elements. Here we found that combined exposure of mercury (Hg) and selenium (Se) triggered endogenous H_2S signal in the roots of *Brasscia rapa*. However, neither Hg nor Se alone worked on it. In roots upon Hg + Se exposure, the defensive role of endogenous H_2S was associated to the decrease in reactive oxygen species (ROS) level, followed by alleviating cell death and recovering root growth. Such findings extend our knowledge of plant H_2S in response to multiple stress conditions.

Keywords: *Brassica rapa*; hydrogen sulfide; mercury; reactive oxygen species; selenium

Citation: Yang, L.; Yang, H.; Bian, Z.; Lu, H.; Zhang, L.; Chen, J. The Defensive Role of Endogenous H_2S in *Brassica rapa* against Mercury-Selenium Combined Stress. *Int. J. Mol. Sci.* **2022**, *23*, 2854. https://doi.org/10.3390/ijms23052854

Academic Editor: Ki-Hong Jung

Received: 7 February 2022
Accepted: 3 March 2022
Published: 5 March 2022

1. Introduction

Heavy metal pollution poses a threat to biological systems, which is a serious problem worldwide. Heavy metal exposure impacts plant growth and development [1]. Hg (mercury) is one of the toxic metals even at low levels. Hg content increases continuously in the environment because of natural release and anthropogenic activities. The average content of soil Hg is 1.1 mg/kg worldwide. The polluted soils contain Hg at various levels even more than 10 mg/kg or higher [2]. Hg^{2+} is the most toxic form of Hg. Hg^{2+} can be readily taken up by plants to induce phytotoxicity. In hydroponic studies, plants exposed to levels of Hg^{2+} (even at 1–5 μM) begin to exhibit toxic symptoms [3,4]. Hg-induced phytotoxicity is associated with the accumulation of ROS (reactive oxygen species), followed by oxidative stress, physiological disturbance, and cell death [3,5]. ROS accumulation is considered as bio-indicator of Hg-induced phytotoxicity [4].

Se (selenium) is a beneficial micronutrient for plant growth. Se at proper concentrations, e.g., 0.5–6.0 mg/L SeO_3^{2-} (about 3.9–47 μM Se) can rescue plant from the toxicity of heavy metals including Hg [6]. However, Se application with excess concentration shows synergistic toxic effect with Hg [7]. Excessive Se is an emerging environmental problem [8]. Se and Hg frequently coexist in the environment [9]. Our previous study also found synergistic toxic effect of Se and Hg on *Brassica rapa*, showing enhanced ROS accumulation and aggravated growth inhibition as compared to Se or Hg alone [10]. However, we still rarely know the plant signaling regulation in response to Hg + Se combined stress.

H_2S (hydrogen sulfide) is an endogenous signaling molecule in plant cells. H_2S can be generated during sulfate assimilation. DES (L-cysteine desulfhydrase, EC4.4.1.1) is considered to be the main enzyme to produce H_2S, regulating various plant physiological processes [11]. H_2S plays important roles in the modulation of plant development and stress responses [12]. Few studies focus on the function of H_2S in plants upon multiple stress conditions at the same time. H_2S regulates plant tolerance against heavy metal stress, such as Cd (cadmium) [13], Al (aluminum) [14], Cr (chromium) [15], Pb (lead) [16], and Hg [17], etc. These reports suggest that H_2S protects plants from metal toxicity by alleviating oxidative injury. We previously found that H_2S was essential for plant combating excessive Se stress [18]. However, whether and how endogenous H_2S regulates plant response upon Se + Hg combination remains elusive.

In this study, we identified the intensified H_2S signaling in the root of *B. rapa* under Hg + Se stress conditions. The results not only help reveal plant signaling regulation of synergistic toxicity of Hg and Se, but also extend our knowledge of H_2S in plant cells in response to multiple stress conditions.

2. Results and Discussion

The roots of *B. rapa* were exposed to $HgCl_2$ (mercury chloride) (1 μM), Na_2SeO_3 (sodium selenite) (4 μM), or their combination for 3 days. Treatment with Hg alone slightly inhibited root growth (Figure 1A). This confirms the phytotoxic effect of Hg even at low levels [19]. Se induces phytotoxicity only at high concentrations [20]. Here we found that treatment with Se alone failed to affect root growth (Figure 1A), suggesting that Se at 4 μM was safe for the growth of *B. rapa* seedlings. However, simultaneous treatment with Hg and Se resulted in remarkable decrease in root length by 48.2% as compared to the control (Figure 1A). Since Hg + Se, but not Hg or Se alone, impacted root growth, we further performed time-course experiment to monitor root length upon Hg + Se exposure. Root growth was significantly inhibited after treatment with Hg + Se for 6 h. Then root growth speed declined with prolonged Hg + Se treatment (Figure 1B). Either Hg or Se was relative safe for the growth of *B. rapa* (Hg showed slight toxicity), but Hg + Se induced severe phytotoxicity. Treatment with Hg and Se together may overwhelm toxic threshold of single element, which further impacting plant growth. Therefore, Se showed synergistic but not antagonistic effect on Hg stress.

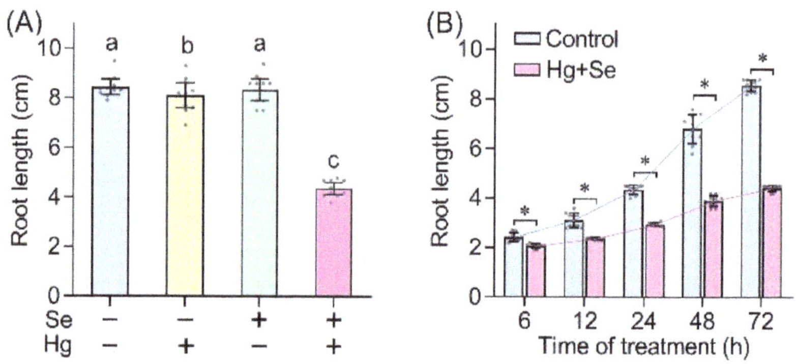

Figure 1. Hg + Se inhibited root growth of *B. rapa* seedlings. (**A**) The effect of Se, Hg, or Se + Hg on root length. Different lowercase letters indicated significant difference among different treatments (one-way analysis of variance, ANOVA; $n = 10$; $p < 0.05$). (**B**) Time-course monitoring of root length upon Se + Hg treatment. Asterisk indicated significant difference between control and Se + Hg (ANOVA; $n = 10$; $p < 0.05$).

The elongation of root tip determines root growth in response to environmental stress [21]. H_2S plays defensive roles in plants against environmental stress [12]. To study the response of H_2S in root tip, we used specific probe WSP-1 (3′-methoxy-3-oxo-3H-spiro[isobenzofuran-1, 9′-xanthen]-6′-yl 2-(pyridin-2-yldisulfanyl)benzoate) to label endogenous H_2S in situ. Neither Hg nor Se treatment alone affected endogenous H_2S level in root tip (Figure 2). Hg + Se treatment enhanced endogenous H_2S remarkably (Figure 2A). WSP-1 fluorescent density increased significantly by 7.71 fold in root tip upon Hg + Se as compared to control (Figure 2B). The changing pattern of endogenous H_2S was associated with root growth inhibition upon stress conditions. Endogenous H_2S were stimulated by Hg + Se (inhibiting growth remarkably), but not Hg or Se alone (without remarkable growth inhibition) (Figures 1A and 2B). The increase in endogenous H_2S level has been found in plant cells in response to growth-inhibited stimuli, such as salinity [22], cadmium (Cd) [23], and chromium (Cr) [24]. Thus, the roots of *B. rapa* may enhance endogenous H_2S to sense Hg + Se stress.

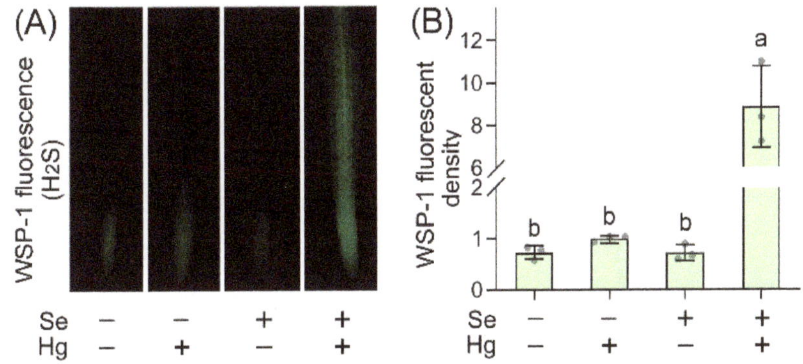

Figure 2. Hg + Se enhanced endogenous H_2S level in the root tip of *B. rapa* seedlings. (**A**) Endogenous H_2S in root tip was indicated by WSP-1 fluorescence. (**B**) Calculated relative WSP-1 fluorescent density in root tips. Different lowercase letters indicated significant difference among different treatments (ANOVA; $n = 3$; $p < 0.05$).

Stress-induced endogenous H_2S production is involved in plant tolerance against abiotic stress [25,26]. Intensified endogenous H_2S signal help sweet potato seedlings against osmotic stress [27]. To further confirm the defensive role of H_2S upon Hg + Se stress, we applied pharmacological experiments to alter endogenous H_2S level in roots. Roots were pretreated with H_2S donor NaHS (sodium hydrosulfide) for 6 h followed by Hg + Se exposure. Applying NaHS promoted root growth under Hg + Se stress in a dose-dependent manner, with maximal effect occurring at 2 mM NaHS (Figure 3A). Pretreatment with H_2S scavenger HT (hypotaurine) or DES inhibitor PAG (DL-propargylglicine) aggravated growth retardation upon Hg + Se stress (Figure 3B,C). In addition, HT (4 μM) or PAG (0.5 μM) counteracted the promoting effect of NaHS (2 mM) on root growth upon Hg + Se stress (Figure 3D). In situ fluorescent detection showed that NaHS further enhanced endogenous H_2S level in root tip treated with Hg + Se, an effect reversed by adding HT or PAG (Figure 3E,F). Hg + Se exposure increased root endogenous H_2S, but it was not sufficient to help plants combat stress condition. Further enhancing endogenous H_2S recovered root growth from subsequent Hg + Se stress. An adverse effect was found by decreasing endogenous H_2S through either scavenging already generated H_2S or impairing H_2S biosynthesis. These results suggested a protective role of plant endogenous H_2S against Hg + Se stress.

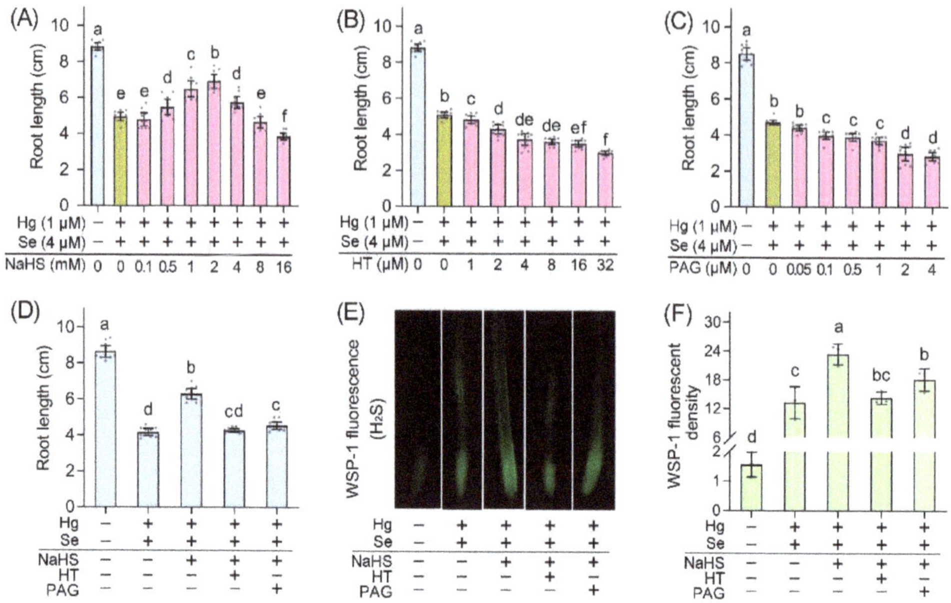

Figure 3. Endogenous H_2S is involved in recovering root growth of *B. rapa* seedlings under Hg + Se stress. (**A**) Effect of pretreatment with NaHS on root growth under Hg + Se stress. (**B**) Effect of pretreatment with HT on root growth under Hg + Se stress. (**C**) Effect of pretreatment with PAG on root growth under Hg + Se stress. (**D**) Effect of combined pretreatment of NaHS, HT, and PAG on root growth under Hg + Se stress. € Effect of combined pretreatment of NaHS, HT, and PAG on endogenous H_2S level indicated by WSP-1 fluorescence in root tips. (**F**) Calculated relative WSP-1 fluorescent density in root tips with respect € (**E**). Different lowercase letters in (**A–D**,**F**) indicated significant difference among different treatments (ANOVA; $n = 10$ or 3; $p < 0.05$).

The next step was to study how endogenous H_2S help plants detoxify Hg + Se. Our previous study found that Hg + Se induced more severe oxidative injury than that of Hg or Se alone in *B. rapa* [10]. We studied the effect of altering endogenous H_2S on oxidative injury in roots upon Hg + Se stress. Plant development needs to maintain the homeostasis of ROS at proper levels as it is an important signaling molecule. ROS overaccumulation is harmful to plants [28]. Total ROS in root tip were detected in vivo with specific fluorescent probe DCFH-DA (2′,7′-dichlorofluorescein diacetate) [29]. Hg + Se exposure led to remarkable increase in ROS level in root tip as compared to the low level of ROS in control. Pretreatment with NaHS significantly decreased ROS level in root tips under Hg + Se exposure. Further adding PAG or HT resulted in ROS accumulation (Figure 4A,B). These results suggested that enhancing endogenous H_2S was able to inhibit ROS accumulation induced by Hg + Se stress.

ROS can cause irreversible cell death in plants in response to stress [30]. Root tip cell death indicated by PI (propidium iodide) showed similar changing pattern to ROS under same treatment conditions (Figure 4C,D), suggesting that enhancing endogenous H_2S attenuated cell death in root treated with (Hg + Se). Root length was negatively correlated to ROS and cell death, respectively, under the condition of altering endogenous H_2S level (Figure 4E,F). Therefore, H_2S-supressed ROS accumulation and cell death may contribute to the recovery of root elongation from Hg + Se stress.

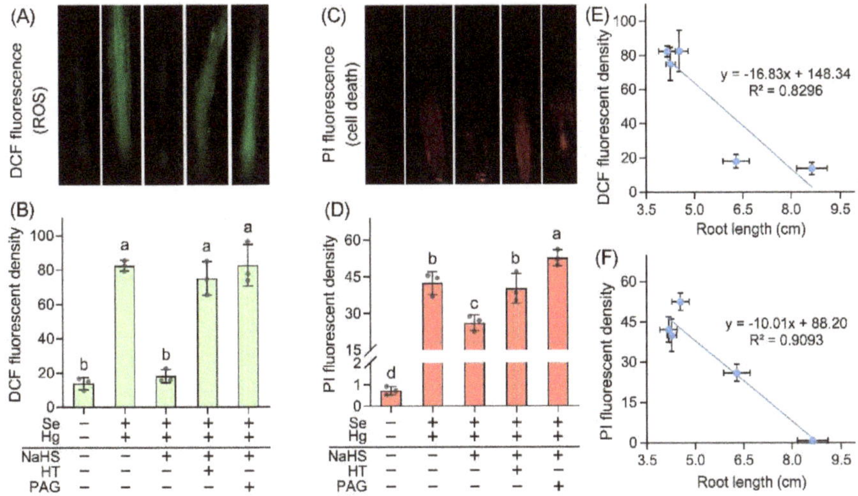

Figure 4. Endogenous H_2S is involved in decreasing ROS accumulation and alleviating cell death in the root tip of *B. rapa* seedlings upon Hg + Se stress. (**A**) In situ detection of ROS accumulation with DCF fluorescence in root tip. (**B**) Calculated relative DCF fluorescent density in root tips. (**C**) In situ detection of ROS accumulation with PI fluorescence in root tip. (**D**) Calculated relative PI fluorescent density in root tips. (**E**) Correlation analysis between root length and DCF fluorescent density. (**F**) Correlation analysis between root length and PI fluorescent density. Different lowercase letters in (**B**,**D**) indicated significant difference among different treatments (ANOVA; $n = 3$; $p < 0.05$).

ROS can attack cell membrane lipid, leading to loss of membrane integrity. This is one of the typical consequences of oxidative injury [31]. We applied histochemical staining to evaluate oxidative injury in roots. Lipid peroxidation (indicated by Schiff's reagent) and loss of membrane integrity (indicated by Evans blue) showed pink and blue, respectively (Figure 5). Hg + Se resulted in extensive staining as compared to light staining in control. Pretreatment with NaHS let to lighter staining in roots treated with Hg + Se, an effect reversed by further adding PAG or HT (Figure 5). These results suggested that endogenous H_2S was important for plants to combat oxidative induced by Hg + Se stress. Linking the above results provides an important clue that endogenous H_2S protects plant from Hg + Se stress by suppressing ROS accumulation followed by the alleviation of oxidative injury and cell death.

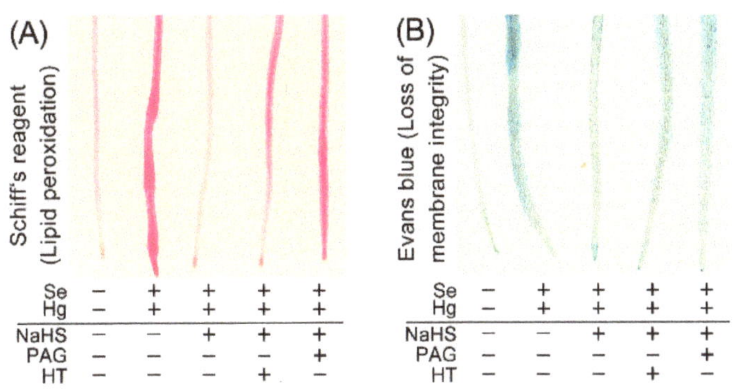

Figure 5. Endogenous H_2S is involved in alleviating oxidative injury in the roots of *B. rapa* seedlings upon Hg + Se stress. (**A**) In vivo detection of lipid peroxidation in roots. (**B**) In vivo detection of loss of membrane integrity in roots.

The antioxidant role of H_2S has been associated with its ability of limiting ROS accumulation in response to abiotic stress [32]. H_2S limits ROS accumulation probably through two ways. First, H_2S can scavenge already generated ROS by activating both antioxidative enzymes (enzymatic system) and antioxidants (non-enzymatic system) [33]. Second, H_2S is able to suppress ROS generation. Superoxide radical and hydrogen peroxide are two typical ROS. NADPH oxidase and PAO (polyamine oxidase) generate superoxide radical and hydrogen peroxide, respectively, in plants upon abiotic stress [34,35]. Excessive Se alone induces PAO- and NADPH oxidase-dependent ROS accumulation in plants [36,37]. Hg stress alone can induce NADPH oxidase-dependent ROS burst in plants as well [38]. It has been reported that H_2S can protect both plant and mammalian cells from oxidative stress by repressing NADPH oxidase-dependent ROS generation [39,40]. Further studies are needed to understand whether H_2S decreases ROS accumulation in plants against Hg + Se stress through the similar mechanisms mentioned above.

The synergistic toxicity of Hg and Se may directly result from the protein disfunction. The -SH in cystine (Cys) play important roles to maintain functional structure of the proteins. Hg^{2+} has strong affinity to -SH groups in functional proteins [41]. Cys in proteins can be easily replaced by SeCys [8]. Both actions are detrimental to the proteins. The combination of Hg and Se stress may aggravate the disturbance of the structure of functional proteins, further leading to negative physiological responses in plants. Recent studies have revealed that H_2S improves protein function via persulfidation of Cys [42,43]. Whether H_2S-mediated protein modification may help overcome the toxicity of Hg + Se needs further studies.

In sum, we found that endogenous H_2S could be triggered as a defensive signal in response to synergistic toxicity of Hg and Se in *B. rapa*. Neither Hg nor Se alone worked on it. Further studies are needed to reveal the molecular mechanism for H_2S-mediated detoxification of Hg + Se, our current results propose a novel role of endogenous H_2S in regulating plant adaption upon multiple stress conditions at the same time.

3. Materials and Methods

3.1. Plant Culture, Treatment, and Chemicals

B. rapa seeds (Lvling) were obtained from Jiangsu Academy of Agricultural Sciences, China. NaClO (sodium hypochlorite) solution (1%) were used to sterilize the surface of seeds for 10 min. After washing with distilled water, the seeds were soaked with distilled water for 3 h, and allowed for germination in darkness for 1 day. Then the seedlings were transferred into 1/4 strength Hoagland solution. The chamber for seedling growth was set at photoperiod of 12 h, photosynthetic active radiation of 200 $\mu mol/m^2/s$, and temperature at 25 °C. [44]. After growing for another 3 days, we selected 30 identical seedlings with root length at 1.5 cm for each treatment. Different chemicals were added to the nutrient solution to treat seedling roots according to different experimental designs. Basically, the seedlings were treated for 3 days. The seedling only had a primary root without appearing lateral roots. So, we evaluated root growth by only measuring the length of primary root.

Na_2SeO_3(sodium selenite) (4 μM) and $HgCl_2$ (mercury chloride) (1 μM) were applied for root treatment according to our previous study [10]. NaHS (sodium sulfide) (0.1–16 mM), PAG (DL-propargylglicine) (0.05–4 μM), and HT (hypotaurine) (1–32 μM) were applied as H_2S donor, H_2S biosynthesis inhibitor, and H_2S scavenger, respectively [18]. All the reagents at analytical purity were obtained from China National Pharmaceutical Group Co., Ltd. (Sinopharm Chemical Reagent Co., Ltd., Beijing, China).

We designed several experiments in this study. First, the roots were exposed to Hg, or Hg + Se, respectively, for 72 h, followed by measuring root length and endogenous H_2S level. Second, the root length were monitored at 6, 12, 24, 48, and 72 h, respectively, upon Hg + Se exposure. Third, roots were pretreated with NaHS, PAG, or HT, respectively, for 6 h. Then roots were exposed to Hg + Se for another 72 h, followed by measuring root length. This helped us to identify the proper concentration of NaHS (2 mM), PAG (0.5 μM), and HT (4 μM) for the following experiment. Fourth, roots were pretreated with

PAG + NaHS or HT + NaHS, respectively, for 6 h. Then roots were exposed to Hg + Se for another 72 h, followed by measuring root length, root tip endogenous H_2S level, root tip endogenous ROS level, and root tip cell death.

3.2. Evaluation of Endogenous H_2S Level in Root Tip

Fluorescent probe WSP-1 (3′-methoxy-3-oxo-3H-spiro[isobenzofuran-1, 9′-xanthen]-6′-yl 2-(pyridin-2-yldisulfanyl)benzoate) was applied to track root tip endogenous H_2S in situ [45,46]. Roots were incubated in WSP-1 solution (15 μM) at 25 °C for 40 min followed by rinsing with distilled water. Then root tips were observed and captured with a fluorescent microscope (ECLIPSE, TE2000-S, Nikon, Melville, NY, USA) with setting of excitation 480–490 nm/emission 515–525 nm.

3.3. Evaluation of Total ROS Level in Root Tip

Fluorescent probe DCFH-DA (2′,7′-dichlorofluorescein diacetate) was applied to detect root tip ROS in situ [29]. Roots were incubated in DCFH-DA solution (10 μM) at 25 °C for 10 min followed by rinsing with distilled water. Then root tips were observed and captured with a fluorescent microscope (ECLIPSE, TE2000-S) with setting of excitation 480–490 nm/emission 515–525 nm.

3.4. Evaluation of Cell Death in Root Tip

Fluorescent probe PI was used to detect root tip cell death in situ [47]. Roots were incubated in PI (propidium iodide) solution (20 μM) at 25 °C for 20 min followed by rinsing with distilled water. Then the root tips were observed and captured with a fluorescent microscope (ECLIPSE, TE2000-S) with setting of excitation 530–540 nm/emission 610–620 nm.

3.5. Evaluation of Oxidative Injury in Root Tip

Schiff's reagent was used to detect root lipid peroxidation in situ [48]. Roots were stained with Schiff's reagent for 20 min followed by rinsing with 0.5% (*w/v*) $K_2S_2O_5$ (prepared in 0.05 M of HCl) until the color became light red. Then roots were photographed with a digital camera.

Evans blue was used to detect loss of membrane integrity in root cells in situ [49]. Roots were stained with Schiff's reagent (0.025%, *w/v*) for 20 min followed by rinsing with distilled water. Then roots were photographed with a digital camera.

3.6. Statistical Analysis

Each result was presented as mean ± SD (standard deviation) of at 3–10 replicates. Least significant difference test (LSD) was performed on data following ANOVA tests to test for significant ($p < 0.05$) differences among treatments. The significant difference at $p < 0.05$ level between two designated treatments was compared using ANOVA with F test.

Author Contributions: Conceptualization, J.C. and L.Y.; methodology, H.Y., Z.B. and L.Z.; formal analysis, J.C. and Z.B.; investigation, L.Y., H.Y. and H.L.; resources, J.C.; writing—original draft preparation, L.Y. and L.Z.; writing—review and editing, J.C.; funding acquisition, L.Y. and J.C. All authors have read and agreed to the published version of the manuscript.

Funding: This research was funded by "National Natural Science Foundation of China, grant number 31771705", "Jiangsu Agricultural Science and Technology Innovation Fund, grant number CX(20)1011" and "China Agriculture Research System, grant number CARS-23-B16".

Institutional Review Board Statement: Not applicable.

Informed Consent Statement: Not applicable.

Data Availability Statement: Not applicable.

Acknowledgments: We would like to thank Ning Wang for her kind help in data processing.

Conflicts of Interest: The authors declare no conflict of interest.

References

1. Nagajyoti, P.C.; Lee, K.D.; Sreekanth, T.V.M. Heavy metals, occurrence and toxicity for plants: A review. *Environ. Chem. Lett.* **2010**, *8*, 199–216. [CrossRef]
2. Gworek, B.; Dmuchowski, W.; Baczewska-Dąbrowska, A.H. Mercury in the terrestrial environment: A review. *Environ. Sci. Eur.* **2020**, *32*, 128. [CrossRef]
3. Zhou, Z.S.; Huang, S.Q.; Guo, K.; Mehta, S.K.; Zhang, P.C.; Yang, Z.M. Metabolic adaptations to mercury-induced oxidative stress in roots of *Medicago sativa* L. *J. Inorg. Biochem.* **2007**, *101*, 1–9. [CrossRef] [PubMed]
4. Zhou, Z.S.; Wang, S.J.; Yang, Z.M. Biological detection and analysis of mercury toxicity to alfalfa (*Medicago sativa*) plants. *Chemosphere* **2008**, *70*, 1500–1509. [CrossRef]
5. Chen, J.; Yang, Z.M. Mercury toxicity, molecular response and tolerance in higher plants. *Biometals* **2012**, *25*, 847–857. [CrossRef]
6. Tran, T.A.T.; Dinh, Q.T.; Zhou, F.; Zhai, H.; Xue, M.; Du, Z.; Bañuelos, G.S.; Liang, D. Mechanisms underlying mercury detoxification in soil–plant systems after selenium application: A review. *Environ. Sci. Pollut. Res.* **2021**, *28*, 46852–46876. [CrossRef]
7. Tran, T.A.T.; Zhou, F.; Yang, W.; Wang, M.; Dinh, Q.T.; Wang, D.; Liang, D. Detoxification of mercury in soil by selenite and related mechanisms. *Ecotoxicol. Environ. Saf.* **2018**, *159*, 77–84. [CrossRef]
8. Hasanuzzaman, M.; Bhuyan, M.H.M.B.; Raza, A.; Hawrylak-Nowak, B.; Matraszek-Gawron, R.; Nahar, K.; Fujita, M. Selenium Toxicity in Plants and Environment: Biogeochemistry and Remediation Possibilities. *Plants* **2020**, *9*, 1711. [CrossRef]
9. Zhang, H.; Feng, X.; Jiang, C.; Li, Q.; Liu, Y.; Gu, C.; Shang, L.; Li, P.; Lin, Y.; Larssen, T. Understanding the paradox of selenium contamination in mercury mining areas: High soil content and low accumulation in rice. *Environ. Pollut.* **2014**, *188*, 27–36. [CrossRef]
10. Bian, Z.-W.; Chen, J.; Li, H.; Liu, D.-D.; Yang, L.-F.; Zhu, Y.-L.; Zhu, W.-L.; Liu, W.; Ying, Z.-Z. The phytotoxic effects of selenium–mercury interactions on root growth in *Brassica rapa* (LvLing). *Hortic. Environ. Biotechnol.* **2016**, *57*, 232–240. [CrossRef]
11. Arif, Y.; Hayat, S.; Yusuf, M.; Bajguz, A. Hydrogen sulfide: A versatile gaseous molecule in plants. *Plant Physiol. Biochem.* **2021**, *158*, 372–384. [CrossRef] [PubMed]
12. Zhang, J.; Zhou, M.; Zhou, H.; Zhao, D.; Gotor, C.; Romero, L.C.; Shen, J.; Ge, Z.; Zhang, Z.; Shen, W.; et al. Hydrogen sulfide, a signaling molecule in plant stress responses. *J. Integr. Plant Biol.* **2021**, *63*, 146–160. [CrossRef] [PubMed]
13. Qiao, Z.; Jing, T.; Liu, Z.; Zhang, L.; Jin, Z.; Liu, D.; Pei, Y. H_2S acting as a downstream signaling molecule of SA regulates Cd tolerance in *Arabidopsis*. *Plant Soil* **2015**, *393*, 137–146. [CrossRef]
14. Chen, J.; Wang, W.-H.; Wu, F.-H.; You, C.-Y.; Liu, T.-W.; Dong, X.-J.; He, J.-X.; Zheng, H.-L. Hydrogen sulfide alleviates aluminum toxicity in barley seedlings. *Plant Soil* **2013**, *362*, 301–318. [CrossRef]
15. Fang, H.; Liu, Z.; Jin, Z.; Zhang, L.; Liu, D.; Pei, Y. An emphasis of hydrogen sulfide-cysteine cycle on enhancing the tolerance to chromium stress in *Arabidopsis*. *Environ. Pollut.* **2016**, *213*, 870–877. [CrossRef]
16. Bharwana, S.A.; Ali, S.; Farooq, M.A.; Ali, B.; Iqbal, N.; Abbas, F.; Ahmad, M.S. Hydrogen sulfide ameliorates lead-induced morphological, photosynthetic, oxidative damages and biochemical changes in cotton. *Environ. Sci. Pollut. Res.* **2014**, *21*, 717–731. [CrossRef]
17. Chen, Z.; Chen, M.; Jiang, M. Hydrogen sulfide alleviates mercury toxicity by sequestering it in roots or regulating reactive oxygen species productions in rice seedlings. *Plant Physiol. Biochem.* **2017**, *111*, 179–192. [CrossRef]
18. Chen, Y.; Mo, H.Z.; Zheng, M.Y.; Xian, M.; Qi, Z.Q.; Li, Y.Q.; Hu, L.B.; Chen, J.; Yang, L.F. Selenium inhibits root elongation by repressing the generation of endogenous hydrogen sulfide in *Brassica rapa*. *PLoS ONE* **2014**, *9*, e110904. [CrossRef]
19. Mei, L.; Zhu, Y.; Zhang, X.; Zhou, X.; Zhong, Z.; Li, H.; Li, Y.; Li, X.; Daud, M.K.; Chen, J.; et al. Mercury-Induced Phytotoxicity and Responses in Upland Cotton (*Gossypium hirsutum* L.) Seedlings. *Plants* **2021**, *10*, 1494. [CrossRef]
20. Kolbert, Z.; Lehotai, N.; Molnar, A.; Feigl, G. "The roots" of selenium toxicity: A new concept. *Plant Signal. Behav.* **2016**, *11*, e1241935. [CrossRef]
21. Ryan, P.R.; Delhaize, E.; Watt, M.; Richardson, A.E. Plant roots: Understanding structure and function in an ocean of complexity. *Ann. Bot.* **2016**, *118*, 555–559. [CrossRef]
22. Christou, A.; Manganaris, G.A.; Papadopoulos, I.; Fotopoulos, V. Hydrogen sulfide induces systemic tolerance to salinity and non-ionic osmotic stress in strawberry plants through modification of reactive species biosynthesis and transcriptional regulation of multiple defence pathways. *J. Exp. Bot.* **2013**, *64*, 1953–1966. [CrossRef]
23. Zhang, L.; Pei, Y.; Wang, H.; Jin, Z.; Liu, Z.; Qiao, Z.; Fang, H.; Zhang, Y. Hydrogen sulfide alleviates cadmium-induced cell death through restraining ROS accumulation in roots of *Brassica rapa* L. ssp. *pekinensis*. *Oxidative Med. Cell. Longev.* **2015**, *2015*, 804603. [CrossRef]
24. Fang, H.; Jing, T.; Liu, Z.; Zhang, L.; Jin, Z.; Pei, Y. Hydrogen sulfide interacts with calcium signaling to enhance the chromium tolerance in *Setaria italica*. *Cell Calcium* **2014**, *56*, 472–481. [CrossRef] [PubMed]
25. Lai, D.; Mao, Y.; Zhou, H.; Li, F.; Wu, M.; Zhang, J.; He, Z.; Cui, W.; Xie, Y. Endogenous hydrogen sulfide enhances salt tolerance by coupling the reestablishment of redox homeostasis and preventing salt-induced K^+ loss in seedlings of *Medicago sativa*. *Plant Sci.* **2014**, *225*, 117–129. [CrossRef]
26. Cui, W.; Chen, H.; Zhu, K.; Jin, Q.; Xie, Y.; Cui, J.; Xia, Y.; Zhang, J.; Shen, W. Cadmium-induced hydrogen sulfide synthesis is involved in cadmium tolerance in *Medicago sativa* by reestablishment of reduced (Homo)glutathione and reactive oxygen species homeostases. *PLoS ONE* **2014**, *9*, e109669. [CrossRef] [PubMed]

27. Zhang, H.; Ye, Y.-K.; Wang, S.-H.; Luo, J.-P.; Tang, J.; Ma, D.-F. Hydrogen sulfide counteracts chlorophyll loss in sweetpotato seedling leaves and alleviates oxidative damage against osmotic stress. *Plant Growth Regul.* **2009**, *58*, 243–250. [CrossRef]
28. Dumanović, J.; Nepovimova, E.; Natić, M.; Kuča, K.; Jaćević, V. The significance of reactive oxygen species and antioxidant defense system in plants: A concise overview. *Front. Plant Sci.* **2021**, *11*, 552969. [CrossRef]
29. Foreman, J.; Demidchik, V.; Bothwell, J.H.; Mylona, P.; Miedema, H.; Torres, M.A.; Linstead, P.; Costa, S.; Brownlee, C.; Jones, J.D.; et al. Reactive oxygen species produced by NADPH oxidase regulate plant cell growth. *Nature* **2003**, *422*, 442–446. [CrossRef]
30. Huang, H.; Ullah, F.; Zhou, D.-X.; Yi, M.; Zhao, Y. Mechanisms of ROS regulation of plant development and stress responses. *Front. Plant Sci.* **2019**, *10*, 800. [CrossRef]
31. Gill, S.S.; Tuteja, N. Reactive oxygen species and antioxidant machinery in abiotic stress tolerance in crop plants. *Plant Physiol. Biochem.* **2010**, *48*, 909–930. [CrossRef] [PubMed]
32. Liu, H.; Wang, J.; Liu, J.; Liu, T.; Xue, S. Hydrogen sulfide (H_2S) signaling in plant development and stress responses. *aBIOTECH* **2021**, *2*, 32–63. [CrossRef] [PubMed]
33. Mostofa, M.G.; Rahman, A.; Ansary, M.M.U.; Watanabe, A.; Fujita, M.; Tran, L.-S.P. Hydrogen sulfide modulates cadmium-induced physiological and biochemical responses to alleviate cadmium toxicity in rice. *Sci. Rep.* **2015**, *5*, 14078. [CrossRef] [PubMed]
34. Gémes, K.; Kim, Y.J.; Park, K.Y.; Moschou, P.N.; Andronis, E.; Valassaki, C.; Roussis, A.; Roubelakis-Angelakis, K.A. An NADPH-oxidase/polyamine oxidase feedback loop controls oxidative burst under salinity. *Plant Physiol.* **2016**, *172*, 1418–1431. [CrossRef]
35. Yu, Z.; Jia, D.; Liu, T. Polyamine oxidases play various roles in plant development and abiotic stress tolerance. *Plants* **2019**, *8*, 184. [CrossRef]
36. Wang, Y.; Ye, X.; Yang, K.; Shi, Z.; Wang, N.; Yang, L.; Chen, J. Characterization, expression, and functional analysis of polyamine oxidases and their role in selenium-induced hydrogen peroxide production in *Brassica rapa*. *J. Sci. Food Agric.* **2019**, *99*, 4082–4093. [CrossRef]
37. Chen, Y.; Mo, H.Z.; Hu, L.; Li, Y.; Chen, J.; Yang, L. The endogenous nitric oxide mediates selenium-induced phytotoxicity by promoting ROS generation in *Brassica rapa*. *PLoS ONE* **2014**, *9*, e110901. [CrossRef]
38. Montero-Palmero, M.B.; Martín-Barranco, A.; Escobar, C.; Hernández, L.E. Early transcriptional responses to mercury: A role for ethylene in mercury-induced stress. *New Phytol.* **2014**, *201*, 116–130. [CrossRef]
39. Zhang, Z.; Jin, S.; Teng, X.; Duan, X.; Chen, Y.; Wu, Y. Hydrogen sulfide attenuates cardiac injury in takotsubo cardiomyopathy by alleviating oxidative stress. *Nitric Oxide* **2017**, *67*, 10–25. [CrossRef]
40. Yu, X.Z.; Chu, Y.P.; Zhang, H.; Lin, Y.J.; Tian, P. Jasmonic acid and hydrogen sulfide modulate transcriptional and enzymatic changes of plasma membrane NADPH oxidases (NOXs) and decrease oxidative damage in *Oryza sativa* L. during thiocyanate exposure. *Ecotoxicology* **2021**, *30*, 1511–1520. [CrossRef]
41. Natasha; Shahid, M.; Khalid, S.; Bibi, I.; Bundschuh, J.; Niazi, N.K.; Dumat, C. A critical review of mercury speciation, bioavailability, toxicity and detoxification in soil-plant environment: Ecotoxicology and health risk assessment. *Sci. Total Environ.* **2020**, *711*, 134749. [CrossRef]
42. Zhou, M.; Zhang, J.; Shen, J.; Zhou, H.; Zhao, D.; Gotor, C.; Romero, L.C.; Fu, L.; Li, Z.; Yang, J.; et al. Hydrogen sulfide-linked persulfidation of ABI4 controls ABA responses through the transactivation of MAPKKK18 in *Arabidopsis*. *Mol. Plant* **2021**, *14*, 921–936. [CrossRef] [PubMed]
43. Chen, S.; Wang, X.; Jia, H.; Li, F.; Ma, Y.; Liesche, J.; Liao, M.; Ding, X.; Liu, C.; Chen, Y.; et al. Persulfidation-induced structural change in SnRK2.6 establishes intramolecular interaction between phosphorylation and persulfidation. *Mol. Plant* **2021**, *14*, 1814–1830. [CrossRef] [PubMed]
44. Hu, L.; Li, H.; Huang, S.; Wang, C.; Sun, W.-J.; Mo, H.-Z.; Shi, Z.Q.; Chen, J. Eugenol confers cadmium tolerance via intensifying endogenous hydrogen sulfide signaling in *Brassica rapa*. *J. Agric. Food Chem.* **2018**, *66*, 9914–9922. [CrossRef] [PubMed]
45. Liu, C.; Pan, J.; Li, S.; Zhao, Y.; Wu, L.Y.; Berkman, C.E.; Whorton, A.R.; Xian, M. Capture and visualization of hydrogen sulfide by a fluorescent probe. *Angew. Chem.* **2011**, *123*, 10511–10513. [CrossRef]
46. Li, Y.J.; Chen, J.; Xian, M.; Zhou, L.G.; Han, F.X.; Gan, L.J.; Shi, Z.Q. In site bioimaging of hydrogen sulfide uncovers its pivotal role in regulating nitric oxide-induced lateral root formation. *PLoS ONE* **2014**, *9*, e90340. [CrossRef]
47. Kellermeier, F.; Chardon, F.; Amtmann, A. Natural variation of Arabidopsis root architecture reveals complementing adaptive strategies to potassium starvation. *Plant Physiol.* **2013**, *161*, 1421–1432. [CrossRef]
48. Wang, Y.S.; Yang, Z.M. Nitric oxide reduces aluminum toxicity by preventing oxidative stress in the roots of *Cassia tora* L. *Plant Cell Physiol.* **2005**, *46*, 1915–1923. [CrossRef]
49. Yamamoto, Y.; Kobayashi, Y.; Matsumoto, H. Lipid peroxidation is an early symptom triggered by aluminum, but not the primary cause of elongation inhibition in pea roots. *Plant Physiol.* **2001**, *125*, 199–208. [CrossRef]

International Journal of *Molecular Sciences*

MDPI

Review

Melatonin Confers Plant Cadmium Tolerance: An Update

Quan Gu [1], Chuyan Wang [1], Qingqing Xiao [1], Ziping Chen [2,*] and Yi Han [3,*]

1 School of Biology, Food and Environment, Hefei University, Hefei 230601, China; guq@hfuu.edu.cn (Q.G.); wangchuyan19@163.com (C.W.); xiaoqq@hfuu.edu.cn (Q.X.)
2 State Key Laboratory of Tea Plant Biology and Utilization, Anhui Agricultural University, Hefei 230036, China
3 National Engineering Laboratory of Crop Stress Resistence Breeding, School of Life Sciences, Anhui Agricultural University, Hefei 230036, China
* Correspondence: zpchen@ahau.edu.cn (Z.C.); yi.han@ahau.edu.cn (Y.H.)

Abstract: Cadmium (Cd) is one of the most injurious heavy metals, affecting plant growth and development. Melatonin (*N*-acetyl-5-methoxytryptamine) was discovered in plants in 1995, and it is since known to act as a multifunctional molecule to alleviate abiotic and biotic stresses, especially Cd stress. Endogenously triggered or exogenously applied melatonin re-establishes the redox homeostasis by the improvement of the antioxidant defense system. It can also affect the Cd transportation and sequestration by regulating the transcripts of genes related to the major metal transport system, as well as the increase in glutathione (GSH) and phytochelatins (PCs). Melatonin activates several downstream signals, such as nitric oxide (NO), hydrogen peroxide (H_2O_2), and salicylic acid (SA), which are required for plant Cd tolerance. Similar to the physiological functions of NO, hydrogen sulfide (H_2S) is also involved in the abiotic stress-related processes in plants. Moreover, exogenous melatonin induces H_2S generation in plants under salinity or heat stress. However, the involvement of H_2S action in melatonin-induced Cd tolerance is still largely unknown. In this review, we summarize the progresses in various physiological and molecular mechanisms regulated by melatonin in plants under Cd stress. The complex interactions between melatonin and H_2S in acquisition of Cd stress tolerance are also discussed.

Keywords: antioxidant defense systems; Cd stress; hydrogen sulfide; melatonin; oxidative stress; transportation and sequestration

Citation: Gu, Q.; Wang, C.; Xiao, Q.; Chen, Z.; Han, Y. Melatonin Confers Plant Cadmium Tolerance: An Update. *Int. J. Mol. Sci.* **2021**, *22*, 11704. https://doi.org/10.3390/ijms222111704

Academic Editors: Yanjie Xie, Francisco J. Corpas and Jisheng Li

Received: 28 September 2021
Accepted: 26 October 2021
Published: 28 October 2021

1. Introduction

Heavy metal pollution is the most widespread contamination resulting from anthropogenic activities in the world [1]. It has raised concerns about its various harmful risks to human health via the metal transfer along the food chain [2]. Among the heavy metals, cadmium (Cd) is a toxic element and poses a hazardous impact to living organisms, such as renal tubular dysfunction and bone disease [3]. In plants, Cd disturbs a range of important biochemical, morphological, physiological, and molecular processes, thus resulting in chlorosis and shunted growth [4,5]. Cd stress deceases the chlorophyll content, net photosynthetic rate, stomatal conductance, intracellular CO_2 concentration, and transpiration rate [4–6]. Cd stress induces the excess accumulation of reactive oxygen species (ROS), mainly due to the imbalance between ROS generation and scavenging [7,8]. Increased concentrations of ROS further induce the lipid peroxidation and oxidative damage, destructing plant membranes, macromolecules, and organelles [7,8]. Additionally, excessive bioaccumulation of Cd in plants inhibits Fe and Zn uptake, and disrupts the uptake and transport of K, Ca, Mg, P, and Mn [9]. In response to Cd stress, plants have evolved the complex biochemical and molecular mechanisms that modulate ROS homeostasis and Cd compartmentation and chelation [7,10–12]. Plant hormones (ethylene, salicylic acid (SA), abscisic acid (ABA), jasmonic acid (JA), auxin, brassinosteroids (BRs), and strigolactones (SLs)) and signaling molecules (nitric oxide (NO), carbon monoxide

(CO), hydrogen sulfide (H_2S), and Ca^{2+}) are involved in plant response to Cd stress [13,14]. Moreover, recent studies have reported that melatonin acts as a master regulator in plant Cd tolerance.

Melatonin (*N*-acetyl-5-methoxytryptamine) was discovered in plants in 1995, and it is since known to act as a pleiotropic molecule to participate in multiple physiological processes, such as plant growth and development, and protection against abiotic and biotic stresses [15,16]. In recent years, numerous studies have focused on the protective role of melatonin against Cd stress in plants [17]. Application of exogenous melatonin increased photosynthetic pigments, and improved relative water content and stomatal conductance in mallow plants upon Cd stress [18]. Many results showed that melatonin could re-establish redox homeostasis by certain enzymatic and non-enzymatic antioxidant defense systems to alleviate Cd-induced oxidative stress [19,20]. In addition, melatonin decreased Cd accumulation via regulating the transcripts of several heavy metal transporter genes to restrict Cd influx, and promote Cd efflux and chelation [19,21]. Moreover, NO and hydrogen peroxide (H_2O_2) signaling, microRNAs, heat shock factor HsfA1a and flavonoids may be involved in melatonin-mediated Cd tolerance in plants [19,22–25].

Apart from NO and H_2O_2, H_2S may also function as a signaling molecule in numerous processes of plants [26–28]. It is produced from the degradation of L-cysteine by L-cysteine desulfhydrase (L-CDes), which is encoded by L-cysteine desulfhydrase (*LCD*), D-cysteine desulfhydrase (*DCD*), and L-cysteine desulfhydrase1 (*DES1*) [29,30]. Exogenous application of H_2S donors regulated plant growth, and conferred tolerance to salinity, heavy metal, heat, and drought stress among others [27,31,32]. H_2S enhanced photosynthesis and antioxidant enzyme activity, and up-regulated the transcripts of PC genes to alleviate Cd stress in tobacco [33]. Moreover, H_2S homeostasis and L-cysteine desulfhydrase activity were involved in melatonin-modulated salt stress tolerance in tomato and cucumber seedlings [31,34,35]. Crosstalk of melatonin and H_2S in alleviating heat stress was also suggested in wheat [36]. However, the involvement of H_2S action in melatonin-mediated abiotic stress tolerance is still largely unknown, especially in Cd stress.

Over the past several years, numerous studies focusing on the role of melatonin in alleviating Cd stress have been steadily increasing in plants. Here, we systematically review and highlight the advanced developments which explore the melatonin-mediated Cd tolerance. For a better understanding of this topic, we also propose and discuss the future studies on the complex interactions between melatonin and H_2S during Cd stress.

2. Role of Melatonin in Plant Abiotic Stress Responses

2.1. Melatonin Biosynthesis and Catabolism

The melatonin metabolic pathway in plants contains two major parts: biosynthesis and catabolism (Figure 1). Melatonin was discovered and confirmed by an isotope tracer study of St. John's wort (*Hypericum perforatum* L. cv. Anthos) seedlings [15,37]. It was found that melatonin is synthesized via four continual enzymatic reactions from tryptophan, requiring at least six enzymes: tryptophan decarboxylase (TDC), tryptophan hydroxylase (TPH), tryptamine 5-hydroxylase (T5H), *N*-acetylserotonin methyltransferase (ASMT), and serotonin *N*-acetyltransferase (SNAT) [17]. T5H-catalyzed hydroxylation of tryptamine is an important step of melatonin biosynthesis in rice (*Oryza sativa*) [38]. In animals, serotonin is initially acetylated to form *N*-acetylserotonin, and then *O*-methylated to form melatonin (named NM pathway) [39]. It has also been found that serotonin is *O*-methylated to form 5-methoxytryptamine, and then acetylated to form melatonin (named MN pathway) [39]. Both NM and MN pathways exist in plants [40].

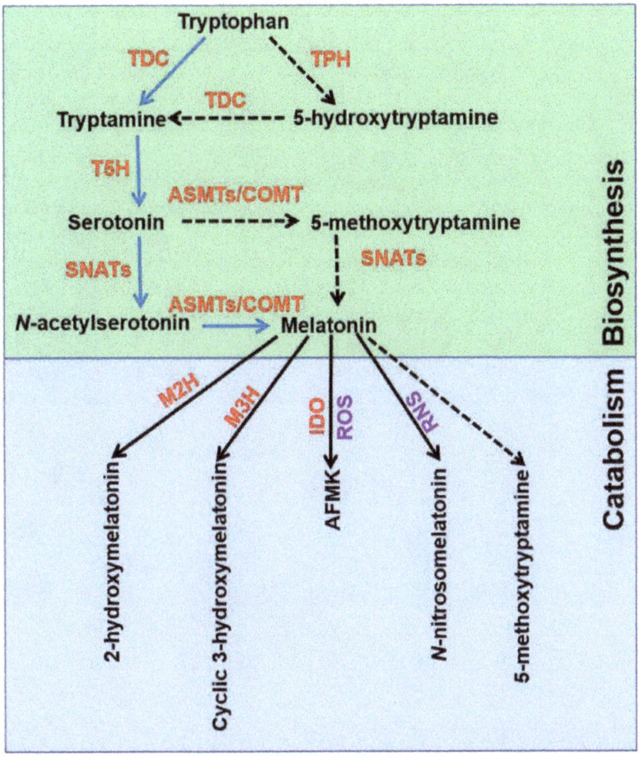

Figure 1. Melatonin biosynthesis and metabolic pathways in plants. TDC, tryptophan decarboxylase; T5H, tryptamine 5-hydroxylase; TPH, tryptophan hydroxylase; SNATs, serotonin N-acetyltransferases; ASMTs, N-acetylserotonin-O-methyltransferases; COMT, caffeic acid O-methyltransferase; M2H, melatonin 2-hydroxylase; M3H, melatonin 3-hydroxylase; IDO, indoleamine 2,3-dioxygenase; AFMK, N^1-acetyl-N^2-formyl-5-methoxykynuramine; ROS, reactive oxygen species; RNS, reactive nitrogen species. The green box indicates melatonin biosynthesis pathways, and blue box indicates melatonin metabolic pathways.

Melatonin can be degraded by two distinct routes: non-enzymatic and enzymatic transformations [17]. Transgenic tomato (*Solanum lycopersicum*) plants expressing the gene encoding indoleamine 2,3-dioxygenase (IDO) in rice showed reduced melatonin levels [41]. Thus, the pathway that melatonin converts to N^1-acetyl-N^2-formyl-5-methoxykynuramine (AFMK) exists in plants. Tan and Reiter speculated that AFMK is the product of melatonin interaction with ROS, which generated during photosynthesis [39]. This might reflect the important role of melatonin in detoxifying ROS accumulation. In addition, melatonin hydroxylation metabolites, 2-hydroxymelatonin (2-OHMel) and cyclic 3-hydroxymelatonin (c3-OHMel), have been identified in plants. Their formation is attributed to melatonin 2-hydroxylase (M2H) and melatonin 3-hydroxylase (M3H), respectively [42–44]. Singh et al. suggested that *N*-nitrosomelatonin (NOmela) likely served as a nitric oxide (NO) carrier that participated in the redox signal transduction [45]. Nevertheless, Mukherjee considered that NOmela served as an intracellular NO reserve in plants was questionable due to its sensitive and unstable characteristics [46]. The processes of NOmela formation and transport are not fully understood and should be thoroughly investigated. In addition, whether 5-methoxytryptamine (5-MT) formed by melatonin deacetylation is of physiological importance remains to be investigated in plants.

2.2. Melatonin Acts as a Master Regulator in Plant Abiotic Stress

As a master regulator, melatonin plays important roles in plant tolerance to abiotic stresses, such as heavy metals, drought, salinity, cold, heat, waterlogging, and pesticides [19,47–52]. This review shows schematically the melatonin-mediated responses to abiotic stresses in plants (Figure 2). Melatonin levels are strongly induced by the above unfavorable conditions. For instance, endogenous melatonin level in *Arabidopsis* wild-type plants was increased in response to salt stress [47]. Loss-of-function mutation *atsnat* in the *AtSNAT* gene showed lower endogenous melatonin content and sensitivity to salinity stress [47]. Cold stress induced melatonin accumulation by upregulating the relative expression of *ClASMT* in watermelon plants [49]. In tomato seedlings, Cd stress induced *COMT1* expression, and thereby improved the accumulation of melatonin [22]. Transcription factor heat shock factor A1a (HsfA1a) bound to the *COMT1* gene promoter and activated the transcription of *COMT1* gene under Cd stress [22]. However, the post-translational regulation of melatonin biosynthesis genes and modification of related proteins still remains largely unknown and should be elucidated in future.

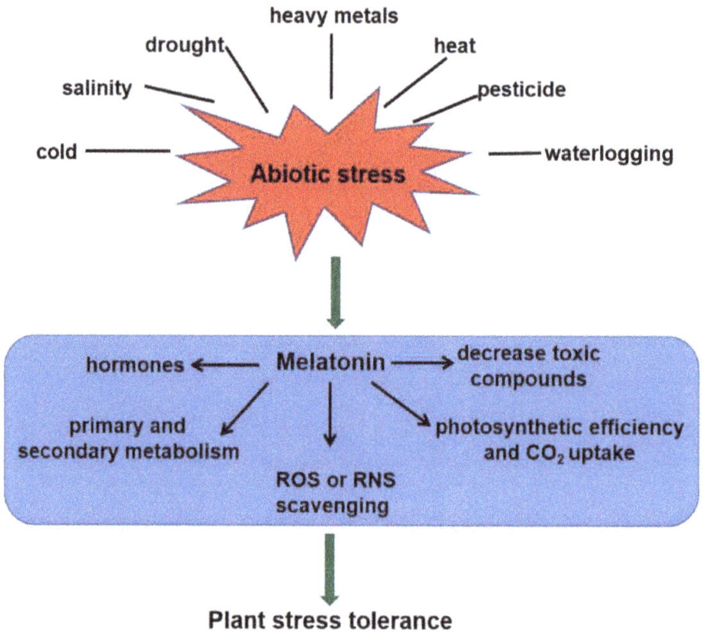

Figure 2. The roles of melatonin in plant tolerance to abiotic stress. Melatonin content of plants increases significantly in responses to abiotic stresses, such as heavy metals, salinity, drought, heat, cold, waterlogging, and pesticides. It confers plant tolerance via multiple mechanisms, including ROS or RNS scavenging, toxic compounds decrease, photosynthetic efficiency increase, interaction with hormones, and secondary metabolite biosynthesis. ROS, reactive oxygen species; RNS, reactive nitrogen species.

Melatonin confers plant tolerance via multiple mechanisms, including photosynthetic efficiency increase, ROS or RNS scavenging, toxic compounds decrease, interaction with hormones, and secondary metabolite biosynthesis (Figure 2). Melatonin stimulated stomatal conductance and improved photosynthesis, thus enhancing tolerance to water-deficient stress in grape cuttings [53]. Another fact is that the photosynthetic efficiency was maximized by higher rates of CO_2 assimilation and stomatal conductance after application of melatonin [54]. Several stresses can induce ROS or RNS accumulation, causing oxidative damage to plants [55]. In this case, melatonin re-establishes the redox balance via activating enzymatic antioxidant defense systems, as well as the ascorbate–glutathione (AsA-GSH) cycle [56]. In plants, the Salt-Overly Sensitive (SOS) pathway mediates ionic homeostasis and contributes to salinity tolerance [57]. This pathway comprises three crucial genes, Salt-Overly Sensitive1 (*SOS1*), Salt-Overly Sensitive2 (*SOS2*) and Salt-Overly Sensitive3 (*SOS3*), which function together to initiate transport of Na^+ out of the cell, or activating other transporters, thus leading to the sequestration of Na^+ in the vacuole [58]. Melatonin reduced ion toxicity and improved salinity tolerance via the SOS pathway [47]. ABA and H_2O_2/NO signaling transduction pathways were also modulated for plant tolerance in response to abiotic stress [47,48,56,59]. In addition, melatonin could increase primary and secondary metabolites including amino acids, organic acids and sugars, and thus improving plant cold tolerance [60].

3. Melatonin Improves Cd Tolerance in Plants

It has been found that Cd affects the ecosystem, causing stress and toxicity in plants. Melatonin acts as a key role in protecting plants from Cd stress. Table 1 summarizes that Cd treatment up-regulates the transcripts of melatonin biosynthesis genes, such as *TDC*, *T5H*, *SNAT*, *ASMT*, and *COMT* in *Arabidopsis thaliana*, *Oryza sativa* L., *Solanum lycopersicum*, *Triticum aestivum* L., *Nicotiana tabacum* L., and *Agaricus campestris* [59,61–67]. Therefore, melatonin contents are significantly increased. Notably, four *M2H* genes, involved in melatonin degradation, were also induced [65]. Byeon et al. suggested that both melatonin degradation and melatonin synthesis occurred in parallel, and 2-hydroxymelatonin of melatonin metabolite also acted as a signaling molecule in plant stress tolerance [65]. As melatonin catabolism is complicated, other pathways and the role of their metabolites should be investigated in plants under Cd stress.

Table 1. Summary table explaining the effect of Cd on genes related to melatonin metabolic pathway.

Plant Species	Cd Stress and Duration	Impact on Genes Related to Melatonin Metabolic Pathway	References
Solanum lycopersicum	100 μM Cd^{2+} for 15 d	*TDC*, *T5H*, *COMT* genes (leaves)	[22]
Oryza sativa L.	500 μM Cd^{2+} for 3 d	*TDC1*, *TDC3*, *SNAT1*, *SNAT2*, *ASMT*, *COMT*, *M2H*, *M3H* genes (seedlings)	[23]
Triticum aestivum L.	200 μM Cd^{2+} for 1 d	*ASMT*, *COMT*, *TDC* genes (root and shoot)	[62]
Nicotiana tabacum L.	10 mg/kg Cd^{2+} for 1, 4, and 7 d	*SNAT1* gene (leaves)	[63]
Agaricus campestris	2, 5, or 8 μM Cd^{2+} for 5 d	*TDC*, *T5H*, *SNAT*, *ASMT*, *COMT* genes	[64]
Oryza sativa L.	200 μM Cd^{2+} for 6, 12, 24, 72 h	*SNAT*, *ASMT*, *COMT*, *TDC*, *T5H* genes (leaves)	[65,67]
Arabidopsis thaliana	300 μM Cd^{2+} for 2, 3, 4 d	*SNAT*, *COMT* genes (leaves)	[66]

TDC1, tryptophan decarboxylase1; *T5H*, tryptamine 5-hydroxylase; *COMT*, caffeic acid O-methyltransferase; *SNAT1*, serotonin N-acetyltransferase1; *SNAT2*, serotonin N-acetyltransferase2; *ASMT*, N-acetylserotonin-O-methyltransferase; *M2H*, melatonin 2-hydroxylase; *M3H*, melatonin 3-hydroxylase.

Most studies showed that melatonin alleviated Cd-induced seedling growth inhibition, including the biomass (fresh weight and dry weight) and root length [19]. Melatonin improved the photosynthesis rate (Pn), transpiration rate (E), intracellular CO_2 concentration and stomatal conductance (Gs) upon Cd stress in *Nicotiana tabacum* L. [6]. That melatonin enhanced stomatal opening and conductance capacity ultimately favored the photosynthesis in plants. Melatonin also prevented the degradation of the chlorophyll and carotenoid molecules in Chinese cabbage seedlings [68]. Similarly, application of melatonin improved chlorophyll and the maximum quantum efficiency of photosystem II (Fv/Fm) levels of wheat plants [20]. In chloroplasts, superoxide anion ($O_2 \cdot^-$) in photosystem I (PSI) is generated by two molecules of O_2 with two electrons from photosystem II (PSII), and disproportionated to H_2O_2 catalyzed with superoxide dismutase (SOD) [69]. The better potential in melatonin treated plants under Cd stress can aid in chlorophyll protection, improve photosynthesis, and mediate redox homeostasis from oxidative damage.

3.1. Melatonin Activates Antioxidant Defense Systems in Response to Cd Stress

Cd stress induces ROS overproduction, containing H_2O_2, $O_2{\cdot}^-$, hydroxyl radical (·OH), and singlet oxygen ($^1O_2$) [70]. These could be formed in photosynthetic cells, mitochondrial respiratory electron transport chain, respiratory processes, and nicotinamide-adenine dinucleotide phosphate (NADPH) oxidases in chloroplasts, mitochondria, peroxisomes, and plasma membrane, respectively [71]. Plants have evolved two antioxidant systems to relieve the ROS-triggered damages, including the enzymatic and non-enzymatic defense systems. Enzymatic defense systems including catalase (CAT), ascorbate peroxidase (APX), guaiacol peroxidase (POD), SOD, glutathione peroxidases (GPX), glutathione reductase (GR), dehydroascorbate reductase (DHAR), peroxiredoxins (PRX), thioredoxins (TRX), and glutaredoxins (GRX) are responsible for ROS scavenging [71]. The non-enzymatic systems, including ascorbate, GSH, flavonoid, anthocyanins, sugars, and carotenoids, are also essential for ROS elimination [71–74].

Melatonin protects plants by enhancing the ROS scavenging efficiency in response to Cd-induced oxidative stress. Application of exogenous melatonin significantly decreased H_2O_2, malondialdehyde (MDA), and $O_2{\cdot}^-$ levels in the tomato leaves/roots under Cd stress [75]. Similar results were also observed in *Triticum aestivum* L., *Nicotiana tabacum* L., *Brassica napus* L., *Catharanthus roseus* (L.), *Malva parviflora*, *Fragaria x ananassa* (Duch.), *Agaricus campestris*, *Carthamus tinctorius* L., *Oryza sativa* L., *Raphanus sativus* L., *Cyphomandra betacea*, *Malachium aquaticum*, and *Zea mays* [18,20,62,64,68,76–86]. In addition, overexpression of *MsSNAT* increased endogenous melatonin level, and reduced ROS accumulation in transgenic *Arabidopsis* plants [19].

Melatonin scavenges the above ROS mainly through two pathways upon Cd stress. Antioxidant enzymes play key roles in melatonin-decreased ROS overproduction, such as APX, CAT, SOD, POD, GPX, GR, DHAR, and monodehydroascorbate reductase (MDHAR). Their functions are confirmed in above plant species. For example, exogenously applied with melatonin counterbalanced the H_2O_2 and MDA accumulation via enhancing APX, CAT, SOD, and POD activities under Cd stress [77]. Enzymes involved in the ascorbate-glutathione (AsA-GSH) cycle, such as DHAR, MDHAR and GR, were also involved in melatonin-mediated ROS balance in sunflower (*Carthamus tinctorius* L.) seedlings [80]. In addition, melatonin interacted with ROS by improving antioxidant levels, including GSH, AsA, and dehydroascorbate (DHA) [80]. Other studies reported melatonin also could increase proline, anthocyanins, flavonoid, and sugars contents in response to Cd-induced oxidative stress [18,64,77,79]. These impacts of melatonin on Cd-induced oxidative stress are summarized in Table 2.

Table 2. Summary table explaining the impacts of melatonin on Cd-induced oxidative stress.

Plant Names	Treatments	Impact on Oxidative Stress Markers and Antioxidative Defense Systems	References
Nicotiana tabacum L.	0, 25, 50, 100, and 250 μM melatonin; 100 μM Cd^{2+} for 7 d	H_2O_2, $O_2{\cdot}^-$; APX, SOD, CAT (leaves)	[6]
Malva parviflora	0, 15, 50, and 100 μM melatonin; 50 μM Cd^{2+} for 8 d	H_2O_2, MDA, SOD, CAT, GPX, PAL, flavonoid, anthocyanins (shoots)	[18]
Medicago sativa L.	0, 10, 50, and 200 μM melatonin; 100 μM Cd^{2+} for 1, 3 d	H_2O_2, $O_2{\cdot}^-$; SOD (roots)	[19]
Triticum aestivum L.	0, 50, and 100 μM melatonin; 100 μM Cd^{2+} for 28 d	H_2O_2, MDA; SOD, CAT, POD (leaves)	[20]
Triticum aestivum L.	0, 50, and 100 μM melatonin; 100 μM Cd^{2+} for 12, 24, 48 h	H_2O_2; APX, SOD, CAT, POD, GSH/GSSG (leaves and roots)	[62]
Agaricus campestris	0, 50, 100, and 200 μM melatonin; 2, 5, and 8 μM Cd^{2+} for 5 d	H_2O_2, MDA; SOD, CAT, POD, APX, GR, proline, sugars	[64]
Solanum lycopersicum	0, 25, 50, 100, 250, and 500 μM melatonin; 100 μM Cd^{2+} for 14 d	H_2O_2, MDA, $O_2{\cdot}^-$; SOD, CAT, GR, POD, APX (leaves)	[75]
Solanum lycopersicum	100 μM melatonin; 100 μM Cd^{2+} for 15 d	H_2O_2; APX, SOD, CAT, POD (leaves and roots)	[76]
Brassica napus L.	0, 50, and 100 μM melatonin; 20 μM Cd^{2+} for 5 d	H_2O_2, MDA; APX, SOD, CAT, POD, proline, anthocyanins (seedlings)	[77]

Table 2. *Cont.*

Plant Names	Treatments	Impact on Oxidative Stress Markers and Antioxidative Defense Systems	References
Catharanthus roseus (L.)	100 μM melatonin; 0, 50, 100, and 200 mg Cd kg^{-1} soil for 30 d	H_2O_2; CAT, POD (leaves)	[78]
Fragaria x ananassa (Duch.)	0, 10, 50, 100, 150, and 200 μM melatonin; 300 mL of 1 mmol·L^{-1} Cd^{2+} for 5, 10 d	MDA; SOD, CAT, POD, APX, soluble protein, anthocyanins (leaves and roots)	[79]
Carthamus tinctorius L.	100 μM melatonin; 100 μM Cd^{2+} for 21 d	H_2O_2, MDA, LOX; ASA, DHA, GSH, GSSG, SOD, APX, DHAR, CAT, GR, MDHAR, Gly (leaves)	[80]
Oryza sativa L.	0, 50, 100, and 200 μM melatonin; 100 μM Cd^{2+} for 10 d	H_2O_2, MDA; SOD, CAT, POD (leaves and roots)	[81]
Zea mays	200 μM melatonin; 150 μM Cd^{2+} for 3 d	MDA; SOD, CAT, POD (root, stem, and leaf)	[82]
Oryza sativa L.	100 μM melatonin	MDA; SOD, CAT, POD (shoots)	[83]
Raphanus sativus L.	0, 10, 25, 50, 100, and 200 μM melatonin; 50 μM Cd^{2+} for 24 h	SOD, CAT, POD, APX, GR (roots and shoots)	[84]
Malachium aquaticum, *Galinsoga parviflora*	0, 50, 100, 150, and 200 μM melatonin; 10 mg/L Cd for 40 d	SOD, POD, CAT (leaves)	[85]
Cyphomandra betacea	0, 50, 100, 150, and 200 μM melatonin; 10 mg/L Cd for 40 d	SOD, POD, CAT (leaves)	[86]

H_2O_2, hydrogen peroxide; MDA, malondialdehyde; $O_2\cdot^-$, superoxide anion; APX, ascorbate peroxidase; SOD, superoxide dismutase; CAT, catalase; GPX, glutathione peroxidase; PAL, phenylalanine ammonia-lyase; POD, guaiacol peroxidase; GSH/GSSG, reduced (GSH)/oxidized (GSSG) glutathione; GR, glutathione reductase; LOX, lipoxygenase; ASA, ascorbate; DHA, dehydroascorbate; DHAR, dehydroascorbate reductase; MDHAR, monodehydroascorbate reductase; Gly, glycine.

3.2. Melatonin Regulates Cadmium Uptake and Translocation

In general, Cd is taken up by plant roots from soil, then transported to shoots through the xylem and phloem, and eventually accumulated in grains [87]. Several processes regulate Cd accumulation, including Cd apoplastic influx, cell wall adsorption, cytoplasm across the membrane, xylem loading, vacuolar sequestration, and energy-driven transport in plants [88]. Natural resistance-associated macrophage protein (*NRAMP*) might be involved in several processes, such as uptake, intracellular transport, translocation, and metal detoxification in various plants [89,90]. Moreover, Cd is also transported through Zn, Fe, and Ca transporters, including Zn transporter proteins (ZRT)- and Fe-regulated transporter (IRT)-like protein (ZIP), yellow strip-like (YS1/YSL), and low-affinity calcium (Ca) transporter 1 (LCT1) [91]. ABC transport (PDR8), metal tolerance proteins (MTPs), cation diffusion facilitators (CDFs), and the P18-type metal transporter ATPase (HMAs) take part in Cd homeostasis [92–94]. Furthermore, GSH and its derivatives, phytochelatins (PCs), bound with Cd, and then transported Cd to vacuoles by ATP-binding cassette subfamily C proteins (ABCCs) [95,96]. HMA3 and CDF transporter family are also involved in the transfer of Cd–PCs complexes into the vacuole [97,98]. Other high-affinity chelators, including metallothioneins (MTs), organic acids, and amino acids play multiple roles in detoxification of Cd [99].

Recent studies have shown that melatonin regulates Cd homeostasis in plants. Exogenous application of melatonin reduced Cd contents in both roots and leaves of *Raphanus sativus* L. and *Brassica pekinensis* (Lour.) Rupr. plants [68,84]. Melatonin significantly decreased Cd contents in the leaves, but not in the roots of *Oryza sativa* L., *Carthamus tinctorius* L., and *Solanum lycopersicum* [61,76,80,81]. However, melatonin increased and decreased Cd contents in roots and shoots of *Malva parviflora*, respectively [18]. These results suggest that the effect of melatonin on translocation factor (Cd content of shoot/root) are different in the above various plants. Melatonin reduced the transcripts of metal transporter-related genes (iron-regulated transporter1 (*OsIRT1*), iron-regulated transporter2 (*OsIRT2*), heavy metal ATPase2 (*OsHMA2*), heavy metal ATPase3 (*OsHMA3*), natural resistance-associated macrophage protein1 (*OsNramp1*), natural resistance-associated macrophage protein5 (*OsNramp5*), and low-affinity cation transporter1 (*OsLCT1*) in leaves, but not in the roots of *Oryza sativa* L. under Cd stress [81]. Expression of *YSLs* and *HMAs* were down-regulated by melatonin, thereby reducing the Cd entering the roots of *Raphanus sativus* L. [84]. In addition, the Metallothionein 1 (*RsMT1*) gene was involved in melatonin-conferred Cd tolerance in transgenic tobacco [84]. In roots of *Brassica pekinensis* (Lour.) Rupr. plants, *IRT1* transcript was down-regulated significantly by melatonin application [68]. Then, Cd content was reduced in root tissues. These impacts of melatonin on Cd uptake and translocation are summarized in Table 3. Therefore, to characterize the biological roles of these metal transporter genes contributes to understanding the melatonin-mediated Cd homeostasis and detoxification.

Table 3. Summary table explaining the impacts of melatonin on Cd uptake and translocation.

Plant Names	Treatments	Impact on Cd in Subcellular Compartment	References
Nicotiana tabacum L.	0, 25, 50, 100, and 250 μM melatonin; 100 μM Cd^{2+} for 7 d	Cd content in leaves; H^+-ATPase activity, *IRT1, IRT2, Nramp1, HMA2, HMA3, HMA4*	[6]
Malva parviflora	0, 15, 50, and 100 μM melatonin; 50 μM Cd^{2+} for 8 d	Cd content in shoots	[18]
Medicago sativa L. *Arabidopsis*	0, 10, 50, and 200 μM melatonin; 100 μM Cd^{2+} for 1, 3 d	Cd content in leaves; *PCR2, Nramp6, PDR8, HMA4*	[19]
Solanum lycopersicum	1 μM melatonin; 100 μM Cd^{2+} for 15 d	Cd content in leaves; GSH and PCs	[61]
Brassica pekinensis (Lour.) *Rupr.*	100 μM melatonin; 20 μM Cd^{2+} for 24 h	Cd contents in roots and leaves; *IRT1/2*	[68]
Solanum lycopersicum	0, 25, 50, 100, 250, and 500 μM melatonin; 100 μM Cd^{2+} for 14 d	Cd content in leaves; H^+-ATPase activity, GSH and PCs; *SlGSH1, SlPCS, SlMT2,* and *SlABC1*	[75]
Solanum lycopersicum	100 μM melatonin; 100 μM Cd^{2+} for 15 d	Cd content in leaves; Cys, γ-glutamyl cysteine, GSH and PCs	[76]
Brassica napus L.	0, 50, and 100 μM melatonin; 20 μM Cd^{2+} for 5 d	Cd content; H^+-ATPase activity	[77]
Carthamus tinctorius L.	100 μM melatonin; 100 μM Cd^{2+} for 21 d	Cd content in roots, stems and leaves; PCs	[80]
Oryza sativa L.	0, 50, 100, and 200 μM melatonin; 100 μM Cd^{2+} for 10 d	Cd content in leaves; *OsIRT1, OsIRT2, OsHMA2, OsHMA3, OsNramp1, OsNramp5,* and *OsLCT1*	[81]
Oryza sativa L.	100 μM melatonin	Cd content in roots and shoots; *Nramp1, Nramp5, IRT1, IRT2, HMA2, HMA3*	[83]
Raphanus sativus L.	0, 10, 25, 50, 100, and 200 μM melatonin; 50 μM Cd^{2+} for 24 h	Cd content in roots and leaves; *PCS*; *MT, CAX4, ZIP12, HMA4, YSL2, YSL7*	[84]
Malachium aquaticum, *Galinsoga parviflora*	0, 50, 100, 150, and 200 μM melatonin; 10 mg/L Cd for 40 d	Cd content in leaves	[85]
Cyphomandra betacea	0, 50, 100, 150, and 200 μM melatonin; 10 mg/L Cd for 40 d	Cd contents in stems, leaves, and shoots	[86]

IRT1, iron-regulated transporter1; *IRT2*, iron-regulated transporter2; *Nramp1*, natural resistance-associated macrophage protein1; *Nramp5*, natural resistance-associated macrophage protein5; *HMA2*, heavy metal ATPase2; *HMA3*, heavy metal ATPase3; *HMA4*, heavy metal ATPase4; *PCR2*, plant cadmium resistance2; *PDR8*, pleiotropic drug resistance8; *GSH1*, glutamate-cysteine ligase; *PCS*, phytochelatin synthase activity; *MT*, metallothionein; *ABC1*, ATP-binding cassette transporter1; *LCT1*, low-affinity cation transporter; *CAX4*, vacuolar cation/proton exchanger4; *ZIP12*, zinc-iron permease12; *YSL2*, yellow stripe-like transporter2; *YSL7*, yellow stripe-like transporter7.

3.3. Other Regulators Are Involved in Melatonin-Mediated Cd Tolerance

It has been widely reported that NO plays a crucial role in regulating various plant physiological processes [100]. Previous studies found that Cd treatment increased NO production, which promoted Cd accumulation by the *IRT1* up-regulation [101,102]. Exogenous melatonin alleviated Cd toxicity by reducing NO accumulation and *IRT1* expression in *Brassica pekinensis* (Lour.) Rupr. [68]. By contrast, melatonin triggered the endogenous NO, and enhanced Cd tolerance via the increase in the activities of antioxidant enzymes in wheat seedlings [20]. Moreover, melatonin can be nitrosated to NOMela by employing four nitrosating entities at the N atom of indole ring [46]. It was suggested that NOMela could release NO. That NO induces *S*-nitrosation is an important redox-based post-translational modification, which is involved in plant responses to abiotic stress [103,104]. Thus, complex interactions between melatonin and NO in Cd resistance should be further investigated. Another important signaling element, salicylic acid (SA), alleviated Cd toxicity by affecting

Cd distribution, the antioxidant defense activities, and photosynthesis [105–107]. Amjadi et al. found that there was a possible synergic interaction between melatonin and SA by reducing Cd uptake and modulating the ascorbate–glutathione cycle and glyoxalase system [80].

4. A Possible Role for H_2S in Melatonin-Mediated Tolerance against Cd Stress

Acting as a signaling molecule, NO interacts with other molecules (H_2O_2, CO, and H_2S) to mediate plant growth and development, as well as abiotic stress responses [100]. Among the molecules, H_2S is also involved in almost all physiological plant processes [27,100]. To date, there is considerable research on the role of NO in melatonin-modulated plant abiotic stress tolerance. However, the functions of H_2S have been largely unknown. It will become a research hotspot to contribute to precise analysis of the collaboration between H_2S and melatonin, and provide deeper insight into melatonin-mitigated signaling mechanisms.

4.1. H_2S Action in Plant Tolerance against Cd Stress

H_2S acts as a signaling molecule in modifying various metabolic processes in plants, especially Cd stress (Figure 3, [27]). Endogenous H_2S production was induced via expression of *LCD*, *DCD*, and *DES1* under Cd stress [108–110]. SA, methane (CH_4), and WRKY DNA-binding protein 13 (*WRKY13*) transcription factor were suggested to be involved in the above process [30,111,112]. H_2S regulated the activities of key enzymes and AsA-GSH cycle involved in ROS homeostasis to alleviate Cd-induced oxidative stress [113–120]. For example, H_2S enhanced the activities of antioxidant enzymes, such as POD, CAT, APX, and SOD, and thereby decreased ROS accumulation [120]. Similarly, it also obviously increased AsA and GSH and the redox status (AsA/DHA and GSH/GSSG) levels to improve rice Cd resistance [114,116].

Increasing evidence demonstrates that H_2S also regulates Cd uptake and translocation in plants [30,117,119,121]. H_2S enhanced the expression of genes encoding metallothionein (MTs) and phytochelatin (PCS) in *Arabidopsis* roots [117]. Therefore, H_2S increased the metal chelators synthesis, contributing to Cd detoxification by binding the trace metal. In addition to enhancing the above genes expression, the protective effect of H_2S was attributed to a decrease in Cd accumulation associated with the expression of Cd transporter genes, such as *PCR1*, *PCR2*, and *PDR8* [30]. Exogenous application of NaHS weakened the expression of *NRAMP1* and *NRAMP6* genes, and intensified the expression of Cd homeostasis-related genes (*CAX2* and *ZIP4*) to enhance Cd tolerance in foxtail millet [122].

A number of studies address that H_2S can interact with other signaling molecules, such as SA, proline, MeJA, Ca, and NO during the responses of plants to Cd stress (Figures 3 and 4; [111,122,123]). H_2S acted as a downstream molecule of SA-transmitted signals to regulate Cd tolerance in *Arabidopsis* [111]. The endogenous production of proline and MeJA enhanced by H_2S donor NaHS responded significantly to Cd stress in foxtail millet [122,123]. H_2S also improved CaM gene expression and controlled the combination of Ca^{2+} and CaM, which act as signal transducers [33].

There exists a complicated and synergistic relationship between H_2S and NO in response to Cd stress in plants (Figure 4; [115,118,124,125]). Exogenous NO and H_2S application increased the Cd tolerance in plants [115,124,126]. Subsequent pharmacological experiments proved that H_2S donor NaHS triggered NO production, which might act as a signal for alleviation of Cd-induced oxidative damage in alfalfa seedling roots [124]. Nevertheless, H_2S production activated by NO is essential in Cd stress response of bermudagrass [115]. As a second messenger, Ca acted both upstream and downstream of NO signal, and crosstalk of Ca and NO regulated the cysteine and H_2S to mitigate Cd toxicity in Vigna radiata [126]. Moreover, application of sodium nitroprusside (SNP), the donor of NO, increased H_2S generation, and thus enhanced Cd stress tolerance in wheat [118]. However, this protective effect was reversed by hypotaurine (HT), the scavenger of H_2S [118]. These results suggested that H_2S and NO can function in a coordinated way under certain signaling cascades in plants under Cd stress.

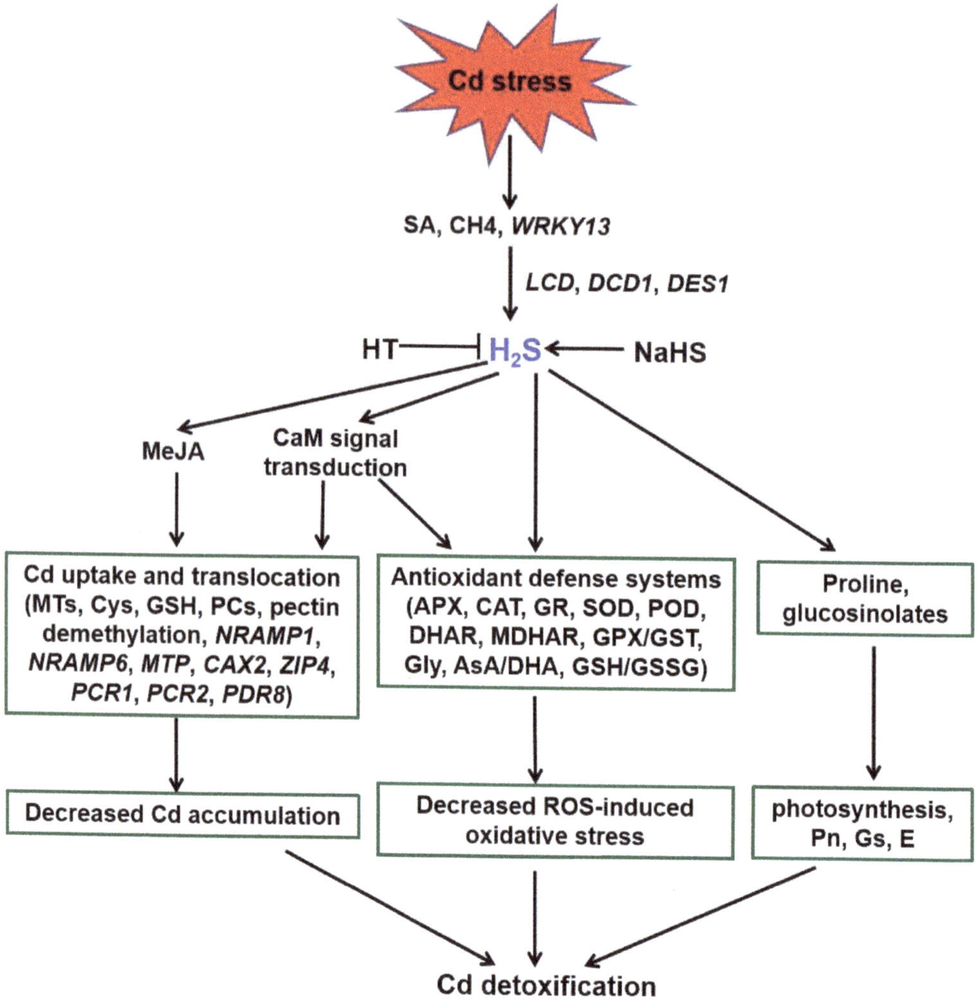

Figure 3. Function of H_2S in plant responses to Cd stress. SA, CH_4, and *WRKY13* are involved in Cd-induced H_2S generation. H_2S enhances the antioxidant defense systems to decrease the ROS accumulation, regulates the transcripts of genes related to Cd uptake and translocation to reduce the Cd accumulation, and increases proline and glucosinolates in response to Cd stress in plants. MeJA and Ca participate in the above regulatory pathways. SA, salicylic acid; CH4, methane; HT, hypotaurine; LCD, L-cysteine desulfhydrase; DCD, D-cysteine desulfhydrase; *DES1*, L-cysteine desulfhydrase 1; MeJA, methyl jasmonate; CaM, calmodulin; *NRAMP1*, natural resistance-associated macrophage protein1; *NRAMP6*, natural resistance-associated macrophage protein6; *MTP*, metal tolerance protein; *CAX2*, vacuolar cation/proton exchanger2; *ZIP4*, zinc-iron permease4; *PCR1*, plant cadmium resistance1; *PCR2*, plant cadmium resistance2; *PDR8*, pleiotropic drug resistance8.

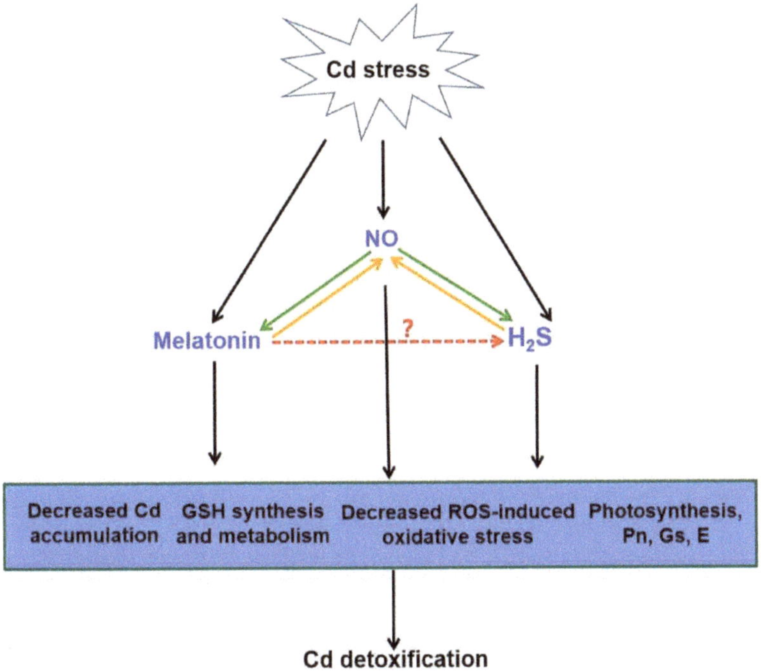

Figure 4. The possible role of H_2S in melatonin-mediated Cd detoxification. NO generation can be induced by Cd stress. Increasing evidence showed that melatonin and H_2S act as the downstream of NO in the responses to Cd stress, respectively (green arrow). It is also suggested that NO acts as a downstream of melatonin or H_2S to improve Cd tolerance (orange arrow). The combination of melatonin, NO and H_2S might be responsible for melatonin-triggered signal transduction in plant Cd tolerance via the decreased Cd accumulation, GSH synthesis and metabolism, decreased ROS-induced oxidative stress and improved photosynthesis. Red arrow, yet largely unknown. Cd, cadmium; NO, nitric oxide; H_2S, hydrogen sulfide; GSH, glutathione; ROS, reactive oxygen species; Pn, photosynthesis rate; Gs, stomatal conductance; E, transpiration.

4.2. Crosstalk of Melatonin and H_2S in Plants

The interaction between melatonin and H_2S plays a beneficial role in abiotic stress response [32]. Exogenous melatonin regulated the endogenous H_2S homeostasis by modulating the L-DES activity in salt-stressed tomato cotyledons [31]. Moreover, an endogenous H_2S-dependent pathway was involved in melatonin-mediated salt stress tolerance in tomato seedling roots [34]. Synergistic effects of melatonin and H_2S regulated K^+/Na^+ homeostasis, and reduced excessive accumulation of ROS by enhancing the activity of antioxidant enzymes. Inhibition of H_2S by HT reversed the melatonin-modulated heat tolerance by inhibiting photosynthesis, carbohydrate metabolism, and the activity of antioxidant enzymes in wheat [36]. Recent investigation has revealed that melatonin-induced pepper tolerance to iron deficiency and salt stress was dependent on H_2S and NO [118]. It was further confirmed that H_2S and NO jointly participated in melatonin-mitigated salt tolerance in cucumber [35]. Thus, these results postulate that H_2S might act as a downstream signaling molecule of melatonin. Combined with the roles of H_2S and melatonin in alleviating Cd stress, it is easy to speculate that H_2S might be involved in melatonin-mediated Cd tolerance in plants (Figure 4).

As mentioned above in Section 3, GSH plays a critical role in plant Cd tolerance. It is synthesized from glutamate, cysteine and glycine by γ-glutamyl cysteine synthetase (γ-ECS, encoded by *GSH1/ECS* gene) and glutathione synthetase (GS, encoded by *GSH2/GS* gene) [127]. The catalysis of GSH1 is the rate-limiting step of GSH biosynthesis [128]. Cd stress induced the transcripts of *GSH1* and *GSH2* in *Arabidopsis*, as well as *ECS* and

GS in *Medicago sativa* [114,129–131]. It was suggested that H_2S could be quickly incorporated into cysteine and subsequently into GSH [132]. Application of NaHS re-established (h)GSH homeostasis by further strengthening the up-regulation of *ECS* and *GS* genes [114]. Similar results were also found in strawberry and cucumber plants [133,134]. Interestingly, exogenous melatonin also increased the GSH content by inducing the transcript levels of *SlGSH1* in tomato [75]. Hence, there might be a certain connection between H_2S and melatonin in regulating the GSH homeostasis at the transcriptional regulatory pathway. This will provide an interesting direction for further research on the complex interactions between melatonin and H_2S in improving Cd tolerance in plants.

5. Conclusions and Future Prospects

Recent studies have strongly indicated that melatonin, a multifunctional molecule, regulates Cd tolerance in plants. To further promote related research in plant Cd tolerance, this review summarizes the regulatory roles and mechanisms of melatonin in response to Cd stress. Melatonin reduces Cd damage mainly through re-establishing the redox homeostasis and decreasing Cd accumulation, but its underlying mechanisms remain to be determined. Intriguingly, melatonin is suggested to be a phytohormone due to the identification of the putative receptor CAND2/PMTR1 [135], although there is still a debate on whether it is a *bona fide* receptor for melatonin [136]. More importantly, more receptor gene(s) should be characterized, which will be critical for precisely understanding the signal transduction pathway of melatonin in plants in response to Cd stress.

Currently, as a signal molecule, the role of NO has been revealed in melatonin-mediated Cd tolerance, likewise H_2S plays a key messenger in plant resistance to Cd stress. That the effects of H_2S have been less explored has prevented precise analysis of the collaboration of H_2S and melatonin. Recently, we presented the underlying mechanisms of H_2S action and its multifaceted roles in plant stress responses [137]. Hence, it would be interesting to fully evaluate the effects of H_2S-based signaling on regulating melatonin-induced Cd tolerance. For directions of future research, biochemical and genetic characterization of H_2S-producing proteins and persulfidation signaling is needed and will shed more light on the integration of H_2S and melatonin signaling during Cd stress.

5.1. *Pharmacological, Genetic and 'Omics' Approach to Understand the Crosstalk of H_2S–Melatonin during Cd Stress*

Various pharmacological, enzyme activity, and gene expression investigations revealed the crosstalk of H_2S–melatonin in response to salt and heat stress in tomato and cucumber [34–36]. Exogenous melatonin induced the H_2S generation by activating the L-cysteine desulfhydrase (L-CDes) activity, which was encoded by *LCD*, *DCD*, and *DES1* [31,35]. Then, the interaction of H_2S and melatonin enhanced the antioxidant defense, and regulated carbohydrate metabolism and ion homeostasis [34–36,118]. Similar pharmacological experiments with an effective concentration range of 1–200 μM in exogenous melatonin, 10–100 μM in 4-Chloro-DL-phenylalanine (p-CPA, melatonin synthesis inhibitor), 10–100 μM in hypotaurine (HT, H_2S inhibitor), and 10–100 μM in NaHS (H_2S donor) could be used to investigate the crosstalk of H_2S–melatonin during Cd stress. Furthermore, genetic modifications with altering melatonin and H_2S levels, such as *snat*, *comt*, *lcd*, *dcd*, and *des1* mutants, should be used to explore their possible roles.

H_2S plays a critical signal mediator in plants in response to Cd stress [111,112,115]. However, there is still an urgent need to elucidate the interactions of H_2S with other signaling molecules in melatonin-mediated Cd tolerance. With the advent of transcriptomic and proteomic analysis, scientists shall reveal the intrinsic regulatory mechanisms of melatonin and H_2S interaction on the regulation of various biological processes. For example, the expression of genes and proteins related to GSH synthesis and metabolism and redox homoeostasis, as well as the hormone biosynthesis pathways, might be used to establish a model system to decipher their signaling interaction.

5.2. The Potential Role of Persulfidation Driven by H_2S in Melatonin-Mediated Cd Tolerance

Recently, it was found that H_2S-mediated post-translational modification (PTM, persulfidation) of protein cysteine residues (RSSH) is an important mechanism in plants to adapt to external environments [27,32]. Protein persulfidation cause various changes in structures, activities, as well as the subcellular localizations of the candidate proteins [138,139]. These proteins are mainly involved in plant growth and development, abiotic stress responses, and carbon/nitrogen metabolism [138]. For example, H_2S production regulated the persulfidation of NADPH oxidase RBOHD at Cys825 and Cys890, leading to improving the ability to produce H_2O_2 signal [140]. It also led to the persulfidation of ABSCISIC ACID INSENSITIVE 4 (ABI4) at Cys250, and persulfidation of SnRK2.6, contributing to reveal the function of H_2S in the complex signal-transduction system [141–143]. By contrast, the residue Cys32 of APX could be persulfidated, thereby enhancing its activity [144]. Therefore, the persulfidation might become a promising direction to investigate the roles of H_2S in melatonin-mediated Cd tolerance in plants. To conclude, the progresses in the various physiological and molecular mechanisms regulated by melatonin are not enough, and future studies along with the above lines should be used to unveil the regulatory mechanism of melatonin and H_2S signaling pathways in plant Cd tolerance.

Author Contributions: Writing—original draft preparation, Q.G., C.W., and Q.X.; writing—review and editing, Z.C. and Y.H. All authors have read and agreed to the published version of the manuscript.

Funding: This research was mainly supported by the Anhui Provincial Natural Science Foundation (2008085MC101), the Natural Science Foundation of China (31300225), Talent Research Fund Project of Hefei University (No. 18-19RC13), and "Borrowing transfer to supplement" Foundation of Hefei (J2019D04).

Data Availability Statement: Not applicable.

Conflicts of Interest: The authors declare no conflict of interest.

References

1. Weissmannová, H.D.; Pavlovský, J. Indices of soil contamination by heavy metals-methodology of calculation for pollution assessment. *Environ. Monit. Assess.* **2017**, *189*, 616. [CrossRef] [PubMed]
2. Clemens, S. Safer food through plant science: Reducing toxic element accumulation in crops. *J. Exp. Bot.* **2019**, *70*, 5537–5557. [CrossRef] [PubMed]
3. Clemens, S.; Aarts, M.G.; Thomine, S.; Verbruggen, N. Plant science: The key to preventing slow cadmium poisoning. *Trends Plant Sci.* **2013**, *18*, 92–99. [CrossRef]
4. DalCorso, G.; Manara, A.; Furini, A. An overview of heavy metal challenge in plants: From roots to shoots. *Metallomics* **2013**, *5*, 1117–1132. [CrossRef]
5. Ismael, M.A.; Elyamine, A.M.; Moussa, M.G.; Cai, M.; Zhao, X.H.; Hu, C.X. Cadmium in plants: Uptake, toxicity, and its interactions with selenium fertilizers. *Metallomics* **2019**, *11*, 255–277. [CrossRef] [PubMed]
6. Wang, M.; Duan, S.; Zhou, Z.; Chen, S.; Wang, D. Foliar spraying of melatonin confers cadmium tolerance in *Nicotiana tabacum* L. *Ecotoxicol. Environ. Saf.* **2019**, *170*, 68–76. [CrossRef]
7. Sharma, S.S.; Dietz, K.J. The relationship between metal toxicity and cellular redox imbalance. *Trends Plant Sci.* **2009**, *14*, 43–50. [CrossRef]
8. Pérez-Chaca, M.V.; Rodríguez-Serrano, M.; Molina, A.S.; Pedranzani, H.E.; Zirulnik, F.; Sandalio, L.M.; Romero-Puertas, M.C. Cadmium induces two waves of reactive oxygen species in *Glycine max* (L.) roots. *Plant Cell Environ.* **2014**, *37*, 1672–1687. [CrossRef]
9. Khaliq, M.A.; James, B.; Chen, Y.H.; Saqib, H.; Li, H.H.; Jayasuriya, P.; Guo, W. Uptake, translocation, and accumulation of Cd and its interaction with mineral nutrients (Fe, Zn, Ni, Ca, Mg) in upland rice. *Chemosphere* **2019**, *215*, 916–924. [CrossRef]
10. Noctor, G.; Mhamdi, A.; Chaouch, S.; Han, Y.; Neukermans, J.; Marquez-Garcia, B.; Foyer, C.H. Glutathione in plants: An integrated overview. *Plant Cell Environ.* **2021**, *35*, 454–484. [CrossRef]
11. Cobbett, C.; Goldsbrough, P. Phytochelatins and metallothioneins: Roles in heavy metal detoxification and homeostasis. *Rev. Plant Biol.* **2002**, *53*, 159–182. [CrossRef]
12. Rizwan, M.; Ali, S.; Adrees, M.; Rizvi, H.; Zia-ur-Rehman, M.; Hannan, F.; Qayyum, M.F.; Hafeez, F.; Ok, Y.S. Cadmium stress in rice: Toxic effects, tolerance mechanisms, and management: A critical review. *Environ. Sci. Pollut. Res. Int.* **2016**, *23*, 17859–17879. [CrossRef]
13. Thao, N.P.; Khan, M.I.R.; Thu, N.B.A.; Hoang, X.L.T.; Asgher, M.; Khan, N.A.; Tran, L.S.P. Role of ethylene and its cross talk with other signaling molecules in plant responses to heavy metal stress. *Plant Physiol.* **2015**, *169*, 73–84. [CrossRef]

14. Rabia, A.; Faiza, M.; Ghulam, K.; Tooba, I.; Maryam, K. Plant signaling molecules and cadmium stress tolerance. *Cadmium Toler. Plants* **2019**, 367–399. [CrossRef]
15. Hattori, A.; Migitaka, H.; Iigo, M.; Itoh, M.; Yamamoto, K.; Ohtani-Kaneko, R.; Hara, M.; Suzuki, T.; Reiter, R.J. Identification of melatonin in plants and its effects on plasma melatonin levels and binding to melatonin receptors in vertebrates. *Biochem. Mol. Biol. Int.* **1995**, *35*, 627–634.
16. Arnao, M.B.; Hernandez-Ruiz, J. Melatonin: A New Plant Hormone and/or a Plant Master Regulator? *Trends Plant Sci.* **2019**, *24*, 38–48. [CrossRef]
17. Sun, C.; Liu, L.; Wang, L.; Li, B.; Jin, C.; Lin, X. Melatonin: A master regulator of plant development and stress responses. *J. Integr. Plant Biol.* **2021**, *63*, 126–145. [CrossRef]
18. Tousi, S.; Zoufan, P.; Ghahfarrokhie, A.R. Alleviation of cadmium-induced phytotoxicity and growth improvement by exogenous melatonin pretreatment in mallow (*Malva parviflora*) plants. *Ecotoxicol. Environ. Saf.* **2020**, *206*, 111403. [CrossRef] [PubMed]
19. Gu, Q.; Chen, Z.; Yu, X.; Cui, W.; Pan, J.; Zhao, G.; Xu, S.; Wang, R.; Shen, W. Melatonin confers plant tolerance against cadmium stress via the decrease of cadmium accumulation and reestablishment of microRNA-mediated redox homeostasis. *Plant Sci.* **2017**, *261*, 28–37. [CrossRef] [PubMed]
20. Kaya, C.; Okant, M.; Ugurlar, F.; Alyemeni, M.N.; Ashraf, M.; Ahmad, P. Melatonin-mediated nitric oxide improves tolerance to cadmium toxicity by reducing oxidative stress in wheat plants. *Chemosphere* **2019**, *225*, 627–638. [CrossRef]
21. He, J.; Zhuang, X.; Zhou, J.; Sun, L.; Wan, H.; Li, H.; Lyu, D. Exogenous melatonin alleviates cadmium uptake and toxicity in apple rootstocks. *Tree Physiol.* **2020**, *40*, 746–761. [CrossRef]
22. Cai, S.Y.; Zhang, Y.; Xu, Y.P.; Qi, Z.Y.; Li, M.Q.; Ahammed, G.J.; Xia, X.; Shi, K.; Zhou, Y.; Reiter, R.; et al. HsfA1a upregulates melatonin biosynthesis to confer cadmium tolerance in tomato plants. *J. Pineal Res.* **2017**, *62*, e12387. [CrossRef] [PubMed]
23. Lee, K.; Choi, G.H.; Back, K. Cadmium-induced melatonin synthesis in rice requires light, hydrogen peroxide, and nitric oxide: Key regulatory roles for tryptophan decarboxylase and caffeic acid O-methyltransferase. *J. Pineal Res.* **2017**, *63*, e12441. [CrossRef] [PubMed]
24. Kyungjin, L.; Jin, H.O.; Reiter, R.J.; Kyoungwhan, B. Flavonoids inhibit both rice and sheep serotonin *N*-acetyltransferases and reduce melatonin levels in plants. *J. Pineal Res.* **2018**, *65*, e12512. [CrossRef]
25. Lu, R.; Liu, Z.; Shao, Y.; Sun, F.; Zhang, Y.; Cui, J.; Zhou, T. Melatonin is responsible for rice resistance to rice stripe virus infection through a nitric oxide-dependent pathway. *Virol. J.* **2019**, *16*, 141. [CrossRef]
26. Zhou, H.; Chen, Y.; Zhai, F.; Zhang, J.; Zhang, F.; Yuan, X.; Xie, Y. Hydrogen sulfide promotes rice drought tolerance via reestablishing redox homeostasis and activation of ABA biosynthesis and signaling. *Plant Physiol. Biochem.* **2020**, *155*, 213–220. [CrossRef]
27. Zhang, J.; Zhou, M.; Zhou, H.; Zhao, D.; Gotor, C.; Romero, L.C.; Shen, J.; Ge, Z.; Zhang, Z.; Shen, W.; et al. Hydrogen sulfide, a signaling molecule in plant stress responses. *J. Integr. Plant Biol.* **2021**, *63*, 146–160. [CrossRef] [PubMed]
28. Zhou, M.; Zhou, H.; Shen, J.; Zhang, Z.; Xie, Y. H_2S action in plant life cycle. *Plant Growth Regul.* **2021**, *94*, 1–9. [CrossRef]
29. Zhou, H.; Guan, W.; Zhou, M.; Shen, J.; Liu, X.; Wu, D.; Yin, X.; Xie, Y. Cloning and characterization of a gene encoding true D-cysteine desulfhydrase from *Oryza sativa*. *Plant Mol. Biol. Rep.* **2020**, *38*, 95–113. [CrossRef]
30. Zhang, Q.; Cai, W.; Ji, T.T.; Ye, L.; Lu, Y.T.; Yuan, T.T. WRKY13 enhances cadmium tolerance by promoting D-cysteine desulfhydrase and hydrogen sulfide production. *Plant Physiol.* **2020**, *183*, 345–357. [CrossRef]
31. Mukherjee, S.; Bhatla, S.C. Exogenous melatonin modulates endogenous H_2S homeostasis and L-cysteine desulfhydrase activity in salt-stressed tomato (*Solanum lycopersicum* L. var. cherry) seedling cotyledons. *J. Plant Growth Regul.* **2020**, *4*, 1–13. [CrossRef]
32. Huang, D.; Huo, J.; Liao, W. Hydrogen sulfide: Roles in plant abiotic stress response and crosstalk with other signals. *Plant Sci.* **2021**, *302*, 110733. [CrossRef]
33. Wang, H.R.; Che, Y.H.; Wang, Z.H.; Zhang, B.N.; Ao, H. The multiple effects of hydrogen sulfide on cadmium toxicity in tobacco may be interacted with CaM signal transduction. *J. Hazard. Mater.* **2021**, *403*, 123651. [CrossRef] [PubMed]
34. Siddiqui, M.; Khan, M.; Mukherjee, S.; Basahi, R.; Alamri, S.; Al-Amri, A.; Alsubaie, Q.; Ali, H.; Al-Munqedhi, B.; Almohisen, I. Exogenous melatonin-mediated regulation of K^+/Na^+ transport, H^+-ATPase activity and enzymatic antioxidative defence operate through endogenous hydrogen sulphide signalling in NaCl-stressed tomato seedling roots. *Plant Biol.* **2021**, *23*, 797–805. [CrossRef] [PubMed]
35. Sun, Y.; Ma, C.; Kang, X.; Zhang, L.; Wang, J.; Zheng, S.; Zhang, T. Hydrogen sulfide and nitric oxide are involved in melatonin-induced salt tolerance in cucumber. *Plant Physiol. Biochem.* **2021**, *167*, 101–112. [CrossRef] [PubMed]
36. Iqbal, N.; Fatma, M.; Gautam, H.; Umar, S.; Khan, N.A. The crosstalk of melatonin and hydrogen sulfide determines photosynthetic performance by regulation of carbohydrate metabolism in wheat under heat stress. *Plants* **2021**, *10*, 1778. [CrossRef]
37. Murch, S.J.; KrishnaRaj, S.; Saxena, P.K. Tryptophan is a precursor for melatonin and serotonin biosynthesis in in vitro regenerated St. John's wort (Hypericum perforatum L. cv. Anthos) plants. *Plant Cell Rep.* **2000**, *19*, 698–704. [CrossRef]
38. Park, S.; Lee, K.; Kim, Y.S.; Back, K. Tryptamine 5-hydroxylase-deficient Sekiguchi rice induces synthesis of 5-hydroxytryptophan and *N*-acetyltryptamine but decreases melatonin biosynthesis during senescence process of detached leaves. *J. Pineal Res.* **2012**, *52*, 211–216. [CrossRef]
39. Tan, D.X.; Reiter, R.J. An evolutionary view of melatonin synthesis and metabolism related to its biological functions in plants. *J. Exp. Bot.* **2020**, *71*, 4677–4689. [CrossRef]

40. Ye, T.; Yin, X.; Yu, L.; Zheng, S.J.; Cai, W.J.; Wu, Y.; Feng, Y.Q. Metabolic analysis of the melatonin biosynthesis pathway using chemical labeling coupled with liquid chromatography-mass spectrometry. *J. Pineal Res.* **2019**, *66*, e12531. [CrossRef]
41. Okazaki, M.; Higuchi, K.; Aouini, A.; Ezura, H. Lowering intercellular melatonin levels by transgenic analysis of indoleamine 2,3-dioxygenase from rice in tomato plants. *J. Pineal Res.* **2010**, *49*, 239–247. [CrossRef]
42. Byeon, Y.; Back, K. Molecular cloning of melatonin 2-hydroxylase responsible for 2-hydroxymelatonin production in rice (*Oryza sativa*). *J. Pineal Res.* **2015**, *58*, 343–351. [CrossRef] [PubMed]
43. Lee, H.J.; Back, K. 2-Hydroxymelatonin promotes the resistance of rice plant to multiple simultaneous abiotic stresses (combined cold and drought). *J. Pineal Res.* **2016**, *61*, 303–316. [CrossRef] [PubMed]
44. Lee, K.; Zawadzka, A.; Czarnocki, Z.; Reiter, R.J.; Back, K. Molecular cloning of melatonin 3-hydroxylase and its production of cyclic 3-hydroxymelatonin in rice (*Oryza sativa*). *J. Pineal Res.* **2016**, *61*, 470–478. [CrossRef]
45. Singh, N.; Kaur, H.; Yadav, S.; Bhatla, S.C. Does N-nitrosomelatonin compete with S-nitrosothiols as a long distance nitric oxide carrier in plants? *Biochem. Anal. Biochem.* **2016**, *5*, 262. [CrossRef]
46. Mukherjee, S. Insights into nitric oxide-melatonin crosstalk and N-nitrosomelatonin functioning in plants. *J. Exp. Bot.* **2019**, *70*, 6035–6047. [CrossRef]
47. Chen, Z.; Xie, Y.; Gu, Q.; Zhao, G.; Zhang, Y.; Cui, W.; Xu, S.; Wang, R.; Shen, W. The AtrbohF-dependent regulation of ROS signaling is required for melatonin-induced salinity tolerance in Arabidopsis. *Free Radical. Biol. Med.* **2017**, *108*, 465–477. [CrossRef] [PubMed]
48. Imran, M.; Shazad, R.; Bilal, S.; Imran, Q.M.; Lee, I.J. Exogenous melatonin mediates the regulation of endogenous nitric oxide in Glycine max L. to reduce effects of drought stress. *Environ. Exp. Bot.* **2021**, *188*, 104511. [CrossRef]
49. Li, H.; Guo, Y.; Lan, Z.; Xu, K.; Chang, J.; Ahammed, G.J.; Ma, J.; Wei, C.; Zhang, X. Methyl jasmonate mediates melatonin-induced cold tolerance of grafted watermelon plants. *Hortic. Res.* **2021**, *8*, 57. [CrossRef]
50. Xia, H.; Zhou, Y.; Deng, H.; Lin, L.; Deng, Q.; Wang, J.; Lv, X.; Zhang, X.; Liang, D. Melatonin improves heat tolerance in Actinidia deliciosa via carotenoid biosynthesis and heat shock proteins expression. *Physiol Plant.* **2021**, *172*, 1582–1593. [CrossRef] [PubMed]
51. Zheng, X.; Zhou, J.; Tan, D.X.; Wang, N.; Wang, L.; Shan, D.; Kong, J. Melatonin improves waterlogging tolerance of *Malus baccata* (Linn.) Borkh. seedlings by maintaining aerobic respiration, photosynthesis and ROS migration. *Front. Plant Sci.* **2017**, *8*, 483. [CrossRef]
52. Pang, Y.W.; Jiang, X.L.; Wang, Y.C.; Wang, Y.Y.; Hao, H.S.; Zhao, S.J.; Du, W.H.; Zhao, X.M.; Wang, L.; Zhu, H.B. Melatonin protects against paraquat-induced damage during in vitro maturation of bovine oocytes. *J. Pineal Res.* **2019**, *66*, e12532. [CrossRef] [PubMed]
53. Meng, J.F.; Xu, T.F.; Wang, Z.Z.; Fang, Y.L.; Xi, Z.M.; Zhang, Z.W. The ameliorative effects of exogenous melatonin on grape cuttings under water-deficient stress: Antioxidant metabolites, leaf anatomy, and chloroplast morphology. *J. Pineal Res.* **2014**, *57*, 200–212. [CrossRef]
54. Wang, P.; Yin, L.; Liang, D.; Li, C.; Ma, F.; Yue, Z. Delayed senescence of apple leaves by exogenous melatonin treatment: Toward regulating the ascorbate-glutathione cycle. *J. Pineal Res.* **2012**, *53*, 11–20. [CrossRef] [PubMed]
55. DalCorso, G.; Farinati, S.; Maistri, S.; Furini, A. How plants cope with cadmium: Staking all on metabolism and gene expression. *J. Integr. Plant Biol.* **2008**, *50*, 1268–1280. [CrossRef] [PubMed]
56. Arnao, M.B.; Hernández-Ruiz, J. Melatonin against environmental plant stressors: A review. *Curr. Protein Pept. Sci.* **2021**, *21*, 413–429. [CrossRef] [PubMed]
57. Zhu, J.K. Plant salt tolerance. *Trends Plant Sci.* **2001**, *6*, 66–71. [CrossRef]
58. Zelm, E.V.; Zhang, Y.; Testerink, C. Salt tolerance mechanisms of plants. *Annu. Rev. Plant Biol.* **2020**, *71*, 403–433. [CrossRef]
59. Lee, H.Y.; Back, K. Melatonin is required for H_2O_2- and NO-mediated defense signaling through MAPKKK3 and OXI1 in Arabidopsis thaliana. *J. Pineal Res.* **2017**, *62*, e12379. [CrossRef]
60. Shi, H.; Jiang, C.; Ye, T.; Tan, D.X.; Reiter, R.J.; Zhang, H.; Liu, R.; Chan, Z. Comparative physiological, metabolomic, and transcriptomic analyses reveal mechanisms of improved abiotic stress resistance in bermudagrass [*Cynodon dactylon* (L). Pers.] by exogenous melatonin. *J. Exp. Bot.* **2015**, *66*, 681–694. [CrossRef]
61. Li, M.Q.; Hasan, M.K.; Li, C.X.; Ahammed, G.J.; Xia, X.J.; Shi, K.; Zhou, Y.; Reiter, R.; Yu, J.; Xu, M.; et al. Melatonin mediates selenium-induced tolerance to cadmium stress in tomato plants. *J. Pineal Res.* **2016**, *61*, 291–302. [CrossRef] [PubMed]
62. Ni, J.; Wang, Q.; Shah, F.A.; Liu, W.; Wang, D.; Huang, S.; Fu, S.; Wu, L. Exogenous melatonin confers cadmium tolerance by counterbalancing the hydrogen peroxide homeostasis in wheat seedlings. *Molecules* **2018**, *23*, 799. [CrossRef]
63. Zhang, J.; Yao, Z.; Zhang, R.; Mou, Z.; Yin, H.; Xu, T.; Zhao, D.; Chen, S. Genome-wide identification and expression profile of the SNAT gene family in tobacco (*Nicotiana tabacum*). *Front. Genet.* **2020**, *11*, 591984. [CrossRef] [PubMed]
64. Gao, Y.; Wang, Y.; Qian, J.; Si, W.; Tan, Q.; Xu, J.; Zhao, Y. Melatonin enhances the cadmium tolerance of mushrooms through antioxidant-related metabolites and enzymes. *Food Chem.* **2020**, *330*, 127263. [CrossRef]
65. Byeon, Y.; Lee, H.Y.; Hwang, O.J.; Lee, H.J.; Back, K. Coordinated regulation of melatonin synthesis and degradation genes in rice leaves in response to cadmium treatment. *J. Pineal Res.* **2015**, *58*, 470–478. [CrossRef]
66. Byeon, Y.; Lee, H.J.; Lee, H.Y.; Back, K. Cloning and functional characterization of the Arabidopsis *N*-acetylserotonin *O*-methyltransferase responsible for melatonin synthesis. *J. Pineal Res.* **2016**, *60*, 65–73. [CrossRef]
67. Byeon, Y.; Lee, H.Y.; Back, K. Cloning and characterization of the serotoninn-acetyltransferase-2 gene (SNAT2) in rice (*Oryza sativa*). *J. Pineal Res.* **2016**, *61*, 198–207. [CrossRef]

68. Wang, T.; Song, J.; Liu, Z.; Liu, Z.; Cui, J. Melatonin alleviates cadmium toxicity by reducing nitric oxide accumulation and IRT1 expression in Chinese cabbage seedlings. *Environ. Sci. Pollut. Res. Int.* **2021**, *28*, 15394–15405. [CrossRef]
69. Asada, K. THE WATER-WATER CYCLE IN CHLOROPLASTS: Scavenging of active oxygens and dissipation of excess photons. *Annu. Rev. Plant Physiol. Plant Mol. Biol.* **1999**, *50*, 601–639. [CrossRef]
70. Considine, M.J.; Foyer, C.H. Oxygen and reactive oxygen species-dependent regulation of plant growth and development. *Plant Physiol.* **2020**, *186*, 79–92. [CrossRef] [PubMed]
71. Kohli, S.K.; Khanna, K.; Bhardwaj, R.; Abd_Allah, E.F.; Ahmad, P.; Corpas, F.J. Assessment of Subcellular ROS and NO Metabolism in Higher Plants: Multifunctional Signaling Molecules. *Antioxidants* **2019**, *8*, 641. [CrossRef]
72. Dai, L.P.; Dong, X.J.; Ma, H.H. Molecular mechanism for cadmium-induced anthocyanin accumulation in Azolla imbricata. *Chemosphere* **2012**, *87*, 319–325. [CrossRef]
73. Alzahrani, Y.; Rady, M.M. Compared to antioxidants and polyamines, the role of maize grain-derived organic biostimulants in improving cadmium tolerance in wheat plants. *Ecotoxicol. Environ. Saf.* **2019**, *182*, 109378. [CrossRef]
74. Kanu, A.S.; Ashraf, U.; Mo, Z.; Sabir, S.; Tang, X. Calcium amendment improved the performance of fragrant rice and reduced metal uptake under cadmium toxicity. *Environ. Sci. Pollut. Res.* **2019**, *26*, 24748–24757. [CrossRef] [PubMed]
75. Hasan, M.K.; Ahammed, G.J.; Yin, L.; Shi, K.; Xia, X.; Zhou, Y.; Yu, J.; Zhou, J. Melatonin mitigates cadmium phytotoxicity through modulation of phytochelatins biosynthesis, vacuolar sequestration, and antioxidant potential in *Solanum lycopersicum* L. *Front. Plant Sci.* **2015**, *6*, 601. [CrossRef] [PubMed]
76. Hasan, M.K.; Ahammed, G.J.; Sun, S.; Li, M.; Yin, H.; Zhou, J. Melatonin inhibits cadmium translocation and enhances plant tolerance by regulating sulfur uptake and assimilation in *Solanum lycopersicum* L. *J. Agric. Food Chem.* **2019**, *67*, 10563–10576. [CrossRef]
77. Sami, A.; Shah, F.A.; Abdullah, M.; Zhou, X.Y.; Zhou, K.J. Melatonin mitigates cadmium and aluminium toxicity through modulation of antioxidant potential in *Brassica napus* L. *Plant Biol.* **2020**, *22*, 679–690. [CrossRef]
78. Nabaei, M.; Amooaghaie, R. Melatonin and nitric oxide enhance cadmium tolerance and phytoremediation efficiency in *Catharanthus roseus* (L.) G. Don. *Environ. Sci. Pollut. Res. Int.* **2020**, *27*, 6981–6994. [CrossRef]
79. Wu, S.; Wang, Y.; Zhang, J.; Gong, X.; Wang, Y. Exogenous melatonin improves physiological characteristics and promotes growth of strawberry seedlings under cadmium stress. *Hortic. Plant J.* **2021**, *7*, 13–22. [CrossRef]
80. Amjadi, Z.; Namdjoyan, S.; Soorki, A.A. Exogenous melatonin and salicylic acid alleviates cadmium toxicity in safflower (Carthamus tinctorius L.) seedlings. *Ecotoxicology* **2021**, *30*, 387–401. [CrossRef] [PubMed]
81. Jiang, M.; Dai, S.; Wang, B.; Xie, Z.; Li, J.; Wang, L.; Li, S.; Tan, Y.; Tian, B.; Shu, Q.; et al. Gold nanoparticles synthesized using melatonin suppress cadmium uptake and alleviate its toxicity in rice. *Environ. Sci. Nano* **2021**, *8*, 1042–1056. [CrossRef]
82. Ma, L.; Huang, Z.; Li, S.; Ashraf, U.; Yang, W.; Liu, H.; Xu, D.; Li, W.; Mo, Z. Melatonin and nitrogen applications modulate early growth and related physio-biochemical attributes in Maize under Cd stress. *J. Soil Sci. Plant Nutr.* **2021**, *21*, 978–990. [CrossRef]
83. Bao, Q.; Bao, W.; Li, Y.; Zhang, S.; Lian, F.; Huang, Y. Silicon combined with foliar melatonin for reducing the absorption and translocation of Cd and As by *Oryza sativa* L. in two contaminated soils. *J. Environ. Manag.* **2021**, *287*, 112343. [CrossRef] [PubMed]
84. Xu, L.; Zhang, F.; Tang, M.; Wang, Y.; Dong, J.; Ying, J.; Chen, Y.; Hu, B.; Li, C.; Liu, L. Melatonin confers cadmium tolerance by modulating critical heavy metal chelators and transporters in radish plants. *J. Pineal Res.* **2020**, *69*, e12659. [CrossRef]
85. Tang, Y.; Lin, L.; Xie, Y.; Liu, J.; Sun, G.; Li, H.; Liao, M.; Wang, Z.; Liang, D.; Xia, H.; et al. Melatonin affects the growth and cadmium accumulation of *Malachium aquaticum* and *Galinsoga parviflora*. *Int. J. Phytoremediation* **2018**, *20*, 295–300. [CrossRef]
86. Lin, L.; Li, J.; Chen, F.; Liao, M.; Tang, Y.; Liang, D.; Xia, H.; Lai, Y.; Wang, X.; Chen, C.; et al. Effects of melatonin on the growth and cadmium characteristics of *Cyphomandra betacea* seedlings. *Environ. Monit. Assess.* **2018**, *190*, 119. [CrossRef] [PubMed]
87. Tanaka, K.; Shu, F.; Fujiwara, T.; Yoneyama, T.; Hayashi, H. Quantitative estimation of the contribution of the phloem in cadmium transport to grains in rice plants (*Oryza sativa* L.). *Soil Sci. Plant Nutr.* **2010**, *53*, 72–77. [CrossRef]
88. Raza, A.; Habib, M.; Kakavand, S.N.; Zahid, Z.; Zahra, N.; Sharif, R.; Hasanuzzaman, M. Phytoremediation of cadmium: Physiological, biochemical, and molecular mechanisms. *Biology* **2020**, *9*, 177. [CrossRef]
89. Nevo, Y.; Nelson, N. The NRAMP family of metal-ion transporters. *Biochim. Biophys. Acta Mol. Cell Res.* **2006**, *1763*, 609–620. [CrossRef]
90. Wu, D.; Yamaji, N.; Yamane, M.; Kashino-Fujii, M.; Sato, K.; Ma, J.F. The HvNramp5 transporter mediates uptake of cadmium and manganese, but not iron. *Plant Physiol.* **2016**, *172*, 1899–1910. [CrossRef]
91. Uraguchi, S.; Kamiya, T.; Sakamoto, T.; Kasai, K.; Sato, Y.; Nagamura, Y.; Yoshida, A.; Kyozuka, J.; Ishikawa, S.; Fujiwara, T. Low-affinity cation transporter (OsLCT1) regulates cadmium transport into rice grains. *Proc. Natl. Acad. Sci. USA* **2011**, *108*, 20959–20964. [CrossRef] [PubMed]
92. Kim, D.Y.; Bovet, L.; Maeshima, M.; Martinoia, E.; Lee, Y. The ABC transporter AtPDR8 is a cadmium extrusion pump conferring heavy metal resistance. *Plant J.* **2007**, *50*, 207–218. [CrossRef] [PubMed]
93. Migocka, M.; Kosieradzka, A.; Papierniak, A.; Maciaszczyk-Dziubinska, E.; Posyniak, E.; Garbiec, A.; Filleur, S. Two metal-tolerance proteins, MTP1 and MTP4, are involved in Zn homeostasis and Cd sequestration in cucumber cells. *J. Exp. Bot.* **2015**, *66*, 1001–1015. [CrossRef] [PubMed]
94. Lekeux, G.; Crowet, J.-M.; Nouet, C.; Joris, M.; Jadoul, A.; Bosman, B.; Carnol, M.; Motte, P.; Lins, L.; Galleni, M. Homology modeling and in vivo functional characterization of the zinc permeation pathway in a heavy metal P-type ATPase. *J. Exp. Bot.* **2019**, *70*, 329–341. [CrossRef] [PubMed]

95. Park, J.; Song, W.Y.; Ko, D.; Eom, Y.; Hansen, T.H.; Schiller, M.; Lee, T.G.; Martinoia, E.; Lee, Y. The phytochelatin transporters AtABCC1 and AtABCC2 mediate tolerance to cadmium and mercury. *Plant J.* **2012**, *69*, 278–288. [CrossRef] [PubMed]
96. Brunetti, P.; Zanella, L.; De Paolis, A.; Di Litta, D.; Cecchetti, V.; Falasca, G.; Barbieri, M.; Altamura, M.M.; Costantino, P.; Cardarelli, M. Cadmium-inducible expression of the ABC-type transporter AtABCC3 increases phytochelatin-mediated cadmium tolerance in Arabidopsis. *J. Exp. Bot.* **2015**, *66*, 3815–3829. [CrossRef] [PubMed]
97. Liu, H.; Zhao, H.; Wu, L.; Liu, A.; Zhao, F.J.; Xu, W. Heavy metal ATPase 3 (HMA3) confers cadmium hypertolerance on the cadmium/zinc hyperaccumulator Sedum plumbizincicola. *New Phytol.* **2017**, *215*, 687–698. [CrossRef]
98. Krämer, U.; Talke, I.N.; Hanikenne, M. Transition metal transport. *FEBS Lett.* **2007**, *581*, 2263–2272. [CrossRef]
99. Rahman, M.F.; Ghosal, A.; Alam, M.F.; Kabir, A.H. Remediation of cadmium toxicity in field peas (*Pisum sativum* L.) through exogenous silicon. *Ecotoxicol. Environ. Saf.* **2017**, *135*, 165–172. [CrossRef] [PubMed]
100. Mishra, V.; Singh, P.; Tripathi, D.K.; Corpas, F.J.; Singh, V.P. Nitric oxide and hydrogen sulfide: An indispensable combination for plant functioning. *Trends Plant Sci.* **2021**, *17*, S1360–S1385. [CrossRef]
101. Besson-Bard, A.; Gravot, A.; Richaud, P.; Auroy, P.; Duc, C.; Gaymard, F.; Taconnat, L.; Renou, J.P.; Pugin, A.; Wendehenne, D. Nitric oxide contributes to cadmium toxicity in Arabidopsis by promoting cadmium accumulation in roots and by up-regulating genes related to iron uptake. *Plant Physiol.* **2009**, *149*, 1302–1315. [CrossRef]
102. Han, B.; Yang, Z.; Xie, Y.; Nie, L.; Jin, C.; Shen, W. Arabidopsis HY1 confers cadmium tolerance by decreasing nitric oxide production and improving iron homeostasis. *Mol. Plant* **2014**, *7*, 388–403. [CrossRef] [PubMed]
103. Sehrawat, A.; Deswal, R. *S-Nitrosylation in Abiotic Stress in Plants and Nitric Oxide Interaction with Plant Hormones*; John Wiley & Sons, Inc.: Hoboken, NJ, USA, 2017; pp. 399–411.
104. Gupta, K.J.; Hancock, J.T.; Petrivalsky, M.; Kolbert, Z.; Lindermayr, C.; Durner, J.; Barroso, J.B.; Palma, J.M.; Brouquisse, R.; Wendehenne, D.; et al. Recommendations on terminology and experimental best practice associated with plant nitric oxide research. *New Phytol.* **2020**, *225*, 1828–1834. [CrossRef]
105. Metwally, A.; Finkemeier, I.; Georgi, M.; Dietz, K.J. Salicylic acid alleviates the cadmium toxicity in barley seedlings. *Plant Physiol.* **2003**, *132*, 272–281. [CrossRef] [PubMed]
106. Krantev, A.; Yordanova, R.; Janda, T.; Szalai, G.; Popova, L. Treatment with salicylic acid decreases the effect of cadmium on photosynthesis in maize plants. *J. Plant Physiol.* **2008**, *165*, 920–931. [CrossRef] [PubMed]
107. Guo, B.; Liang, Y.C.; Zhu, Y.G.; Zhao, F.J. Role of salicylic acid in alleviating oxidative damage in rice roots (*Oryza sativa*) subjected to cadmium stress. *Environ. Pollut.* **2007**, *147*, 743–749. [CrossRef]
108. Zhang, L.; Pei, Y.; Wang, H.; Jin, Z.; Liu, Z.; Qiao, Z.; Fang, H.; Zhang, Y. Hydrogen sulfide alleviates cadmium-induced cell death through restraining ROS accumulation in roots of *Brassica rapa* L. ssp. pekinensis. *Oxidative Med. Cell. Longev.* **2015**, *2015*, 804603. [CrossRef]
109. Hu, L.; Li, H.; Huang, S.; Wang, C.; Sun, W.J.; Mo, H.Z.; Shi, Z.Q.; Chen, J. Eugenol confers cadmium tolerance via intensifying endogenous hydrogen sulfide signaling in *Brassica rapa*. *J. Agric. Food Chem.* **2018**, *66*, 9914–9922. [CrossRef]
110. Zhang, J.; Zhou, M.J.; Ge, Z.L.; Shen, J.; Zhou, C.; Gotor, C.; Romero, L.C.; Duan, X.L.; Liu, X.; Wu, D.L.; et al. Abscisic acid-triggered guard cell L-cysteine desulfhydrase function and in situ hydrogen sulfide production contributes to heme oxygenase-modulated stomatal closure. *Plant Cell Environ.* **2020**, *43*, 624–636. [CrossRef]
111. Qiao, Z.; Tao, J.; Liu, Z.; Zhang, L.; Jin, Z.; Liu, D.; Pei, Y. H_2S acting as a downstream signaling molecule of SA regulates Cd tolerance in Arabidopsis. *Plant Soil* **2015**, *393*, 137–146. [CrossRef]
112. Yang, X.; Kong, L.; Wang, Y.; Su, J.; Shen, W. Methane control of cadmium tolerance in alfalfa roots requires hydrogen sulfide. *Environ. Pollut.* **2021**, *284*, 117123. [CrossRef]
113. Sun, J.; Wang, R.; Zhang, X.; Yu, Y.; Zhao, R.; Li, Z.; Chen, S. Hydrogen sulfide alleviates cadmium toxicity through regulations of cadmium transport across the plasma and vacuolar membranes in Populus euphratica cells. *Plant Physiol. Biochem.* **2013**, *65*, 67–74. [CrossRef] [PubMed]
114. Cui, W.; Chen, H.; Zhu, K.; Jin, Q.; Xie, Y.; Cui, J.; Xia, Y.; Zhang, J.; Shen, W. Cadmium-induced hydrogen sulphide synthesis is involved in cadmium tolerance in Medicago sativa by reestablishment of reduced (homo) glutathione and reactive oxygen species homeostases. *PLoS ONE* **2014**, *9*, e109669. [CrossRef] [PubMed]
115. Shi, H.; Ye, T.; Chan, Z. Nitric oxide-activated hydrogen sulfide is essential for cadmium stress response in bermudagrass (Cynodon dactylon (L). pers.). *Plant Physiol. Biochem.* **2014**, *74*, 99–107. [CrossRef]
116. Mostofa, M.G.; Rahman, A.; Ansary, M.; Watanabe, A.; Fujita, M.; Tran, L.P. Hydrogen sulfide modulates cadmium-induced physiological and biochemical responses to alleviate cadmium toxicity in rice. *Sci. Rep.* **2015**, *5*, 14078. [CrossRef] [PubMed]
117. Jia, H.; Wang, X.; Dou, Y.; Liu, D.; Si, W.; Fang, H.; Zhao, C.; Chen, S.; Xi, J.; Li, J. Hydrogen sulfide-cysteine cycle system enhances cadmium tolerance through alleviating cadmium-induced oxidative stress and ion toxicity in Arabidopsis roots. *Sci. Rep.* **2016**, *6*, 39702. [CrossRef]
118. Kaya, C.; Ashraf, M.; Alyemeni, M.N.; Ahmad, P. Responses of nitric oxide and hydrogen sulfide in regulating oxidative defence system in wheat plants grown under cadmium stress. *Physiol. Plantarum* **2020**, *168*, 345–360. [CrossRef] [PubMed]
119. Lv, H.; Xu, J.; Bo, T.; Wang, W. Comparative transcriptome analysis uncovers roles of hydrogen sulfide for alleviating cadmium toxicity in Tetrahymena thermophila. *BMC Genom.* **2021**, *22*, 21. [CrossRef]

120. Li, G.; Shah, A.A.; Khan, W.U.; Yasin, N.A.; Ahmad, A.; Abbas, M.; Ali, A.; Safdar, N. Hydrogen sulfide mitigates cadmium induced toxicity in Brassica rapa by modulating physiochemical attributes, osmolyte metabolism and antioxidative machinery. *Chemosphere* **2021**, *263*, 127999. [CrossRef]
121. Jia, H.; Wang, X.; Shi, C.; Guo, J.; Ma, P.; Ren, X.; Wei, T.; Liu, H.; Li, J. Hydrogen sulfide decreases Cd translocation from root to shoot through increasing Cd accumulation in cell wall and decreasing Cd^{2+} influx in Isatis indigotica. *Plant Physiol. Biochem.* **2020**, *155*, 605–612. [CrossRef]
122. Tian, B.; Zhang, Y.; Jin, Z.; Liu, Z.; Pei, Y. Role of hydrogen sulfide in the methyl jasmonate response to cadmium stress in foxtail millet. *Front. Biosci.* **2017**, *22*, 530–538. [CrossRef]
123. Tian, B.; Qiao, Z.; Zhang, L.; Li, H.; Pei, Y. Hydrogen sulfide and proline cooperate to alleviate cadmium stress in foxtail millet seedlings. *Plant Physiol. Biochem.* **2016**, *109*, 293–299. [CrossRef] [PubMed]
124. Li, L.; Wang, Y.; Shen, W. Roles of hydrogen sulfide and nitric oxide in the alleviation of cadmium-induced oxidative damage in alfalfa seedling roots. *Biometals* **2012**, *25*, 617–631. [CrossRef] [PubMed]
125. Khan, M.N.; Siddiqui, M.H.; AlSolami, M.A.; Alamri, S.; Hu, Y.; Ali, H.M.; Al-Amri, A.A.; Alsubaie, Q.D.; Al-Munqedhi, B.M.A.; Al-Ghamdi, A. Crosstalk of hydrogen sulfide and nitric oxide requires calcium to mitigate impaired photosynthesis under cadmium stress by activating defense mechanisms in Vigna radiata. *Plant Physiol. Biochem.* **2020**, *156*, 278–290. [CrossRef] [PubMed]
126. Fang, L.; Ju, W.; Yang, C.; Jin, X.; Liu, D.; Li, M.; Yu, J.; Zhao, W.; Zhang, C. Exogenous application of signaling molecules to enhance the resistance of legume-rhizobium symbiosis in Pb/Cd-contaminated soils. *Environ. Pollut.* **2020**, *265*, 114744. [CrossRef]
127. Graham, N.; Arisi, A.M.; Lise, J.; Kunert, K.J.; Heinz, R.; Foyer, C.H. Glutathione: Biosynthesis, metabolism and relationship to stress tolerance explored in transformed plants. *J. Exp. Bot.* **1998**, *49*, 623–647. [CrossRef]
128. Jozefczak, M.; Remans, T.; Vangronsveld, J.; Cuypers, A. Glutathione is a key player in metal-induced oxidative stress defenses. *Int. J. Mol. Sci.* **2012**, *13*, 3145–3175. [CrossRef]
129. Chen, J.; Yang, L.; Yan, X.; Liu, Y.; Wang, R.; Fan, T.; Ren, Y.; Tang, X.; Xiao, F.; Liu, Y.; et al. Zinc-finger Transcription Factor ZAT6 positively regulates cadmium tolerance through the glutathione-dependent pathway in Arabidopsis. *Plant Physiol.* **2016**, *171*, 707–719. [CrossRef]
130. Shen, J.; Su, Y.; Zhou, C.; Zhang, F.; Yuan, X. A putative rice L-cysteine desulfhydrase encodes a true L-cysteine synthase that regulates plant cadmium tolerance. *Plant Growth Regul.* **2019**, *89*, 217–226. [CrossRef]
131. Gu, Q.; Chen, Z.; Cui, W.; Zhang, Y.; Hu, H.; Yu, X.; Wang, Q.; Shen, W. Methane alleviates alfalfa cadmium toxicity via decreasing cadmium accumulation and reestablishing glutathione homeostasis. *Ecotoxicol. Environ. Saf.* **2018**, *147*, 861–871. [CrossRef]
132. Kok, L.; Bosma, W.; Maas, F.M.; Kuiper, P. The effect of short-term H_2S fumigation on water-soluble sulphydryl and glutathione levels in spinach. *Plant Cell Environ.* **1985**, *8*, 189–194. [CrossRef]
133. Anastasis, C.; Manganaris, G.A.; Ioannis, P.; Vasileios, F. Hydrogen sulfide induces systemic tolerance to salinity and non-ionic osmotic stress in strawberry plants through modification of reactive species biosynthesis and transcriptional regulation of multiple defence pathways. *J. Exp. Bot.* **2013**, *7*, 1953–1966. [CrossRef]
134. Liu, F.; Zhang, X.; Cai, B.; Pan, D.; Ai, X. Physiological response and transcription profiling analysis reveal the role of glutathione in H_2S-induced chilling stress tolerance of cucumber seedlings. *Plant Sci.* **2020**, *291*, 110363. [CrossRef]
135. Wei, J.; Li, D.X.; Zhang, J.R.; Shan, C.; Rengel, Z.; Song, Z.B.; Chen, Q. Phytomelatonin receptor PMTR1-mediated signaling regulates stomatal closure in Arabidopsis thaliana. *J. Pineal Res.* **2018**, *65*, e12500. [CrossRef]
136. Lee, H.Y.; Back, K. The phytomelatonin receptor (PMRT1) Arabidopsis Cand2 is not a bona fide G protein-coupled melatonin receptor. *Melatonin Res.* **2020**, *3*, 177–186. [CrossRef]
137. Chen, T.; Tian, M.; Han, Y. Hydrogen sulfide: A multi-tasking signal molecule in the regulation of oxidative stress responses. *J. Exp. Bot.* **2020**, *71*, 2862–2869. [CrossRef]
138. Aroca, A.; Benito, J.M.; Gotor, C.; Romero, L.C. Persulfidation proteome reveals the regulation of protein function by hydrogen sulfide in diverse biological processes in Arabidopsis. *J. Exp. Bot.* **2017**, *68*, 4915–4927. [CrossRef]
139. Filipovic, M.R.; Zivanovic, J.; Alvarez, B.; Banerjee, R. Chemical biology of H_2S signaling through persulfidation. *Chem. Rev.* **2018**, *118*, 1253–1337. [CrossRef] [PubMed]
140. Shen, J.; Zhang, J.; Zhou, M.; Zhou, H.; Cui, B.; Gotor, C.; Romero, L.C.; Fu, L.; Yang, J.; Foyer, C.H.; et al. Persulfidation-based modification of cysteine desulfhydrase and the NADPH oxidase RBOHD controls guard cell abscisic acid signaling. *Plant Cell* **2020**, *32*, 1000–1017. [CrossRef]
141. Zhou, M.; Zhang, J.; Shen, J.; Zhou, H.; Zhao, D.; Gotor, C.; Romero, L.C.; Fu, L.; Li, Z.; Yang, J.; et al. Hydrogen sulfide-linked persulfidation of ABI4 controls ABA responses through the transactivation of MAPKKK18 in Arabidopsis. *Mol. Plant* **2021**, *14*, 921–936. [CrossRef]
142. Chen, S.S.; Jia, H.L.; Wang, X.F.; Shi, C.; Wang, X.; Ma, P.; Wang, J.; Wang, M.J.; Li, J. Hydrogen sulfide positively regulates abscisic acid signaling through persulfidation of SnRK2.6 in guard cells. *Mol. Plant* **2020**, *13*, 732–744. [CrossRef] [PubMed]
143. Chen, J.; Zhou, H.; Xie, Y. SnRK2.6 phosphorylation/persulfidation: Where ABA and H_2S signaling meet. *Trends Plant Sci.* **2021**. [CrossRef] [PubMed]
144. Aroca, Á.; Serna, A.; Gotor, C.; Romero, L.C. S-sulfhydration: A cysteine posttranslational modification in plant systems. *Plant Physiol.* **2015**, *168*, 334–342. [CrossRef] [PubMed]

International Journal of *Molecular Sciences*

MDPI

Review

Crosstalk between Melatonin and Reactive Oxygen Species in Plant Abiotic Stress Responses: An Update

Quan Gu [1], Qingqing Xiao [1], Ziping Chen [2,*] and Yi Han [3,*]

[1] School of Biological Food and Environment, Hefei University, Hefei 230601, China; guq@hfuu.edu.cn (Q.G.); xiaoqq@hfuu.edu.cn (Q.X.)
[2] State Key Laboratory of Tea Plant Biology and Utilization, Anhui Agricultural University, Hefei 230036, China
[3] National Engineering Laboratory of Crop Stress Resistance Breeding, School of Life Sciences, Anhui Agricultural University, Hefei 230036, China
* Correspondence: zpchen@ahau.edu.cn (Z.C.); yi.han@ahau.edu.cn (Y.H.)

Abstract: Melatonin acts as a multifunctional molecule that takes part in various physiological processes, especially in the protection against abiotic stresses, such as salinity, drought, heat, cold, heavy metals, etc. These stresses typically elicit reactive oxygen species (ROS) accumulation. Excessive ROS induce oxidative stress and decrease crop growth and productivity. Significant advances in melatonin initiate a complex antioxidant system that modulates ROS homeostasis in plants. Numerous evidences further reveal that melatonin often cooperates with other signaling molecules, such as ROS, nitric oxide (NO), and hydrogen sulfide (H_2S). The interaction among melatonin, NO, H_2S, and ROS orchestrates the responses to abiotic stresses via signaling networks, thus conferring the plant tolerance. In this review, we summarize the roles of melatonin in establishing redox homeostasis through the antioxidant system and the current progress of complex interactions among melatonin, NO, H_2S, and ROS in higher plant responses to abiotic stresses. We further highlight the vital role of respiratory burst oxidase homologs (RBOHs) during these processes. The complicated integration that occurs between ROS and melatonin in plants is also discussed.

Keywords: reactive oxygen species; nitric oxide; hydrogen sulfide; melatonin; RBOHs; signaling networks; abiotic stress

Citation: Gu, Q.; Xiao, Q.; Chen, Z.; Han, Y. Crosstalk between Melatonin and Reactive Oxygen Species in Plant Abiotic Stress Responses: An Update. *Int. J. Mol. Sci.* **2022**, *23*, 5666. https://doi.org/10.3390/ijms23105666

Academic Editors: Yanjie Xie, Francisco J. Corpas and Jisheng Li

Received: 21 April 2022
Accepted: 16 May 2022
Published: 18 May 2022

1. Introduction

In nature, many plants are constantly challenged by various abiotic environmental conditions, such as salinity, cold, heat, drought, heavy metals, and nutrient deficiencies. These stresses have important impacts on crop growth, development, and productivity [1]. Abiotic stresses affect multiple aspects of plant physiology and cause widespread damages to cellular processes [2]. Plants have evolved complex regulatory pathways to sense and respond to these stresses in a timely manner [3]. In general, abiotic stress often causes oxidative stress and cell damage through inducing excess reactive oxygen species (ROS) generation, such as superoxide anion ($O_2^{\bullet-}$), hydrogen peroxide (H_2O_2), hydroxyl radical ($\cdot OH$), and singlet oxygen ($^1O_2$) [4–6]. Plants have evolved sophisticated antioxidant mechanisms to modulate ROS homeostasis in response to oxidative stress [4,5]. For example, plants recruit abundant enzymatic and non-enzymatic antioxidants. Among these, superoxide dismutase (SOD), guaiacol peroxidase (POD), catalase (CAT), ascorbate peroxidase (APX), thioredoxins (TRX), peroxiredoxins (PRX), glutathione peroxidase (GPX), and glutathione reductase (GR) comprise the enzymatic antioxidant system to regulate ROS accumulation [7,8]. Non-enzymatic antioxidants, such as ascorbic acid (ASC), glutathione (GSH), tocopherol (vitamin E), and flavonoids, are also responsible for keeping ROS balanced at a basal non-toxic level [7,8]. The ascorbate-glutathione cycle (AsA-GSH cycle) is regarded as an important part of the redox hub [5].

Melatonin (*N*-acetyl-5-methoxytryptamine), known as phytomelatonin, is a modulatory agent of plant growth and stress responses, such as lateral root growth, salinity, drought, heat, heavy metals, and defense against UV-B irradiation and bacterial pathogen infection [9–17]. Interestingly, endogenous melatonin levels are increased rapidly by the induction of synthetic genes under the above unfavorable conditions in plants [11,12,14,15]. Melatonin acts as an antioxidant in the control of ROS levels via regulating redox enzymes (including SOD, POD, CAT, APX, GR, etc.) and metabolites (including ASC, GSH, flavonoid, anthocyanins, etc.) [11,12,17–20]. Several studies have revealed that melatonin regulates the primary and secondary metabolism via mediating the master factors of metabolic processes [10,19–22].

Moreover, any deviation from ROS balance can be thought of as a reaction to ROS signaling [8,23]. Numerous studies have revealed that ROS signaling plays a dual role, and is also beneficial to plants at specific cellular compartments during the abiotic stress process [8,23]. Multiple enzymatic systems, such as POD and plasma membrane-bound NADPH oxidase, generate ROS [24]. The respiratory burst oxidase homolog (RBOH) NADPH oxidases are the primary source of ROS production at the apoplast [25]. Superoxide is generated by NADPH oxidase and dismutated to H_2O_2 [26]. It has been observed that *AtrbohD* and *AtrbohF* can regulate sodium (Na) and potassium (K) transport, thus limiting Na concentrations and enhancing salinity tolerance [27,28]. *AtrbohF* also plays a vital role in mediating cadmium (Cd) uptake, chelation, and translocation [29]. Moreover, the ROS wave is required for a plant's high light, cold, and heat tolerance as well [30–32].

To date, several papers have confirmed the crosstalk between melatonin and signaling molecules, such as nitric oxide (NO), hydrogen sulfide (H_2S), and hydrogen gas (H_2); therefore, the present paper does not elucidate it in detail [9,10,18,19,33–35]. Here, we systematically review the updated literature on the crosstalk between melatonin and ROS in plants upon abiotic stresses, highlight the role of RBOHs, and give perspectives for future research.

2. Melatonin Acts as an Antioxidant to Establish Redox Homeostasis through the Antioxidant System in Plants under Abiotic Stresses

As a master regulator, melatonin plays an important role in plant tolerance to abiotic stresses, such as salinity, cold, heat, drought, and heavy metals [9–17]. Our previous review systematically summarized the melatonin biosynthesis and catabolism in plants [19]. We also showed that Cd stress strongly induced melatonin accumulation via regulating the expression of genes encoding tryptophan decarboxylase (TDC), tryptamine 5-hydroxylase (T5H), *N*-acetylserotonin methyltransferase (ASMT), caffeic acid O-methyltransferase (COMT), and serotonin *N*-acetyltransferase (SNAT, also called arylakylamine *N*-acetyltransferase (AANAT)) [21]. Moreover, salinity stress up-regulated the expression *SNAT* genes and improved melatonin levels in *Arabidopsis* and *Brassica napus* [35,36]. In cucumber, cold treatment induced the expression of *TDC, T5H, SNAT*, and *COMT* genes, and thus enhanced melatonin levels [37]. Heat stress also improved the transcripts of *T5H* and *ASMT* genes and melatonin levels in tomato seedlings [38]. Similarly, drought stress up-regulated the expression of *TDC1, T5H, AANAT2*, and *ASMT1* in *Malus hupehensis* and maize plants [39]. These studies mainly found that melatonin levels were significantly induced via up-regulating the transcriptional level of melatonin biosynthesis genes in response to abiotic stress in plants. Interestingly, MzASMT9 protein levels were enhanced by salinity stress in leaves of *Malus zumi* [40]. Moreover, abiotic stress also tightly regulated the activities of melatonin biosynthesis enzymes, such as *T5H, TDC, SNAT*, and *ASMT* enzymes [41,42]. For example, high temperature elevated SNAT and ASMT activity, and increased melatonin levels in rice seedlings [41]. Salinity stress induced serotonin accumulation and *N*-acetylserotonin O-methyltransferase (HIOMT) activity in vascular bundles and the cortex, leading to melatonin accumulation in sunflower (*Helianthus annuus*) plants [42].

In general, these stresses caused endogenous melatonin accumulation, indicating that melatonin might be involved in a plant's tolerance to abiotic stress. A series of studies found that the application of exogenous melatonin increased the level of endogenous melatonin, thereby improving plant tolerance to abiotic stress [20,21,36,43,44]. For example, the endogenous melatonin content was increased in maize by application of exogenous melatonin upon both control and Al stress conditions [43]. Moreover, this increase significantly mitigated Al-induced oxidative stress [43]. Similar results were found in the role of exogenous melatonin application in the alleviation of Cd-induced growth inhibition of mallow (*Malva parviflora*) plants [20]. Pharmacological studies also revealed that exogenous melatonin application improved resistance to salinity and drought stresses via the modulation of photosynthesis and starch/sucrose metabolism in soybean [21]. The application of melatonin partly counteracted salinity-induced seedling growth inhibition in rapeseed (*Brassica napus* L.) [36]. The role of exogenous melatonin in reducing the severity induced by heat stress in wheat seedlings was also evaluated [44].

Recently, several general genetic studies have been conducted in plants [10–12,35,45–47]. In these studies, it was demonstrated that both the *TDC*-silenced mutant and *COMT1*-silenced mutant showed a lower level of melatonin in tomato plants [10]. In Arabidopsis, we found that the *atsnat* mutant showed a low content of melatonin, and appeared hypersensitive to salinity stress in comparison with the wild-type seedlings [12,35]. It was also observed that both *atsnat-1* and *atsnat-1* showed sensitivity to high levels of light [46]. The transgenic *Arabidopsis* seedlings overexpressing alfalfa *SNAT* enhanced melatonin accumulation and exhibited more resistance to Cd stress than wild-type plants [11]. In tomato plants, overexpressing *AANAT* or *HIOMT* enhanced melatonin accumulation and improved drought tolerance [45]. Heterologous expression of *HIOMT* in apple leaves showed higher melatonin levels and improved salinity stress tolerance [46]. Nevertheless, there is still much to be learned about the post-translational modulation of melatonin biosynthesis genes and the regulation of related proteins, which should be further studied in the future.

Abiotic stresses cause endogenous ROS (mainly $O_2^{\bullet-}$, H_2O_2, and MDA) accumulation in plants, which generates in different various organelles including chloroplast, peroxisome, mitochondria, and the cell membrane during abiotic stresses (Figure 1) [4]. For example, $O_2^{\bullet-}$ acts as the by-product of oxygen reduction by the electron transport chain (ETC) in chloroplast and mitochondria [48,49]. They also generate by photorespiration and fatty acid oxidation in peroxisome [50,51]. Then, H_2O_2 is produced from $O_2^{\bullet-}$ by the activity of SOD or glycolate oxidases. Furthermore, NADPH oxidases, cell-wall-bound peroxidases (POX), and polyamine oxidases (PAO) result in ROS generation in the cell membrane, cell wall, and apoplast, respectively [4,52]. As toxic byproducts, ROS could cause damages to the RNA, DNA, and proteins of plants (oxidative stress situations) [8].

Melatonin acts as a potential antioxidant against abiotic stresses in plants. Afterwards, melatonin enhances the tolerance via up- or down-regulating downstream regulating elements within the physiological environments of various plants (Figure 1, Table 1). The increased melatonin decreases $O_2^{\bullet-}$ and H_2O_2 accumulation via the enhanced antioxidant enzyme activities and antioxidant levels [9,10,18,19]. Some examples of the various roles of melatonin in the regulation of redox homeostasis in plants under abiotic stresses are illustrated in Table 1. Salinity stress is one of the serious threats to crop growth and development. Many studies indicate that melatonin enhances tolerance to salinity stress in various plant species, including *Arabidopsis*, *Brassica napus*, rice, wheat, tomato, cucumber, *Malus domestica*, *Limonium bicolor*, sunflower, and olive [12,35,36,40,42,47,53–59]. Within these studies, melatonin regulated ion homeostasis, especially Na^+ and K^+ homeostasis, thus alleviating the salinity damage. Melatonin treatment up-regulated the expression of *SOS1*, *NHX1*, and/or *AKT1*, and then maintained K^+/Na^+ homeostasis in *Arabidopsis* and rice [12,35,36,53]. Moreover, to reestablish the redox homeostasis, melatonin also enhanced the expression of genes encoding antioxidant enzymes (such as *APX1*, *APX2*, *CAT1*, *FSD1*, *CuZnSOD*, and *MnSOD*), and improved the activities of APX, SOD, CAT, POD, Δ1-pyrroline-5-carboxylate synthesis (P5CS), as well as the levels of antioxidants (ASC, GSH, proline, and total soluble carbohydrates) [12,35,36,40,42,47,54–59] (Table 1).

Similarly, it was well established that melatonin boosted the activities of many antioxidant enzymes (including SOD, POD, APX, CAT, DHAR, GST, GR, MDHAR, and PPO) and the levels of antioxidants (including ASC, DHA, GSH, proline, flavonoid, carotenoid, and phenolic compounds), thus reducing ROS levels and improving tolerance to drought stress in plants, such as maize, tomato, citrus, soybean, *Malus*, or kiwifruit plants [14,39,60–64] (Table 1). Melatonin also acted as a priming agent to improve *Medicago sativa* tolerance to drought stress via the nitro-oxidative homeostasis [65]. Both cold and heat stress can induce ROS accumulation and alter the redox homeostasis. The increase in melatonin alleviated the inhibition of germination and growth of plants, such as *Arabidopsis*, rice, watermelon, *Camellia sinensis*, cucumber, tomato, soybean, *Chrysanthemum*, *Actinidia deliciosa*, etc. [15,37,38,44,66–77] (Table 1). Similar to salinity and drought stresses, cold or heat stress induced severe oxidative stress, and melatonin increased APX, SOD, CAT, POD, GPX, GR, Gly I, and Gly II activity, as well as GSH, ASC, proline, flavonoid, and proline contents. Furthermore, melatonin treatment positively modulated *ZAT10* and *ZAT12*, which encode transcriptional regulators of ROS-related antioxidant genes [66]. Several studies suggested that heat shock proteins (HSPs) were also involved in melatonin-regulated heat tolerance in plants, such as *Arabidopsis*, tomato, and kiwifruit [74,76,77]. In addition, heavy metal pollutants were shown to induce serious stress and toxicity in plants. Melatonin protects plants upon heavy metal stress, such as Cd, aluminum (Al), lead (Pb), mercury (Hg), copper (Cu), vanadium (V), and arsenic (As) [11,19,20,43,78–80] (Table 1). In these studies, melatonin improved Cd-triggered redox imbalance through changes in Cu/ZnSOD genes, which are regulated by miR398a and miR398b [11]. Seeds of red cabbage with melatonin pretreatment conferred Cu tolerance by blocking the membrane peroxidation and DNA damages [78]. Treatment of exogenous melatonin or improvement of endogenous melatonin by overexpressing the melatonin synthetic-related genes stimulated the activities of antioxidant enzymes (including APX, SOD, CAT, POD, GPX, GR, and PAL) and increased antioxidant levels (including DHA, GSH, proline, flavonoid, and anthocyanins), thus inhibiting ROS production in plants, such as tomato, *Nicotiana tabacum* L., rice, maize, wheat, *Azolla imbricata*, and watermelon seedlings under the stress of heavy metals [11,19,20,43,78–80] (Table 1). Melatonin also improved the efficiency of PSII and regulated amino acids, sugar alcohols, and carotenoids metabolism to enhance plant tolerance to abiotic stress (Figure 1).

Table 1. Some examples of the roles of melatonin in regulating redox homeostasis in plants under abiotic stresses.

Abiotic Stressors	Impact on Oxidative Stress Markers and Antioxidative Defense Systems (Enzymes and Related Genes)	Plant Species	References
Salinity stress	H_2O_2, $O_2^{\bullet-}$, MDA, ·OH, and EL; APX, SOD, CAT, POD, Δ1-pyrroline-5-carboxylate synthetase, ASC, GSH, proline, and total soluble carbohydrates; *APX1*, *APX2*, *CAT1*, *FSD1*, *CuZnSOD*, and *MnSOD*	*Arabidopsis*, *Brassica napus*, *Malus domestica*, olive, tomato, wheat, cucumber, rice, *Limonium bicolor*	[12,35,36,40,42, 47,53–59]
Drought stress	H_2O_2, MDA, $O_2^{\bullet-}$, and EL; APX, SOD, CAT, POD, DHAR, GST, GR, MDHAR, PPO, ASC, DHA, GSH, proline, flavonoid, carotenoid, and phenolic compounds; *Cu/ZnSOD*, *Fe/MnSOD*, *APX*, *CAT*, *GR*, *POD*, *GST*, *DHAR*, and *MDHAR*	maize, tomato, citrus, soybean, kiwifruit, Malus	[14,39,60–65]
Cold stress	H_2O_2, $O_2^{\bullet-}$, MDA, and EL; APX, SOD, CAT, POD, GR, GSH, ASC, proline, polyamine; *APX*, *CAT*, *SOD*, *GR*, *ZAT10*, and *ZAT12*	*Arabidopsis*, watermelon, *Camellia sinensis*, rice, cucumber, tomato	[15,37,66–70]
Heat stress	H_2O_2, $O_2^{\bullet-}$, MDA, and EL; APX, SOD, CAT, POD, GPX, GR, Gly I, Gly II, GSH, ASC, proline, flavonoid, proline, polyamine, and carotenoid; *APX*, *CAT*, *SOD*, *POD*, *HsfA2*, and *Hsp90*	rice, soybean, maize, *Chrysanthemum*, *Actinidia deliciosa*	[38,44,71–77]

Table 1. *Cont.*

Abiotic Stressors	Impact on Oxidative Stress Markers and Antioxidative Defense Systems (Enzymes and Related Genes)	Plant Species	References
Heavy metals stress	H_2O_2, $O_2^{\bullet -}$, MDA, and EL; APX, SOD, CAT, POD, GPX, GR, PAL, ASC, DHA, GSH, proline, flavonoid, anthocyanins *APX, CAT, POD, SOD, GR, GSH1, PCS*	*Tomato, Nicotiana tabacum* L., *Brassica napus* L., rice, maize, wheat, alfalfa, Azolla imbricata, watermelon	[11,19,20,43,78–80]

H_2O_2, hydrogen peroxide; $O_2^{\bullet -}$, superoxide anion; MDA, malondialdehyde; ·OH, hydroxyl radical; EL, electrolytic leakage; APX, ascorbate peroxidase; SOD, superoxide dismutase; CAT, catalase; POD, guaiacol peroxidase; DHAR, dehydroascorbate reductase; GST, glutathione S-transferase; GR, glutathione reductase; MDHAR, monodehydroascorbate reductase, PPO, polyphenol oxidase; ACS, ascorbate; GSH, reduced glutathione; DHA, dehydroascorbate; GPX, glutathione peroxidase; ZAT, ROS-related responsive elements; Gly, glyoxalase; HsfA, heat-shock factor; HSP, heat-shock protein; PAL, phenylalanine ammonia-lyase.

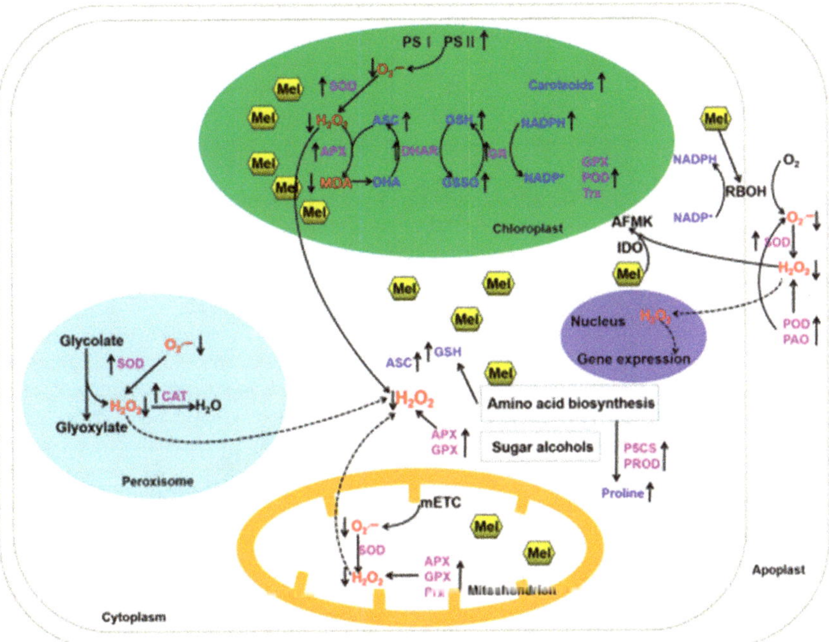

Figure 1. The relationship between melatonin and ROS in plant responses to abiotic stresses. Abiotic stresses, such as salinity, heat, cold, drought, and heavy metals, induce melatonin (Mel) and ROS (mainly $O_2^{\bullet -}$, H_2O_2, and MDA) accumulation. ROS generates in excess in chloroplast, peroxisome, mitochondria, the cell membrane, and apoplast. Melatonin further regulates the activity of several antioxidant enzymes and the contents of antioxidants. Moreover, melatonin modulates the RBOH involved in H_2O_2 accumulation, and thereby acts as a signaling molecule to regulate gene expression in the nucleus. It is also suggested that ROS interact with melatonin to form AFMK via IDO enzyme. Melatonin regulates the amino acid biosynthesis, sugar alcohols, and carotenoids to alleviate abiotic stresses. Mel, melatonin; PS, photosystem; mETC, the electron transport chain in mitochondria; ROS, reactive oxygen species; $O_2^{\bullet -}$, superoxide anion; H_2O_2, hydrogen peroxide; SOD, superoxide dismutase; APX, ascorbate peroxidase; POD, guaiacol peroxidase; CAT, catalase; GR, glutathione reductase; DHAR, dehydroascorbate reductase; GPX, glutathione peroxidase; Trx, thioredoxins; Prx, peroxiredoxins; P5CS, pyrroline-5-carboxylate synthetase; PROD, proline dehydrogenase; IDO, indoleamine 2,3-dioxygenase; PAO, polyamine oxidase; ASC, ascorbic acid; DHA, dehydroascorbate; GSH, reduced glutathione; GSSG, oxidized glutathione; AFMK, N^1-acetyl-N^2-formyl-5-methoxykynuramine; RBOH, respiratory burst oxidase.

3. Plant Abiotic Stress Tolerance Is Mediated by the Crosstalk between Melatonin and Signal Molecules (NO, H_2S, and ROS)

Melatonin was shown to be a crucial regulator occupying extensive roles in many physiological and biochemical processes throughout plant life, especially plant responses to abiotic stress [9,10,19]. Furthermore, melatonin was shown to function with other signal molecules in order to manipulate environmental damages, such as NO, H_2S, ROS, and H_2 [9,10,18,19,33–35]. The latest reviews systematically revealed the inter-relationship between melatonin and gasotransmitters (including NO, CO, H_2S, CH_4, and H_2) in resistance to plant abiotic stress [10,81–83]. For example, melatonin altered the endogenous NO accumulation, and reduced reactive nitrogen species (RNS) (ONOO–, and peroxynitrous acid), which was generated by stress [81,82]. Nevertheless, melatonin regulated the expression of the nitric oxide synthase (*NOS*) gene, and triggered endogenous NO accumulation [84]. Ample evidence manifested that NO acted as the downstream signal of melatonin to regulate plant tolerance to salinity, drought, heat, cold, Cd, and Al stresses [33,34,36,85–88]. Zhao et al. found that melatonin enhanced rapeseed seedling tolerance via NO signaling against salinity stress, and similar results were obtained for sunflower seedlings as well [36,85]. Melatonin down-regulated the NO accumulation, thus promoting soybean tolerance to drought stress [86]. Moreover, positive and antagonistic interactions between melatonin and NO might exist in plant responses to stress caused by heavy metals [19,87,88].

It was also shown that H_2S plays a vital role in enhancing plant tolerance to abiotic stress and alleviating its detrimental effects [89–92]. Melatonin increased the activities of the H_2S-produced enzymes (*D*-cysteine desulfhydrase (DCD), *L*-cysteine desulfhydrase (LCD)), thus improving H_2S accumulation [93–95]. Further, application of hypotaurine (HT, H_2S scavenger) reversed the contribution of melatonin in alleviation of the salinity and heat damages by reestablishing the redox homeostasis in tomato, cucumber, and wheat seedlings [44,94,95]. Kaya et al. found that the interactive effect of NO and H_2S improved the wheat's resistance to Cd stress via enhancing the antioxidative defense system and reducing the damage induced by oxidative stress [96]. Moreover, the H_2S and NO jointly were involved in melatonin-regulated salinity tolerance in cucumbers [95]. They were also involved in melatonin-mediated resistance to iron deficiency and salinity stress in pepper seedlings [97]. Until now, the interactions among melatonin, NO, and H_2S in plant responses to abiotic stress were not largely explored.

In recent years, apart from NO and H_2S, great efforts were made in studies conducted on ROS-directed plant abiotic stress responses [4–8]. In this review, Section 2 shows that ROS are inevitably produced by adverse environmental conditions, and thereby significantly cause damages to the structural and functional integrity of the whole plant seedling. More importantly, ROS instantly produced in chloroplast, peroxisome, mitochondria, and cell membrane organelles often modulate signaling pathways when maintained at a moderate concentration [4–8]. Recent studies have further shed new light on the role of ROS in melatonin-regulated tolerance to abiotic stresses in plants. In early responses to cold stress, melatonin was found to stimulate H_2O_2 accumulation in watermelon [98]. Chen et al. found that endogenous melatonin rapidly induced ROS accumulation under short-term salinity treatment in *Arabidopsis* [12]. Then, ROS triggered SOS-mediated Na^+ efflux and intensified the increased antioxidant defense [12]. Similarly, melatonin triggered an ROS burst that enhanced the expression of K^+ uptake transporters to enable K^+ retention under salinity stress in rice [53]. H_2O_2 scavengers negated the effects of melatonin-mitigated abiotic stress, such as drought, heat, and cold stress in tomato plants [99]. Collectively, these studies preliminarily revealed that ROS signaling acts downstream of melatonin in alleviation of abiotic stress in plants.

Several articles have also shown the mechanisms underlying the complexity of ROS with NO and/or H_2S signaling in plant tolerance against abiotic stress [2–4,89,90,92]. Zeng et al. reviewed the crosstalk among melatonin, NO, and ROS in plant tolerance to bacterial, fungal, and viral diseases [100]. This phenomenon was also shown to promote fruit ripening [9]. However, more advances should be made to provide new insights on the understanding of the crosstalk among melatonin, NO, H_2S, and ROS in plant abiotic tolerance using genetic, pharmacological, genomic, and proteomic approaches.

4. The Roles of RBOH-Involved ROS Signaling in Melatonin-Modulated Plant Processes

ROS produced in several organs (the cell membrane, chloroplast, peroxisome, mitochondria, and apoplast) are implicated in signaling pathways (Figure 1). Respiratory burst oxidase homolog (RBOH) proteins are the NADPH oxidases localized on the plasma membrane [101]. They are the key proteins associated with the signal transduction event [101]. There is a C-terminal FAD/NADP(H)-binding domain, a N-terminal regulatory domain, six transmembrane domains, and several potential phosphorylation sites in *RBOHs* [24,25]. The NADPH oxidases are modulated by phosphorylation and/or binding of calcium ions at the cytosol, and then produce $O_2^{\bullet-}$ at the apoplast (Figure 1). $O_2^{\bullet-}$ is converted into H_2O_2 via the action of SOD, or spontaneously [8,101]. Afterwards, H_2O_2 could enter various types of cells and trigger different signaling responses. There are ten or eight *RBOH* genes encoding NADPH oxidase in Arabidopsis or tomato, respectively [24,25]. In recent years, many studies have shed light on ROS-directed plant growth and stress responses in *Arabidopsis* and other crops. In this review, we emphasize the roles of RBOH-involved ROS signaling in melatonin-mediated plant abiotic stress tolerance.

Recently, the roles of Rbohs were analyzed in *Arabidopsis*, tomato, tobacco, rice, cucumber, alfalfa, etc. [12,53,102–106]. Most of these genes played important roles in resistance to salinity, high/low temperature, heavy metals, and biological stress [12,53,102–106]. It was reported that *AtRbohC* regulated the stability of *SOS1* mRNA to improve salinity tolerance [102]. *AtrbohC*, *AtrbohD*, and *AtrbohF* also inhibited calcium (Ca), zinc (Zn), and iron (Fe) translocation [29]. *SlRbohB*-involved ROS signaling was required for tolerance to drought stress in tomatoes [103]. Meanwhile, *RBOH1*-produced H_2O_2 induced the expression of downstream genes to enhance tomato's resistance to cold, salinity, and salinity–alkalinity stresses [104–106]. In tobacco, the *NtRbohE*-derived ROS signaling pathway improved salinity tolerance [107]. Moreover, RBOH-mediated ROS production was involved in lateral root growth, and AtRbohC promoted root hair budding in Arabidopis [108,109]. *AtrbohF*-mediated H_2O_2 signaling acted as a key mediator in stomatal closure in guard cells [110]. Hence, *RBOHs*-involved ROS signaling plays a vital role in plant growth and abiotic stress tolerance.

Much more is now known about how *RBOH* genes are regulated in response to abiotic stresses. Recent evidence suggests that ROS function as signaling molecules in relation to hormone responses, and the mechanistic bases are complicated. Clearly, melatonin has complex crosstalks with signaling molecules and other phytohormones [19,111]. In our studies, melatonin-triggered lateral root formation was H_2O_2-dependent in alfalfa [13]. Further, melatonin induced PAO and *RBOH*-derived ROS accumulation to facilitate the lateral root development in tomatoes [112,113]. The roles of *PuRBOHF*-dependent H_2O_2 were also essential for melatonin-induced anthocyanin accumulation in red pear fruit [114,115]. Moreover, a *RbohF*-dependent ROS burst was required for melatonin-triggered salinity tolerance in *Arabidopsis* and rice [12,53]. So far, the ways that most *RBOHs* function in melatonin-regulated processes in response to abiotic stress are still elusive, and it should be clarified in future studies.

5. How Melatonin Directs with the RBOH-Regulated ROS Signaling in Plant Tolerance to Abiotic Stress

Several studies take the important stance that the transmembrane receptor of melatonin (PMTR1/CAND2) is found in *Arabidopsis*, tobacco, alfalfa, and maize [116–119]. It is also located in the plasma membrane, and can interact with G-protein α subunits, thereby activating *RBOHs* to promote stomatal closure in Arabidopsis (Figure 2) [116,119]. Melatonin inhibits endogenous NO accumulation and reduces the S-nitrosylation of RBOH to activate the ROS signaling pathway [99]. ROS signaling induces the expression of defensive genes to enhance plant tolerance to oxidative stress [120,121]. However, whether or how the interaction between PMTR1/CAND2 and G-protein α subunits directly regulates RBOHs in plant response to abiotic stress remains to be deciphered.

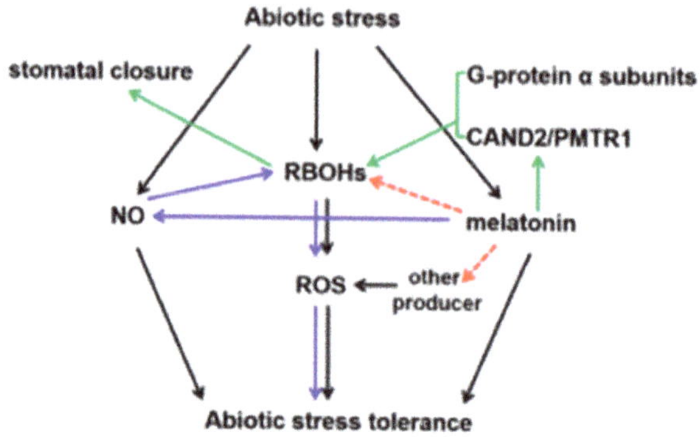

Figure 2. Probable integrative model of ROS with melatonin regulator in plant responses to abiotic stress. Increasing evidence shows that melatonin induces NO generation and enhances RBOH activity through denitrosylation, thereby activating ROS signaling in tomatoes (blue arrow). Interaction between CAND2/PMTR1 (the melatonin receptor) and G-protein α subunits activates *RBOHs*, resulting in stomatal closure (green arrow). The direct relationship between the melatonin and RBOHs in plant tolerance to abiotic stress is still largely unknown (red arrow, yet largely unknown). NO, nitric oxide; ROS, reactive oxygen species; RBOHs, respiratory burst oxidase homologs.

Recent data imply that melatonin connects with the multiple elements, including the hormone and signaling molecules. Moreover, it was found that H_2S modulates the post-translational modification of protein cysteine residues in plants [122–124]. Our previous review further suggested that ROS interacted with H_2S by regulating transcriptional or post-translational modifications in response to oxidative stress [125]. Therefore, interaction of melatonin with ROS and H_2S in regulating abiotic stress also has significant importance and remains to be identified in future.

6. Conclusions and Perspectives

Much more is now known about the regulatory mechanism of melatonin-mediated tolerance to abiotic stresses, especially the cooperation between melatonin and ROS, NO, and/or H_2S. To further promote this research in plants, our review summarizes the ROS-involved regulatory roles and mechanisms of melatonin-mediated abiotic stress resistance. Melatonin confers oxidative stress tolerance mainly through the reestablishment of redox homeostasis. Moreover, ROS act as signaling molecules that regulate melatonin-modulated protective effects. In particular, the vital role of RBOHs during these processes was shown. However, there are still many questions that should be characterized to understand the signal transduction pathway of melatonin in plants in response to abiotic stress. For example, it is necessary to focus more attention on the signaling role of ROS produced by photosystem II (PSII) and photorespiration in melatonin-alleviated abiotic stress in future studies.

As melatonin is an important regulatory element of phytohormones, it collaborates with multiple elements (such as the discovered signaling molecules NO, H_2S, and ROS) and hormones (such as auxin, ethylene, salicylic acid, gibberellin, and abscisic acid signaling). Importantly, most studies do not provide solid in vivo evidence. Future studies using related mutants produced by gene editing technology and plastid transformation technology [126,127] should aim to illustrate how melatonin functions with these signaling molecules in plants under stressful situations.

Author Contributions: Writing—original draft preparation, Q.G. and Q.X.; writing—review and editing, Z.C. and Y.H. All authors have read and agreed to the published version of the manuscript.

Funding: This research was mainly supported by the Natural Science Foundation of Anhui Province (2008085MC101), Talent Research Fund Project of Hefei University (No. 18-19RC13), Postdoctoral Projects of Anhui Province (Grant No. 2020A396).

Institutional Review Board Statement: Not applicable.

Informed Consent Statement: Not applicable.

Data Availability Statement: Not applicable.

Conflicts of Interest: The authors declare no conflict of interest.

References

1. Bailey-Serres, J.; Parker, J.E.; Ainsworth, E.A.; Oldroyd, G.E.D.; Schroeder, J.I. Genetic strategies for improving crop yields. *Nature* **2019**, *575*, 109–118. [CrossRef] [PubMed]
2. Zhang, H.; Zhao, Y.; Zhu, J.K. Thriving under stress: How plants balance growth and the stress response. *Dev. Cell* **2020**, *55*, 529–543. [CrossRef] [PubMed]
3. Zhang, H.; Zhu, J.; Gong, Z.; Zhu, J. Abiotic stress responses in plants. *Nat. Rev. Genet.* **2022**, *23*, 104–111. [CrossRef]
4. Sachdev, S.; Ansari, S.A.; Ansari, M.I.; Fujita, M.; Hasanuzzaman, M. Abiotic stress and reactive oxygen species: Generation, signaling, and defense mechanisms. *Antioxidants* **2021**, *10*, 277. [CrossRef] [PubMed]
5. Foyer, C.H.; Noctor, G. Oxidant and antioxidant signalling in plants: A re-evaluation of the concept of oxidative stress in a physiological context. *Plant Cell Environ.* **2005**, *28*, 1056–1071. [CrossRef]
6. Qi, J.; Song, C.P.; Wang, B.; Zhou, J.; Kangasjärvi, J.; Zhu, J.K.; Gong, Z. Reactive oxygen species signaling and stomatal movement in plant responses to drought stress and pathogen attack. *J. Integr. Plant Biol.* **2018**, *60*, 805–826. [CrossRef]
7. Apel, K.; Hirt, H. Reactive oxygen species: Metabolism, oxidative stress, and signal transduction. *Annu. Rev. Plant Biol.* **2004**, *55*, 373–399. [CrossRef]
8. Mittler, R. ROS are good. *Trends Plant Sci.* **2017**, *22*, 11–19. [CrossRef]
9. Sun, C.; Liu, L.; Wang, L.; Li, B.; Jin, C.; Lin, X. Melatonin: A master regulator of plant development and stress responses. *J. Integr. Plant Biol.* **2021**, *63*, 126–145. [CrossRef]
10. Arnao, M.B.; Cano, A.; Hernández-Ruiz, J. Phytomelatonin: An unexpected molecule with amazing performances in plants. *J. Exp. Bot.* **2022**, erac009. [CrossRef]
11. Gu, Q.; Chen, Z.; Yu, X.; Cui, W.; Pan, J.; Zhao, G.; Xu, S.; Wang, R.; Shen, W. Melatonin confers plant tolerance against cadmium stress via the decrease of cadmium accumulation and reestablishment of microRNA-mediated redox homeostasis. *Plant Sci.* **2017**, *261*, 28–37. [CrossRef] [PubMed]
12. Chen, Z.; Xie, Y.; Gu, Q.; Zhao, G.; Zhang, Y.; Cui, W.; Xu, S.; Wang, R.; Shen, W. The AtrbohF-dependent regulation of ROS signaling is required for melatonin-induced salinity tolerance in Arabidopsis. *Free Radic. Biol. Med.* **2017**, *108*, 465–477. [CrossRef] [PubMed]
13. Chen, Z.; Gu, Q.; Yu, X.; Huang, L.; Xu, S.; Wang, R.; Shen, W.; Shen, W. Hydrogen peroxide acts downstream of melatonin to induce lateral root formation. *Ann. Bot.* **2018**, *121*, 1127–1136. [CrossRef] [PubMed]
14. Sun, L.; Song, F.; Guo, J.; Zhu, X.; Liu, S.; Liu, F.; Li, X. Nano-ZnO-induced drought tolerance is associated with melatonin synthesis and metabolism in maize. *Int. J. Mol. Sci.* **2020**, *21*, 782. [CrossRef]
15. Li, H.; Guo, Y.; Lan, Z.; Xu, K.; Chang, J.; Ahammed, G.J.; Ma, J.; Wei, C.; Zhang, X. Methyl jasmonate mediates melatonin-induced cold tolerance of grafted watermelon plants. *Hortic. Res.* **2021**, *8*, 57. [CrossRef]
16. Haskirli, H.; Yilmaz, O.; Ozgur, R.; Uzilday, B.; Turkan, I. Melatonin mitigates UV-B stress via regulating oxidative stress response, cellular redox and alternative electron sinks in Arabidopsis thaliana. *Phytochemistry* **2021**, *182*, 112592. [CrossRef]
17. Lee, H.Y.; Back, K. Mitogen-activated protein kinase pathways are required for melatonin-mediated defense responses in plants. *J. Pineal Res.* **2016**, *60*, 327–335. [CrossRef]
18. Arnao, M.B.; Hernandez-Ruiz, J. Melatonin: A new plant hormone and/or a plant master regulator? *Trends Plant Sci.* **2019**, *24*, 38–48. [CrossRef]
19. Gu, Q.; Wang, C.; Xiao, Q.; Chen, Z.; Han, Y. Melatonin confers plant cadmium tolerance: An update. *Int. J. Mol. Sci.* **2021**, *22*, 11704. [CrossRef]
20. Tousi, S.; Zoufan, P.; Ghahfarrokhie, A.R. Alleviation of cadmium-induced phytotoxicity and growth improvement by exogenous melatonin pretreatment in mallow (*Malva parviflora*) plants. *Ecotox. Environ. Safe.* **2020**, *206*, 111403. [CrossRef]
21. Wei, W.; Li, Q.T.; Chu, Y.N.; Reiter, R.J.; Yu, X.M.; Zhu, D.H.; Zhang, W.K.; Ma, B.; Lin, Q.; Zhang, J.S.; et al. Melatonin enhances plant growth and abiotic stress tolerance in soybean plants. *J. Exp. Bot.* **2015**, *66*, 695–707. [CrossRef] [PubMed]
22. Teng, Z.; Yu, Y.; Zhu, Z.; Hong, S.B.; Yang, B.; Zang, Y. Melatonin elevated Sclerotinia sclerotiorum resistance via modulation of ATP and glucosinolate biosynthesis in *Brassica rapa* ssp. pekinensis. *J. Proteom.* **2021**, *243*, 104264. [CrossRef] [PubMed]

23. Choudhury, F.K.; Rivero, R.M.; Blumwald, E.; Mittler, R. Reactive oxygen species, abiotic stress and stress combination. *Plant J.* **2017**, *90*, 856–867. [CrossRef] [PubMed]
24. Marino, D.; Dunand, C.; Puppo, A.; Pauly, N. A burst of plant NADPH oxidases. *Trends Plant Sci.* **2012**, *17*, 9–15. [CrossRef] [PubMed]
25. Sagi, M.; Fluhr, R. Production of reactive oxygen species by plant NADPH oxidases. *Plant Physiol.* **2006**, *141*, 336–340. [CrossRef] [PubMed]
26. Torres, M.A.; Jones, J.D.; Dangl, J.L. Pathogen-induced, NADPH oxidase-derived reactive oxygen intermediates suppress spread of cell death in Arabidopsis thaliana. *Nat. Genet.* **2005**, *37*, 1130–1134. [CrossRef]
27. Jiang, C.; Belfield, E.J.; Mithani, A.; Visscher, A.; Ragoussis, J.; Mott, R.; Smith, J.A.; Harberd, N.P. ROS-mediated vascular homeostatic control of root-to-shoot soil Na delivery in Arabidopsis. *EMBO J.* **2012**, *31*, 4359–4370. [CrossRef]
28. Ma, L.; Zhang, H.; Sun, L.; Jiao, Y.; Zhang, G.; Miao, C.; Hao, F. NADPH oxidase AtrbohD and AtrbohF function in ROS-dependent regulation of Na^+/K^+ homeostasis in Arabidopsis under salt stress. *J. Exp. Bot.* **2012**, *63*, 305–317. [CrossRef]
29. Gupta, D.K.; Pena, L.B.; Romero-Puertas, M.C.; Hernández, A.; Inouhe, M.; Sandalio, L.M. NADPH oxidases differentially regulate ROS metabolism and nutrient uptake under cadmium toxicity. *Plant Cell Environ.* **2017**, *40*, 509–526. [CrossRef]
30. Wang, L.; Guo, Y.; Jia, L.; Chu, H.; Zhou, S.; Chen, K.; Wu, D.; Zhao, L. Hydrogen peroxide acts upstream of nitric oxide in the heat shock pathway in Arabidopsis seedlings. *Plant Physiol.* **2014**, *164*, 2184–2196. [CrossRef]
31. Zhou, J.; Wang, J.; Shi, K.; Xia, X.J.; Zhou, Y.H.; Yu, J.Q. Hydrogen peroxide is involved in the cold acclimation-induced chilling tolerance of tomato plants. *Plant Physiol. Biochem.* **2012**, *60*, 141–149. [CrossRef] [PubMed]
32. Zandalinas, S.I.; Fichman, Y.; Devireddy, A.R.; Sengupta, S.; Azad, R.K.; Mittler, R. Systemic signaling during abiotic stress combination in plants. *Proc. Natl. Acad. Sci. USA* **2000**, *117*, 13810–13820. [CrossRef] [PubMed]
33. Mukherjee, S. Insights to nitric oxide-melatonin crosstalk and N-nitrosomelatonin functioning in plants: Where do we stand? *J. Exp. Bot.* **2019**, *70*, 6035–6047. [CrossRef] [PubMed]
34. Zhu, Y.; Gao, H.; Lu, M.; Hao, C.; Pu, Z.; Guo, M.; Hou, D.; Chen, L.Y.; Huang, X. Melatonin-nitric oxide crosstalk and their roles in the redox network in plants. *Int. J. Mol. Sci.* **2019**, *20*, 6200. [CrossRef] [PubMed]
35. Su, J.; Yang, X.; Shao, Y.; Chen, Z.; Shen, W. Molecular hydrogen-induced salinity tolerance requires melatonin signalling in Arabidopsis thaliana. *Plant Cell Environ.* **2021**, *44*, 476–490. [CrossRef] [PubMed]
36. Zhao, G.; Zhao, Y.; Yu, X.; Kiprotich, F.; Han, H.; Guan, R.; Wang, R.; Shen, W. Nitric oxide is required for melatonin-enhanced tolerance against salinity stress in rapeseed (*Brassica napus* L.) seedlings. *Int. J. Mol. Sci.* **2018**, *19*, 1912. [CrossRef]
37. Yang, N.; Sun, K.; Wang, X.; Wang, K.; Kong, X.; Gao, J.; Wen, D. Melatonin participates in selenium-enhanced cold tolerance of cucumber seedlings. *Front. Plant Sci.* **2021**, *12*, 786043. [CrossRef]
38. Jahan, M.S.; Guo, S.; Sun, J.; Shu, S.; Wang, Y.; El-Yazied, A.A.; Alabdallah, N.M.; Hikal, M.; Mohamed, M.H.M.; Ibrahim, M.F.M.; et al. Melatonin-mediated photosynthetic performance of tomato seedlings under high-temperature stress. *Plant Physiol. Biochem.* **2021**, *167*, 309–320. [CrossRef]
39. Li, C.; Tan, D.X.; Liang, D.; Chang, C.; Jia, D.; Ma, F. Melatonin mediates the regulation of ABA metabolism, free-radical scavenging, and stomatal behaviour in two Malus species under drought stress. *J. Exp. Bot.* **2015**, *66*, 669–680. [CrossRef]
40. Zheng, X.; Tan, D.X.; Allan, A.C.; Zuo, B.; Zhao, Y.; Reiter, R.J.; Wang, L.; Wang, Z.; Guo, Y.; Zhou, J.; et al. Chloroplastic biosynthesis of melatonin and its involvement in protection of plants from salt stress. *Sci. Rep.* **2017**, *7*, 41236. [CrossRef]
41. Byeon, Y.; Back, K. Molecular cloning of melatonin 2-hydroxylase responsible for 2-hydroxymelatonin production in rice (*Oryza sativa*). *J. Pineal Res.* **2015**, *58*, 343–351. [CrossRef] [PubMed]
42. Mukherjee, S.; David, A.; Yadav, S.; Baluška, F.; Bhatla, S.C. Salt stress-induced seedling growth inhibition coincides with differential distribution of serotonin and melatonin in sunflower seedling roots and cotyledons. *Physiol. Plant.* **2014**, *152*, 714–728. [CrossRef] [PubMed]
43. Ren, J.; Yang, X.; Zhang, N.; Feng, L.; Ma, C.; Wang, Y.; Yang, Z.; Zhao, J. Melatonin alleviates aluminum-induced growth inhibition by modulating carbon and nitrogen metabolism, and reestablishing redox homeostasis in *Zea mays* L. *J. Hazard. Mater.* **2022**, *423*, 127159. [CrossRef] [PubMed]
44. Iqbal, N.; Fatma, M.; Gautam, H.; Umar, S.; Khan, N.A. The crosstalk of melatonin and hydrogen sulfide determines photosynthetic performance by regulation of carbohydrate metabolism in wheat under heat stress. *Plants* **2021**, *10*, 1778. [CrossRef]
45. Wang, L.; Zhao, Y.; Reiter, R.J.; He, C.; Liu, G.; Lei, Q.; Zuo, B.; Zheng, X.; Li, Q.; Kong, J. Changes in melatonin levels in transgenic 'Micro-Tom' tomato overexpressing ovine AANAT and ovine HIOMT genes. *J. Pineal Res.* **2014**, *56*, 134–142. [CrossRef] [PubMed]
46. Yang, S.J.; Huang, B.; Zhao, Y.Q.; Hu, D.; Chen, T.; Ding, C.B.; Chen, Y.E.; Yuan, S.; Yuan, M. Melatonin enhanced the tolerance of Arabidopsis thaliana to high light through improving anti-oxidative system and photosynthesis. *Front. Plant Sci.* **2021**, *12*, 752584. [CrossRef]
47. Tan, K.; Zheng, J.; Liu, C.; Liu, X.; Liu, X.; Gao, T.; Song, X.; Wei, Z.; Ma, F.; Li, C. Heterologous expression of the melatonin-related gene HIOMT improves salt tolerance in *Malus domestica*. *Int. J. Mol. Sci.* **2021**, *22*, 12425. [CrossRef]
48. Pandey, V.; Dixit, V.; Shyam, R. Chromium effect on ROS generation and detoxification in pea (*Pisum sativum*) leaf chloroplasts. *Protoplasma* **2009**, *236*, 85–95. [CrossRef]
49. Chang, R.; Jang, C.J.; Branco-Price, C.; Nghiem, P.; Bailey-Serres, J. Transient MPK6 activation in response to oxygen deprivation and reoxygenation is mediated by mitochondria and aids seedling survival in Arabidopsis. *Plant Mol. Biol.* **2012**, *78*, 109–122. [CrossRef]

50. Corpas, F.J.; Del Río, L.A.; Palma, J.M. Plant peroxisomes at the crossroad of NO and H_2O_2 metabolism. *J. Integr. Plant Biol.* **2019**, *61*, 803–816.
51. Hofmann, A.; Wienkoop, S.; Harder, S.; Bartlog, F.; Lüthje, S. Hypoxia-responsive class III peroxidases in maize roots: Soluble and membrane-bound isoenzymes. *Int. J. Mol. Sci.* **2020**, *21*, 8872. [CrossRef] [PubMed]
52. Wang, L.; Mu, X.; Chen, X.; Han, Y. Hydrogen sulfide attenuates intracellular oxidative stress via repressing glycolate oxidase activities in Arabidopsis thaliana. *BMC Plant Biol.* **2022**, *22*, 98. [CrossRef] [PubMed]
53. Liu, J.; Shabala, S.; Zhang, J.; Ma, G.; Chen, D.; Shabala, L.; Zeng, F.; Chen, Z.H.; Zhou, M.; Venkataraman, G. Melatonin improves rice salinity stress tolerance by NADPH oxidase-dependent control of the plasma membrane K^+ transporters and K^+ homeostasis. *Plant Cell Environ.* **2020**, *43*, 2591–2605. [CrossRef] [PubMed]
54. Zahedi, S.M.; Hosseini, M.S.; Fahadi, H.N.; Gholami, R.; Abdelrahman, M.; Tran, L.P. Exogenous melatonin mitigates salinity-induced damage in olive seedlings by modulating ion homeostasis, antioxidant defense, and phytohormone balance. *Physiol. Plant.* **2021**, *173*, 1682–1694. [CrossRef]
55. Li, C.; Wang, P.; Wei, Z.; Liang, D.; Liu, C.; Yin, L.; Jia, D.; Fu, M.; Ma, F. The mitigation effects of exogenous melatonin on salinity-induced stress in *Malus hupehensis*. *J. Pineal Res.* **2012**, *53*, 298–306. [CrossRef] [PubMed]
56. Siddiqui, M.H.; Alamri, S.; Al-Khaishany, M.Y.; Khan, M.N.; Al-Amri, A.; Ali, H.M.; Alaraidh, I.A.; Alsahli, A.A. Exogenous melatonin counteracts NaCl-induced damage by regulating the antioxidant system, proline and carbohydrates metabolism in tomato seedlings. *Int. J. Mol. Sci.* **2019**, *20*, 353. [CrossRef] [PubMed]
57. Zhang, H.J.; Zhang, N.; Yang, R.C.; Wang, L.; Sun, Q.Q.; Li, D.B.; Cao, Y.Y.; Weeda, S.; Zhao, B.; Ren, S.; et al. Melatonin promotes seed germination under high salinity by regulating antioxidant systems, ABA and GA_4 interaction in cucumber (*Cucumis sativus* L.). *J. Pineal Res.* **2014**, *57*, 269–279. [CrossRef]
58. Yan, F.; Wei, H.; Ding, Y.; Li, W.; Liu, Z.; Chen, L.; Tang, S.; Ding, C.; Jiang, Y.; Li, G. Melatonin regulates antioxidant strategy in response to continuous salt stress in rice seedlings. *Plant Physiol. Biochem.* **2021**, *165*, 239–250. [CrossRef]
59. Li, J.; Liu, Y.; Zhang, M.; Xu, H.; Ning, K.; Wang, B.; Chen, M. Melatonin increases growth and salt tolerance of Limonium bicolor by improving photosynthetic and antioxidant capacity. *BMC Plant Biol.* **2022**, *22*, 16. [CrossRef]
60. Altaf, M.A.; Shahid, R.; Ren, M.X.; Naz, S.; Altaf, M.M.; Khan, L.U.; Tiwari, R.K.; Lal, M.K.; Shahid, M.A.; Kumar, R.; et al. Melatonin improves drought stress tolerance of tomato by modulating plant growth, root architecture, photosynthesis, and antioxidant defense system. *Antioxidants* **2022**, *11*, 309. [CrossRef]
61. Jafari, M.; Shahsavar, A. The effect of foliar application of melatonin on changes in secondary metabolite contents in two Citrus species under drought stress conditions. *Front. Plant Sci.* **2021**, *12*, 692735. [CrossRef] [PubMed]
62. Shamloo-Dashtpagerdi, R.; Aliakbari, M.; Lindlöf, A.; Tahmasebi, S. A systems biology study unveils the association between a melatonin biosynthesis gene, O-methyl transferase 1 (OMT1) and wheat (*Triticum aestivum* L.) combined drought and salinity stress tolerance. *Planta* **2022**, *255*, 99. [CrossRef] [PubMed]
63. Imran, M.; Latif, K.A.; Shahzad, R.; Aaqil, K.M.; Bilal, S.; Khan, A.; Kang, S.M.; Lee, I.J. Exogenous melatonin induces drought stress tolerance by promoting plant growth and antioxidant defence system of soybean plants. *AoB Plants* **2021**, *13*, plab026. [CrossRef] [PubMed]
64. Xia, H.; Ni, Z.; Hu, R.; Lin, L.; Deng, H.; Wang, J.; Tang, Y.; Sun, G.; Wang, X.; Li, H.; et al. Melatonin alleviates drought stress by a non-enzymatic and enzymatic antioxidative system in kiwifruit seedlings. *Int. J. Mol. Sci.* **2020**, *21*, 852. [CrossRef]
65. Antoniou, C.; Chatzimichail, G.; Xenofontos, R.; Pavlou, J.J.; Panagiotou, E.; Christou, A.; Fotopoulos, V. Melatonin systemically ameliorates drought stress-induced damage in *Medicago sativa* plants by modulating nitro-oxidative homeostasis and proline metabolism. *J. Pineal Res.* **2017**, *62*, e12401. [CrossRef]
66. Bajwa, V.S.; Shukla, M.R.; Sherif, S.M.; Murch, S.J.; Saxena, P.K. Role of melatonin in alleviating cold stress in Arabidopsis thaliana. *J. Pineal Res.* **2014**, *56*, 238–245. [CrossRef]
67. Li, X.; Wei, J.P.; Scott, E.R.; Liu, J.W.; Guo, S.; Li, Y.; Zhang, L.; Han, W.Y. Exogenous melatonin alleviates cold stress by promoting antioxidant defense and redox homeostasis in *Camellia sinensis* L. *Molecules* **2018**, *23*, 165. [CrossRef]
68. Han, Q.H.; Huang, B.; Ding, C.B.; Zhang, Z.W.; Chen, Y.E.; Hu, C.; Zhou, L.J.; Huang, Y.; Liao, J.Q.; Yuan, S.; et al. Effects of melatonin on anti-oxidative systems and photosystem II in cold-stressed rice seedlings. *Front. Plant Sci.* **2017**, *8*, 785. [CrossRef]
69. Marta, B.; Szafrańska, K.; Posmyk, M.M. Exogenous melatonin improves antioxidant defense in cucumber seeds (*Cucumis sativus* L.) germinated under chilling stress. *Front. Plant Sci.* **2016**, *7*, 575. [CrossRef]
70. Ding, F.; Liu, B.; Zhang, S. Exogenous melatonin ameliorates cold-induced damage in tomato plants. *Sci. Horticult.* **2017**, *219*, 264–271. [CrossRef]
71. Yu, Y.; Deng, L.; Zhou, L.; Chen, G.; Wang, Y. Exogenous melatonin activates antioxidant systems to increase the ability of rice seeds to germinate under high temperature conditions. *Plants* **2022**, *11*, 886. [CrossRef] [PubMed]
72. Imran, M.; Aaqil, K.M.; Shahzad, R.; Bilal, S.; Khan, M.; Yun, B.W.; Khan, A.L.; Lee, I.J. Melatonin ameliorates thermotolerance in soybean seedling through balancing redox homeostasis and modulating antioxidant defense, phytohormones and polyamines biosynthesis. *Molecules* **2021**, *26*, 5116. [CrossRef] [PubMed]
73. Xing, X.; Ding, Y.; Jin, J.; Song, A.; Chen, S.; Chen, F.; Fang, W.; Jiang, J. Physiological and transcripts analyses reveal the mechanism by which melatonin alleviates heat stress in *Chrysanthemum* seedlings. *Front. Plant Sci.* **2021**, *12*, 673236. [CrossRef] [PubMed]

74. Xia, H.; Zhou, Y.; Deng, H.; Lin, L.; Deng, Q.; Wang, J.; Lv, X.; Zhang, X.; Liang, D. Melatonin improves heat tolerance in *Actinidia deliciosa* via carotenoid biosynthesis and heat shock proteins expression. *Physiol. Plant.* **2021**, *172*, 1582–1593. [CrossRef]
75. Li, Z.G.; Xu, Y.; Bai, L.K.; Zhang, S.Y.; Wang, Y. Melatonin enhances thermotolerance of maize seedlings (*Zea mays* L.) by modulating antioxidant defense, methylglyoxal detoxification, and osmoregulation systems. *Protoplasma* **2019**, *256*, 471–490. [CrossRef]
76. Shi, H.; Tan, D.X.; Reiter, R.J.; Ye, T.; Yang, F.; Chan, Z. Melatonin induces class A1 heat-shock factors (HSFA1s) and their possible involvement of thermotolerance in Arabidopsis. *J. Pineal Res.* **2015**, *58*, 335–342. [CrossRef]
77. Xu, W.; Cai, S.Y.; Zhang, Y.; Wang, Y.; Ahammed, G.J.; Xia, X.J.; Shi, K.; Zhou, Y.H.; Yu, J.Q.; Reiter, R.J. Melatonin enhances thermotolerance by promoting cellular protein protection in tomato plants. *J. Pineal Res.* **2016**, *61*, 457–469. [CrossRef]
78. Posmyk, M.M.; Kuran, H.; Marciniak, K.; Janas, K.M. Presowing seed treatment with melatonin protects red cabbage seedlings against toxic copper ion concentrations. *J. Pineal Res.* **2008**, *45*, 24–31. [CrossRef]
79. Sun, C.; Lv, T.; Huang, L.; Liu, X.; Jin, C.; Lin, X. Melatonin ameliorates aluminum toxicity through enhancing aluminum exclusion and reestablishing redox homeostasis in roots of wheat. *J. Pineal Res.* **2020**, *68*, e12642. [CrossRef]
80. Nawaz, M.A.; Jiao, Y.; Chen, C.; Shireen, F.; Zheng, Z.; Imtiaz, M.; Bie, Z.; Huang, Y. Melatonin pretreatment improves vanadium stress tolerance of watermelon seedlings by reducing vanadium concentration in the leaves and regulating melatonin biosynthesis and antioxidant-related gene expression. *J. Plant Physiol.* **2018**, *220*, 115–127. [CrossRef]
81. Wang, Y.; Cheng, P.; Zhao, G.; Li, L.; Shen, W. Phytomelatonin and gasotransmitters: A crucial combination for plant physiological functions. *J. Exp. Bot.* **2022**, *17*, erac159. [CrossRef] [PubMed]
82. Tan, D.X.; Manchester, L.C.; Terron, M.P.; Flores, L.J.; Reiter, R.J. One molecule, many derivatives: A never-ending interaction of melatonin with reactive oxygen and nitrogen species? *J. Pineal Res.* **2007**, *42*, 28–42. [CrossRef] [PubMed]
83. He, H.; He, L.F. Crosstalk between melatonin and nitric oxide in plant development and stress responses. *Physiol. Plant.* **2020**, *170*, 218–226. [CrossRef] [PubMed]
84. Aghdam, M.S.; Luo, Z.; Jannatizadeh, A.; Sheikh-Assadi, M.; Sharafi, Y.; Farmani, B.; Fard, J.R.; Razavi, F. Employing exogenous melatonin applying confers chilling tolerance in tomato fruits by upregulating *ZAT2/6/12* giving rise to promoting endogenous polyamines, proline, and nitric oxide accumulation by triggering arginine pathway activity. *Food Chem.* **2019**, *275*, 549–556. [CrossRef] [PubMed]
85. Arora, D.; Bhatla, S.C. Melatonin and nitric oxide regulate sunflower seedling growth under salt stress accompanying differential expression of Cu/ZnSOD and MnSOD. *Free Radic Biol. Med.* **2017**, *106*, 315–328. [CrossRef] [PubMed]
86. Imran, M.; Shazad, R.; Bilal, S.; Imran, Q.M.; Lee, I.J. Exogenous melatonin mediates the regulation of endogenous nitric oxide in *Glycine max* L. to reduce effects of drought stress. *Environ. Exp. Bot.* **2021**, *188*, 104511. [CrossRef]
87. Pardo-Hernández, M.; López-Delacalle, M.; Rivero, R.M. ROS and NO regulation by melatonin under abiotic stress in plants. *Antioxidants* **2020**, *9*, 1078. [CrossRef]
88. Wang, T.; Song, J.; Liu, Z.; Liu, Z.; Cui, J. Melatonin alleviates cadmium toxicity by reducing nitric oxide accumulation and IRT1 expression in Chinese cabbage seedlings. *Environ. Sci. Pollut. Res. Int.* **2021**, *28*, 15394–15405. [CrossRef]
89. Zhang, J.; Zhou, M.; Zhou, H.; Zhao, D.; Gotor, C.; Romero, L.C.; Shen, J.; Ge, Z.; Zhang, Z.; Shen, W.; et al. Hydrogen sulfide, a signaling molecule in plant stress responses. *J. Integr. Plant Biol.* **2021**, *63*, 146–160. [CrossRef]
90. Chen, T.; Tian, M.; Han, Y. Hydrogen sulfide: A multi-tasking signal molecule in the regulation of oxidative stress responses. *J. Exp. Bot.* **2020**, *71*, 2862–2869. [CrossRef]
91. Jia, H.; Wang, X.; Dou, Y.; Liu, D.; Si, W.; Fang, H.; Zhao, C.; Chen, S.; Xi, J.; Li, J. Hydrogen sulfide-cysteine cycle system enhances cadmium tolerance through alleviating cadmium-induced oxidative stress and ion toxicity in *Arabidopsis* roots. *Sci. Rep.* **2016**, *6*, 39702. [CrossRef] [PubMed]
92. Huang, D.; Huo, J.; Liao, W. Hydrogen sulfide: Roles in plant abiotic stress response and crosstalk with other signals. *Plant Sci.* **2021**, *302*, 110733. [CrossRef] [PubMed]
93. Mukherjee, S.; Bhatla, S.C. Exogenous melatonin modulates endogenous H_2S homeostasis and L-cysteine desulfhydrase activity in salt-stressed tomato (*Solanum lycopersicum* L. var. cherry) seedling cotyledons. *J. Plant Growth Regul.* **2020**, *40*, 2502–2514. [CrossRef]
94. Siddiqui, M.; Khan, M.; Mukherjee, S.; Basahi, R.; Alamri, S.; Al-Amri, A.; Alsubaie, Q.; Ali, H.; Al-Munqedhi, B.; Almohisen, I. Exogenous melatonin-mediated regulation of K^+/Na^+ transport, H^+-ATPase activity and enzymatic antioxidative defence operate through endogenous hydrogen sulphide signalling in NaCl-stressed tomato seedling roots. *Plant Biol.* **2021**, *23*, 797–805. [CrossRef]
95. Sun, Y.; Ma, C.; Kang, X.; Zhang, L.; Wang, J.; Zheng, S.; Zhang, T. Hydrogen sulfide and nitric oxide are involved in melatonin-induced salt tolerance in cucumber. *Plant Physiol. Biochem.* **2021**, *167*, 101–112. [CrossRef]
96. Kaya, C.; Ashraf, M.; Alyemeni, M.N.; Ahmad, P. Responses of nitric oxide and hydrogen sulfide in regulating oxidative defence system in wheat plants grown under cadmium stress. *Physiol. Plant.* **2020**, *168*, 345–360. [CrossRef]
97. Kaya, C.; Higgs, D.; Ashraf, M.; Alyemeni, M.N.; Ahmad, P. Integrative roles of nitric oxide and hydrogen sulfide in melatonin-induced tolerance of pepper (*Capsicum annuum* L.) plants to iron deficiency and salt stress alone or in combination. *Physiol. Plant.* **2020**, *168*, 256–277. [CrossRef]

98. Chang, J.; Guo, Y.; Li, J.; Su, Z.; Wang, C.; Zhang, R.; Wei, C.; Ma, J.; Zhang, X.; Li, H. Positive interaction between H_2O_2 and Ca^{2+} mediates melatonin-induced CBF pathway and cold tolerance in watermelon (*Citrullus lanatus* L.). *Antioxidants* **2021**, *10*, 1457. [CrossRef]
99. Gong, B.; Yan, Y.; Wen, D.; Shi, Q. Hydrogen peroxide produced by NADPH oxidase: A novel downstream signaling pathway in melatonin induced stress tolerance in *Solanum lycopersicum*. *Physiol. Plant.* **2017**, *160*, 396–409. [CrossRef]
100. Zeng, H.; Bai, Y.; Wei, Y.; Reiter, R.J.; Shi, H. Phytomelatonin as a central molecule in plant disease resistance. *J. Exp. Bot.* **2022**, *17*, erac111. [CrossRef]
101. Suzuki, N.; Miller, G.; Morales, J.; Shulaev, V.; Torres, M.A.; Mittler, R. Respiratory burst oxidases: The engines of ROS signaling. *Curr. Opin. Plant Biol.* **2011**, *14*, 691–699. [CrossRef] [PubMed]
102. Chung, J.S.; Zhu, J.K.; Bressan, R.A.; Hasegawa, P.M.; Shi, H. Reactive oxygen species mediate Na-induced SOS1 mRNA stability in Arabidopsis. *Plant J.* **2008**, *53*, 554–565. [CrossRef] [PubMed]
103. Li, X.H.; Zhang, H.J.; Tian, L.M.; Huang, L.; Liu, S.X.; Li, D.Y.; Song, F.M. Tomato SlRbohB, a member of the NADPH oxidase family, is required for disease resistance against *Botrytis cinerea* and tolerance to drought stress. *Front. Plant Sci.* **2015**, *6*, 463. [CrossRef] [PubMed]
104. Li, X.; Li, Y.; Ahammed, G.J.; Zhang, X.N.; Ying, L.; Zhang, L.; Yan, P.; Zhang, L.P.; Li, Q.Y.; Han, W.Y. RBOH1-dependent apoplastic H_2O_2 mediates epigallocatechin- 3-gallate-induced abiotic stress tolerance in *Solanum lycopersicum* L. *Environ. Exp. Bot.* **2019**, *161*, 357–366. [CrossRef]
105. Liu, T.; Ye, X.L.; Li, M.; Li, J.M.; Hu, X.H. H_2O_2 and NO are involved in trehalose-regulated oxidative stress tolerance in cold-stressed tomato plants. *Environ. Exp. Bot.* **2020**, *171*, 103961. [CrossRef]
106. Xu, J.; Kang, Z.; Zhu, K.; Zhao, D.; Yuan, Y.; Yang, S.; Zhen, W.; Hu, X. RBOH1-dependent H_2O_2 mediates spermine-induced antioxidant enzyme system to enhance tomato seedling tolerance to salinity-alkalinity stress. *Plant Physiol. Biochem.* **2021**, *164*, 237–246. [CrossRef]
107. Liu, D.; Li, Y.Y.; Zhou, Z.C.; Xiang, X.; Liu, X.; Wang, J.; Hu, Z.R.; Xiang, S.P.; Li, W.; Xiao, Q.Z.; et al. Tobacco transcription factor bHLH123 improves salt tolerance by activating NADPH oxidase *NtRbohE* expression. *Plant Physiol.* **2021**, *186*, 1706–1720. [CrossRef]
108. Orman-Ligeza, B.; Parizot, B.; de Rycke, R.; Fernandez, A.; Himschoot, E.; Van Breusegem, F.; Bennett, M.J.; Périlleux, C.; Beeckman, T.; Draye, X. RBOH-mediated ROS production facilitates lateral root emergence in Arabidopsis. *Development* **2016**, *143*, 3328–3339. [CrossRef]
109. Mangano, S.; Denita-Juarez, S.P.; Choi, H.S.; Marzol, E.; Hwang, Y.; Ranocha, P.; Velasquez, S.M.; Borassi, C.; Barberini, M.L.; Aptekmann, A.A.; et al. Molecular link between auxin and ROS-mediated polar growth. *Proc. Natl. Acad. Sci. USA* **2017**, *114*, 5289–5294. [CrossRef]
110. Jiang, C.; Belfield, E.J.; Cao, Y.; Smith, J.A.; Harberd, N.P. An Arabidopsis soil-salinity-tolerance mutation confers ethylene-mediated enhancement of sodium/potassium homeostasis. *Plant Cell* **2013**, *25*, 3535–3552. [CrossRef]
111. Arnao, M.B.; Hernández-Ruiz, J. Regulatory role of melatonin in the redox network of plants and plant hormone relationship in stress. In *Hormones and Plant Response. Plant in Challenging Environments*; Gupta, D.K., Corpas, F.J., Eds.; Springer: Cham, Switzerland, 2021; Volume 2, pp. 235–272.
112. Chen, J.; Li, H.; Yang, K.; Wang, Y.; Yang, L.; Hu, L.; Liu, R.; Shi, Z. Melatonin facilitates lateral root development by coordinating PAO-derived hydrogen peroxide and Rboh-derived superoxide radical. *Free Radic. Biol. Med.* **2019**, *143*, 534–544. [CrossRef] [PubMed]
113. Bian, L.; Wang, Y.; Bai, H.; Li, H.; Zhang, C.; Chen, J.; Xu, W. Melatonin-ROS signal module regulates plant lateral root development. *Plant Signal. Behav.* **2021**, *16*, 1901447. [CrossRef] [PubMed]
114. Sun, H.; Cao, X.; Wang, X.; Zhang, W.; Li, W.; Wang, X.; Liu, S.; Lyu, D. RBOH-dependent hydrogen peroxide signaling mediates melatonin-induced anthocyanin biosynthesis in red pear fruit. *Plant Sci.* **2021**, *313*, 111093. [CrossRef] [PubMed]
115. Corpas, F.J.; Rodríguez-Ruiz, M.; Muñoz-Vargas, M.A.; González-Gordo, S.; Reiter, R.J.; Palma, J.M. Interactions of melatonin, ROS and NO during fruit ripening: An update and prospective view. *J. Exp. Bot.* **2022**, erac128. [CrossRef]
116. Wei, J.; Li, D.X.; Zhang, J.R.; Shan, C.; Rengel, Z.; Song, Z.B.; Chen, Q. Phytomelatonin receptor PMTR1-mediated signaling regulates stomatal closure in *Arabidopsis thaliana*. *J. Pineal Res.* **2018**, *65*, e12500. [CrossRef]
117. Kong, M.; Sheng, T.; Liang, J.; Ali, Q.; Gu, Q.; Wu, H.; Chen, J.; Liu, J.; Gao, X. Melatonin and its homologs induce immune responses via receptors trP47363-trP13076 in *Nicotiana benthamiana*. *Front. Plant Sci.* **2021**, *12*, 691835. [CrossRef]
118. Wang, L.F.; Li, T.T.; Zhang, Y.; Guo, J.X.; Lu, K.K.; Liu, W.C. CAND2/PMTR1 is required for melatonin-conferred osmotic stress tolerance in Arabidopsis. *Int. J. Mol. Sci.* **2021**, *22*, 4014. [CrossRef]
119. Wang, L.F.; Lu, K.K.; Li, T.T.; Zhang, Y.; Guo, J.X.; Song, R.F.; Liu, W.C. Maize PHYTOMELATONIN RECEPTOR1 functions in plant osmotic and drought stress tolerance. *J. Exp. Bot.* **2021**, erab553. [CrossRef]
120. Han, Y.; Mhamdi, A.; Chaouch, S.; Noctor, G. Regulation of basal and oxidative stress-triggered jasmonic acid-related gene expression by glutathione. *Plant Cell Environ.* **2013**, *36*, 1135–1146. [CrossRef]
121. Han, Y.; Chaouch, S.; Mhamdi, A.; Queval, G.; Zechmann, B.; Noctor, G. Functional analysis of Arabidopsis mutants points to novel roles for glutathione in coupling H2O2 to activation of salicylic acid accumulation and signaling. *Antioxid. Redox Signal.* **2013**, *18*, 2106–2121. [CrossRef]

122. Chen, J.; Zhou, H.; Xie, Y. SnRK2.6 phosphorylation/persulfidation: Where ABA and H_2S signaling meet. *Trends Plant Sci.* **2021**, *26*, 1207–1209. [CrossRef] [PubMed]
123. Zhou, M.; Zhang, J.; Shen, J.; Zhou, H.; Zhao, D.; Gotor, C.; Romero, L.C.; Fu, L.; Li, Z.; Yang, J.; et al. Hydrogen sulfide-linked persulfidation of ABI4 controls ABA responses through the transactivation of MAPKKK18 in Arabidopsis. *Mol. Plant* **2021**, *14*, 921–936. [CrossRef] [PubMed]
124. Chen, S.S.; Jia, H.L.; Wang, X.F.; Shi, C.; Wang, X.; Ma, P.; Wang, J.; Wang, M.J.; Li, J. Hydrogen sulfide positively regulates abscisic acid signaling through persulfidation of SnRK2.6 in guard cells. *Mol. Plant* **2020**, *13*, 732–744. [CrossRef] [PubMed]
125. de Bont, L.; Mu, X.; Wei, B.; Han, Y. Abiotic stress-triggered oxidative challenges: Where does H_2S act? *J. Genet. Genom.* 2022, *in press*. [CrossRef]
126. Xu, S.; Chen, T.; Tian, M.; Rahantaniaina, M.S.; Zhang, L.; Wang, R.; Xuan, W.; Han, Y. Genetic manipulation of ROS homeostasis utilizing CRISPR/Cas9-based gene editing in rice. *Methods Mol. Biol.* 2022, *in press*. [CrossRef]
127. Li, S.; Shen, P.; Wang, B.; Mu, X.; Tian, M.; Chen, T.; Han, Y. Modification of chloroplast antioxidant capacity by plastid transformation. *Methods Mol. Biol.* 2022, *in press*. [CrossRef]

International Journal of *Molecular Sciences*

MDPI

Review

Hydrogen Sulfide in Plants: Crosstalk with Other Signal Molecules in Response to Abiotic Stresses

Chunlei Wang, Yuzheng Deng, Zesheng Liu and Weibiao Liao *

College of Horticulture, Gansu Agricultural University, Lanzhou 730070, China; wangchunlei@gsau.edu.cn (C.W.); dengyz0830@163.com (Y.D.); lzs0724@163.com (Z.L.)
* Correspondence: liaowb@gsau.edu.cn; Tel.: +86-13893287942 or +86-931-7632155

Abstract: Hydrogen sulfide (H_2S) has recently been considered as a crucial gaseous transmitter occupying extensive roles in physiological and biochemical processes throughout the life of plant species. Furthermore, plenty of achievements have been announced regarding H_2S working in combination with other signal molecules to mitigate environmental damage, such as nitric oxide (NO), abscisic acid (ABA), calcium ion (Ca^{2+}), hydrogen peroxide (H_2O_2), salicylic acid (SA), ethylene (ETH), jasmonic acid (JA), proline (Pro), and melatonin (MT). This review summarizes the current knowledge within the mechanism of H_2S and the above signal compounds in response to abiotic stresses in plants, including maintaining cellular redox homeostasis, exchanging metal ion transport, regulating stomatal aperture, and altering gene expression and enzyme activities. The potential relationship between H_2S and other signal transmitters is also proposed and discussed.

Keywords: hydrogen sulfide; nitric oxide; abscisic acid; Ca^{2+}; hydrogen peroxide; abiotic stresses; signal transmitters; stomatal movement

Citation: Wang, C.; Deng, Y.; Liu, Z.; Liao, W. Hydrogen Sulfide in Plants: Crosstalk with Other Signal Molecules in Response to Abiotic Stresses. *Int. J. Mol. Sci.* **2021**, *22*, 12068. https://doi.org/10.3390/ijms222112068

Academic Editors: Yanjie Xie, Francisco J. Corpas, Jisheng Li and Jozef Kovacik

Received: 4 September 2021
Accepted: 5 November 2021
Published: 8 November 2021

1. Introduction

Several abiotic stresses such as salt, drought, flooding, heat, cold, and freezing easily result in the loss of crop production and a drop in economy in the world. Furthermore, with ongoing industrialization and pesticides application, plants are more likely subjected to some abiotic stresses including salinity and heavy metal (aluminum (Al); cadmium (Cd); chromium (Cr); lead (Pb); cobalt (Co); arsenic (As); nickel (Ni)) stresses [1,2]. In order to survive, plants must make a series of adjustments in morphology and physiological and biochemical metabolism when they are subjected to abiotic stresses. There are many kinds of mechanisms for plants to respond to abiotic stresses, including plant hormones, osmotic regulators, active oxygen scavenging systems, genes, and proteins. When plants are subjected to adversity stress, a series of changes will occur in the hormone levels, thereby initiating or regulating certain physiological and biochemical processes related to stress resistance to complete the response to adversity. Moreover, some inorganic and organic osmotic substances such as Na^+, K^+, Cl^-, proline (Pro), and soluble sugars may accumulate when plants encounter stresses. Further, under normal circumstances, the reactive oxygen species (ROS) are tightly controlled in plants, because plants have a reactive oxygen scavenging system, which keeps the production and removal of reactive oxygen species in a dynamic balance. This ROS includes hydrogen peroxide (H_2O_2), superoxide anion ($O_2{\cdot}^-$), singlet oxygen (${\cdot}O_2$), and hydroxyl radical (${\cdot}OH$) [3]. Under the condition of adversity, this balance is broken, and a large amount of active oxygen is produced. Active oxygen attacks the membrane system, causing changes in membrane lipid components and conformation of various enzymes on the membrane, loss of membrane selective permeability, leakage of electrolytes and certain small molecular organic substances, and disorder of mitochondria and chloroplast functions [1,4]. The active oxygen scavenging system mainly includes two types of substances: one is an enzymatic protection system composed of superoxide dismutase (SOD), peroxidase (POD), and catalase (CAT), etc.; the other is non-enzymatic antioxidants including reduced glutathione (GSH), carotenoids (Car), vitamin E, and other antioxidants [5]. Last but not least, some proteins such as NAC, WRKY, basic region/leucine zipper motif (bZTP), and salt overly sensitive1 (SOS1) participate in plant response to abiotic stresses [2,6].

Hydrogen sulfide (H_2S) is a colorless, combustible, and hydrosoluble gas with an obvious smell of rotten eggs, which has been widely considered as the third gasotransmitter molecule besides nitric oxide (NO) and carbon monoxide (CO) [7]. The emission of H_2S was studied a long time ago. In 1978, Wilson et al. (1978) firstly observed the emission of H_2S in the leaves of cucumber (*Cucumis sativus* L.), squash and pumpkin (*Cucurbita pepo* L.), cantaloupe (*Cucumis melo* L.), maize, soybean (*Glycine max* L. Merr), and cotton (*Gossypium hirsutum* L.) [8]. Current studies show that H_2S can be biosynthesized through a variety of enzymes such as cysteine synthase (CS), β-cyanoalanine synthase (CAS), L-cysteine desulfhydrase (LCD), D-cysteine desulfhydrase (DCD), and sulfite reductase (SiR) in mitochondria, cytosol, and chloroplast [9,10]. In mitochondria, H_2S can be produced by CAS in the course of cyanide detoxification. The generation of H_2S mainly occurs by inducing the activities of LCD and DCD from cysteine (Cys) in the cytosol, which is also accompanied by the formation of pyruvate and ammonia. SiR is the reaction catalyst in the photosynthetic sulfate-assimilation pathway which induces the release of H_2S in the chloroplast [11,12]. Thus, endogenous H_2S can be produced under the catalysis of the corresponding enzymes [8–11]. The changes in endogenous H_2S level can influence cellular metabolisms, enzyme activities, and gene expressions, and thus modulate plant growth and development [5,13]. Therefore, H_2S is widely considered as a signaling molecule within organic cells.

In the last few decades, increasing evidence has shown that H_2S plays a vital role in the treatment of diseases for animals and humans, including cancer [13], burns [14], neurodegenerative diseases [15], and inflammation [16]. In addition, it is involved in many processes of growth and development in plants. It can influence the seed germination, root organogenesis, photosynthesis, stomatal movement, leaf senescence, fruit ripening and nodulation, and nitrogen fixation [17]. H_2S can also enhance the plant's tolerance to diverse biotic and abiotic stresses, such as bacterial and fungal pathogens, salinity, drought, heat, hyperosmotic, oxidative and heavy metal stresses, etc. [5,17–19].

As a gaseous signaling molecule, H_2S can interact with other signal molecules to influence the growth and development of, and respond to abiotic stresses in, plants. Plenty of research demonstrates that H_2S is involved in NO-alleviated salt stress and heavy metal stresses in the seedling roots of pea (*Pisum sativum* L. cv. Azad P-1) and barley (*Hordeum vulgare* L.), as well as the seeds of alfalfa (*Medicago sativa* L. cv. Victoria) [20–22]. Besides, some plant hormones such as abscisic acid (ABA), salicylic acid (SA), ethylene (ETH), jasmonic acid (JA), and melatonin (MT) could alleviate abiotic stresses together with H_2S in the process of plant growth and development. Some ionic signals such as calcium ion (Ca^{2+}) and H_2S are interrelated under stresses [23]. Meanwhile, H_2O_2 and proline (Pro) have been reported to have a relationship with H_2S under abiotic stresses during the process of plant growth [17,24,25]. Here, we comprehensively review the crosstalk between H_2S and other signal molecules in response to abiotic stresses. Also, new research directions and future prospects in this area will be discussed in this review (Figure 1).

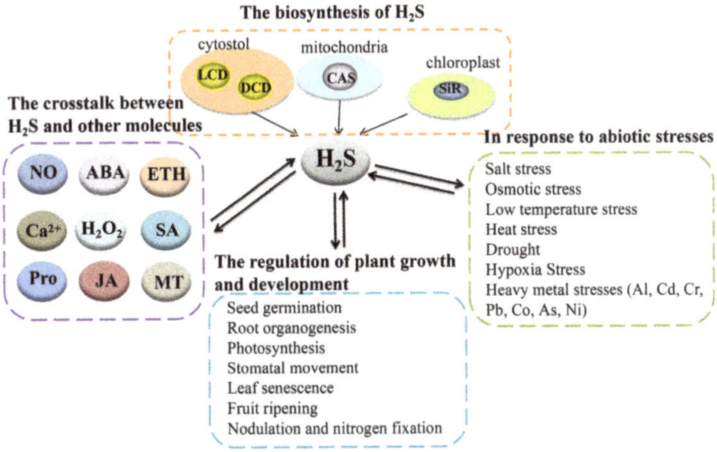

Figure 1. The summary of the biosynthesis of H_2S, the crosstalk between H_2S and other molecules, the regulation of plant growth and development, and the response to abiotic stresses by H_2S. H_2S, hydrogen sulfide; LCD, L-cysteine desulfhydrase; DCD, D-cysteine desulfhydrase; CAS, β-cyanoalanine synthase; SiR, sulfite reductase; NO, nitric oxide; ABA, abscisic acid; Ca^{2+}, calcium ion; H_2O_2, hydrogen peroxide; SA, salicylic acid; JA, jasmonic acid; Pro, proline; MT, melatonin; Al, aluminum; Cd, cadmium; Cr, chromium; Pb, lead; Co, cobalt; As, arsenic; Ni, nickel.

2. Crosstalk between H_2S and NO in Response to Abiotic Stresses

NO is widely recognized as a gas transmitter in the regulation of seed germination, dormancy, stomatal aperture, adventitious root development, and photosynthesis in plants [26,27]. NO also takes part in many stress alleviation processes, such as heavy metal, extreme temperature, drought, salt, and UV-B radiation [4,28]. Moreover, the relationship between H_2S and NO under different stress conditions has been explored at both the physiological and molecular levels, which remains a hot topic in plant science research in recently years. The obtained achievements in this field were collected and shown below.

2.1. Crosstalk between H_2S and NO in Response to Heavy Metal Stress

There is considerable research on how H_2S and NO interplay with each other in plants under heavy metal stress. In pea seedlings, As (V) reduced growth, photosynthesis capacity, and nitrogen content [29]. An application of exogenous NaHS alleviated As (V) toxicity by inducing H_2S and NO generation. These results suggest a vital role of H_2S in As (V) stress tolerance. Also, exogenous H_2S and NO could reduce the influence of Cr (VI) toxicity in maize (*Zea mays* L.) in a similar manner [30]. Furthermore, H_2S donor NaHS and NO donor sodium nitroprusside (SNP), rather than other derivatives, were found to specifically ameliorate Cd-induced oxidative damage in the root tissues of alfalfa seedlings [31]. This work further confirms that both H_2S and NO may participate in alleviating heavy metal stress. In addition, the alleviation effects of NaHS and SNP were reversed by NO scavenger 2-(4-carboxyphenyl)-4,4,5,5-tetramethylimidazoline-1-oxyl-3-oxide potassium salt (cPTIO) [31], illustrating crosstalk between H_2S and NO during the response to Cd stress. Another study in wheat (*Triticum aestivum* L.) obtained similar results that exogenous H_2S might correlate with NO to enhance Co tolerance [32]. The above studies show that H_2S may cooperate with the NO signal in managing different heavy metal stresses in plants.

The pharmacological method of introducing specific scavengers into different experimental conditions was further employed to research the relationship between H_2S and NO under heavy metal ion stress in plants. Cd stress was shown to induce a burst of endogenous NO and H_2S in bermudagrass [*Cynodon dactylon* (L). Pers.] [33]. Moreover, exogenous NO donor SNP and H_2S donor NaHS could improve Cd stress tolerance, while the positive roles of SNP and NaHS were specifically blocked by H_2S scavenger hypotaurine (HT, $C_2H_7NO_2S$), but not by NO scavenger cPTIO and H_2S inhibitors potassium pyruvate (PP, $C_3H_3KO_3$) and hydroxylamine (HA, H_3NO). PP is regarded as the substrate of dehydrogenase. H_2S could interact with the dehydrogenase. HA is an alkaline inorganic amine, which can react with the acid gas H_2S; thereby, PP and HA are able to inhibit the production of endogenous H_2S [33]. Thus, NO could activate the H_2S signal in response to Cd stress, and maybe H_2S is downstream of the NO signal. This phenomenon was further proved by the study of Al stress in soybean roots, in which NO modulated *GmMATE13* and *GmMATE47* gene expressions to enhance citrate secretion, and regulated PM H^+-ATPase activity through regulating H_2S biosynthesis and degradation [34]. H_2S and NO improved Pb tolerance in *Sesamum indicum*, while the H_2S-induced response was completely eliminated by NO scavenger cPTIO [35]. Meanwhile, only part of the effect conducted by NO was weakened by H_2S scavenger HT. It seems that NO acts downstream of H_2S or independent of H_2S in conferring plant tolerance to Pd stress. More recently, the downstream role of NO in cooperation with H_2S was also discovered in pepper (*Capsicum annuum* L.) and wheat under Cd stress [36,37]. From the numerous studies of H_2S and NO, a hypothesis may be drawn that there exists a two-side signal cascades mechanism between H_2S and NO in mediating heavy metal damage (Figure 2).

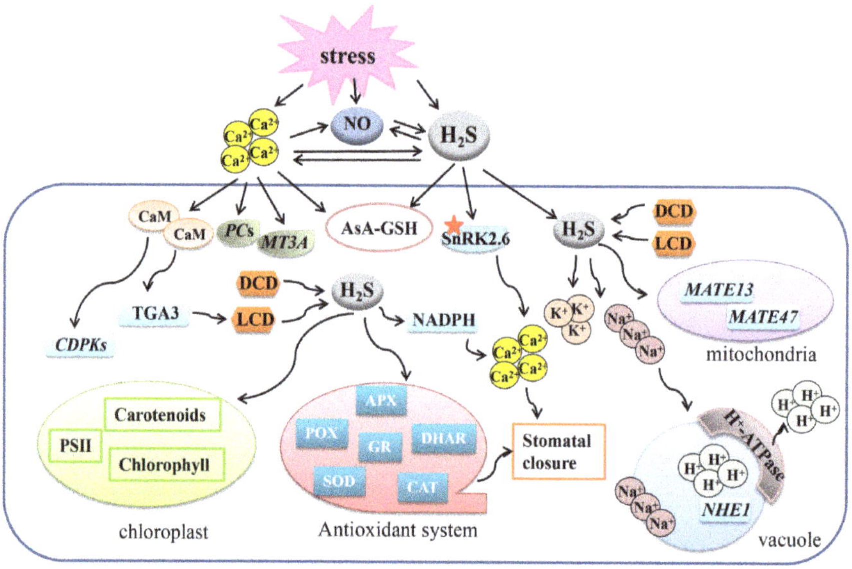

Figure 2. Overview for the mechanisms of the crosstalk between Ca^{2+}, NO, and H_2S to regulate plant response to abiotic stresses. A protein marked with a red asterisk means that the protein can be persulfided. Ca^{2+}, calcium ion; NO, nitric oxide; H_2S, hydrogen sulfide; LCD, L-cysteine desulfhydrase; DCD, D-cysteine desulfhydrase; APX, ascorbate peroxidase; SOD, superoxide dismutase; GR, glutathione reductase; POD, peroxidase; CAT, catalase; CaM, calmodulin; PCs, phytochelatin synthase; MT3A, metallothionein-like type 3; CDPKs, Ca^{2+}-dependent protein kinases; AsA-GSH, ascorbate-glutathione cycle; DHAR, dehydroascorbate reductase; POD, peroxidase; CAT, catalase.

2.2. Crosstalk between H_2S and NO in Response to Salt Stress

It has long been recognized that H_2S and NO participate in alleviating salt stress in different plant species. Salt treatment (conducted by NaCl) could increase endogenous H_2S and NO generation in the leaves of *Nicotiana tabacum* L. cv. Havana by increasing L-Cys and L-Arg contents and enhancing H_2S and NO biosynthesis enzyme activities [38].

Then, H_2S and NO help plants to cope with oxidative stress induced by salinity. These results suggest that both H_2S and NO contribute to enhancing salt tolerance. Moreover, H_2S donor NaHS and NO donor SNP relieved the inhibition of seed germination under salt stress in alfalfa through reestablishing ion homeostasis and maintaining activities of antioxidant enzymes [39]. The attenuation effect of salinity damage by H_2S was reversed by NO scavenger cPTIO, suggesting that H_2S enhanced salt tolerance through the NO pathway [39]. Another report discovered a similar relationship between H_2S and NO in rescuing salt-induced inhibition of plant growth by regulating ion homeostasis [22].

The relationship between H_2S and NO in salt resistance is still puzzled. It has been found that NO accumulation occurred ahead of H_2S, however, H_2S could not stimulate NO accumulation during the initial stage in salt-treated tomato (*Solanum lycopersicum*) roots [40]. The results above illustrate that H_2S acts downstream of NO under salt stress, and may further induce NO production to strengthen the signal cascade in a feedback manner (Figure 2). In addition, H_2S and NO may act downstream of MT to alleviate salt stress in pepper seedlings [41].

2.3. Crosstalk between H_2S and NO in Response to Other Stresses

There also exists multiple pieces of evidence that H_2S and NO cooperate with each other in heat, drought, osmotic, and flooding stresses. The pretreatment of exogenous NO enhanced the survival rate of maize seedlings under heat stress, and NO increased H_2S content [42]. Furthermore, NO-induced heat tolerance was eliminated by H_2S synthesis inhibitors and a H_2S scavenger [42], indicating that H_2S may act downstream of the NO signal in NO-induced heat tolerance. Later, another study discovered that SNP treatment facilitated the survival of submerged maize by enhancing the antioxidant system and regulating ROS content, elevating intracellular Ca^{2+} content and ADH activity, and increasing expressions of hypoxia-induced genes in maize seedling roots [43]. Moreover, SNP induced endogenous H_2S generation, and H_2S increased the NO-enhanced acquisition of tolerance to flooding-induced hypoxia in maize seedling roots [43], suggesting an analogical pattern of H_2S and NO signal cascades in relieving heat and hypoxia stresses.

H_2S may act as a downstream component of NO in ethylene-induced stomatal closure in *Vicia faba* L. [44]. Also, NO represented downstream of H_2S in ABA-triggered stomatal closure, which may suggest a paradoxical relationship between H_2S and NO under drought condition [45]. As for osmotic stress in wheat seedlings, the application of exogenous NO markedly improved H_2S synthesis enzymes LCD and DCD, as well as enhancing the activity of *O*-acetylserine (thiol)lyase (OAS-TL) to modulate Cys homeostasis [46]. On the other hand, NO scavenger cPTIO and H_2S scavenger HT invalidated the effect of NO on endogenous H_2S levels and Cys homeostasis in wheat [46]. Thus, both H_2S and NO could contribute to reinforcing osmotic tolerance and direct stomatal closure, though the concrete mechanism is largely unknown.

The H_2S donor GYY4137 released a less severe H_2S shock and a more prolonged H_2S flux; however, it decreased NO accumulation in guard cells of *A. thaliana* leaves, in accordance with another type of H_2S donor, NaHS [47]. In *Medicago sativa*, pretreatment with NOSH or NOSH-aspirin, the novel donors, which can donate NO and H_2S simultaneously to plants, could enhance plant tolerance to drought stress and improve the recovery phenotype followed by rewatering [48]. Considering the cooperative relationship between H_2S and NO, acting as signal molecules in retarding environmental damages, NOSH or NOSH-aspirin seems to be more favorable compared with NaHS and GYY4137 when used in plant guard cells, however, the effect and dosage have yet to be demonstrated (Figure 2).

3. Crosstalk between H_2S and ABA in Response to Abiotic Stresses

ABA has long been recognized as a significant phytohormone with the function of regulating plant growth, development processes, and responses to diverse environmental stresses [49]. Within drought stress, ABA may take a central role in endogenous physiological processes, including stomatal movement [50,51]. Stomata are pores of plant aerial tissues and consist of a pair of guard cells. The stomatal aperture can be modulated by these specialized cells to respond to external and internal stimuli [52]. Within the past 10 years, the research of H_2S and ABA crosstalk in augmenting plant tolerance to abiotic stresses has always come along with the regulation mechanism of stomatal movement.

3.1. Crosstalk between H_2S and ABA in Response to Abiotic Stresses through Regulating Stomatal Closure

H_2S cooperates with ABA in modulating the stomatal aperture, which has long been reported since [53] found that exogenous H_2S regulated stomatal movement and enhanced leaf relative water content (RWC) to strengthen plant drought tolerance in *Arabidopsis thaliana*. Furthermore, scavenging H_2S by HT or inhibiting H_2S biosynthesis partially blocked ABA-dependent stomatal closure through regulating ATP-binding cassette transporters [53]. Similarly, pretreatment with H_2S could considerably enhance rice's tolerance to drought stress by decreasing lipid peroxidation, maintaining antioxidant system activation, and improving ABA biosynthesis [54]. The results above affirm a role of H_2S in ABA signaling under environmental stresses. Furthermore, the stomatal aperture was enlarged in *lcd* mutant plants, causing a sensitive drought phenotype [55]. In addition, *LCD* expression and H_2S generation were down-regulated in ABA-related mutants *aba3* and *abi1*, and NaHS application increased stomatal closure in these mutants [55]. Thus, H_2S may regulate stomatal aperture in an ABA-dependent manner, and ABA may induce H_2S biosynthesis under drought stress. Simultaneously, another report revealed that pretreatment of exogenous H_2S enhanced wheat seedling tolerance to drought conditions through reinforcing antioxidant capacity [56]. Besides, the application of H_2S modulated ABA metabolic pathway genes and up-regulated ABA receptors, indicating again that H_2S alleviates drought stress, at least in part, through the ABA signaling pathway. Furthermore, exogenous ABA induced the endogenous H_2S content under drought stress [56], illustrating a complex relationship between H_2S and ABA signals in modulating drought stress.

Mitogen-activated protein kinases (MAPKs) belong to a crucial signaling molecule family which adjusts plants to multiple environmental stimuli [49]. In *A. thaliana*, drought stress fortified H_2S generation and gene expression of MAPK, however, the induced MAPK expression was abolished in H_2S synthesis double mutants *lcd des1* [57]. Further, the contributions of ABA to stomatal movements were also inhibited in *lcd des1* and *mpk4* mutants. In addition, H_2S-enhanced stomatal closure was impaired in *slac1-3* mutants [57], in which SLAC1 is an S-type anion channel that responds to ABA signaling in stomatal closure [58]. A previous report announced that H_2S could activate S-type anion currents via SLAC1 to induce stomatal closure [59]. In all, it could be proposed that H_2S is involved in ABA-stimulated stomatal closure. Thus, MPK4 may act downstream of H_2S, and H_2S-MPK4 signal cascade is involved in ABA-stimulated stomatal closure in alleviating drought stress [57].

Osmotic stress adversely causes internal environmental disorder on account of the overproduction of ROS, which leads to a decrease in plant growth and productivity. Usually, plants resist osmotic stress by enhancing the antioxidant system and stimulating signal transductions [60]. Wheat could adjust itself to resisting osmotic stress by enhancing antioxidant systems and inducing H_2S biosynthesis [61]. Furthermore, exogenous ABA induced AsA-GSH cycle activity, but H_2S scavenger HT and synthesis inhibitor aminooxy acetic acid (AOA) reversed the activities mentioned above [61]. These results suggest that H_2S induced by exogenous ABA is a signal that triggers the up-regulation of the AsA-GSH cycle under osmotic stress. Obviously, H_2S takes part in ABA-related stomatal closure in response to different environmental stresses; however, the relationship between them is complicated (Figure 3).

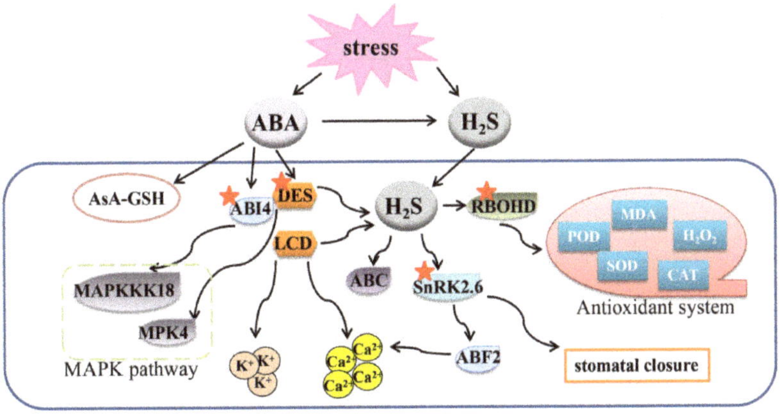

Figure 3. Overview of the mechanisms of the crosstalk between ABA and H_2S to regulate plant response to abiotic stresses. A gene or protein marked with a red asterisk means that the protein can be persulfided. H_2S, hydrogen sulfide; ABA, abscisic acid; ABF2, ABA response element-binding factor 2; AsA-GSH, ascorbate-glutathione cycle; SnRK2.6, snf1-related protein kinase 2.6; RBOHD, respiratory burst oxidase homolog protein d; MDA, malondialdehyde, ABI4, abscisic acid insensitive 4; MAPK, mitogen-activated protein kinase.

3.2. Crosstalk between DES1/H_2S and ABA in Response to Drought Stress through Regulating Protein Persulfidation

ABA could stimulate H_2S generation under stresses, but how H_2S synthesis enzyme DES1 contributes to the crosstalk between ABA and H_2S is puzzled. Recently, by creating transgenic lines that expressed *DES1* in a tissue-specific pattern, it was found that the guard cell-specific DES1 was involved in ABA-induced physiological molecular responses [62]. ABA-induced DES1 expression and H_2S production in guard cells were inhibited by H_2S scavenger and restored by H_2S donor [62]. The above genetic and pharmacological evidence further confirmed the hypothesis that DES1 is a unique component in ABA signaling in guard cells, and guard cell in situ DES1, together with H_2S, participates in ABA-guided stomatal closure [63].

Excitingly, another report discovered that the ABA signal was, in turn, commanded by H_2S-induced persulfidation of Open stomata 1 (OST1)/Snf1-related protein kinase 2.6 (SnRK2.6) on Cys131 and Cys137 residues in *A. thaliana* [64]. The persulfidated SnRK2.6 then interacted with ABA response element-binding factor 2 (ABF2), an ABA downstream protein, to modulate stomatal movement. Also, ABA was detected to induce DES1 and DCD expressions within 5–30 min previously [63,65], which suggests that the accumulation of H_2S by ABA is ahead of the occurrence of protein persulfidation. Together with the works above, a hypothesis that ABA induces H_2S accumulation, which further persulfidates SnRK2.6 continuously to promote ABA signaling in guard cells, would be proposed. The persulfidated SnRK2.6 then enhanced ABA- and H_2S-induced Ca^{2+} influx, which subsequently caused stomatal closure through the inhibition of inward K^+ channels and activation of outward anion channels [66]. To be encouraged continually, the DES1/H_2S-triggered persulfidation mechanism in ABA-regulated stomatal movement has been confirmed in another two reports [67,68]. One of their works found that ABA triggered DES1 accumulation, and DES1 auto-presulfidated at Cys44 and Cys205 in a redox-dependent fashion, causing a trigger of transient H_2S overproduction in guard cells [67]. They also found that the sustained DES1/H_2S drove persulfidation of the NADPH oxidase respiratory burst oxidase homolog protein d (RBOHD) at Cys825 and Cys890 to strengthen its ability to introduce a ROS burst, which in turn induced stomatal closure [67]. Together, this work suggests that H_2S-guided persulfidantion of DES1 and RBOHD may form a negative feedback loop that fine-tunes guard cell redox homeostasis and ABA signaling.

Abscisic acid insensitive 4 (ABI4) could also be persulfidated by DES1 at Cys250 in vitro and in vivo, and served as a downstream target of H_2S in plant's response to ABA under stress conditions [68]. In addition, DES1-linked persulfication of ABI4 induced *MPAKKK18* transactivation through binding to the CE1 motif in the *MAPKKK18* promoter, which further enlarged the MAPK signaling cascade induced by ABA. Meanwhile, ABI4 could bind to the *DES1* promoter and, in turn, activate its transcription, forming a DES1-ABI4 loop to fine-tune ABA-MAPK signals [68]. The results above illustrate a redox-based protein persulfidation mechanism within the crosstalk between H_2S- and ABA-involved stomatal movement [69]. Further work may focus on the molecular mechanisms of persulfidation and other post-translational modification events in H_2S-regulated ABA signaling in guard cells (Figure 3).

4. Crosstalk between H_2S and Ca^{2+} in Response to Abiotic Stresses

Ca^{2+} is another well-known second messenger in plant cells with the function of regulating intracellular physiological and biochemical processes, including alleviating abiotic stresses. Calmodulin (CaM) is a receptor protein in calcium signal transduction, and its main function is to perceive the volatility of intracellular calcium ions [10,70]. Recent studies uncovered a new signal transduction pattern in which Ca^{2+} and H_2S cooperate to help plants resist environmental stresses.

4.1. Crosstalk between H_2S and Ca^{2+} in Response to Heavy Metal Stress

Ca^{2+} influx was found to participate in restraining heavy metal contamination together with H_2S signal cascade. H_2S synthesis inhibitor and Ca^{2+} chelators aggravated the toxic phenotypes of foxtail millet (*Setaria italica*) exposed to Cr (VI) damage, demonstrating the involvement of H_2S and Ca^{2+} signals during this process [71]. Furthermore, Ca^{2+} enhanced the expressions of heavy metal chelator biosynthesis genes *Metallothionein-like type 3* (*MT3A*) and *Phytochelatin Synthase* (*PCS*) and activated the antioxidant system, which was partially dependent on the H_2S signal [71], indicating a downstream role of H_2S in Ca^{2+} signaling. A later report in *A. thaliana* further discovered that the expression of H_2S synthesis enzyme LCD was increased through a Ca^{2+}/calmodulin 2 (CaM2)-directed pathway, which may explain the generation of H_2S in the defense of plants against the Cr (VI) toxic condition [72,73]. The detailed mechanism was that the extracellular Cr (VI) stimulated Ca^{2+} influx, and the CaM2 protein then bound Ca^{2+} and interacted with the bZIP transcription factor TGA3, which further reinforced *LCD* gene expression and enhanced H_2S production [72]. Ca^{2+} and H_2S donor NaHS induced AsA-GSH cycle, redox homeostasis, and *Ca^{2+}-dependent protein kinase* (*CDPK*) and *Phytochelatins* (*PCs*) genes expressions under Ni toxicity in zucchini seedlings [74]. In addition, H_2S scavenger HT inhibited H_2S accumulation induced by Ca^{2+}, and Ca^{2+} chelator ethylene glycol-bis(b-aminoethylether)-N,N,N',N'-tetra-acetic acid (EGTA) eliminated the impacts of seed priming induced by NaHS [74]. Thus, Ca^{2+} and H_2S may manifest a two-side crosstalk in inoculating plants against heavy metal conditions. The relationship between NO and H_2S has been discussed in another part of the present article, and it was put forward that Ca^{2+}, in association with NO and H_2S, improved chlorophyll metabolism, photosynthesis, carbohydrate accumulation, and maintained redox homeostasis in *Vigna radiata* under Cd stress [32]. The study also discovered that NO scavenger cPTIO could reduce Ca^{2+} content, and that EGTA reduced H_2S content and altered Ca^{2+}-dependent LCD and DCD enzyme activities, but that HT could not considerably reduce Ca^{2+} content [32]. Therefore, Ca^{2+}, as a downstream signal of NO, may act in a two-side crosstalk pattern with H_2S during plants' adjustment to heavy mental contamination (Figure 2).

4.2. Crosstalk between H_2S and Ca^{2+} in the Regulation of Stomatal Closure

Stomatal closure is an important physiological process under stress conditions; thus, the role of Ca^{2+} in stomatal closure was also summarized here. As mentioned above, H_2S contributed to regulate S-type anion channel activation in guard cells, and this process was correlated with the SnRK2.6 function and the level of cytosolic free Ca^{2+} [59]. Further, H_2S induced the Ca^{2+} influx in guard cells by stimulating the accumulation of ROS [75]. H_2S triggered the persulfidation of SnRK2.6, and the persulfidated SnRK2.6 enhanced ABA- and H_2S-induced Ca^{2+} influx, which subsequently caused stomatal closure [64]. Therefore, Ca^{2+} may function downstream of H_2S-driven stomatal closure in a redox- and post-translational persulfidation-dependent manner (Figure 2).

4.3. Crosstalk between H_2S and Ca^{2+} in Response to Other Stresses

As signal messengers, the crosstalk between H_2S and Ca^{2+} has also been validated in many kinds of other stress conditions. Pretreating with H_2S enhanced the heat tolerance of tobacco (*Nicotiana tabacum* L.) suspension-cultured cells by inhibiting electrolyte leakage and MDA accumulation, and exogenous Ca^{2+} and its ionophore A23187 intensified these effects [76]. However, H_2S-induced heat tolerance was restrained by the application of Ca^{2+} chelator EGTA, as well as CaM antagonists chlorpromazine (CPZ) and trifluoperazine (TFP), illustrating a role of Ca^{2+} and CaM in H_2S-triggered heat tolerance [76]. Afterward, another study announced that exogenous H_2S enhanced the heat resistance of wheat coleoptiles through strengthening antioxidant enzyme activities in a Ca^{2+}-dependent manner [77]. Thus, Ca^{2+} and CaM participate in H_2S-induced heat tolerance in plants.

As for K^+ deficiency under NaCl stress in *Vigna radiata* seedlings, Ca^{2+} increased endogenous H_2S generation, and Ca^{2+} and H_2S then cooperated with each other to induce an Na^+/H^+ antiport system and antioxidant defense [78]. Considering another result that adding of Ca^{2+}-chelator EGTA and H_2S scavenger HT reversed the effects of Ca^{2+} [78], a hypothesis may be drawn that H_2S acts downstream during Ca^{2+}-mediated plant adaptive responses to NaCl stress (Figure 2).

5. Crosstalk between H_2S and H_2O_2 in Response to Abiotic Stresses

H_2O_2 is a colorless transparent liquid and crucial signaling molecule. Various studies have shown that H_2O_2 plays important roles in seed germination, stomatal movement, shoot and root development, pollination, and fruit ripening [79]. Also, it can modulate the plant growth and development under abiotic stresses [80]. The crosstalk between H_2S and H_2O_2 under stress has been studied in recent years.

5.1. Crosstalk between H_2S and H_2O_2 in Response to Heavy Metal Stress

Cd stress could regulate the homeostasis of ROS and promote oxidative injury, which may cause cell death [81]. Cd could decrease vacuolar H^+-ATPase activity, which was able to generate a proton gradient across the vacuolar membrane [82]. Under high Cd concentration stress, H_2O_2 and $O_2{\cdot}^-$ significantly enhanced and triggered the oxidative injury, thus resulting in cell death in *Brassica rapa* root tips [81]. However, when *B. rapa* was exposed to low concentration Cd stress, the transcript levels of H_2S biosynthesis-related genes *LCD* and *DCD* were significantly increased. Simultaneously, H_2O_2 had a remarkable increase and $O_2{\cdot}^-$ went down, whereas H_2S biosynthesis inhibitor or H_2S scavenger reversed the positive effects, indicating a role of H_2S in alleviating low Cd stress by adjusting the balance between H_2O_2 and $O_2{\cdot}^-$ [81]. H_2S donor NaHS treatment increased the photosynthetic fluorescence parameters in cotyledons of cucumber (*C. sativus* L. var. Wisconsin) seedling roots exposed to 100 μM $CdCl_2$ for 24 h [82]. In addition, both the enhancement of H_2O_2 content and the decline in H_2S content in roots decreased vacuolar H^+-ATPase activity under Cd stress. Further, the increase in H_2S content in root tissue by exogenous H_2O_2 had nothing to do with the desulfurization enzyme activity. Exogenous H_2S remarkably enhanced the NADPH oxidase activity and the relative gene expression; however, it did not have an effect on the accumulation of H_2O_2 in cucumber roots under Cd stress [82]. Hence, H_2S content might be partially enhanced through the H_2O_2/NADPH oxidase-induced pathway, independent of desulfhydrase activity (Figure 4).

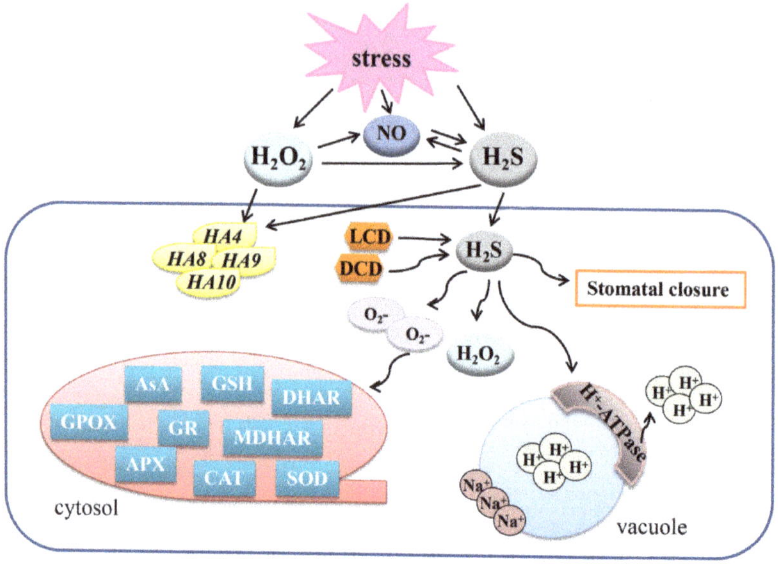

Figure 4. Overview of the mechanisms of the crosstalk between H_2S, NO, and H_2O_2 to regulate plant response to abiotic stresses. H_2S, hydrogen sulfide; H_2O_2, hydrogen peroxide; NO, nitric oxide; LCD, L-cysteine desulfhydrase; DCD, D-cysteine desulfhydrase; AsA, ascorbic acid; GSH, glutathione; GR, glutathione reductase; APX, ascorbate peroxidase; GPOX, guaiacol peroxidase; CAT, catalase; SOD, superoxide dismutase; DHAR, dehydroascorbate reductase; MDHAR, monodehydroascorbate reductase.

5.2. Crosstalk between H_2S and H_2O_2 in Response to Salt Stress

H_2S donor NaHS could enhance the activity of PM H^+-ATPase under salt or low-temperature stress in cucumber, and the transcript levels of the plasma membrane proton pump-related genes including *CsHA2*, *CsH4*, *CsH8*, *CsH9*, and *CsHA10* were also increased [83]. However, NO and H_2O_2 only enhanced the expression of *CsHA1*. Therefore, H_2S, NO, and H_2O_2 could resist the salt stress by regulating the plasma membrane proton pump at different standards. Usually, salt stress could induce stomata closure. However, the H_2S scavengers HT, AOA, hydroxylamine (NH_2OH), potassium pyruvate ($C_3H_3KO_3$), ammonia (NH_3), H_2O_2, ascorbic acid (AsA), CAT, and diphenyl iodide (DPI) suppressed the closure of stomata in *V. faba* L. [44], suggesting that both H_2S and H_2O_2 could regulate stomatal movement under salt stress. Furthermore, endogenous H_2S and H_2O_2 accumulation and the activities of LCD and DCD were enhanced by salt treatment in guard cells. Nevertheless, these effects were inhibited by H_2O_2 and H_2S scavengers. Exogenous H_2O_2 scavengers prevented the increase in endogenous H_2S level as well as the stomatal closure; however, H_2O_2 generation was barely influenced with the application of H_2S scavengers in guard cells responding to salt stress [44]. Hence, H_2S may act as the downstream of H_2O_2-alleviated salt stress (Figure 4).

5.3. Crosstalk between H_2S and H_2O_2 in Response to Drought Stress

Drought stress is one of the most serious abiotic stresses in the world. Treatment by spermidine (Spd) remarkably enhanced H_2S production and activities of antioxidant enzymes [SOD, CAT, guaiacol peroxidase (GPOX), APX, GR, dehydroascorbate reductase (DHAR), and monodehydroascorbate reductase (MDHAR)] in white clover (*Trifolium repens*) under dehydration conditions [84]. Furthermore, NO and H_2S scavengers could not reduce the generation of H_2O_2 induced by Spd, but H_2O_2 scavengers could effectively inhibit the increase of NO and H_2S induced by Spd. The H_2S signal induced by Spd was also significantly inhibited by NO scavenger [84]. Hence, in response to dehydration, H_2S may be the downstream signaling molecule to interact with NO and H_2O_2 (Figure 4).

5.4. Crosstalk between H_2S and H_2O_2 in Response to Other Stresses

UV-B is a common stress in practical agricultural production. When plants encounter the UV-B stress, the levels of electrolyte leakage, MDA, and ultraviolet absorbing compounds decreased, and the activities of antioxidant enzymes, GSH, and AsA also declined [85]. However, exogenous H_2S, H_2O_2, and putrescence (Put) could alleviate the negative effects of UV-B stress. The protective role of Put in UV-B radiation damage was reduced by the inhibitors of H_2S, H_2O_2, and Put [86]. Moreover, the level of H_2O_2 was increased by exogenous H_2S, and the enhanced H_2O_2 promoted the accumulation of UV absorbing compounds in hulless barley (*H. vulgare* L. var. nude, Kunlun-12) seedlings, thus preserving the steady state of oxidation-reduction under UV-B stress and improving its UV-B tolerance [86].

In addition, extreme temperature is a key factor which influences plant growth and development. H_2S, NO, and H_2O_2 had a significant impact in response to low temperature (10 °C) by modulating the plasma membrane proton pump in cucumber roots [83]. Moreover, H_2O_2 treatment could improve the heat resistance in maize (*Z. mays* L., Huidan No. 4) seedlings, and this effect could be strengthened by NO and H_2S donors but abolished by NO and H_2S scavengers or synthesis inhibitors [87]. It seems that NO and H_2S act downstream of H_2O_2 in the acquisition of heat resistance in plants (Figure 4).

6. Crosstalk between H_2S and Other Signal Molecules in Response to Abiotic Stresses

In recent years, many kinds of signal transmitters have emerged to regulate plant growth and development, and to acclimate to environment changes. The protective role of H_2S related to these signal molecules such as SA, ETH, JA, Pro, and MT (mentioned in another part of the article) under toxic environment in plants has also been explored to some extent.

6.1. Crosstalk between H_2S and SA in Response to Abiotic Stresses

SA has long been recognized as a pivotal signal messenger, manifesting multiple functions in defending plant disease and adverse environmental conditions. Endogenous SA biosynthesis is mainly proceeded in the cytoplasm through the phenylalanine route by phenylalanine ammonia lyase (PAL) and benzoic-acid-2-hydroxylase (BA2H) [10,88,89]. SA and H_2S enhanced heat tolerance by strengthening the activities of antioxidant enzymes and increasing osmolyte content in maize seedlings [90]. Further, SA induced endogenous H_2S generation by enhancing the activity of H_2S synthesis enzyme DES [91]. While the increase in SA production and the relative enzyme activities of PAL and BA2H were rarely influenced by H_2S, this downstream role of H_2S in SA-induced stress responses was also similarly reported in Cd tolerance in *A. thaliana* [92]. Thus, the positive role of SA under the stress condition is partially dependent on H_2S. Pb stress accelerated endogenous H_2S production [35]. Moreover, SA improved enzyme activities of the AsA-GSH cycle system in pepper under Pb stress [93]. In addition, exogenous SA enhanced the H_2S content, which was further reinforced by H_2S donor NaHS. It seems that SA triggers endogenous H_2S accumulation, which further regulates the AsA-GSH cycle to resist Pb toxicity (Figure 5).

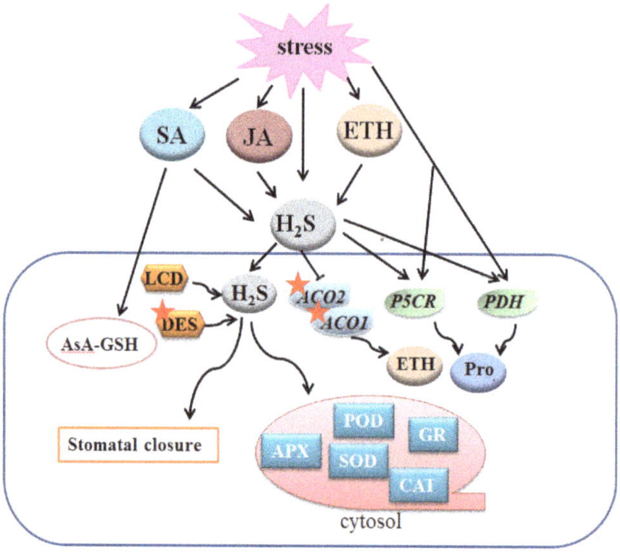

Figure 5. Overview of the mechanisms of the crosstalk between H_2S and JA, SA, ETH, and Pro to regulate plant response to abiotic stresses. A gene or protein marked with a red asterisk means that the protein can be persulfided. H_2S, hydrogen sulfide; LCD, L-cysteine desulfhydrase; DES, desulfhydrase; SA, salicylic acid; JA, jasmonic acid; ETH, ethylene; Pro, proline; APX, ascorbate peroxidase; SOD, superoxide dismutase; GR, glutathione reductase; POD, peroxidase; CAT, catalase; ACO1, 1-aminocyclopropane-1-carboxylic acid oxidase 1; ACO2, 1-aminocyclopropane-1-carboxylic acid oxidase 2; P5CR, proline-5-carboxylate reductase; PDH, proline dehydrogenase.

6.2. Crosstalk between H_2S and ETH in Response to Abiotic Stresses

Ethylene induced H_2S biosynthesis in guard cells in tomatoes under osmotic stress [94]. Moreover, the effect of ethylene on resisting osmotic stress was reversed by H_2S scavenger HT or H_2S synthetic inhibitor PAG, suggesting a downstream component of H_2S in ethylene-triggered stomatal closure under osmotic stress. Further, H_2S induced the persulfidation of 1-aminocyclopropane-1-carboxylic acid oxidase1 (ACO1) and ACO2, and restrained their expressions. As a result, H_2S negatively regulated ethylene generation in response to osmotic stress [94]. These results are parallel with a recently published mechanism of waterlogging damage resistance in peach (*Prunus persica* L. Batsch) seedlings [95], in which H_2S restrained over-synthesis of ethylene as well as inhibited oxidative damage under waterlogging stress (Figure 5).

6.3. Crosstalk between H_2S and JA in Response to Abiotic Stresses

JA is another phytohormone kind signal transmitter with extensive modulation functions in plant root elongation [96], anthocyanin accumulation and trichome initiation [97], stamen development and flowing [98], leaf senescence [99], and stress resistance [100]. A recent study announced a critical role of JA in inhibiting stomatal development in *A. thaliana* [101]. Furthermore, JA positively modified LCD activity and H_2S production. The JA-deficient mutants represented a high stomatal density phenotype, which could be reversed by exogenous H_2S, whereas the H_2S synthesis-deficient mutants *lcd* displayed similar stomatal development phenotype as the JA-deficient mutants, which could be rescued by H_2S donor NaHS but not by JA [102]. Thus, H_2S may act as a downstream member of JA in stomatal development (Figure 5).

6.4. Crosstalk between H_2S and Pro in Response to Abiotic Stresses

Pro is a kind of organic osmolyte with a wide distribution in plant cells. Previous studies have demonstrated the increase of Pro after the application of signal transmitter agents in defense of abiotic stresses [89,103,104]. Pretreatment with exogenous H_2S increased endogenous Pro content, and the activities and transcription levels of proline-5-carboxylate reductase (P5CR) and proline dehydrogenase (PDH) in foxtail millet, whereas H_2S scavenger or inhibitor reduced the above effects [105]. Moreover, the combined application of H_2S and Pro resulted in preferable growth status, stomatal movement, and oxidative remission under stress conditions. These results indicate a cooperation of Pro and H_2S under adverse environments (Figure 5).

7. Conclusions and Outlook

The disadvantageous environment conditions cause oxidative damage, ionic imbalance, and osmotic stress to plants, resulting in a weakened growth and development status. H_2S can reinforce plant tolerance to these stresses through constructing a luxuriant crosstalk with other signal molecules, such as NO, ABA, Ca^{2+}, H_2O_2, SA, ETH, JA, Pro, and MT. The genes regulated by H_2S and other molecules under abiotic stress conditions are displayed in Table 1. There exists a legible clue that environmental stresses and various signal transmitters stimulate endogenous H_2S generation and improve the activities of H_2S synthesis enzymes under the stress condition. Meanwhile, H_2S represents a feedback manner to enhance the signal cascades in inducing the accumulation of some signal messengers, especially NO, ABA, and Ca^{2+}. In addition, the existence of DES1-related auto-persulfidation and persulfidation may be the reason for the extensive inspiration of its enzyme activity in different stress conditions. In summary, H_2S acts as a downstream signal member in cooperation with ABA, H_2O_2, SA, ETH, JA, and MT, but an upstream signal member of Pro under stress condition. Nevertheless, the crosstalk between H_2S, NO, and Ca^{2+} represents a two-side signal cascades manner, whereas relationships between H_2S and other signal molecules vary on account of the specific stress pattern.

Multiple types of research need to be done to explore the point-to-point mechanism within the crosstalk between H_2S and one single signal transducer under abiotic stress conditions. Firstly, the feedback molecular mechanism of H_2S and NO, and the interactions within protein persulfidation, *S*-sulfhydration, and *S*-nitrosylation, remain unclear. Next, more post-translational modification proteins need to be discovered and identified that are triggered by H_2S in ABA- or NO-dependent signal pathways under stress condition. Finally, new signal messengers related to H_2S activity are waiting to be discovered.

Table 1. Genes regulated by H_2S and other molecules under abiotic stress conditions.

Crosstalk between H_2S and other Molecules	Stresses	Plant Species	Tissue	Regulated Genes	References
H_2S and NO	salt stress	*Medicago sativa*	seeds	*APX-1*, *APX-2*, and *Cu/Zn-SOD*	[39]
		Hordeum vulgare L.	seedlings	*HvHA*, *HvVHA-β*, *HvSOS1*, *HvVNHX2*, *HvAKT1* and *HvHAK4*	[22]
		Solanum lycopersicum	seedlings	*SlL-DES*, *SlCAS* and *SlCS*	[40]
	drought	*M. sativa* L.	leaves	*GST17*, *Cu/ZnSOD*, *FeSOD*, *NR*, c*APX*, *PIP*	[48]
	hypoxia stress	*Zea mays* L.	seedlings	*P4H*, *ADH*, *CRT1*, *GS*, *CYP51* and *ME*	[43]
	cadmium stress	*M. sativa* L.	seedlings	*Cu/Zn–SOD*, *APX* and *POD*	[31]
	cobalt stress	*Triticum aestivum* L.	seedlings	*RbcL*	[32]
	aluminum stress	*Glycine max* L.	seedlings	*MATE13*, *MATE47*, *MATE58*, *MATE74*, *MATE79*, *MATE84*, and *MATE87*	[34]
H_2S and ABA	drought	*Oryza sativa* L.	seedlings	*NCED2*, *NCED3*, *NCED5*, *AREB1*, *AREB8*, *bZIP23* and *LEA3*	[54]
		Arabidopsis	seedlings	*TPC1*, *GORK*, *SKOR*, *KCO1*, *MYP5*, *ACA9*, *ACA11*, *CAX1*, *SLAC1*, *AKT1A*, *KT2*, *KC1* and *KAT1*	[55]
		T. aestivum L.	leaves and roots	*TaZEP*, *TaNCED*, *TaAAO* and *TaSDR*	[56]
		Arabidopsis thaliana	-	*MAPKs*	[57]
	chromium stress	*A. thaliana*	seedlings	*LCD*	[72]
	nickel stress	*Cucurbita pepo* L.	seedlings	*CDPK* and *PCS1*	[74]
H_2S and Ca^{2+}	chromium stress	*Setaria italica*	seedlings	*MT3A*, *PCS*, *CaM*, *CBL* and *CDPK*	[71]
H_2S-H_2O_2	cadmium stress	*Brassica rapa.*	seedlings	*Br_UPB1A*, *Br_UPB1B*↑; *Bra035235*, *Bra033551*, *Bra006423*, ra023639	[89]
	cadmium stress	*Cucumis sativus* L.	roots	*CsVHA-A*, *CsVHA-B*, *CsVHA-a1*, *CsVHA-a2*, *CsVHA-a3*, *CsVHA-c1*, *CsVHA-c2* and *CsVHA-c3*	[82]

Table 1. *Cont.*

Crosstalk between H_2S and other Molecules	Stresses	Plant Species	Tissue	Regulated Genes	References
H2S, NO and H_2O_2	salt or low temperature	*C. sativus* L.	roots	*CsHA1, CsHA2, CsH4, CsH8, CsH9* and *CsHA10*	[83]
	dehydration	*Trifolium repens*	seedlings	*bZIP37, bZIP107, DREB2, DREB4* and *WRKY108715*	[84]
H_2S and ETH	osmotic stress	*S. lycopersicum*	seedlings	*LeACO1 and LeACO2*	[94]
H_2S and Pro	cadmium stress	Foxtail millet	seedlings	*PDH* and *P5CR*	[105]

APX, ascorbate peroxidase; SOD, superoxide dismutase; HA, H^+-ATPase; VNHX2, vacuolar Na^+/H^+ antiporter; VHA-β, H^+-ATPase subunit β; HAK4, high-affinity K^+ uptake system; L-DES, L-cysteine desulfhydrase; CAS, β-cyanoalanine synthase; CS, L-cysteine synthase; P4H, prolyl 4-hydroxylase; ADH, alcohol dehydrogenase; CRT1, calcium binding protein; CYP51, cytochrome P450 14a-sterol demethylase; GS, glutamate synthase 1; ME, NADP-dependent malic enzyme; POD, peroxidase; rbcL, rubisco large subunit; NCED, 9′-cis-epoxycarotenoid dioxygenase; TPC1, two pore segment channel 1; GORK, guard cell outward-rectifying Kþ channel; SKOR, SKI family transcriptional corepressor; KCO, outward-rectifying K^+ channel; ACA, adenylyl cyclase-associated protein; CAX, calcium exchanger; SLAC1, slow anion channel associated 1; AKT, *Arabidopsis* potassium transporter; KC1, potassium channel 1; KAT1, potassium channel in *Arabidopsis thaliana* 1; ZEP, zeaxanthin epoxidase; AAO, abscisic aldehyde oxidase; SDR, short-chain dehydrogenase; MAPK, mitogen-activated protein kinase; LCD, L-cysteine desulfhydrase; CDPK, Ca^{2+}-dependent protein kinase; PCS, phytochelatin; CaM, calmodulin; CBL, calcineurin B-like; ACO, 1-aminocyclopropane-1-carboxylic oxidase; PDH, proline dehydrogenase; P5CR, proline-5-carboxylate reductase.

Author Contributions: Conceptualization W.L. and C.W.; Collection and analysis of bibliography, Z.L. and Y.D.; Writing original draft, W.L., C.W., Y.D., and Z.L. All authors have read and agreed to the published version of the manuscript.

Funding: This research was funded by the Scientific research start-up funds for openly-recruited doctors (GAU-KYQD-2017RCZX-29); the National Natural Science Foundation of China (Nos. 32102370, 32072559, 31860568, 31560563, and 31160398); the National Key Research and Development Program of China (Grant No. 2018YFD1000800); the Natural Science Foundation of Gansu Province, China (No. 20JR5RA027); the Fuxi Young Talents Fund of Gansu Agricultural University (No. Gaufx-03Y07).

Conflicts of Interest: The authors declare no conflict of interest.

References

1. Kazan, K. Diverse roles of jasmonates and ethylene in abiotic stress tolerance. *Trends Plant Sci.* **2015**, *20*, 219–229. [CrossRef]
2. Emamverdian, A.; Ding, Y.; Mokhberdoran, F.; Xie, Y. Heavy metal stress and some mechanisms of plant defense response. *Sci. World J.* **2015**, *2015*, 756120. [CrossRef] [PubMed]
3. Choudhury, F.K.; Rivero, R.M.; Blumwald, E.; Mittler, R. Reactive oxygen species, abiotic stress and stress combination. *Plant J.* **2017**, *90*, 856–867. [CrossRef]
4. Devireddy, A.R.; Zandalinas, S.I.; Fichman, Y.; Mittler, R. Integration of reactive oxygen species and hormone signaling during abiotic stress. *Plant J.* **2021**, *105*, 459–476. [CrossRef] [PubMed]
5. Fotopoulos, V.; Christou, A.; Manganaris, G.A. Hydrogen sulfide as a potent regulator of plant responses to abiotic stress factors. *Mol. Approaches Plant Abiotic Stress* **2013**, 353–373. [CrossRef]
6. Li, W.; Pang, S.; Lu, Z.; Jin, B. Function and mechanism of WRKY transcription factors in abiotic stress responses of plants. *Plants* **2020**, *9*, 1515. [CrossRef] [PubMed]
7. Calderwood, A.; Kopriva, S. Hydrogen sulfide in plants: From dissipation of excess sulfur to signaling molecule. *Nitric oxide.* **2014**, *41*, 72–78. [CrossRef]
8. Wilson, L.G.; Bressan, R.A.; Filner, P. Light-dependent Emission of Hydrogen Sulfide from Plants. *Plant Physiol.* **1978**, *61*, 184–189. [CrossRef]
9. Papenbrock, J.; Riemenschneider, A.; Kamp, A.; Schulz-Vogt, H.N.; Schmidt, A. Characterization of Cysteine-Degrading and H_2S-Releasing Enzymes of Higher Plants-From the Field to the Test Tube and Back. *Plant Biol.* **2007**, *9*, 582–588. [CrossRef]
10. Huang, D.; Huo, J.; Liao, W. Hydrogen sulfide: Roles in plant abiotic stress response and crosstalk with other signals. *Plant Sci.* **2021**, *302*, 110733. [CrossRef]
11. Takahashi, H.; Kopriva, S.; Giordano, M.; Saito, K.; Hell, R. Sulfur assimilation in photosynthetic organisms: Molecular functions and regulations of transporters and assimilatory enzymes. *Annu. Rev. Plant Biol.* **2011**, *62*, 157–184. [CrossRef] [PubMed]
12. Singh, S.; Kumar, V.; Kapoor, D.; Kumar, S.; Singh, S.; Dhanjal, D.S.; Datta, S.; Samuel, J.; Dey, P.; Wang, S.; et al. Revealing on hydrogen sulfide and nitric oxide signals coordination for plant growth under stress conditions. *Physiol. Plant.* **2019**, *168*, 301–317. [CrossRef] [PubMed]
13. Lee, Z.W.; Deng, L.W. Role of H_2S Donors in Cancer Biology. *Handb. Exp. Pharmacol.* **2015**, *230*, 243–265. [CrossRef] [PubMed]
14. Akter, F. The role of hydrogen sulfide in burns. *Burns* **2016**, *42*, 519–525. [CrossRef]
15. Tabassum, R.; Jeong, N.; Jung, J. Therapeutic importance of hydrogen sulfide in age-associated neurodegenerative diseases. *Neural. Regen. Res.* **2020**, *15*, 653–662. [CrossRef]
16. Bhatia, M. H_2S and inflammation: An overview. *Handb. Exp. Pharmacol.* **2015**, *230*, 165–180. [CrossRef]

17. Mukherjee, S.; Corpas, F.J. Crosstalk among hydrogen sulfide (H_2S), nitric oxide (NO) and carbon monoxide (CO) in root-system development and its rhizosphere interactions: A gaseous interactome. *Plant Physiol. Bioch.* **2020**, *155*, 800–814. [CrossRef]
18. Aroca, A.; Gotor, C.; Romero, L.C. Hydrogen Sulfide Signaling in Plants: Emerging Roles of Protein Persulfidation. *Front. Plant Sci.* **2018**, *9*, 1369. [CrossRef]
19. Zhang, J.; Zhou, H.; Zhou, M.; Ge, Z.; Zhang, F.; Foyer, C.H.; Yuan, X.; Xie, Y. The coordination of guard-cell autonomous ABA synthesis and DES1 function in situ regulates plant water deficit responses. *J. Adv. Res.* **2020**, *27*, 191–197. [CrossRef]
20. Singh, V.P.; Tripathi, D.K.; Fotopoulos, V. Hydrogen sulfide and nitric oxide signal integration and plant development under stressed/non-stressed conditions. *Physiol. Plant.* **2020**, *168*, 239–240. [CrossRef] [PubMed]
21. Corpas, F.J.; Barroso, J.B.; González-Gordo, S.; Muñoz-Vargas, M.A.; Palma, J.M. Hydrogen sulfide: A novel component in *Arabidopsis* peroxisomes which triggers catalase inhibition. *J. Integr. Plant Biol.* **2019**, *61*, 871–883. [CrossRef] [PubMed]
22. Chen, J.; Wang, W.H.; Wu, F.H.; He, E.M.; Liu, X.; Shangguan, Z.P.; Zheng, H.L. Hydrogen sulfide enhances salt tolerance through nitric oxide-mediated maintenance of ion homeostasis in barley seedling roots. *Sci. Rep.* **2015**, *5*, 12516. [CrossRef] [PubMed]
23. Xuan, L.; Li, J.; Wang, X.; Wang, C. Crosstalk between hydrogen sulfide and other signal molecules regulates plant growth and development. *Int. J. Mol. Sci.* **2020**, *21*, 4593. [CrossRef]
24. Ahmad, R.; Ali, S.; Rizwan, M.; Dawood, M.; Farid, M.; Hussain, A.; Wijaya, L.; Alyemeni, M.N.; Ahmad, P. Hydrogen sulfide alleviates chromium stress on cauliflower by restricting its uptake and enhancing antioxidative system. *Physiol. Plant.* **2020**, *168*, 289–300. [CrossRef] [PubMed]
25. Zhang, Y.; Cheng, P.; Wang, Y.; Li, Y.; Su, J.; Chen, Z.; Yu, X.; Shen, W. Genetic elucidation of hydrogen signaling in plant osmotic tolerance and stomatal closure via hydrogen sulfide. *Free Radic. Bio. Med.* **2020**, *161*, 1–14. [CrossRef]
26. Deng, Y.; Wang, C.; Wang, N.; Wei, L.; Li, W.; Yao, Y.; Liao, W. Roles of Small-Molecule Compounds in Plant Adventitious Root Development. *Biomolecules* **2019**, *9*, 420. [CrossRef]
27. Corpas, F.J. Nitric oxide and hydrogen sulfide in higher plants under physiological and stress conditions. *Antioxidants* **2019**, *8*, 457. [CrossRef]
28. Palma, J.M.; Mateos, R.M.; López-Jaramillo, J.; Rodríguez-Ruiz, M.; González-Gordo, S.; Lechuga-Sancho, A.M.; Corpas, F.J. Plant catalases as NO and H_2S targets. *Redox Biol.* **2020**, *34*, 101525. [CrossRef] [PubMed]
29. Singh, V.P.; Singh, S.; Kumar, J.; Prasad, S.M. Hydrogen sulfide alleviates toxic effects of arsenate in pea seedlings through up-regulation of the ascorbate-glutathione cycle: Possible involvement of nitric oxide. *J. Plant Physiol.* **2015**, *181*, 20–29. [CrossRef]
30. Kharbech, O.; Houmani, H.; Chaoui, A.; Corpas, F.J. Alleviation of Cr(VI)-induced oxidative stress in maize (*Zea mays* L.) seedlings by NO and H_2S donors through differential organ-dependent regulation of ROS and NADPH-recycling metabolisms. *J. Plant Physiol.* **2017**, *219*, 71–80. [CrossRef]
31. Li, L.; Wang, Y.; Shen, W. Roles of hydrogen sulfide and nitric oxide in the alleviation of cadmium-induced oxidative damage in alfalfa seedling roots. *Biometals* **2012**, *25*, 617–631. [CrossRef]
32. Ozfidan-Konakci, C.; Yildiztugay, E.; Elbasan, F.; Kucukoduk, M.; Turkan, I. Hydrogen sulfide (H_2S) and nitric oxide (NO) alleviate cobalt toxicity in wheat (*Triticum aestivum* L.) by modulating photosynthesis, chloroplastic redox and antioxidant capacity. *J. Hazard. Mater.* **2020**, *388*, 122061. [CrossRef]
33. Shi, H.; Ye, T.; Chan, Z. Nitric oxide-activated hydrogen sulfide is essential for cadmium stress response in bermudagrass (*Cynodon dactylon* (L). Pers.). *Plant Physiol. Bioch.* **2014**, *74*, 99–107. [CrossRef] [PubMed]
34. Wang, H.; Ji, F.; Zhang, Y.; Hou, J.; Liu, W.; Huang, J.; Liang, W. Interactions between hydrogen sulphide and nitric oxide regulate two soybean citrate transporters during the alleviation of aluminium toxicity. *Plant Cell Environ.* **2019**, *42*, 2340–2356. [CrossRef]
35. Amooaghaie, R.; Zangene-Madar, F.; Enteshari, S. Role of two-sided crosstalk between NO and H_2S on improvement of mineral homeostasis and antioxidative defense in Sesamum indicum under lead stress. *Ecotox. Environ. Saf.* **2017**, *139*, 210–218. [CrossRef] [PubMed]
36. Kaya, C.; Akram, N.A.; Ashraf, M.; Alyemeni, M.N.; Ahmad, P. Exogenously supplied silicon (Si) improves cadmium tolerance in pepper (*Capsicum annuum* L.) by up-regulating the synthesis of nitric oxide and hydrogen sulfide. *J. Biotechnol.* **2020**, *316*, 35–45. [CrossRef] [PubMed]
37. Kaya, C.; Ashraf, M.; Alyemeni, M.N.; Ahmad, P. Responses of nitric oxide and hydrogen sulfide in regulating oxidative defence system in wheat plants grown under cadmium stress. *Physiol. Plant.* **2019**, *168*, 345–360. [CrossRef] [PubMed]
38. Da Silva, C.J.; Batista Fontes, E.P.; Modolo, L.V. Salinity-induced accumulation of endogenous H_2S and NO is associated with modulation of the antioxidant and redox defense systems in Nicotiana tabacum L. cv. Havana. *Plant Sci.* **2017**, *256*, 148–159. [CrossRef] [PubMed]
39. Wang, Y.; Li, L.; Cui, W.; Xu, S.; Shen, W.; Wang, R. Hydrogen sulfide enhances alfalfa (*Medicago sativa*) tolerance against salinity during seed germination by nitric oxide pathway. *Plant Soil* **2012**, *351*, 107–119. [CrossRef]
40. Da-Silva, C.J.; Mollica, D.C.F.; Vicente, M.H.; Peres, L.E.P.; Modolo, L.V. NO, hydrogen sulfide does not come first during tomato response to high salinity. *Nitric Oxide* **2018**, *76*, 164–173. [CrossRef]
41. Kaya, C.; Higgs, D.; Ashraf, M.; Alyemeni, M.N.; Ahmad, P. Integrative roles of nitric oxide and hydrogen sulfide in melatonin-induced tolerance of pepper (*Capsicum annuum* L.) plants to iron deficiency and salt stress alone or in combination. *Physiol. Plant.* **2020**, *168*, 256–277. [CrossRef] [PubMed]
42. Li, Z.; Yang, S.; Long, W.; Yang, G.; Shen, Z. Hydrogen sulphide may be a novel downstream signal molecule in nitric oxide-induced heat tolerance of maize (*Zea mays* L.) seedlings. *Plant Cell Environ.* **2013**, *36*, 1564–1572. [CrossRef] [PubMed]

43. Peng, R.; Bian, Z.; Zhou, L.; Cheng, W.; Hai, N.; Yang, C.; Yang, T.; Wang, X.; Wang, C. Hydrogen sulfide enhances nitric oxide-induced tolerance of hypoxia in maize (*Zea mays* L.). *Plant Cell Rep.* **2016**, *35*, 2325–2340. [CrossRef] [PubMed]
44. Liu, J.; Hou, Z.; Liu, G.; Hou, L.; Liu, X. Hydrogen sulfide may function downstream of nitric oxide in ethylene-induced stomatal closure in *Vicia faba* L. *J. Integr. Agric.* **2012**, *11*, 1644–1653. [CrossRef]
45. Mishra, V.; Singh, P.; Tripathi, D.K.; Corpas, F.J.; Singh, V.P. Nitric oxide and hydrogen sulfide: An indispensable combination for plant functioning. *Trends Plant Sci.* **2021**, 1360–1385. [CrossRef]
46. Khan, M.N.; Mobin, M.; Abbas, Z.K.; Siddiqui, M.H. Nitric oxide-induced synthesis of hydrogen sulfide alleviates osmotic stress in wheat seedlings through sustaining antioxidant enzymes, osmolyte accumulation and cysteine homeostasis. *Nitric Oxide* **2017**, *68*, 91–102. [CrossRef]
47. Lisjak, M.; Srivastava, N.; Teklic, T.; Civale, L.; Lewandowski, K.; Wilson, I.; Wood, M.E.; Whiteman, M.; Hancock, J.T. A novel hydrogen sulfide donor causes stomatal opening and reduces nitric oxide accumulation. *Plant Physiol. Bioch.* **2010**, *48*, 931–935. [CrossRef]
48. Antoniou, C.; Xenofontos, R.; Chatzimichail, G.; Christou, A.; Kashfi, K.; Fotopoulos, V. Exploring the potential of nitric oxide and hydrogen sulfide (NOSH)-releasing synthetic compounds as novel priming agents against drought stress in *Medicago sativa* plants. *Biomolecules* **2020**, *10*, 120. [CrossRef]
49. Urhonen, T.; Lie, A.; Aamodt, G. Associations between long commutes and subjective health complaints among railway workers in Norway. *Prev. Med. Rep.* **2016**, *4*, 490–495. [CrossRef]
50. Zhu, J. Abiotic Stress Signaling and Responses in Plants. *Cell* **2016**, *167*, 313–324. [CrossRef]
51. He, J.; Duan, Y.; Hua, D.; Fan, G.; Wang, L.; Liu, Y.; Chen, Z.; Han, L.; Qu, L.; Gong, Z. DEXH box RNA helicase-mediated mitochondrial reactive oxygen species production in *Arabidopsis* mediates crosstalk between abscisic acid and auxin signaling. *Plant Cell* **2012**, *24*, 1815–1833. [CrossRef] [PubMed]
52. Liu, H.; Xue, S. Interplay between hydrogen sulfide and other signaling molecules in the regulation of guard cell signaling and abiotic/biotic stress response. *Plant Commun.* **2021**, *2*, 100179. [CrossRef] [PubMed]
53. García-Mata, C.; Lamattina, L. Hydrogen sulphide, a novel gasotransmitter involved in guard cell signalling. *New Phytol.* **2010**, *188*, 977–984. [CrossRef]
54. Zhou, H.; Chen, Y.; Zhai, F.; Zhang, J.; Zhang, F.; Yuan, X.; Xie, Y. Hydrogen sulfide promotes rice drought tolerance via reestablishing redox homeostasis and activation of ABA biosynthesis and signaling. *Plant Physiol. Bioch.* **2020**, *155*, 213–220. [CrossRef] [PubMed]
55. Jin, Z.; Xue, S.; Luo, Y.; Tian, B.; Fang, H.; Li, H.; Pei, Y. Hydrogen sulfide interacting with abscisic acid in stomatal regulation responses to drought stress in *Arabidopsis*. *Plant Physiol. Bioch.* **2013**, *62*, 41–46. [CrossRef] [PubMed]
56. Ma, D.; Ding, H.; Wang, C.; Qin, H.; Han, Q.; Hou, J.; Lu, H.; Xie, Y.; Guo, T. Alleviation of drought stress by hydrogen sulfide is partially related to the abscisic acid signaling pathway in wheat. *PLoS ONE* **2016**, *11*, e163082. [CrossRef] [PubMed]
57. Du, X.; Jin, Z.; Zhang, L.; Liu, X.; Yang, G.; Pei, Y. H_2S is involved in ABA-mediated stomatal movement through MPK4 to alleviate drought stress in *Arabidopsis thaliana*. *Plant Soil* **2019**, *435*, 295–307. [CrossRef]
58. Hua, D.; Wang, C.; He, J.; Liao, H.; Duan, Y.; Zhu, Z.; Guo, Y.; Chen, Z.; Gong, Z. A plasma membrane receptor kinase, GHR1, mediates abscisic acid- and hydrogen peroxide-regulated stomatal movement in *Arabidopsis*. *Plant Cell* **2012**, *24*, 2546–2561. [CrossRef] [PubMed]
59. Wang, L.; Wan, R.; Shi, Y.; Xue, S. Hydrogen sulfide activates S-type anion channel via OST1 and Ca^{2+} Modules. *Mol. Plant* **2016**, *9*, 489–491. [CrossRef]
60. Golldack, D.; Lüking, I.; Yang, O. Plant tolerance to drought and salinity: Stress regulating transcription factors and their functional significance in the cellular transcriptional network. *Plant Cell Rep.* **2011**, *30*, 1383–1391. [CrossRef]
61. Shan, C.; Zhang, S.; Zhou, Y. Hydrogen sulfide is involved in the regulation of ascorbate-glutathione cycle by exogenous ABA in wheat seedling leaves under osmotic stress. *Cereal Res. Commun.* **2017**, *45*, 411–420. [CrossRef]
62. Zhang, J.; Zhou, M.; Ge, Z.; Shen, J.; Zhou, C.; Gotor, C.; Romero, L.C.; Duan, X.; Liu, X.; Wu, D.; et al. Abscisic acid-triggered guard cell L-cysteine desulfhydrase function and in situ hydrogen sulfide production contributes to heme oxygenase-modulated stomatal closure. *Plant Cell Environ.* **2020**, *43*, 624–636. [CrossRef] [PubMed]
63. Scuffi, D.; álvarez, C.; Laspina, N.; Gotor, C.; Lamattina, L.; García-Mata, C. Hydrogen sulfide generated by L-cysteine desulfhydrase acts upstream of nitric oxide to modulate abscisic acid-dependent stomatal closure. *Plant Physiol.* **2014**, *166*, 2065–2076. [CrossRef] [PubMed]
64. Chen, S.; Jia, H.; Wang, X.; Shi, C.; Wang, X.; Ma, P.; Wang, J.; Ren, M.; Li, J. Hydrogen sulfide positively regulates abscisic acid signaling through persulfidation of SnRK2.6. *Guard Cells* **2020**, *13*, 732–744. [CrossRef]
65. Hou, Z.; Wang, L.; Liu, J.; Hou, L.; Liu, X. Hydrogen sulfide regulates ethylene-induced stomatal closure in *Arabidopsis thaliana*. *J. Integr. Plant Biol.* **2013**, *55*, 277–289. [CrossRef] [PubMed]
66. Kim, T.H.; Böhmer, M.; Hu, H.; Nishimura, N.; Schroeder, J.I. Guard cell signal transduction network: Advances in understanding abscisic acid, CO_2, and Ca^{2+} signaling. *Annu. Rev. Plant Biol.* **2010**, *61*, 561–591. [CrossRef]
67. Shen, J.; Zhang, J.; Zhou, M.; Zhou, H.; Cui, B.; Gotor, C.; Romero, L.C.; Fu, L.; Yang, J.; Foyer, C.H.; et al. Persulfidation-based modification of cysteine desulfhydrase and the NADPH oxidase RBOHD controls guard cell abscisic acid signaling. *Plant Cell* **2020**, *32*, 1000–1017. [CrossRef]

68. Zhou, M.; Zhang, J.; Shen, J.; Zhou, H.; Zhao, D.; Gotor, C.; Romero, L.C.; Fu, L.; Li, Z.; Yang, J.; et al. Hydrogen sulfide-linked persulfidation of ABI4 controls ABA responses through the transactivation of MAPKKK18. *Arabidopsis* **2021**, *14*, 921–936. [CrossRef]
69. Zhou, H.; Zhang, J.; Shen, J.; Zhou, M.; Yuan, X.; Xie, Y. Redox-based protein persulfidation in guard cell ABA signaling. *Plant Signal. Behav.* **2020**, *15*, 1741987. [CrossRef]
70. Peiter, E. The plant vacuole: Emitter and receiver of calcium signals. *Cell Calcium* **2011**, *50*, 120–128. [CrossRef]
71. Fang, H.; Jing, T.; Liu, Z.; Zhang, L.; Jin, Z.; Pei, Y. Hydrogen sulfide interacts with calcium signaling to enhance the chromium tolerance in *Setaria italica*. *Cell Calcium* **2014**, *56*, 472–481. [CrossRef]
72. Fang, H.; Liu, Z.; Long, Y.; Liang, Y.; Jin, Z.; Zhang, L.; Liu, D.; Li, H.; Zhai, J.; Pei, Y. The Ca^2/calmodulin2-binding transcription factor TGA3 elevates *LCD* expression and H_2S production to bolster Cr^{6+} tolerance in *Arabidopsis*. *Plant J.* **2017**, *91*, 1038–1050. [CrossRef] [PubMed]
73. He, H.; Li, Y.; He, L. The central role of hydrogen sulfide in plant responses to toxic metal stress. *Ecotox. Environ. Saf.* **2018**, *157*, 403–408. [CrossRef] [PubMed]
74. Valivand, M.; Amooaghaie, R.; Ahadi, A. Seed priming with H_2S and Ca^{2+} trigger signal memory that induces cross-adaptation against nickel stress in zucchini seedlings. *Plant Physiol. Biochem.* **2019**, *143*, 286–298. [CrossRef] [PubMed]
75. Gong, C.; Shi, C.; Ding, X.; Liu, C.; Li, J. Hydrogen sulfide induces Ca^{2+} signal in guard cells by regulating reactive oxygen species accumulation. *Plant Signal. Behav.* **2020**, *15*, 1805228. [CrossRef] [PubMed]
76. Li, Z.; Gong, M.; Xie, H.; Yang, L.; Li, J. Hydrogen sulfide donor sodium hydrosulfide-induced heat tolerance in tobacco (*Nicotiana tabacum* L) suspension cultured cells and involvement of Ca^{2+} and calmodulin. *Plant Sci.* **2012**, *185*, 185–189. [CrossRef]
77. Kolupaev, Y.E.; Firsova, E.N.; Yastreb, T.O.; Lugovaya, A.A. The participation of calcium ions and reactive oxygen species in the induction of antioxidant enzymes and heat resistance in plant cells by hydrogen sulfide donor. *Appl. Biochem. Microbiol.* **2017**, *53*, 573–579. [CrossRef]
78. Khan, M.N.; Siddiqui, M.H.; Mukherjee, S.; Alamri, S.; Al-Amri, A.A.; Alsubaie, Q.D.; Al-Munqedhi, B.; Ali, H.M. Calcium-hydrogen sulfide crosstalk during K^+-deficient NaCl stress operates through regulation of Na^+/H^+ antiport and antioxidative defense system in mung bean roots. *Plant Physiol. Biochem.* **2021**, *159*, 211–225. [CrossRef] [PubMed]
79. Waszczak, C.; Carmody, M.; Kangasjarvi, J. Reactive oxygen species in plant signaling. *Annu. Rev. Plant Biol.* **2018**, *69*, 209–236. [CrossRef]
80. Černý, M.; Habánová, H.; Berka, M.; Luklová, M.; Brzobohatý, B. Hydrogen peroxide: Its role in plant biology and crosstalk with signalling networks. *Int. J. Mol. Sci.* **2018**, *19*, 2812. [CrossRef]
81. Lv, W.; Yang, L.; Xu, C.; Shi, Z.; Shao, J.; Xian, M.; Chen, J. Cadmium disrupts the balance between hydrogen peroxide and superoxide radical by regulating endogenous hydrogen sulfide in the root tip of *Brassica rapa*. *Front. Plant Sci.* **2017**, *8*, 232. [CrossRef] [PubMed]
82. Kabała, K.; Zboińska, M.; Głowiak, D.; Reda, M.; Jakubowska, D.; Janicka, M. Interaction between the signaling molecules hydrogen sulfide and hydrogen peroxide and their role in vacuolar H^+-ATPase regulation in cadmium-stressed cucumber roots. *Physiol. Plant.* **2019**, *166*, 688–704. [CrossRef] [PubMed]
83. Janicka, M.; Reda, M.; Czyżewska, K.; Kabała, K. Involvement of signalling molecules NO, H_2O_2 and H_2S in modification of plasma membrane proton pump in cucumber roots subjected to salt or low temperature stress. *Funct. Plant Biol.* **2018**, *45*, 428–439. [CrossRef] [PubMed]
84. Li, Z.; Zhu, Y.; He, X.; Yong, B.; Peng, Y.; Zhang, X.; Ma, X.; Yan, Y.; Huang, L.; Nie, G. The hydrogen sulfide, a downstream signaling molecule of hydrogen peroxide and nitric oxide, involves spermidine-regulated transcription factors and antioxidant defense in white clover in response to dehydration. *Environ. Exp. Bot.* **2019**, *161*, 255–264. [CrossRef]
85. Zhao, C.; Wang, X.; Wang, X.; Wu, K.; Li, P.; Chang, N.; Wang, J.; Wang, F.; Li, J.; Bi, Y. Glucose-6-phosphate dehydrogenase and alternative oxidase are involved in the cross tolerance of highland barley to salt stress and UV-B radiation. *J. Plant Physiol.* **2015**, *181*, 83–95. [CrossRef]
86. Li, Q.; Wang, Z.; Zhao, Y.; Zhang, X.; Zhang, S.; Bo, L.; Wang, Y.; Ding, Y.; An, L. Putrescine protects hulless barley from damage due to UV-B stress via H_2S- and H_2O_2-mediated signaling pathways. *Plant Cell Rep.* **2016**, *35*, 1155–1168. [CrossRef] [PubMed]
87. Li, Z.; Luo, L.; Sun, Y. Signal crosstalk between nitric oxide and hydrogen sulfide may be involved in hydrogen peroxide-induced thermotolerance in maize seedlings. *Russ. J. Plant Physiol.* **2015**, *62*, 507–514. [CrossRef]
88. Janda, M.; Ruelland, E. Magical mystery tour: Salicylic acid signalling. *Environ. Exp. Bot.* **2015**, *114*, 117–128. [CrossRef]
89. Zhang, J.; Zhou, M.; Zhou, H.; Zhao, D.; Gotor, C.; Romero, L.C.; Shen, J.; Ge, Z.; Zhang, Z.; Shen, W.; et al. Hydrogen sulfide, a signaling molecule in plant stress responses. *J. Integr. Plant Biol.* **2021**, *63*, 146–160. [CrossRef]
90. Li, Z. Synergistic effect of antioxidant system and osmolyte in hydrogen sulfide and salicylic acid crosstalk-induced heat tolerance in maize (*Zea mays* L.) seedlings. *Plant Signal. Behav.* **2015**, *10*, e1051278. [CrossRef]
91. Li, Z.; Xie, L.; Li, X. Hydrogen sulfide acts as a downstream signal molecule in salicylic acid-induced heat tolerance in maize (*Zea mays* L.) seedlings. *J. Plant Physiol.* **2015**, *177*, 121–127. [CrossRef]
92. Qiao, Z.; Jing, T.; Liu, Z.; Zhang, L.; Jin, Z.; Liu, D.; Pei, Y. H_2S acting as a downstream signaling molecule of SA regulates Cd tolerance in *Arabidopsis*. *Plant Soil.* **2015**, *393*, 137–146. [CrossRef]
93. Kaya, C. Salicylic acid-induced hydrogen sulphide improves lead stress tolerance in pepper plants by upraising the ascorbate-glutathione cycle. *Physiol. Plant.* **2020**, *173*, 8–19. [CrossRef] [PubMed]

94. Jia, H.; Chen, S.; Liu, D.; Liesche, J.; Shi, C.; Wang, J.; Ren, M.; Wang, X.; Yang, J.; Shi, W.; et al. Ethylene-induced hydrogen sulfide negatively regulates ethylene biosynthesis by persulfidation of ACO in tomato under osmotic stress. *Front. Plant Sci.* **2018**, *9*, 1517. [CrossRef] [PubMed]
95. Xiao, Y.; Wu, X.; Sun, M.; Peng, F. Hydrogen sulfide alleviates waterlogging-induced damage in peach seedlings via enhancing antioxidative system and inhibiting ethylene synthesis. *Front Plant Sci.* **2020**, *11*, 696. [CrossRef] [PubMed]
96. Pauwels, L.; Barbero, G.F.; Geerinck, J.; Tilleman, S.; Grunewald, W.; Pérez, A.C.; Chico, J.M.; Bossche, R.V.; Sewell, J.; Gil, E.; et al. NINJA connects the co-repressor TOPLESS to jasmonate signalling. *Nature* **2010**, *464*, 788–791. [CrossRef] [PubMed]
97. Qi, T.; Song, S.; Ren, Q.; Wu, D.; Huang, H.; Chen, Y.; Fan, M.; Peng, W.; Ren, C.; Xie, D. The Jasmonate-ZIM-domain proteins interact with the WD-Repeat/bHLH/MYB complexes to regulate Jasmonate-mediated anthocyanin accumulation and trichome initiation in *Arabidopsis thaliana*. *Plant Cell* **2011**, *23*, 1795–1814. [CrossRef]
98. Qi, T.; Huang, H.; Song, S.; Xie, D. Regulation of Jasmonate-mediated stamen development and seed production by a bHLH-MYB complex in *Arabidopsis*. *Plant Cell* **2015**, *27*, 1620–1633. [CrossRef]
99. Qi, T.; Wang, J.; Huang, H.; Liu, B.; Gao, H.; Liu, Y.; Song, S.; Xie, D. Regulation of Jasmonate-Induced Leaf Senescence by Antagonism between bHLH Subgroup IIIe and IIId Factors in Arabidopsis. *Plant Cell* **2015**, *27*, 1634–1649. [CrossRef]
100. Verma, V.; Ravindran, P.; Kumar, P.P. Plant hormone-mediated regulation of stress responses. *BMC Plant Biol.* **2016**, *16*, 1–10. [CrossRef]
101. Han, X.; Hu, Y.; Zhang, G.; Jiang, Y.; Chen, X.; Yu, D. Jasmonate negatively regulates stomatal development in *Arabidopsis* cotyledons. *Plant Physiol.* **2018**, *176*, 2871–2885. [CrossRef] [PubMed]
102. Deng, G.; Zhou, L.; Wang, Y.; Zhang, G.; Chen, X. Hydrogen sulfide acts downstream of jasmonic acid to inhibit stomatal development in *Arabidopsis*. *Planta* **2020**, *251*, 42. [CrossRef]
103. He, H.; He, L. Regulation of gaseous signaling molecules on proline metabolism in plants. *Plant Cell Rep.* **2018**, *37*, 387–392. [CrossRef]
104. Yao, Y.; Yang, Y.; Li, C.; Huang, D.; Zhang, J.; Wang, C.; Li, W.; Wang, N.; Deng, Y.; Liao, W. Research progress on the functions of gasotransmitters in plant responses to abiotic stresses. *Plants* **2019**, *8*, 605. [CrossRef] [PubMed]
105. Tian, B.; Qiao, Z.; Zhang, L.; Li, H.; Pei, Y. Hydrogen sulfide and proline cooperate to alleviate cadmium stress in foxtail millet seedlings. *Plant Physiol. Bioch.* **2016**, *109*, 293–299. [CrossRef] [PubMed]

International Journal of *Molecular Sciences*

MDPI

Review

The Interplay between Hydrogen Sulfide and Phytohormone Signaling Pathways under Challenging Environments

Muhammad Saad Shoaib Khan [1,†], Faisal Islam [1,†], Yajin Ye [2], Matthew Ashline [3], Daowen Wang [4], Biying Zhao [1], Zheng Qing Fu [3,*] and Jian Chen [1,*]

1 International Genome Center, Jiangsu University, Zhenjiang 212013, China; yusufzai.pathan786@hotmail.com (M.S.S.K.); faysal224@yahoo.com (F.I.); zhaoby@ujs.edu.cn (B.Z.)
2 Key Laboratory of Forest Genetics and Biotechnology, Ministry of Education of China, Co-Innovation Center for the Sustainable Forestry in Southern China, Nanjing Forestry University, Nanjing 210037, China; yajinye@njfu.edu.cn
3 Department of Biological Sciences, University of South Carolina, Columbia, SC 29208, USA; mashline@email.sc.edu
4 State Key Laboratory of Wheat and Maize Crop Science, College of Agronomy, Henan Agricultural University, Zhengzhou 450002, China; dwwang@henau.edu.cn
* Correspondence: zfu@mailbox.sc.edu (Z.Q.F.); jianchen@ujs.edu.cn (J.C.)
† These authors contributed equally to this work.

Abstract: Hydrogen sulfide (H_2S) serves as an important gaseous signaling molecule that is involved in intra- and intercellular signal transduction in plant–environment interactions. In plants, H_2S is formed in sulfate/cysteine reduction pathways. The activation of endogenous H_2S and its exogenous application has been found to be highly effective in ameliorating a wide variety of stress conditions in plants. The H_2S interferes with the cellular redox regulatory network and prevents the degradation of proteins from oxidative stress via post-translational modifications (PTMs). H_2S-mediated persulfidation allows the rapid response of proteins in signaling networks to environmental stimuli. In addition, regulatory crosstalk of H_2S with other gaseous signals and plant growth regulators enable the activation of multiple signaling cascades that drive cellular adaptation. In this review, we summarize and discuss the current understanding of the molecular mechanisms of H_2S-induced cellular adjustments and the interactions between H_2S and various signaling pathways in plants, emphasizing the recent progress in our understanding of the effects of H_2S on the PTMs of proteins. We also discuss future directions that would advance our understanding of H_2S interactions to ultimately mitigate the impacts of environmental stresses in the plants.

Keywords: hydrogen sulfide; biotic stress; abiotic stress; salicylic acid; abscisic acid; jasmonic acid; ethylene; auxin; phytohormones

Citation: Khan, M.S.S.; Islam, F.; Ye, Y.; Ashline, M.; Wang, D.; Zhao, B.; Fu, Z.Q.; Chen, J. The Interplay between Hydrogen Sulfide and Phytohormone Signaling Pathways under Challenging Environments. *Int. J. Mol. Sci.* **2022**, *23*, 4272. https://doi.org/10.3390/ijms23084272

Academic Editor: Francisco J. Corpas

Received: 20 February 2022
Accepted: 11 April 2022
Published: 12 April 2022

1. Introduction

The in-depth understanding of mechanisms/processes involved in plant growth and development is critical for improving crop quality and productivity, as well as the development of more stable and climate-resilient crops. Due to their sessile nature, plants have evolved several adaptive mechanisms for survival. Among them, phytohormones are complex signaling factors that regulate a myriad of physio-biochemical processes to maintain optimum growth, development, and performance [1]. The synthesis and level of hormones could vary significantly in different plant tissues, during different developmental stages, and under different environmental conditions [2]. Furthermore, there is less knowledge about the coordination of the spatial and temporal distribution of plant hormones and how these dynamic processes trigger diverse responses in plants [3].

Recently, numerous investigations have revealed hydrogen sulfide (H_2S) as one of the critical components in various acclimation processes in plants under normal and stressful conditions (Figure 1). H_2S is a colorless, lipophilic, toxic, volatile, inflammable, and water-soluble gas with a pungent odor, similar to that of rotten eggs. Amidst the emergence of life on Earth approximately 3.8 billion years ago, H_2S acted as a major energy source; however, H_2S-dependent organisms disappeared after a burst of oxygen [4]. Nevertheless, the biogeochemical sulfur cycle was preserved in organisms and is presently limited to some vital metabolic and signaling events [5,6]. H_2S receives extensive attention in the animal field due to its multiple physiological and pathophysiological functions in different organs due to clear and well-established experimental models/approaches [7]. However, it was not until recently that the roles of H_2S in plants have gained the attention of scientists due to the involvement of H_2S in adverse stress conditions via regulation of gene expression, post-translational modifications (PTMs), and crosstalk with other gaseous signals and phytohormones [8,9].

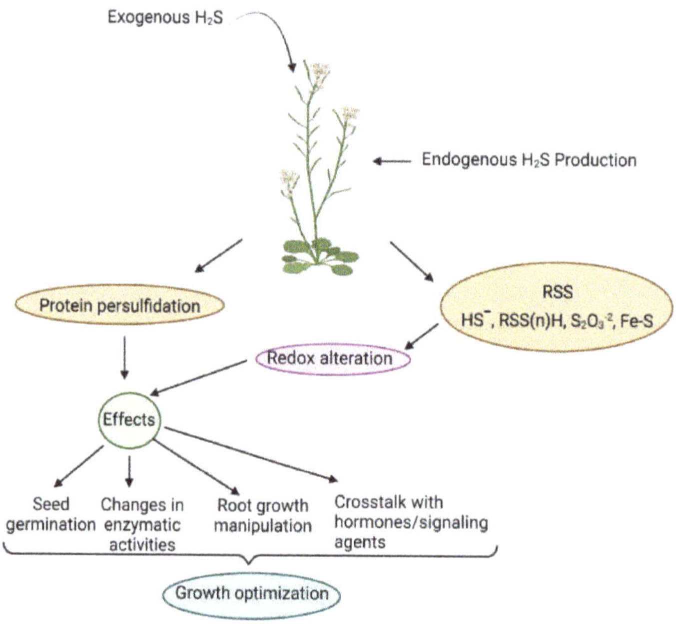

Figure 1. Overview of hydrogen sulfide (H_2S) production and the regulation of several physiological, metabolic, and morphological processes by H_2S to optimize growth in plants.

The fine-tuned interaction of H_2S with other gaseous signaling biomolecules and hormones orchestrates molecular, metabolic, and physiological adaptive responses and permits the plants to respond properly to changing environmental conditions. In this review article, we will explain the central role of H_2S in the regulation of various physiological and molecular processes. We will also discuss how hormonal homeostasis plays a crucial role in stress conditions and how H_2S synergistically/antagonistically regulates the biosynthesis and degradation of the associated plant hormones and modulates their signaling to generate adaptive responses in plants.

2. H_2S Biosynthesis in Different Organelles and Associated Enzymes

Plant roots absorb sulfate (SO_4^{2-}), which is reduced into H_2S via the action of APS reductase (adenosine-5′-phosphoryl sulfate reductase) and SiR (sulfite reductase). H_2S is later transformed into cysteine amino acid via catalysis of O-Acetylserine (thiol) lyase (OASTL), as a final step of sulfate assimilation in plants (Figure 2). In *A. thaliana*, cytosolic OAS-A1 (At4g14880), the plastid OAS-B (At2g43750), and the mitochondrial OAS-C (At3g59760) are considered true OASTL because they incorporate an O-acetylserine (OAS) and sulfide into cysteine synthesis [10–12]. The presence of functional OASTL was also identified in pollen [13]. Additionally, plant cells contain nutritional sulfur (SO_4^{2-}) and SO_2 (collected from the atmosphere) that is consequently converted into SO_3^{2-} and is used to produce H_2S in the presence of ferredoxin and APS reductase [14,15]. In salt-stressed tobacco plants, malfunction of SiR leads to decreased H_2S production, correlating with less availability of SO_2 on account of stomatal closure. This series represents the functional role of SiR in H_2S metabolism under stress conditions [16].

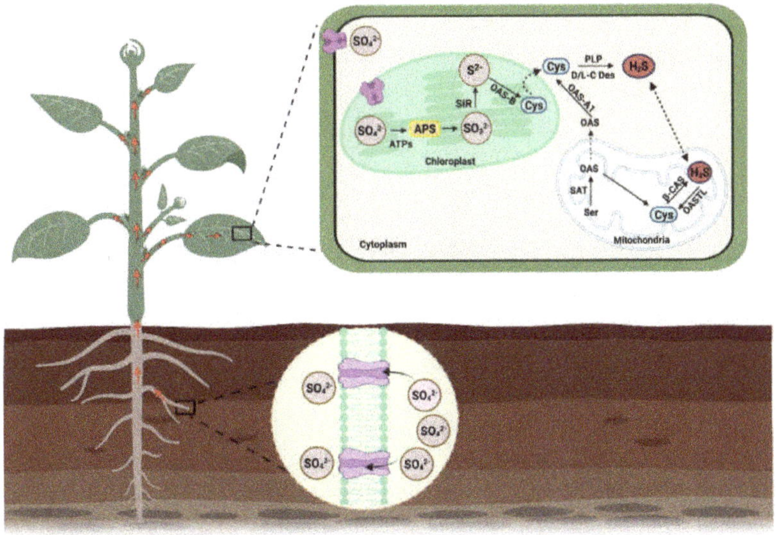

Figure 2. H_2S biosynthesis in plants. In plants, sulfate (SO_4^{2-}) is transported from the roots, which is then distributed to all parts of the plant through the xylem vessels. SO_4^{2-} entering the cells is assimilated in the chloroplasts and mitochondria. In chloroplast, SO_4^{2-} is reduced to sulfite (SO_3^{2-}) by APS reductase after it is activated to APS. Under the catalysis of SiR, the sulfite is then reduced to sulfide (S^{2-}) using six electrons transferred from ferredoxin. As a result, sulfide is produced, which is used to produce cysteine. The OASTL enzyme catalyzes the synthesis of cysteine along with O-acetylserine. The enzyme CDes and pyridoxal 5-phosphate (PLP) participate in degrading cysteine to generate H_2S. In mitochondria, serine acetyltransferase (SAT) catalyzes the conversion of serine (Ser) into OAS and produces cysteine, which is converted to H_2S via the catalytic activity of β-cyanoalanine synthase (β-CAS).

H_2S is also synthesized in the chloroplasts and mitochondria when cysteine is reduced by cysteine desulfhydrase (CDes) and β-cyanoalanine synthase (β-CAS), respectively (Figure 2). Genetic and molecular evidence indicated that mitochondrial isoforms of CAS are CYS-C1 (At3g61440) and OAS-C (At3g59760), and chloroplastic isoforms of CAS are OAS-B (At2g43750) and SCS (At3g03630) [17]. The cytosolic release of H_2S is dependent upon the functioning of D/L cysteine desulfhydrases (L/D-CDes). Several L-CDes of the Arabidopsis plant are well characterized and are involved in the breakdown of L-cysteine to sulfide, NH_3, and pyruvate [18–20]. However, D-CDes are completely different proteins and belong to the pyridoxal 5-phosphate (PLP)-dependent enzyme superfamily, and its activity is PLP dependent [21,22]. The model plant Arabidopsis contains two putative *D-cysteine desulfhydrases* (*D-CDes*) genes (At1g48420 and At3g26115) [21–23], while two *D-CDes* are also functionally characterized in rice (*OsDCD1* and *OsLCD2*) and some other crops [24,25]. The D-cysteine desulfhydrases 2 carry out the decomposition of both L- and D-Cystine into H_2S. Accumulating evidence signifies that NifS-like L-CDes are also involved in the generation of H_2S. The presence of H_2S in plant peroxisomes and its interaction with catalase is also observed; however, the synthesis mechanisms and involved enzymes are still unknown [26].

The mitochondria play a vital role in the catabolism of H_2S and maintain its steady-state levels in cells. In mitochondria, H_2S is generated during cyanide detoxification through the catalysis of β-CAS. The functional mitochondria isoform of CAS is CYS-C1 (At3g61440), which catalyzes the conversion of cysteine and cyanide into hydrogen sulfide and β-CAS and maintains optimum levels of cyanide to prevent phytotoxicity [27]. This yielded H_2S is converted back into cysteine via mitochondrial OASTL (OAS-C, At3g59760), which will again be used in the detoxification of cyanide. This process is considered a cyclic pathway of cysteine generation via H_2S consumption in mitochondria [28]. Under stress conditions, excess accumulation of H_2S raises the pH of mitochondria, leading to the conversion of H_2S into hydrosulfide ions (HS^-). Excess accumulation of H_2S also prevents the loss of H_2S from mitochondrial membranes and maintains H_2S homeostasis (Figure 2). The environmental cues also modulate the endogenous H_2S biosynthesis by stimulating desulfhydrase activities in plant cells [18].

In plastids, the reduction of sulfate to sulfide and its incorporation into the OAS is executed as an entry point of reduced sulfur to plant metabolism for growth and development via a photosynthetic sulfate assimilation pathway [18,29]. The OAS interaction with serine acetyltransferase (SAT) forms a cysteine synthase complex (CSC), which generates demand-driven synthesis of cystine in plant cells [30,31]. Subsequently, the breakdown of cysteine in the chloroplast generates H_2S due to the catalysis of DES1 and L/D-cysteine desulfhydrase (Figure 2). The generation of H_2S in chloroplasts acts as a signaling molecule because it substantially impacts cellular metabolism by limiting the rate of photosynthesis.

The peroxisome is an essential single membrane-bound organelle involved in the metabolism of reactive nitrogen species (RNS), including H_2S [26,32,33]. Recent studies demonstrated the presence of H_2S in plant peroxisomes [34]. Some studies speculated that peroxisomes have the capacity to transform sulfite to sulfate under the catalysis of Sulfite oxidase (At3g01910) in *A. thaliana*. Presently, no enzymatic source for H_2S metabolism has been observed in the peroxisome of Arabidopsis, and tomato [34–36]; the mechanism of H_2S production in peroxisome is still obscure. The H_2S characterization study in *Solanum lycopersicum* showed the localization of OASTL9 in the peroxisome, which exhibited upregulation under different developmental stages and pathogenic bacterial treatments [36].

In the plant, several additional enzymes are also involved in H_2S synthesis, and most of the H_2S in the cell is produced during the necessary consumption of cysteine. For example, At5g28030 encodes a cysteine synthase (CS)-like protein that degrades L-cysteine and produces H_2S [28]. This protein is also localized in the cytoplasm as AtDES1 (desulfhydrase). The homolog of this protein in *Brassica napus* (BnDES) is also involved in the breakdown of cysteine [37]. However, AtDES1 homolog in rice (OsLCD2) exhibits cysteine biosynthesis activity [38]. The Arabidopsis nitrogen fixation-like 1 and 2 (At5g65720; At1g08490) also use L-cysteine as a substrate and produce H_2S during the synthesis of L-alanine in the cytosol [20,28,39]. This diversity in enzymatic functioning and discrepancies in their substrates' catalyzation may allow the plants to calibrate endogenous H_2S levels according to their requirements and external prompts.

3. Role of H_2S in the Modulation of Abiotic Stress Responses

H_2S plays a vital role in protecting plants against several abiotic stressors. Environmental stress factors such as salinity, drought, waterlogging, high temperature, excessive light, heavy metals, and chilling could adversely affect plant growth and development (Figure 3) [40–43]. Generally, under most stress conditions, plants reduce uptake of CO_2 due to the closure of stomata and limiting CO_2 fixation. This condition causes alternation in cell metabolism due to restricted photosynthetic capacity that leads to the generation of reactive oxygen/nitrogen species (ROS/RNS) [44–50]. H_2S directly regulates the cysteine (Cys) residues' persulfidation via posttranslational modification (PTM), allowing the H_2S

to regulate protein functioning through persulfidation [51,52]. For example, APX protein was persulfidated in different compartments of cells (cytosol, chloroplasts, mitochondria, and peroxisomes) in Arabidopsis [26,53–55]. These findings indicate that the ROS-induced toxicity in stressed plants is regulated by H_2S-mediated persulfidation post-translationally via triggering the ROS scavenging enzyme activities [56].

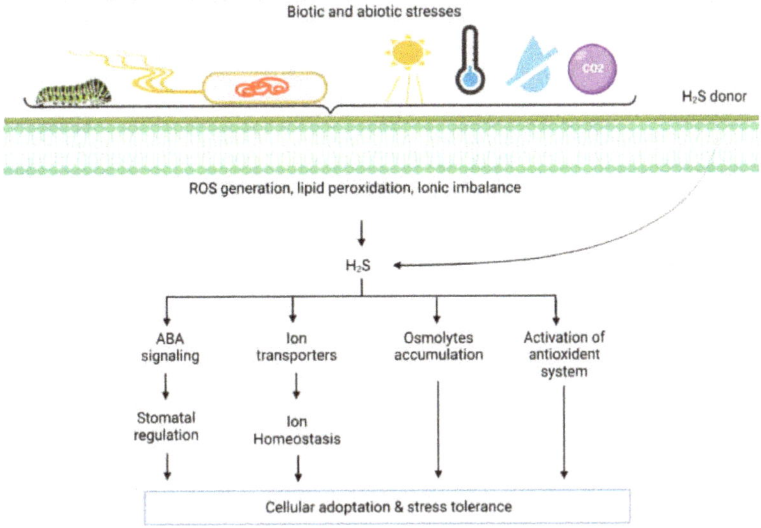

Figure 3. Multiple environmental stressors can induce endogenous hydrogen sulfide (H_2S) production in plants. The H_2S production mediates the different physiological processes in plants by undergoing interaction with plant hormones and other cellular entities to maintain homeostasis under normal and stressful conditions.

3.1. Application of H_2S in Plant Drought Responses

During osmotic stress, improved water status of plants is a vital survival strategy that is achieved via accumulating osmolytes to maintain normal hydration levels. Exposure to drought stress or PEG-induced osmotic stress in plants enhances the accumulation of osmolytes such as proline and glycine betaine to maintain normal water status in stressed plants. However, sometimes the accumulation of these osmolytes fails to maintain adequate water status due to the severity of osmotic stress [50,52]. The endogenous stimulation of H_2S regulates the proline synthesizing enzyme via stimulating the expression of 1-pyrroline-5-carboxylate synthetase, and by inhibiting the activity of the proline-degrading enzyme. On the other hand, H_2S also triggers the activity of glycine betaine biosynthesis enzymes (aldehyde dehydrogenase), which reduce the osmotic stress and assist the plants in enhancing osmotic pressures to improve water uptake and relative water content in vital tissues [57,58]. The pre-exposure of SO_2 to drought-stressed wheat plants showed a pronounced increase in endogenous H_2S. This inflation may be caused by the conversion of SO_2 into the SO_3^{2-} and decomposition of L-/D-Cys, which generates enough H_2S to initiate drought adaptive responses in the stressed seedling. However, when hypotaurine (HT; H_2S scavenger) was applied on SO_2-pretreated seedlings, reduced content of H_2S and severe symptoms of drought toxicity appeared in seedlings. In addition, endogenous generation of H_2S via pretreatment of SO_2/NaHS, fully activated the antioxidant enzymes (SOD, CAT, and POD) and reduced the production of H_2O_2 and MDA content in drought-stressed plants [41,47,59,60]. The endogenous H_2S modulation in plants also activated the expression of transcription factors (TFs) such as *ERF1*, *NAC69*, and *MYB30* [41,61]. The findings of several studies indicated that TF NAC69 could confer resistance in drought-stressed plants via the H_2S mediated ABA signaling pathway.

Additionally, the upregulation of TFs such as *ERF1* and *MYB30* may activate signal transduction pathways and regulate stress-responsive gene expression profiling under drought stress conditions [62–64]. Since the application of H_2S scavengers inhibited the transcript abundance of *ERF1, NAC69,* and *MYB30* in wheat plants under drought stress conditions, there must be direct involvement of H_2S in the regulation of stress-related TFs in response to drought stress [61,65,66]. Some studies also recognized that H_2S signaling in response to drought stress influences the functioning of ABA biosynthesis genes such as *NCED2, NCED3,* and *NCED5* and suppresses the ABA catabolic genes (*ABA8ox1, ABA8ox2,* and *ABA8ox3*), which is consistent with ABA accumulation in drought-stressed plants [47,50].

3.2. Role of H_2S in the Alleviation of Metal Stress

Under metal toxicity, plants modulate several metal/metalloid ions from toxic to less toxic forms, such as reduction of arsenate (AsV) to arsenite (ASIII), and *hexavalent chromium* ($Cr^{(VI)}$) to less toxic trivalent $Cr^{(III)}$, and sequester these metal ions via thiols (GSH) and phytochelatins (PCs) ligands [64]. These metabolites (GSH and PCs) actively participate in the intracellular redox balance and metal tolerance capacity of crop plants and prevent the cells from entering programmed cell death or necrosis phases [67,68]. Due to metal-induced oxidative stress, the intracellular redox becomes oxidized, decreasing levels of reduced molecules such as NADH/NADPH and allowing apoptosis or necrosis to be initiated. The endogenous production of H_2S or exogenous application of H_2S donors assists in maintaining the levels of GSH and phytochelatins in the plant to sustain optimum redox balance and the sequestration of toxic metal ions into the vacuoles [41,69]. The GSH and PCs are sulfur enriched compounds, whereas, in sulfur metabolism, metabolites such as sulfite, H_2S, cysteine, and GSH are highly interconnected, and depletion of GSH during metal toxicity could potentially accelerate cysteine breakdown and ultimately enhance the GSH and H_2S supply to the cell [16,67]. In several published studies, it is observed that the mitigation effects of H_2S under different abiotic stresses and metal excess conditions are related to the upregulation or superior maintenance of redox-active compounds such as ASA-, GSH, and PCs [40,67,69,70]. This finding of these studies provides compelling evidence that modulation of endogenous H_2S during stressful conditions could help the plant to maintain or reduce the loss of intracellular glutathione, which supports the overall redox positive state of the cell and verifies that H_2S has an important influence on cell functions under stressful conditions [41,70,71].

H_2S not only overcomes ROS-induced toxicity in metal exposed plants but also plays an effective role in the inhibition of metal transport and absorption. H_2S has the ability to alter chemical forms of metal ions into insoluble phosphate compounds, which decreases metal toxicity and movements [72]. However, the metal reduction capacity of H_2S is much lower than GSH, cysteine, phytochelatins, and metallothioneins [73]. H_2S mediated reduction in metal transport/immobilization is usually associated with downregulation of metal transporters or secretion of chelating compounds to prevent the further translocation of metal ions to the sensitive tissues or uptake from the root zone. For example, in several crop plants, exogenous application of H_2S intensifies the citrate secretion and expression of citrate transporters, so the non-toxic complexes of citrate with Al^{3+} could be formed in the rhizosphere [74–76]. Similarly, H_2S also suppresses pectin methyl esterase activity, which suppresses Al^{3+} binding sites by reducing negative charge in root cells, which has direct implications for Al^{3+} tolerance [77,78]. In the case of Cd metal, H_2S triggers the expression of phytochelatin synthase (*PCS*) and the Cd-ATPase gene to effectively chelate and transport metal ions into the vacuoles through the help of HMT transmembrane transporter channels [79]. The L-DC-mediated H_2S accumulation modulates root pectin content with a lower degree of methylation to facilitate the binding of Cd^{2+} to the cell wall, which ultimately diminishes its further translocation from root to shoot and toxicity symptoms in exposed plants [80]. In Arabidopsis, exogenous application of H_2S activated the generation of Cr^{6+} binding peptides, such as phytochelatins and metallothioneins, to carry toxic Cr^{6+} to insensitive regions mediated by compartmentalization [81,82]. Based on these studies, we infer that H_2S plays a pivotal role in the chelation of heavy metals for inactivation and later sequesters them into the vacuole to increase the metal stress tolerance of plants.

3.3. Effect of H_2S on Plant Salt Tolerance

Salinity is a major constraint limiting agriculture productivity due to poor irrigation practices and continuous climate fluctuations [83]. Saline stress imposes both osmotic stress and ionic toxicity, which retard plant growth and productivity. The unregulated accumulation of sodium (Na^+) hinders water and nutrient uptake and induces water deficit conditions for plants. Furthermore, an excessive amount of Na^+ and chloride (Cl^-) accumulation in plants disturbs ionic homeostasis. The depolarization of membranes leads to the loss of potential stress mitigating ions such as K^+ and Ca^{2+} and induces changes in transpiration rate, photosynthesis, oxidative stress, etc. [84–86]. Saline stress in plants reinforces several physiological, molecular, and metabolic disorders that completely inhibit plant growth [87–89]. The maintenance of ionic homeostasis and a lower cytosolic Na^+/K^+ ratio is critical for salt adaptation and tolerance. It is observed that several Na^+/K^+ ion transporters and stress-responsive gene activation pathways are interconnected with plant hormones because stress and growth hormones are spatially involved in mediating salt-stress signaling and maintaining the balance between stress responses and growth in plants [83,87,88]. In this regard, H_2S biosynthesis and signaling are implicated in saline stress tolerance in plants. [90–93]. Several studies demonstrated that exogenous application of H_2S reduces the uptake of Na^+ and increases the accumulation of K^+ that untimely preserves an optimal Na^+/K^+ ratio for the plant's vital functioning. [90–93]. It is proven via pharmacological studies that when H_2S scavengers were applied to the salt-stressed plants, the depletion of endogenous H_2S aggravated the saline stress symptoms and increased the Na^+/K^+ ratio and cytosolic concentration of Na^+ in studied plants. These studies also highlighted that H_2S application significantly maintains K^+ homeostasis in plants by preventing K^+ leakage by reducing oxidative stress-mediated lipid peroxidation and membrane depolarization. [90–93]. At the molecular level, it was observed that H_2S regulated the activity of SKOR (outward rectifying K^+ channel) by inhibiting its expression and preventing the loss of K^+ into the xylem under saline stress conditions. However, when H_2S scavengers (DL-propargylglycine or HT) were applied to the plants, SKOR expression was not compromised. [90–93]. Similarly, the K^+ retention during saline stress conditions normalizes H^+-ATPase, because H^+ gradient-mediated H^+-ATPase activity repolarizes the PM to accelerate potassium influx and sodium efflux [90–93]. This repolarization occurs because H_2S is involved in the stimulation of gene expression and phosphorylation-mediated upregulation of H^+-ATPase activity under salinity [94,95]. This observation suggests that H_2S shows the implication of K^+ uptake and its homeostasis via upregulating the K^+/Na^+ antiport system through modulating H^+-ATPase activity [42,91,92,95]. Besides this, AKT1 (inward rectifying potassium channels) is located in root epidermal tissue [96], and HAK5 (potassium transporter) gene is located in the tonoplast and the PM [96]. These genes are also coupled with maintaining K^+ and plant resistance to salt. The exogenous application of H_2S donors improved the transcript expression of *AKT1* and *HAK5* and total K content in the salt-challenged *Brassica napus* plant [97]. Similarly, NaHS induced H_2S promoted the expression of *HvAKT1* and *HvHAK5* in roots of barley seedlings under salinity [8]. All these findings advocate that the potential increase in H_2S and its signaling is a positive regulator of K^+ homeostasis and maintenance of the Na^+/K^+ ratio during saline stress in plants [8,95,98,99].

The (SOS) pathway is critical for the exclusion of Na^+ under saline stress conditions. SOS1 is involved in the long-distance transport of Na^+ from roots to shoots [95,100]. The increase in transcript abundance of *SOS1* favors the accumulation of SOS1 proteins in the PM, which triggers the exclusion of Na^+ from cells and minizines the Na^+ load in the cytosol [60]. The H_2S application under alkaline and normal salt stress conditions stabilizes the mRNA level of *SOS1*, which leads to the reduced Na^+ content in the roots of cultivated apple plants [101]. SOS1 is regulated by the H^+ gradient provided by PM H^+-ATPase. Several studies identified that H_2S positively influences the gene expression and phosphorylation of PM H^+-ATPase under salinity [102]. In pharmacological experiments where endogenous H_2S production was inhibited, the expression level of *SOS1* and related Na^+ antiporters were downregulated, and salinity tolerance of plants was compromised due to unregulated accumulation of Na^+ in sensitive tissues [100]. The PM H^+-ATPase on the membranes of vacuoles also regulates the expression and activation of the Na^+/H^+ antiporter, because the compartmentalization of Na^+ ions into the vacuoles is an alternative solution to decrease the Na^+ induced toxicity in cells [42,103]. The H_2S application greatly induces the transcript accumulation of *NHX2* and *VHA-β* genes (Na^+/H^+ antiporter) in salt-exposed plants. This finding also advocates that Na^+ caging in vacuoles is influenced by H_2S signaling [8,85]. Meanwhile, for the regulation of Na^+/K^+ homeostasis, H_2S also controls the H_2O_2 mediated activity of PM-bound NADPH oxidases [104]. For instance, PM NADPH oxidase inhibitor (diphenyleneiodonium chloride) suppressed the H_2S mediated increase in H_2O_2 in the root of Arabidopsis under salinity. The application of ROS scavenger (N,N′-Dimethylthiourea) abolished the H_2S mediated H_2O_2 production in salt stress plants due to the Na^+ uptake being high in salt-stressed plants from the absence of H_2S mediated activation of NADPH oxidase [104]. This conclusion indicates that H_2O_2 might act as a downstream signal for H_2S-mediated Na^+/K^+ homeostasis [85,104,105]. The findings of these studies demonstrate that H_2S regulated signaling influences the activity of H^+-ATPase and the expression of PM Na^+/H^+ antiporter that enhances the salt tolerance by maintaining Na^+/K^+ homeostasis in plants [85,106].

4. Crosstalk of H_2S with Signaling/Phytohormones under Changing Environmental Conditions

Phytohormones, or plant growth regulators (PGRs), are the most significant signaling molecules, synthesized in specific locations within plants, and can be translocated to different parts to regulate stress responses [106]. PGR such as abscisic acid (ABA), auxins (IAA), brassinosteroids (BRs), cytokinins (CK), gibberellins (GA), jasmonic acid (JA), and salicylic acid (SA) help the plants to overcome numerous biotic/abiotic adversities by triggering physiological and molecular responses [107,108]. H_2S, which acts as an endogenous gasotransmitter, is recognized in relevance with other signaling molecules such as NO [109], ROS [110], H_2O_2 [111], CO [112], and plant hormones such as ABA [113], JA [114], GA [115] and ethylene.

H_2S in plants exhibits a dual role, either disseminated as pernicious cellular repercussion or as credible signaling molecules depending upon stress conditions. A study discovered that H_2S operates downstream of NO and helps decrease oxidative stress during salt stress in tomatoes. H_2S helps minimize postharvest ripening and senescence in bananas because it inhibits ethylene signaling as well as mitigating oxidative stress [115]. Additional studies revealed that H_2S regulates NADPH oxidase (RBOH) activity, leading to ROS accumulation [116]. Simultaneously, the concentration of phosphatidic acid generated via phospholipase D [117,118] is also modulated by H_2S, which helps further to inhibit the cellular signaling pathway [1]. In Arabidopsis, H_2S operates upstream of the MAPKs pathway, and both of these work parallelly under cold stress conditions [119]. Various developmental processes such as organogenesis, seed germination, and the advent of senescence are spurred by H_2S produced from sodium hydrosulfide (NaHS) and morpholin-4-ium 4-methoxyphenyl (morpholino) phosphinodithiolate (GYY4137) [119,120]. As a signaling molecule, H_2S participates in several cross-talk networks amid H_2O_2, NO, CO, and phytohormone ABA during different stress conditions [121]. It is evident that signaling molecules such as H_2S interplay an essential role in several stages of plant development because of the interaction between H_2S and numerous phytohormones. In the future, genes involved in governing the new signaling molecules such as H_2S could be targeted to develop a genetically improved crop.

4.1. Crosstalk of H_2S and Abscisic Acid (ABA)

Plants modify ABA levels continually in response to changing physiological and environmental conditions, while bioactive ABA levels are sustained through a fine balance between generation and catabolism [45,86,87]. Several ABA receptors are involved in signal perception and transduction [45]. Earlier studies revealed that the interaction of H_2S with ABA receptor genes implied that H_2S regulates ABA signaling via influencing ABA receptors [45,122,123]. H_2S application in drought-stressed plants upregulated the expression of potential ABA receptors such as *RCAR* (*The regulatory component of ABA*), *ABAR* (*abscisic acid receptor*), *PYR1* (*pyrabactin resistant protein*), *GTG1* (*GPCR-type G proteins*), and *CHLH* (*H subunit of the Mg-chelatase*) [45,124]. Some studies point out that ABA regulates many physiological processes, and H_2S sometimes regulates these responses in a similar way [45,113,124]. Exogenous application of ABA triggers the endogenous production of H_2S, suggesting complex crosstalk between two signaling molecules exists under drought stress conditions [45]. Similarly, under heat stress, ABA could trigger the accumulation of endogenous H_2S and act as a new downstream gaseous signaling molecule that regulates ABA-induced stress responses in heat-stressed plants [45].

In plants, stomatal closure or opening is regulated by guard cells. The plant hormone ABA regulates the function of several ion channels in an ABA-dependent manner to control stomatal closure and opening [124–128]. A wealth of literature provides ample evidence that H_2S regulates stomatal aperture in various plant species, and it may have implications for ABA-dependent stomatal closures in plants under stressful conditions [124]. The earlier study of Wang et al. [129] illuminated this underlying mechanism and revealed that exogenous application of H_2S activates the S-type anion currents in guard cells of Arabidopsis. Concurrently, the elevated level of free Ca^{2+} is a prerequisite for its activation [129]. H_2S triggers Ca^{2+} waves in guard cells. In guard cells, Ca^{2+} sensing is perceived by a heterotrimeric G-protein β-subunit (AGB1) that collaborates in Ca^{2+} induced stomatal closure in Arabidopsis [130]. Ca^{2+} ions also activate SLAC1 by stimulating CPK (calcium-dependent protein kinase) activity. It was observed that lower concentrations of ABA partially impaired stomatal closure in *CPK* quadruple mutant plants; however, higher concentrations of ABA effectively close stomata. The application of Ca^{2+} chelator (1,2-bis(*o*-aminophenoxy) ethane-*N*,*N*,*N*,*N*-tetraacetic acid (BAPTA) completely inhibited the ABA-mediated activation of anion channel in guard cells and prevented the ABA-induced stomatal closure [131,132]. These studies showed that H_2S and ABA are signaling components in stomatal closure in plants.

A recent study demonstrated that H_2S mediated persulfidation of SnRK2.6/OST1 in response to ABA signaling initiated stomatal closure (Figure 4). In guard cells, SnRK2.6/OST1 acts as a core component of ABA signaling that controls stomatal movements, and its function is tightly regulated by H_2S-mediated PTMs. Under certain physiological conditions, ABA induces the generation of H_2S by activating DES1 in the guard cell. The accumulation of H_2S persulfidates SnRK2.6 on Cyc131 and Cys137, which are close to the catalytic loop and near to Ser175 residues, which is vital for the phosphorylation of SnRK2.6 [133–137]. The Cys137 can also undergo *S*-nitrosylation and could inhibit the activity of SnRK2.6 [9,136]. However, persulfidation promotes SnRK2.6 activity, and it is believed that persulfidation occurs earlier than *S*-nitrosylation [9,137]. Due to Cyc131/137 persulfidation induced changes, Ser175 affinity for ATP-γ-phosphate proton acceptor site (Asp140) increases, which leads to the robust autophosphorylation of Ser175 and triggers efficient interaction of SnRK2.6 with its target. This observation confirms that H_2S-mediated persulfidation positively impacts the function of SnRK2.6 in ABA-mediated stomatal closure in guard cells [9,135]. Likewise, Shen et al. [138] reported that during drought stress, ABA signaling in guard cells is promoted by H_2S interaction with ABA. The drought stress mediates the accumulation of ABA, which stimulates persulfidation of DES1 in a redox-dependent manner. At the physiological level, enhanced accumulation of H_2S in the guard cell leads to the persulfidation of H_2O_2 producing enzymes, such as NADPH oxidase, which triggers the generation of H_2O_2 in the guard cell that reinforces ABA signaling and the closure of stomata [138]. Another study revealed that abscisic acid insensitive 4 (ABI4) is involved in the facilitation of ABA and H_2S crosstalk at the transcriptional level (Figure 4). ABI4 is a vital TF in the ABA signaling cascade, and little was known about the PTMs that regulate its activity in response to ABA/H_2S interaction in plants. The ABA accumulation triggers a massive generation of H_2S that leads to the persulfidation of ABI4, which allows the binding of ABI4 to the E1 motif of the *MAPKKK18* (*mitogen-activated protein kinase kinase kinase 18*) promoter to activate *DES1* transcription to close stomata under the ABA-dependent signaling cascade [43]. This study provides compelling evidence that the DES1/H_2S-ABI4 module acts downstream of ABA signaling to regulate stomatal closure [43,139] (Figure 4).

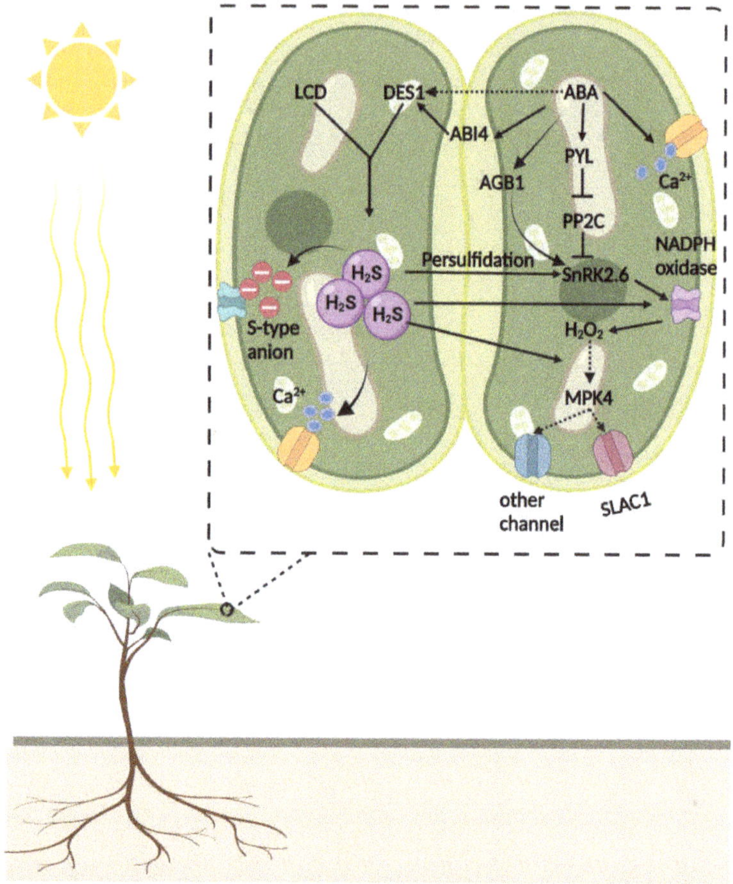

Figure 4. Under normal conditions, ABA receptors (PYR/PYL/RCAR) bind to the PP2Cs and inhibit the activity of SnRK2.6, which deactivates NADPH oxidase, SALC1, and other ion channels to reinforce the normal functioning of stomata. Under water-stressed conditions, ABA signaling stimulates ABA receptors (PYR/PYL/RCAR) that lead to the activation of SnRK2.6, which triggers SLAC1 and NADPH oxidase to produce H_2O_2 and regulate stomatal movements. During drought stress, ABA signaling increases the biosynthesis of H_2S via persulfidation of ABI4-mediated activation of DES1 transcription. The burst of H_2S in guard cells activates the S-type anion and spikes the Ca^{2+} wave alongside strong persulfidation of SnRK2.6. The persulfidated SnRK2.6 robustly phosphorylates SALAC1 and NADPH oxidase to produce a long-lasting burst of ROS to modulate water efflux in guard cells to close stomata, similarly to the way that ABA induces stomatal closure.

In some of the recently published reports, it was also revealed that H_2S might be involved in the biosynthesis of ABA in guard cells [140]. The H_2S promotes the synthesis of cysteine, which is a substrate of ABA3 (molybdenum cofactor sulfurase) enzymes that regulate the activation of AAO3 (abscisic aldehyde oxidase 3) [141]. The higher accumulation of cysteine stimulates the activity of AAO (in vivo) and favors the synthesis of ABA [39] by stimulating the transcript abundance of *NCED3* (9-*cis*-epoxycarotenoid dioxygenase 3). It was revealed that H_2S could boost ABA synthesis, because in a cysteine-biosynthesis-depleted mutant with the disrupted ABA biosynthesis, the H_2S was unable to induce stomatal closure [135,136]. All these studies point out the involvement/crosstalk of H_2S with SnRK2.6, CPK6, MAPKKK18, ABI1, NADPH oxidase, Ca^{2+}, and ROS in ABA-mediated signaling for stomatal movements in plants [135–137].

4.2. Nitric Oxide (NO) and H_2S: Two Interacting Gaseous Molecules Essential for Plant Functioning

Nitric oxide (NO) is also a lipophilic gaseous hormone that could diffuse into inter- or intra cellular spaces without the need for any carrier or transport channel. NO is also involved in PTMs via tyrosine nitration, metal nitrosylation, and S-nitrosylation, whereas H_2S mediated-PTM is associated with persulfidation. However, all these reactions led to the modification of structure, localization, and function of target proteins. Several studies have shown that H_2S interacts with NO and other signaling molecules to modulate plant development and stress responses [7,26,32,34,142]. Earlier reports indicate that the interaction of H_2S towards NO is complementary or inhibitory [55,143–146]. The positive or negative interaction of these two gaseous signaling molecules may be dependent upon the dosage of exogenous H_2S or NO application. For instance, the level of NO was reduced in plant tissues that were treated with H_2S modulator (NaSH) [126,147]. However, crosstalk of NO-H_2S showed synergistic interaction during abiotic stresses and inhibition of ethylene-induced fruit ripening, whereas antagonistic interaction of H_2S-NO-ethylene is also reported [16,148–150]. The discrepancy in H_2S and NO interaction may depend upon the specific location of these gaseous molecules in the cell that deicide their signaling behavior [151]. There is also a possibility that both gaseous molecules may compete for the same targeting protein in the cell. For example, SnRK2.6 is a target of both NO and H_2S biomolecules, and S-nitrosylation of SnRK2.6 via NO inhibits its activity while persulfidation enhances its activity and mediate stomatal movements [135,137]. Additionally, H_2S and NO could react among themselves to produce nitrosothiol compounds that are also involved in signaling responses. The crosstalk of ROS with H_2S–NO cascades also modulates their interactions in positive or negative ways [152]. Taken together, the nature of the interaction between NO and H_2S may vary for different physiological functions based upon their location and concentration in the cell.

NO and H_2S belong to the family of reactive nitrogen and sulfur species (RNS and RSS), and their positive combinations regulate various important physiological and molecular processes in plants. For example, the interaction of H_2S with NO and Ca^{2+} regulate lateral root (LR) formation in tomato plants. The exogenous application of NO triggers the accumulation of H_2S in tomato roots due to the upregulation of H_2S biosynthesis enzymes, which induce later root formation [6]. However, when H_2S inhibitor/scavengers were applied, LRs' formation was partially arrested. These findings indicate that NO-induced H_2S synthesis governs the later root formation [6,153].

Stomatal movements are regulated by many endogenous signaling molecules; among them, H_2S and NO crosstalk are also responsible for stomatal closure. In a recent study, with the employment of pharmacological, spectrophotographic, and fluorescence microscope techniques, the coordinated action of H_2S and NO in the presence of 2,4-epibrassinolide (EBR) was involved in stomatal regulation [154,155]. The authors demonstrated the application of EBR-induced stomatal closure in a dose and time-dependent manner via modifying the levels of NO, and H_2S in *Vicia faba*. The application of EBR upregulated the activity of L-/D-cysteine desulfhydrase and enhanced the endogenous levels of H_2S together with H_2O_2 and NO generation in guard cells. The application of the H_2S inhibitor significantly reduced L-/D-cysteine desulfhydrase activity and H_2S endogenous production, which in turn abolished the EBR mediated stomatal closure effect [154]. The H_2S scavengers/inhibitors did not affect the NO and H_2O_2 levels in guard cells. However, the application of NO and H_2O_2 inhibitors/modulators significantly affected the endogenous production of H_2S and its biosynthesis enzymes and compromised the EBR-induced stomatal closure [154]. Similarly, Jing et al. [156] found that H_2S may function downstream of NO in ethylene-induced stomatal closure in *V. faba*. These results indicate that H_2S and NO participate in EBR-mediated stomatal closure response and H_2S signifies an essential constituent downstream of H_2O_2 and NO in EBR-induced stomatal closure in *V. faba* [154,157]. Previous studies demonstrated that H_2S inhibits ABA-mediated NO generation in Arabidopsis and *Capsicum annuum* guard cells. Conversely, H_2S increased NO levels in alfalfa seedlings [55,147], while H_2S induces NO generation in Arabidopsis guard cells. Conversely, NO scavenger inhibited H_2S-induced stomatal closure [145]. However, investigation of H_2S-mediated guard cell signaling in Arabidopsis revealed that the H_2S induced signaling cascade for stomatal closure is NO-dependent [128], and both H_2S and NO equally contribute to the production of 8-mercapto-cGMP, which triggers stomatal closure. In the same way, H_2S and NO collaborate in ethylene induce stomatal closure responses in Arabidopsis plants, and H_2S generation is mediated by NO, which suggests that H_2S acts as a downstream signaling agent in ethylene induce stomatal closure [158].

The crosstalk of H_2S and NO in the alleviation of metal toxicity is also reported, but these studies focused more on stress physiology and lacked underlying molecular mechanisms of crosstalk [159]. The exogenous application of H_2S donor alleviated Cd stress in alfalfa plants by triggering the synthesis of NO. The interaction mechanism between H_2S and NO improved the Cd stress tolerance by reducing Cd accumulation and lowering the lipid peroxidation in stressed plants [136]. Another study, where H_2S and NO scavenger and inhibitor were applied to Cd stressed bermudagrass plants, revealed that depletion of NO makes them more vulnerable to metal toxicity. Furthermore, through pharmacological experiments, it was demonstrated that NO-activated H_2S was essential for cadmium stress responses in bermudagrass [160]. In *Pisum sativum*, positive interaction of NO and H_2S was also explored under arsenate stress [109]. The application of H_2S donor triggered endogenous H_2S and NO accumulation in *P. sativum*, which led to the strengthening of the antioxidant defense system, reduced arsenate accumulation, and maintained the redox balance of *P. sativum* plant under metal toxicity [109]. Similarly, the crosstalk of NO and H_2S reduced oxidative stress and increased salinity tolerance in alfalfa, while barley seedlings under H_2S application regulate ion homeostasis under salinity via maintaining the NO signaling pathway [8,146]. Most of the published studies on the interaction of NO and H_2S in the context of metal toxicity/salinity proposed that crosstalk of these gaseous molecules ameliorates stress-induced toxicity in exposed plants via (i) improving the antioxidant defense to prevent oxidative stress, (ii) reducing the metal uptake, and (iii) by modulating the expression of associated metal transporter genes [159].

In short, H_2S and NO are both gaseous biomolecules with common signaling pathways, and it seems that one pathway controls the functions of the other [159]. The persulfidation promoted by H_2S reacts with thiol groups in the same way as NO does in modification through S-nitrosation [159,161]. However, there is still a need to investigate the interaction of H_2S and NO in different plant species, tissues, and diverse environmental conditions to unveil the regulatory mechanism of the NO–H_2S signaling cascade in plants.

4.3. H_2S-Mediated Manipulation of Auxin Signaling in Plants

The development of roots, including lateral and adventitious roots, is incredibly important for normal plant growth and the successful completion of the life cycle. Plant root architecture is mainly based on the LR that is generated from pericycle founder cells [155]. The plant hormone auxin and environmental factors (i.e., water and nutrient availability) are key influencers in lateral root formation [162,163]. Since auxin is a master regulator of root development in plants, there have always been complex crosstalks of auxin with other signaling agents in the root development [162,164,165].

Several studies have reported that H_2S and auxin interact with each other to regulate root growth; however, mechanistic insight remains to be elucidated [120,154,166]. The earlier studies demonstrated that the application of exogenous H_2S on the sweet potato seedling stimulated the numbers and length of adventitious roots by modulating the IAA levels in a dose-dependent manner [154]. It was also noted that pretreatment of H_2S donor upregulated the transcript abundance of the auxin-dependent Cyclin-Dependent Kinases gene (*CDKA1*) and a cell cycle regulatory gene (*CYCA2*) [153,165]. The activity of both of these genes was inhibited either by auxin blocker or H_2S inhibitor, which illustrated that H_2S mediated LR development is dependent upon the IAA signaling via influencing the regulation of *CDKA1* and *CYCA2* [153,165]. Similarly, when higher doses of H_2S donor (1 mM) were applied, *the RBOH1* (respiration burst oxidase homologous) transcript was significantly upregulated and ROS accumulation triggered the later root formation [115] (Figure 5). The pharmacological studies revealed that H_2S triggered the expression activity of *RBOH1*, which stimulated an H_2O_2-mediated increase in IAA signaling via regulation of *CDKA1*, *CYCA2*, and *Kip-Related Protein 2* (*KRP2*), to activate LR formation [115]. A transcriptomic study revealed that exogenous application of H_2S impacted the regulation of various auxin pathway-related genes. The accumulation of auxin biosynthesis genes (*TAA1* and *UGT74B1*) was correlated with the increase in auxin levels in roots. The genes involved in auxin polar subcellular distribution, such as *PIN2*, *ABCB1*, *ABCB19*, *PILS3*, and *PILS7*, were differentially expressed, while *PIN1c* appeared as a hub gene on the basis of WGCNA analysis. This study provides sufficient evidence that H_2S induced root development emanates from regulating the genes involved in transcriptional control and synthesis of auxin [166] (Figure 5).

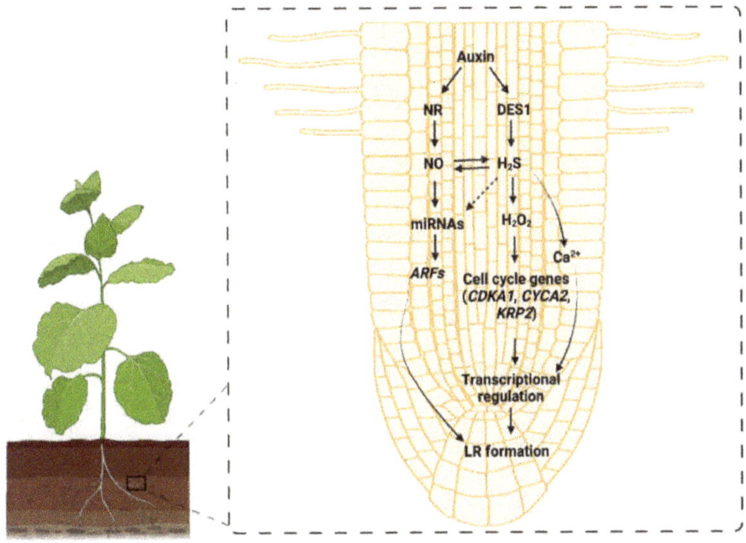

Figure 5. Schematic representation of the signaling pathways involving auxin, DES (cysteine desulfhydrase), NO (Nitric oxide), and hydrogen sulfide (H_2S) interaction during lateral root formation in plants. The interaction between H_2S and NO under the influence of auxin participates in the development of the lateral root via modulating the expressions and activities of different effector genes or proteins in a framework of regulatory pathways to permit root growth. miRNA: Micro RNA; ARFs: Auxin Response Factors; *CDKA1*: *Cyclin-Dependent Kinases* gene; *CYCA2*: *cell cycle regulatory* gene; *KRP2*: Kip-Related Protein 2; NR: Nitrate reductase; LR: lateral root.

In some studies, the application of higher dosages of H_2S showed changes in root development and inhibition of auxin transport due to the alteration in the polar subcellular distribution of the PIN proteins [166]. The polar subcellular movement of auxin in root cells is an actin-dependent process, and H_2S is involved in the regulation of actin dynamics due to the persulfidation and depolymerization of F actin [167]. Furthermore, during root hair development, the H_2S fine-tuned polar auxin transport via persulfidation and actin filament growth [167,168]. In the root developmental process, actin-binding proteins work downstream of the H_2S signal transduction pathway because actin-binding proteins are involved in the depolymerization of F-actin in root cells, which regulate the distribution and transport of auxin [168]. Auxin affects the patterning and organization of the actin cytoskeleton in root cells during cellular growth [169,170]. Conversely, the actin cytoskeleton modulates the directional transport of auxin by altering auxin efflux carriers [171,172]. This finding indicates that overproduction of H_2S significantly increases the S-sulfhydration level of actin-2 and decreases the distribution of actin cytoskeleton in root cells, thereby reducing auxin's polar transport, which restricts the LR and the root hair growth [44,167,168].

The exposure of plants to CH_4 strongly induces H_2S production and affects the root growth, adventitious root numbers, and root length in cucumber explants [106,173]. At the transcriptional level, it was observed that H_2S modulated auxin-signaling genes (*Aux22D-like* and *Aux22B-like*) reinforce the CH_4-induced cucumber adventitious rooting network [111,173–175]. Similarly, in tomato plants, LRs formation was also triggered by the CH_4-mediated H_2S signaling cascade. It was hypothesized that the possible involvement of auxin transport and auxin signaling in CH_4-induced LR formation is involved [176]. However, more biochemical and genetic investigations are required to analyze the detailed targets and their functions in root organogenesis under CH_4-H_2S-Auxin crosstalks [173,176].

The signaling pathways of H_2S and auxin interaction under the chilling stress were recently explored in cucumber plants [177–179] (Figure 6). The study demonstrated that chilling stress in cucumber arrested photosynthesis and induced oxidative stress; however, deleterious effects were alleviated due to exogenous application of H_2S donor or IAA application [179]. The expression of *YUCCA2* (auxin biosynthesis gene) and auxin contents were very high in chilling-exposed cucumber seedlings. This result may be due to the inhibition of polar transport of IAA in long-term chilling stress, which increases auxin concentration in leaves and inhibits plant growth. The complex interaction of H_2S and IAA under chilling stress improved the activities and gene expression of key enzymes of the Calvin–Benson cycle (Ribulose-1,5-bisphosphatecarboxylase, fructose bisphosphatase, sedoheptulose-1,7-bisphosphatase, fructose-1,6-bisphosphate aldolase, and transketolase) and strengthened the photosynthetic carbon assimilation capacity [179] (Figure 6). The results also indicated that auxin is a downstream signal for the protective effects induced by H_2S under chilling-induced tolerance in cucumber plants [179]. Furthermore, the overexpression of *auxin response factor 5* (*ARF5*) in cucumber unveiled the molecular mechanism of cold tolerance. In transgenic plants overexpressing *ARF5* under cold stress, ARF5 directly activates the expression of *dehydration-responsive element-binding protein 3* (*DREB3*) for the reinforcement of auxin signaling to improve cold stress tolerance in cucumber in response to H_2S application [180] (Figure 6). Previously, it was observed that *auxin response factors* (*ARFs*) and miR390 formed an auxin-responsive regulatory network (miR390-TAS3-ARF2/ARF3/ARF4) that strengthens auxin signaling in plants [181].

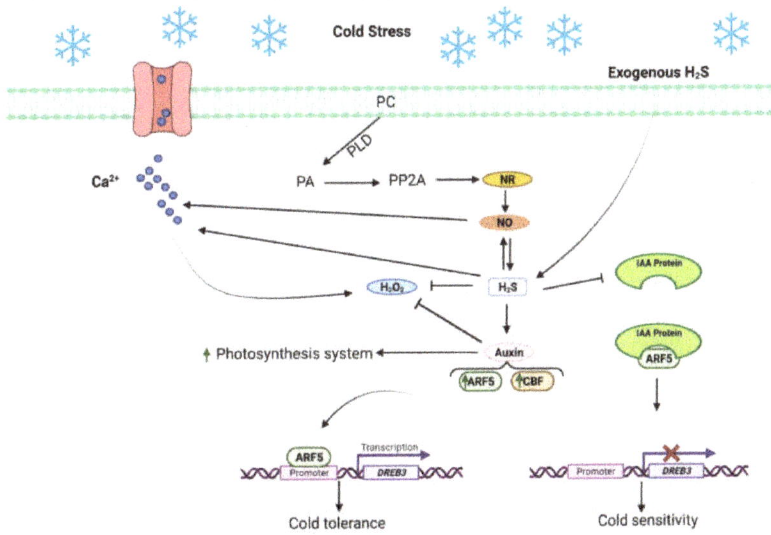

Figure 6. A regulatory model elucidates the role of hydrogen sulfide (H_2S) in mediating the cold stress response in plants via auxin signaling. In the presence of cold stress, the phospholipase D (PLD) is activated and degrades the phosphatidylcholine (PC) phospholipid of the cell membrane. As a result, phosphatidic acid (PA) is produced, which further regulates protein phosphatase 2A (PP2A), nitrate reductase (NR), nitric oxide (NO), and finally H_2S. In the absence of H_2S, auxin distribution, photosynthesis, and carbon assimilation are inhibited in plants under exposure to cold stress. The exogenous application or endogenous H_2S mediate auxin redistribution in plants and activate the antioxidant defense system along with improved photosynthesis to restore the normal function of the plant at physiological levels. On the other hand, C-repeat binding factors (CBFs) and ARF (auxin-responsive proteins) promote the dehydration-responsive element-binding (DREB) and other related proteins to promote cold tolerance at molecular levels under H_2S-mediated signaling.

4.4. Interaction between H_2S and Gibberellic Acid

Gibberellic acid (GA) is a phytohormone that substantially influences the seed germination and growth of seedlings. Imbibition of barley grains in 0.25 mM NaHS solution caused an upsurge in antioxidant enzymes such as CAT, POD, APX, and SOD in the aleurone layer [182]. In tomato plants, boron stress reduced dry weight, photosynthetic rate, water content, chlorophyll content, and increased H_2O_2, MDA, and endogenous H_2S. GA foliar spray reduced the harmful effects of boron by raising endogenous H_2S, Ca^{2+}, and K^+, as well as lowering the levels of H_2O_2, MDA, and boron, as well as membrane leakage. Surprisingly, NaHS further increased GA-induced boron tolerance, whereas H_2S scavengers prevented it (HT). These findings indicate that H_2S plays a signaling role downstream of GA in the development of boron stress tolerance in tomato plants. During cadmium stress, the NaHS treatment stimulated the activities of amylase and antioxidant enzymes in cucumber hypocotyls and radicles, which might be connected to H_2S-induced Cd stress tolerance.

Moreover, GA can cause programmed cell death (PCD); however, NaHS application can prevent PCD by lowering L-cysteine desulfhydrase (LCD) activity and accumulating endogenous H_2S in wheat aleurone layers [49]. GA-induced PCD is reduced in the aleurone layer in the NaHS-treated seeds by diminishing the endogenous GSH levels. H_2S concentration regulates the GSH levels, which upsurges expression of the *HEME OXYGENASE-1* (*HO-1*) gene, resulting in the alleviation of apoptosis in the aleurone layer and an overall decrease in PCD. Hence, in the aleurone layer, there are regulatory interactions between GA, H_2S, GSH, and HO-1. Intriguingly, NaHS pretreatment slowed Arabidopsis seed germination, but Arabidopsis *des1* mutant seedlings were more susceptible to ABA than the wild-type. These findings suggest that H_2S interacts with GA in plants to control seed germination under normal and stressful circumstances.

4.5. Interaction between H_2S and Melatonin

Melatonin (N-acetyl-5-methoxytryptamine) is a multifaceted phytohormone involved in germination, ripening, flowering, photosynthesis, and defense mechanisms [183]. In plants, melatonin alters the permeability of the cell layer governed by ion transporters, which control stomatal opening and closure. Studies have shown that melatonin can increase the photosynthetic capacity of plants, which leads to greater levels of nitrogen and chlorophyll. In tomato and wheat, increased transcription of stress-responsive genes was induced by melatonin, resulting in better tolerance to high temperature [184,185]. Furthermore, melatonin cross-talks with various plant hormones and signaling molecules. It was also discovered that H_2S and melatonin conjointly helped alleviate salt stress-induced growth reduction in tomatoes, and exogenous melatonin treatment assisted in regulating early H_2S signaling [186]. In wheat, the heat stress-induced oxidative damage was mitigated by exogenous melatonin and further increased the H_2S production, suggesting that melatonin-mediated H_2S was involved in alleviating the oxidative stress. However, the melatonin function was attenuated when H_2S was inhibited by its inhibitor, indicating that the cross-talk between H_2S and melatonin, and possibly melatonin, regulates heat stress signaling by acting upstream of H_2S [187].

5. H_2S-Plant Hormone Cross-Talk under Pathogen Attack

In plants, the dual roles of H_2S in interactions with phytohormones determine the biological roles of H_2S in plant growth, development, and responses to biotic stresses. In response to biotic stresses, the crosstalk between H_2S and phytohormones, as well as several other signaling molecules, has been studied less; however, some critical molecular insights have been found in the recent past. In the following paragraph we discuss the H_2S–phytohormone interplay under biotic stress.

5.1. Interaction between H_2S and Salicylic Acid

Salicylic acid (SA) is a phytohormone that triggers a defense response in plants against biotrophic and hemibiotrophic phytopathogens. SA activates a large number of defense-related genes, especially those that encode pathogenesis-related (PR) proteins [188,189]. Susceptibility to virulent and avirulent pathogens develops as a result of mutations that impede SA production. In *Nicotiana tabacum* cv. Xanthi-nc, acetyl SA (aspirin) confers resistance to tobacco mosaic virus [190]. Previously, it was found that the expression of multiple WRKY transcription factors (TFs) is modulated by pathogen attack or SA treatment [191]. A subsequent study has shown that the mutation in *WRKY18*, *WRKY40*, and *WRKY60* resulted in the up-regulation of *LCD*, *DES*, *DCD1*, and higher production of H_2S in Arabidopsis [192]. In Arabidopsis, the expression level of a *PR* gene-regulating transcription factor *WRKY54* was elevated in *des1* mutants and decreased in *oas-a1* mutants [193]. Furthermore, *des1* mutants had lower levels of L-glutathione oxidation than *oas-a1* mutants, and lesser intracellular redox potential was caused by higher L-Cys levels in *des1* mutants, which may help boost plant resistance to pathogen invasion [193]. Later, Alvarez et al. [194] demonstrated that Arabidopsis *des1* mutants have increased amounts of SA and developed more resilience against *Pseudomonas syringae* pv. *tomato* (*Pst*) DC3000 *avrRpm1*, while *oas-a1* mutants were more vulnerable to this pathogen [194]. The *des1* mutants exhibited all the constitutive systemic acquired resistance characteristics, including high resistance against biotrophic and necrotrophic pathogens, accumulation of salicylic acid, and induction of *WRKY54* and *PR1* [194]. In contrast to the *oas-a1* mutants, Arabidopsis *cad2-1* mutants showed lower levels of L-glutathione but a non-significant change in the L-Cys levels. In *cad2-1* mutants, repression of *WRKY54* was also not observed, which suggests that lower expression of *PR* genes in *oas-a1* mutants might be due to reduced L-Cys level [192]. In order to determine if L-Cys is involved in plant immunity, researchers exposed *oas-a1* mutants to the bacterial pathogen *Pst* DC3000, which releases effectors that suppress PAMP-triggered immunity (PTI). The Arabidopsis *oas-a1* mutant plants were shown to be more susceptible to infection by this pathogen [195]. Thus, the results from the previously mentioned studies suggest that higher L-Cys decreases cytoplasmic redox potential, which may play a key role in pathogen defense in Arabidopsis and other plant species. Still, more research is needed in Arabidopsis and other plant species.

Among SA-biosynthesis genes in Arabidopsis, the phytoalexin deficient (*PAD*) genes (*PAD1*, *PAD2*, *PAD3*, and *PAD4*) encode regulatory proteins that function against the eukaryotic biotroph *Peronospora parasitica* and promote resistance to downy mildew [196]. Increased sensitivity to the bacterial pathogen *Pst* DC3000 has been observed in the *pad1*, *pad2*, and *pad4* mutants [196]. Enhanced disease susceptibility1 (*EDS1*) gene codes for a lipases-like protein that acts in resistance (*R*) gene-dependent effector-triggered immunity and contributes to basal defense in plants. EDS1 is also required for pathogen-induced *PAD4* mRNA accumulation [197]. The *PAD4* and *EDS1* genes involved in SA biosynthesis were found to be constitutively activated in Arabidopsis plants with high H_2S concentrations but found to be reduced in plants with low H_2S levels (Figure 7) [58]. NPR1 plays an essential function in SA signaling because it binds SA and initiates a SAR response [198–200]. Other similar molecules such as methyl salicylate (MeSA) or gentisic acid promote *PR1* expression in addition to SA [201]. The deposition of SA is required for triggering the expression of SA-mediated genes, such as *PR*s [189]. Plants with greater H_2S levels showed increased expression of SA-mediated *PR* genes, which improved pathogen resistance, and vice versa (Figure 7).

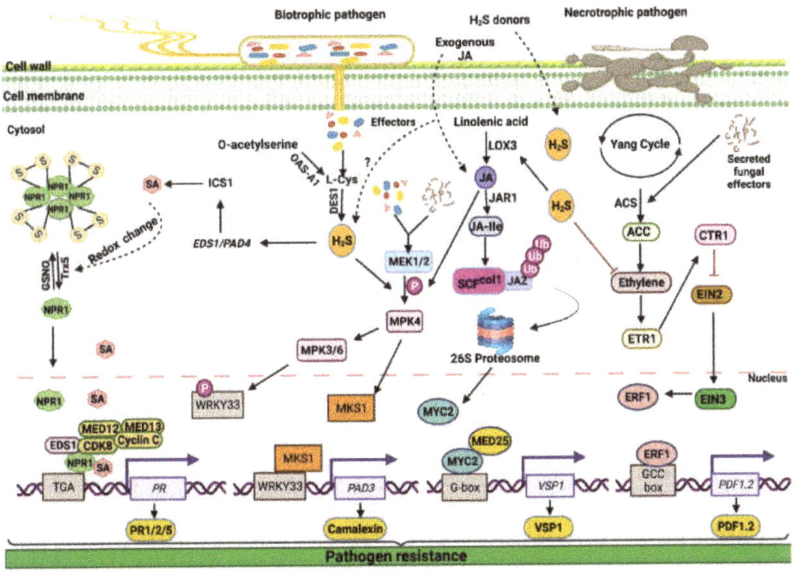

Figure 7. A schematic model of the cross-talks between H_2S and salicylic acid (SA), jasmonic acid (JA), and ethylene (ET) in plant defense against pathogens. The biotrophic pathogen attacks plants and secretes effectors into plant cells. The conversion of O-Acetylserine into L-Cysteine (L-Cys) is catalyzed by Anthranilate synthase (OAS-A1). Similarly, effectors also induce the biosynthesis of L-Cys. The plant cytosol contains the enzyme L-cysteine desulfhydrase (DES1), which is responsible for L-Cys decomposition and endogenous H_2S production. The higher concentration of H_2S triggers the upregulation of SA biosynthesis-related genes (*PAD4*/*EDS1*). The enzyme ICS1 catalyzes the conversion

of chorismite into isochorismate, which is then exported to the cytosol by EDS5. The L-glutamate is converted into isochorismate-9-glutamate in the cytosol by PBS3. Subsequently, SA is produced from isochorismate-9-glutamate through spontaneous decay. By acting as an isochorismate A pyruvoyl-glutamate lyase (IPGL), EPS1 also degrades N-pyruvoyl-L-glutamate to create SA. The *NPR1* gene expression is aided by SA due to the interaction of WRKY transcription factors with NPR1, which promotes the recruitment of CDK8 to the *NPR1* promoter's W-box. Pathogen-induced defense signals enhance the accumulation of salicylic acid (SA) in plants by enhancing the expression of *Isochorismate Synthase* (*ICS*) genes. In addition, SA promotes redox reactions that lead to the reduction of NPR1 oligomers to monomers. The monomeric NPR1 molecules move from the cytosol to the nucleus, where they form a protein complex with transcription factor (TGA), EDS1, SA, and CDK8, resulting in the transcription of *PR* genes. A higher concentration of H_2S upregulates the JA biosynthetic gene *LOX3*. Moreover, the exogenous application of JA also increases the endogenous H_2S and JA. The secreted effectors by biotrophic and necrotrophic pathogens trigger the pattern recognition receptors (PRR), which further activate the plant mitogen-activated protein kinase (MEK1/2) cascades. H_2S and JA participate in phosphorylation of MEK1/2, subsequently triggering MPK4. The MPK4 activates the MPK3/MPK6 and MKS1 (the substrate of MPK4). WRKY33 is involved in the biosynthesis of camalexin (a phytoalexin). MPK3/MPK6 phosphorylate the WRKY33 and increase its transactivation activity. The WRKY33 forms a complex with MKS1 for the transcription of *PAD3*, which activates the biosynthesis of camalexin. In the elicited cells, JA-Ile COI1, an F-box protein in the SCF ubiquitin E3 ligase complex, recognizes JA-Ile and facilitates the binding between COI1 and the JAZ family of repressor proteins, resulting in JAZs being ubiquitinated. The 26S proteasome then degrades the ubiquitinated JAZs. JAZ degradation promotes downstream JA responses by releasing the target transcription factor (MYC2) from inhibition. The Mediator25 binds to the MYC2 to enhance the transcriptional activity of wound-responsive gene *VSP1*. H_2S molecules work as a repressor for ethylene signaling. In response to the effectors of the necrotrophic pathogen, the ethylene biosynthesis genes 1-aminocyclopropane-1-carboxylic acid synthase (ACS), 1-aminocyclopropane-l-carboxylic acid (ACC) are activated, resulting in the formation of ethylene. Under normal growth conditions with low ethylene levels, the Ethylene receptor 1 (ERT1) remains in the active state and associates with CTR1, which, in turn, inhibits the downstream signaling pathway. The ethylene binding inactivates its receptors and in turn deactivates the Raf-like kinase CTR1. Consequentially, EIN2 can function and signal positively downstream to the ethylene insensitive 3 (EIN3) of transcription factors situated in the nucleus. EIN3 drives the expression of ethylene response factor (*ERF1*). Subsequently, the ERF1 binds to the GCC box and invokes the *PDF1.2* defense gene.

5.2. Interaction between H_2S and Jasmonic Acid

Jasmonic acid (JA) is a lipid-derived signaling molecule that plays a significant role in many biological processes in plant cells. Herbivorous insects chewing on the leaves or necrotrophic diseases trigger the JA response pathway. Plants have evolved to remember these attacks and employ this pre-conditioned situation effectively and to their benefit in a mechanism termed induced systemic resistance (ISR). Interestingly, the biological pathways of JA and SA have been reported to function antagonistically [202]. JA and SA enhance plant defense against nematodes such as *M. incognita* [203]. This pathogen causes plants to trigger SA pathways and prevent JA in leaves to permit successful invasion of the pathogen. Furthermore, JA showed a higher concentration in roots following the nematodic infection that is subsequently transferred to leaves, helping plants to defend themselves against pathogens [204]. In another study, when Arabidopsis was deprived of the sulfur element, it led to activation of the JA and SA metabolism; but the plant showed susceptibility to necrotrophic *Botrytis cinerea* [205]. This discovery suggests that the presence of sulfur-containing compound H_2S is essential for plant defense mechanisms through its interaction with SA and JA.

H_2S interacts with JA to promote pathogen resistance in plants (Figure 7). The redox state of ascorbate is shown to be regulated in the leaves of *A. thaliana* by the interaction between H_2S and mitogen-activated protein kinase (MEK1/2) (Figure 7) [206]. In Arabidopsis, the exogenous application of JA resulted in a significant increase in endogenous

H_2S generation, MEK1/2 phosphorylation, and a lower ascorbate to dehydroascorbate ratio (AsA/DHA) [195]. The increase in the phosphorylation level of MEK1/2, endogenous H_2S generation, and the AsA/DHA ratio in wild-type hosts was shown to be caused by hypotaurine (HT), an H_2S scavenger, resulting in a decrease in JA. The application of sodium hydrosulfide, which acts as an H_2S donor in mutant *A. thaliana* plants, was observed to enhance these indicators. When these mutant plants were given an application of NaHS after being treated with HT and JA, the effects of hypotaurine on those JA-induced indicators were not reversed.

5.3. Interaction between H_2S and Ethylene

Phytohormones play a critical role in the defense mechanism in plants against various pathogens. SA often controls biotrophic and hemibiotrophic pathogen defense responses, but ethylene and JA promote defense responses to necrotrophic pathogens. However, sometimes hormone signal transduction pathways that conferred resistance and vulnerability were found to be diametrically opposed. Plant resistance was shown to be associated with an increase in SA signaling, whereas susceptibility was found to be associated with an increase in the ethylene pathway and a decrease in SA and cytokinin signaling. According to Foucher et al. [207], two *Phaseolus vulgaris* L. genotypes (resistant and susceptible) were screened against common bacterial blight caused by *Xanthomonas phaseoli* pv. *phaseoli*. The transcriptomic study revealed that resistance was associated with an increase in the SA pathway and a decrease in photosynthetic activity as well as sugar metabolism. Susceptibility was associated with an increase in the ethylene pathway and genes that modify cell walls, as well as a decrease in the downregulation of resistance genes [207].

Pathogenic bacteria cannot form merism when exposed to exogenous NaHS, which helps plants recover from infection [208]. Fumigation with H_2S has been shown to suppress spore germination, mycelial growth, and pathogenicity of *Monilinia fructicola* in peach fruit, as well as *Aspergillus niger* and *Penicillium expansum* in pear [209]. These findings show that H_2S can promote a plant's resistance to pathogen infection, and that immunological signals and exogenous sulfide can both trigger the production of endogenous H_2S. Exogenous H_2S reversed the impacts of ETH by reducing the activity of enzymes involved in cell wall modification (cellulase and polygalacturonase) via transcription suppression rather than direct post-translational modification (sulfhydration) by H_2S [210]. H_2S also controlled the expression of *SlIAA3*, *SlIAA4*, *ILR-L3*, and *ILR-L4* (all of which are involved in auxin signaling), which suppressed petiole abscission by controlling the amount of free auxin in tomato abscission zone cells. In rose and lily plants, similar findings were observed in floral organ abscission and anther dehiscence [210]. These findings suggest that H_2S interacts with ethylene and auxin during plant organ abscission.

Exogenous ethylene donor (ethephon) stimulated the activities of LCD and DCD in Arabidopsis and *Vicia faba* plants, resulting in H_2S production in guard cells and stomatal closure, whereas H_2S-synthesis inhibitors (PAG) reversed ethylene-induced stomatal closure, indicating H_2S-mediated ethylene-induced stomatal closure [211]. Furthermore, early leaf senescence was seen in Arabidopsis *des1* mutants (due to reduced endogenous H_2S content), whereas NaHS treatment reversed the senescence and extended the vase life of cut flowers by elevating endogenous H_2S levels. In addition, by reducing ethylene synthesis, H_2S-delayed senescence was seen in green leafy crops [212]. These findings demonstrate that ethylene promotes stomatal closure and organ senescence in plants by independently increasing and suppressing endogenous H_2S generation.

6. Conclusions and Future Prospects

For a long time, H_2S was considered an undesirable by-product of sulfur metabolism, which could adversely affect plant cells. However, this perception was altered after it was discovered that H_2S could have signaling properties. H_2S is involved in many plant processes and can interact with other phytohormones to mitigate stress in plants. However, most research is focused on the H_2S interaction with phytohormones under abiotic stress.

In contrast, there is very limited research progress on the interaction of H_2S with SA, JA, and especially, ethylene in plants under biotic stresses. The exogenous ethylene donor (ethephon) stimulated the activities of LCD and DCD in Arabidopsis and *V. faba* plants, resulting in H_2S production in guard cells and stomatal closure, whereas H_2S-synthesis inhibitors (PAG) reversed ethylene-induced stomatal closure, indicating H_2S mediates ethylene-induced stomatal closure [211]. Since ethylene promotes stomatal closure, it might prevent the invasion of pathogens. Therefore, it is likely the crosstalk between H_2S and ethylene plays a pivotal role in the regulation of stomatal closure during plant defense against pathogen invasion, which warrants further investigation.

In plants, the H_2S-mediated persulfation can significantly impact protein function, altering protein conformation and regulating protein activity under stress response. According to Chen et al. [135], H_2S positively regulates abscisic acid signaling by sulfidating SnRK2.6 in guard cells. H_2S has also been reported to persulfidate MAPK in Arabidopsis to alleviate cold stress [213]. Numerous studies have been conducted to understand H_2S-mediated persulfation of proteins in plants under abiotic stress; however, H_2S-mediated persulfation is not studied sufficiently in plant–pathogen interaction. H_2S can also be involved in protein functions through trans-persulfidation and regulating cellular redox state in other unexplored H_2S-related molecules in the plant metabolism such as glutathione persulfide (GSSH) and cysteine persulfide (CysSSH).

In future studies, more fundamental research is required to investigate the fate and regulation of endogenous H_2S production, and its subsequent interaction with and regulation of different plant processes under laboratory as well as in field conditions. However, the exogenous application of H_2S on plants in controlled conditions has generated plenty of experimental results that have explained at least some of the underlying mechanisms of actions driven by H_2S molecules in plants. In the animal field, several exogenous sources of H_2S have been utilized that can slowly release H_2S in media (mimicking the natural generation of H_2S). However, for plants, NaHS and inorganic sodium polysulfides (Na_2S_n) such as Na_2S_2, Na_2S_3, and Na_2S_4 are currently used in various research reports to study the H_2S impacts in plants. The NaHS and related H_2S generation compounds are usually short-lived donors and do not mimic the slow release of H_2S in in-vivo conditions. Recently, dialkyldithiophosphate demonstrated the potential to release H_2S slowly and enhance the maize plant biomass upon application [214]. In addition, more precise and advanced methods of H_2S application to the plants under various growth stages and environmental stresses, and H_2S suitable dosages for different crop species are also required.

Author Contributions: M.S.S.K. and F.I.: original draft preparation; M.S.S.K.: figure preparation; J.C. and Z.Q.F.: supervision and conceptualization; Y.Y., M.A., D.W., B.Z., Z.Q.F. and J.C.: reviewing and editing; B.Z. and J.C.: funding acquisition. All authors have read and agreed to the published version of the manuscript.

Funding: This work is supported by grants from Jiangsu University High-level Talent Funding (20JDG34), Natural Science Foundation of Jiangsu Province (BK20211319), and National Natural Science Foundation of China (32000201) to J.C., and National Natural Science Foundation of China (No. 31800386) and Chinese Postdoctoral Science Found (No. 2019M651721) to Z.B.

Institutional Review Board Statement: Not applicable.

Informed Consent Statement: Not applicable.

Data Availability Statement: Not applicable.

Acknowledgments: We would like to acknowledge BIORender.com, which was used to create all figures.

Conflicts of Interest: The authors declare no conflict of interest.

References

1. Kour, J.; Khanna, K.; Sharma, P.; Singh, A.D.; Sharma, I.; Arora, P.; Kumar, P.; Devi, K.; Ibrahim, M.; Ohri, P. Hydrogen sulfide and phytohormones crosstalk in plant defense against abiotic stress. In *Hydrogen Sulfide in Plant Biology*; Elsevier: Amsterdam, The Netherlands, 2021; pp. 267–302. [CrossRef]
2. Swarup, R.; Perry, P.; Hagenbeek, D.; Van Der Straeten, D.; Beemster, G.T.; Sandberg, G.r.; Bhalerao, R.; Ljung, K.; Bennett, M.J. Ethylene upregulates auxin biosynthesis in Arabidopsis seedlings to enhance inhibition of root cell elongation. *Plant Cell* **2007**, *19*, 2186–2196. [CrossRef] [PubMed]
3. Isoda, R.; Yoshinari, A.; Ishikawa, Y.; Sadoine, M.; Simon, R.; Frommer, W.B.; Nakamura, M. Sensors for the quantification, localization and analysis of the dynamics of plant hormones. *Plant J.* **2021**, *105*, 542–557. [CrossRef] [PubMed]
4. Olson, K.R.; Straub, K.D. The role of hydrogen sulfide in evolution and the evolution of hydrogen sulfide in metabolism and signaling. *Physiology* **2016**, *31*, 60–72. [CrossRef] [PubMed]
5. Fike, D.A.; Bradley, A.S.; Rose, C.V. Rethinking the ancient sulfur cycle. *Annu. Rev. Earth Planet. Sci.* **2015**, *43*, 593–622. [CrossRef]
6. Li, Y.J.; Chen, J.; Xian, M.; Zhou, L.G.; Han, F.X.; Gan, L.J.; Shi, Z.Q. In site bioimaging of hydrogen sulfide uncovers its pivotal role in regulating nitric oxide-induced lateral root formation. *PLoS ONE* **2014**, *9*, e90340. [CrossRef] [PubMed]
7. Yamasaki, H.; Cohen, M.F. Biological consilience of hydrogen sulfide and nitric oxide in plants: Gases of primordial earth linking plant, microbial and animal physiologies. *Nitric Oxide* **2016**, *55*, 91–100. [CrossRef]
8. Chen, J.; Wang, W.H.; Wu, F.H.; He, E.M.; Liu, X.; Shangguan, Z.P.; Zheng, H.L. Hydrogen sulfide enhances salt tolerance through nitric oxide-mediated maintenance of ion homeostasis in barley seedling roots. *Sci. Rep.* **2015**, *5*, 12516. [CrossRef]
9. Chen, S.; Wang, X.; Jia, H.; Li, F.; Ma, Y.; Liesche, J.; Liao, M.; Ding, X.; Liu, C.; Chen, Y. Persulfidation-induced structural change in SnRK2. 6 establishes intramolecular interaction between phosphorylation and persulfidation. *Mol. Plant* **2021**, *14*, 1814–1830. [CrossRef]
10. Bonner, E.R.; Cahoon, R.E.; Knapke, S.M.; Jez, J.M. Molecular basis of cysteine biosynthesis in plants: Structural and functional analysis of O-acetylserine sulfhydrylase from Arabidopsis thaliana. *J. Biol. Chem.* **2005**, *280*, 38803–38813. [CrossRef]
11. Heeg, C.; Kruse, C.; Jost, R.; Gutensohn, M.; Ruppert, T.; Wirtz, M.; Hell, R.D. Analysis of the Arabidopsis O-acetylserine (thiol) lyase gene family demonstrates compartment-specific differences in the regulation of cysteine synthesis. *Plant Cell* **2008**, *20*, 168–185. [CrossRef]
12. Jez, J.M.; Dey, S. The cysteine regulatory complex from plants and microbes: What was old is new again. *Curr. Opin. Struct. Biol.* **2013**, *23*, 302–310. [CrossRef] [PubMed]
13. Birke, H.; Heeg, C.; Wirtz, M.; Hell, R. Successful fertilization requires the presence of at least one major O-acetylserine (thiol) lyase for cysteine synthesis in pollen of Arabidopsis. *Plant Physiol.* **2013**, *163*, 959–972. [CrossRef] [PubMed]
14. Nakayama, M.; Akashi, T.; Hase, T. Plant sulfite reductase: Molecular structure, catalytic function and interaction with ferredoxin. *J. Inorgan. Biochem.* **2000**, *82*, 27–32. [CrossRef]
15. Li, Z.G.; Xie, L.R.; Li, X.J. Hydrogen sulfide acts as a downstream signal molecule in salicylic acid-induced heat tolerance in maize (*Zea mays* L.) seedlings. *J. Plant Physiol.* **2015**, *177*, 121–127. [CrossRef] [PubMed]
16. Da-Silva, C.J.; Modolo, L.V. Hydrogen sulfide: A new endogenous player in an old mechanism of plant tolerance to high salinity. *Acta Bot. Bras.* **2017**, *32*, 150–160. [CrossRef]
17. Hatzfeld, Y.; Maruyama, A.; Schmidt, A.; Noji, M.; Ishizawa, K.; Saito, K. β-Cyanoalanine synthase is a mitochondrial cysteine synthase-like protein in spinach and Arabidopsis. *Plant Physiol.* **2000**, *123*, 1163–1172. [CrossRef]
18. Gotor, C.; García, I.; Aroca, Á.; Laureano-Marín, A.M.; Arenas-Alfonseca, L.; Jurado-Flores, A.; Moreno, I.; Romero, L.C. Signaling by hydrogen sulfide and cyanide through post-translational modification. *J. Exp. Bot.* **2019**, *70*, 4251–4265. [CrossRef]
19. Garcia-Arriaga, V.; Alvarez-Ramirez, J.; Amaya, M.; Sosa, E. H_2S and O_2 influence on the corrosion of carbon steel immersed in a solution containing 3 M diethanolamine. *Corros. Sci.* **2010**, *52*, 2268–2279. [CrossRef]
20. Shen, J.; Xing, T.; Yuan, H.; Liu, Z.; Jin, Z.; Zhang, L.; Pei, Y. Hydrogen sulfide improves drought tolerance in Arabidopsis thaliana by microRNA expressions. *PLoS ONE* **2013**, *8*, e77047. [CrossRef]
21. Riemenschneider, A.; Bonacina, E.; Schmidt, A.; Papenbrock, J. Isolation and characterization of a second D-cysteine desulfhydrase-like protein from Arabidopsis. In *Sulfur Transport and Assimilation in Plants in the Post Genomic Era*; Backhuys Publishers: Leiden, The Netherlands, 2005; pp. 103–106.
22. Riemenschneider, A.; Wegele, R.; Schmidt, A.; Papenbrock, J. Isolation and characterization of ad-cysteine desulfhydrase protein from Arabidopsis thaliana. *FEBS J.* **2005**, *272*, 1291–1304. [CrossRef]
23. Papenbrock, J.; Riemenschneider, A.; Kamp, A.; Schulz-Vogt, H.; Schmidt, A. Characterization of cysteine-degrading and H_2S-releasing enzymes of higher plants-from the field to the test tube and back. *Plant Biol.* **2007**, *9*, 582–588. [CrossRef] [PubMed]
24. Zhou, H.; Guan, W.; Zhou, M.; Shen, J.; Liu, X.; Wu, D.; Yin, X.; Xie, Y. Cloning and Characterization of a gene Encoding True D-cysteine Desulfhydrase from Oryza sativa. *Plant Mol. Biol. Rep.* **2020**, *38*, 95–113. [CrossRef]
25. Khan, M.N.; AlZuaibr, F.M.; Al-Huqail, A.A.; Siddiqui, M.H.; Ali, H.M.; Al-Muwayhi, M.A.; Al-Haque, H.N. Hydrogen sulfide-mediated activation of O-Acetylserine (thiol) Lyase and L/D-Cysteine desulfhydrase enhance dehydration tolerance in Eruca sativa mill. *Int. J. Mol. Sci.* **2018**, *19*, 3981. [CrossRef] [PubMed]
26. Corpas, F.J.; Del Río, L.A.; Palma, J.M. Plant peroxisomes at the crossroad of NO and H_2O_2 metabolism. *J. Integr. Plant Biol.* **2019**, *61*, 803–816. [CrossRef]

27. García, I.; Castellano, J.M.; Vioque, B.; Solano, R.; Gotor, C.; Romero, L.C. Mitochondrial β-cyanoalanine synthase is essential for root hair formation in Arabidopsis thaliana. *Plant Cell* **2010**, *22*, 3268–3279. [CrossRef]
28. Álvarez, C.; García, I.; Romero, L.C.; Gotor, C. Mitochondrial sulfide detoxification requires a functional isoform O-acetylserine (thiol) lyase C in Arabidopsis thaliana. *Mol. Plant* **2012**, *5*, 1217–1226. [CrossRef]
29. García, I.; Romero, L.C.; Gotor, C. *Cysteine Homeostasis*; CABI Publishing: Sao Paulo, Brazil, 2015; Chapter 12; pp. 219–233.
30. Feldman-Salit, A.; Wirtz, M.; Lenherr, E.D.; Throm, C.; Hothorn, M.; Scheffzek, K.; Hell, R.; Wade, R.C. Allosterically gated enzyme dynamics in the cysteine synthase complex regulate cysteine biosynthesis in Arabidopsis thaliana. *Structure* **2012**, *20*, 292–302. [CrossRef]
31. Wirtz, M.; Birke, H.; Heeg, C.; Müller, C.; Hosp, F.; Throm, C.; König, S.; Feldman-Salit, A.; Rippe, K.; Petersen, G. Structure and function of the hetero-oligomeric cysteine synthase complex in plants. *J. Biol. Chem.* **2010**, *285*, 32810–32817. [CrossRef]
32. Corpas, F.J.; González-Gordo, S.; Cañas, A.; Palma, J.M. Nitric oxide and hydrogen sulfide in plants: Which comes first? *J. Exp. Bot.* **2019**, *70*, 4391–4404. [CrossRef]
33. Corpas, F.J.; González-Gordo, S.; Muñoz-Vargas, M.A.; Rodríguez-Ruiz, M.; Palma, J.M. The modus operandi of hydrogen sulfide (H_2S)-dependent protein persulfidation in higher plants. *Antioxidants* **2021**, *10*, 1686. [CrossRef]
34. Corpas, F.J.; Barroso, J.B.; González-Gordo, S.; Muñoz-Vargas, M.A.; Palma, J.M. Hydrogen sulfide: A novel component in Arabidopsis peroxisomes which triggers catalase inhibition. *J. Integr. Plant Biol.* **2019**, *61*, 871–883. [CrossRef]
35. Corpas, F.J.; Palma, J.M. H_2S signaling in plants and applications in agriculture. *J. Adv. Res.* **2020**, *24*, 131–137. [CrossRef] [PubMed]
36. Choudhary, A.; Singh, S.; Khatri, N.; Gupta, R. Hydrogen sulphide: An emerging regulator of plant defence signalling. *Plant Biol.* **2021**. [CrossRef]
37. Xie, Y.; Lai, D.; Mao, Y.; Zhang, W.; Shen, W.; Guan, R. Molecular cloning, characterization, and expression analysis of a novel gene encoding L-cysteine desulfhydrase from Brassica napus. *Mol. Biotechnol.* **2013**, *54*, 737–746. [CrossRef] [PubMed]
38. Shen, J.; Su, Y.; Zhou, C.; Zhang, F.; Zhou, H.; Liu, X.; Wu, D.; Yin, X.; Xie, Y.; Yuan, X.A. *Putative* rice L-cysteine desulfhydrase encodes a true L-cysteine synthase that regulates plant cadmium tolerance. *Plant Growth Regul.* **2019**, *89*, 217–226. [CrossRef]
39. González-Gordo, S.; Palma, J.M.; Corpas, F.J. Appraisal of H_2S metabolism in Arabidopsis thaliana: In silico analysis at the subcellular level. *Plant Physiol. Biochem.* **2020**, *155*, 579–588. [CrossRef]
40. Cao, M.J.; Wang, Z.; Zhao, Q.; Mao, J.L.; Speiser, A.; Wirtz, M.; Hell, R.; Zhu, J.K.; Xiang, C.B. Sulfate availability affects ABA levels and germination response to ABA and salt stress in Arabidopsis thaliana. *Plant J.* **2014**, *77*, 604–615. [CrossRef]
41. Shan, C.J.; Zhang, S.; Li, D.F.; Zhao, Y.Z.; Tian, X.; Zhao, X.; Wu, Y.X.; Wei, X.Y.; Liu, R.Q. Effects of exogenous hydrogen sulfide on the ascorbate and glutathione metabolism in wheat seedlings leaves under water stress. *Acta Physiol. Plant.* **2011**, *33*, 2533. [CrossRef]
42. Zhao, N.; Zhu, H.; Zhang, H.; Sun, J.; Zhou, J.; Deng, C.; Zhang, Y.; Zhao, R.; Zhou, X.; Lu, C. Hydrogen sulfide mediates K+ and Na+ homeostasis in the roots of salt-resistant and salt-sensitive poplar species subjected to NaCl stress. *Front. Plant Sci.* **2018**, *9*, 1366. [CrossRef]
43. Zhou, M.; Zhang, J.; Shen, J.; Zhou, H.; Zhao, D.; Gotor, C.; Romero, L.C.; Fu, L.; Li, Z.; Yang, J. Hydrogen sulfide-linked persulfidation of ABI4 controls ABA responses through the transactivation of MAPKKK18 in Arabidopsis. *Mol. Plant* **2021**, *14*, 921–936. [CrossRef]
44. Zhang, J.; Zhou, H.; Zhou, M.; Ge, Z.; Zhang, F.; Foyer, C.H.; Yuan, X.; Xie, Y. The coordination of guard-cell autonomous ABA synthesis and DES1 function in situ regulates plant water deficit responses. *J. Adv. Res.* **2021**, *27*, 191–197. [CrossRef] [PubMed]
45. Zhou, H.; Zhou, Y.; Zhang, F.; Guan, W.; Su, Y.; Yuan, X.; Xie, Y. Persulfidation of Nitrate Reductase 2 Is Involved in l-Cysteine Desulfhydrase-Regulated Rice Drought Tolerance. *Int. J. Mol. Sci.* **2021**, *22*, 12119. [CrossRef]
46. Xuan, L.; Li, J.; Wang, X.; Wang, C. Crosstalk between hydrogen sulfide and other signal molecules regulates plant growth and development. *Int. J. Mol. Sci.* **2020**, *21*, 4593. [CrossRef] [PubMed]
47. Ma, D.; Ding, H.; Wang, C.; Qin, H.; Han, Q.; Hou, J.; Lu, H.; Xie, Y.; Guo, T. Alleviation of drought stress by hydrogen sulfide is partially related to the abscisic acid signaling pathway in wheat. *PLoS ONE* **2016**, *11*, e0163082. [CrossRef] [PubMed]
48. Xie, R.; Deng, L.; Jing, L.; He, S.; Ma, Y.; Yi, S.; Zheng, Y.; Zheng, L. Recent advances in molecular events of fruit abscission. *Biol. Plant* **2013**, *57*, 201–209. [CrossRef]
49. Xie, Z.; Shi, M.; Xie, L.; Wu, Z.Y.; Li, G.; Hua, F.; Bian, J.S. Sulfhydration of p66Shc at cysteine59 mediates the antioxidant effect of hydrogen sulfide. *Antioxid. Redox. Signal.* **2014**, *21*, 2531–2542. [CrossRef]
50. Zhou, H.; Chen, Y.; Zhai, F.; Zhang, J.; Zhang, F.; Yuan, X.; Xie, Y. Hydrogen sulfide promotes rice drought tolerance via reestablishing redox homeostasis and activation of ABA biosynthesis and signaling. *Plant Physiol. Biochem.* **2020**, *155*, 213–220. [CrossRef]
51. Ziogas, V.; Tanou, G.; Filippou, P.; Diamantidis, G.; Vasilakakis, M.; Fotopoulos, V.; Molassiotis, A. Nitrosative responses in citrus plants exposed to six abiotic stress conditions. *Plant Physiol. Biochem.* **2013**, *68*, 118–126. [CrossRef]
52. Ziogas, V.; Tanou, G.; Belghazi, M.; Filippou, P.; Fotopoulos, V.; Grigorios, D.; Molassiotis, A. Roles of sodium hydrosulfide and sodium nitroprusside as priming molecules during drought acclimation in citrus plants. *Plant Mol. Biol.* **2015**, *89*, 433–450. [CrossRef]
53. Zhou, M.; Zhang, J.; Zhou, H.; Zhao, D.; Duan, T.; Wang, S.; Yuan, X.; Xie, Y. Hydrogen Sulfide-Linked Persulfidation Maintains Protein Stability of abscisic acid-insensitive 4 and Delays Seed Germination. *Int. J. Mol. Sci.* **2022**, *23*, 1389. [CrossRef]

54. Begara-Morales, J.C.; López-Jaramillo, F.J.; Sánchez-Calvo, B.; Carreras, A.; Ortega-Muñoz, M.; Santoyo-González, F.; Corpas, F.J.; Barroso, J.B. Vinyl sulfone silica: Application of an open preactivated support to the study of transnitrosylation of plant proteins by S-nitrosoglutathione. *BMC Plant Biol.* **2013**, *13*, 61. [CrossRef] [PubMed]
55. Begara-Morales, J.C.; Sánchez-Calvo, B.; Chaki, M.; Valderrama, R.; Mata-Pérez, C.; López-Jaramillo, J.; Padilla, M.N.; Carreras, A.; Corpas, F.J.; Barroso, J.B. Dual regulation of cytosolic ascorbate peroxidase (APX) by tyrosine nitration and S-nitrosylation. *J. Exp. Bot.* **2014**, *65*, 527–538. [CrossRef] [PubMed]
56. Li, J.; Shi, C.; Wang, X.; Liu, C.; Ding, X.; Ma, P.; Wang, X.; Jia, H. Hydrogen sulfide regulates the activity of antioxidant enzymes through persulfidation and improves the resistance of tomato seedling to copper oxide nanoparticles (CuO NPs)-induced oxidative stress. *Plant Physiol. Biochem.* **2020**, *156*, 257–266. [CrossRef] [PubMed]
57. Naz, R.; Batool, S.; Shahid, M.; Keyani, R.; Yasmin, H.; Nosheen, A.; Hassan, M.N.; Mumtaz, S.; Siddiqui, M.H. Exogenous silicon and hydrogen sulfide alleviates the simultaneously occurring drought stress and leaf rust infection in wheat. *Plant Physiol. Biochem.* **2021**, *166*, 558–571. [CrossRef]
58. Shi, H.; Ye, T.; Han, N.; Bian, H.; Liu, X.; Chan, Z. Hydrogen sulfide regulates abiotic stress tolerance and biotic stress resistance in Arabidopsis. *J. Integr. Plant Biol.* **2015**, *57*, 628–640. [CrossRef]
59. Li, L.; Wang, Y.; Shen, W. Roles of hydrogen sulfide and nitric oxide in the alleviation of cadmium-induced oxidative damage in alfalfa seedling roots. *Biometals* **2012**, *25*, 617–631. [CrossRef]
60. Shi, H.; Ye, T.; Chan, Z. Exogenous application of hydrogen sulfide donor sodium hydrosulfide enhanced multiple abiotic stress tolerance in bermudagrass (*Cynodon dactylon* (L). Pers.). *Plant Physiol. Biochem.* **2013**, *71*, 226–234. [CrossRef]
61. Christou, A.; Manganaris, G.A.; Papadopoulos, I.; Fotopoulos, V. Hydrogen sulfide induces systemic tolerance to salinity and non-ionic osmotic stress in strawberry plants through modification of reactive species biosynthesis and transcriptional regulation of multiple defence pathways. *J. Exp. Bot.* **2013**, *64*, 1953–1966. [CrossRef]
62. Zhang, L.; Zhao, G.; Xia, C.; Jia, J.; Liu, X.; Kong, X. A wheat R2R3-MYB gene, TaMYB30-B, improves drought stress tolerance in transgenic Arabidopsis. *J. Exp. Bot.* **2012**, *63*, 5873–5885. [CrossRef]
63. Xu, Z.S.; Xia, L.Q.; Chen, M.; Cheng, X.G.; Zhang, R.Y.; Li, L.C.; Zhao, Y.X.; Lu, Y.; Ni, Z.Y.; Liu, L. Isolation and molecular characterization of the Triticum aestivum L. ethylene-responsive factor 1 (TaERF1) that increases multiple stress tolerance. *Plant Mol. Biol.* **2007**, *65*, 719–732. [CrossRef]
64. Joshi, R.; Wani, S.H.; Singh, B.; Bohra, A.; Dar, Z.A.; Lone, A.A.; Pareek, A.; Singla-Pareek, S.L. Transcription factors and plants response to drought stress: Current understanding and future directions. *Front. Plant Sci.* **2016**, *7*, 1029. [CrossRef] [PubMed]
65. Baillo, E.H.; Kimotho, R.N.; Zhang, Z.; Xu, P. Transcription factors associated with abiotic and biotic stress tolerance and their potential for crops improvement. *Genes* **2019**, *10*, 771. [CrossRef] [PubMed]
66. Xue, G.P.; Way, H.M.; Richardson, T.; Drenth, J.; Joyce, P.A.; McIntyre, C.L. Overexpression of TaNAC69 leads to enhanced transcript levels of stress up-regulated genes and dehydration tolerance in bread wheat. *Mol. Plant* **2011**, *4*, 697–712. [CrossRef] [PubMed]
67. Li, L.H.; Yi, H.L.; Liu, X.P.; Qi, H.X. Sulfur dioxide enhance drought tolerance of wheat seedlings through H_2S signaling. *Ecotoxicol. Environ. Saf.* **2021**, *207*, 111248. [CrossRef] [PubMed]
68. Kaya, C.; Ashraf, M. Nitric oxide is required for aminolevulinic acid-induced salt tolerance by lowering oxidative stress in maize (*Zea mays*). *J. Plant Growth Regul.* **2021**, *40*, 617–627. [CrossRef]
69. Schafer, F.Q.; Buettner, G.R. Redox environment of the cell as viewed through the redox state of the glutathione disulfide/glutathione couple. *Free Radic. Biol. Med.* **2001**, *30*, 1191–1212. [CrossRef]
70. Hancock, J.T. Hydrogen sulfide and environmental stresses. *Environ. Exp. Bot.* **2019**, *161*, 50–56. [CrossRef]
71. Foyer, C.H.; Theodoulou, F.L.; Delrot, S. The functions of inter-and intracellular glutathione transport systems in plants. *Trend. Plant Sci.* **2001**, *6*, 486–492. [CrossRef]
72. Noctor, G.; Reichheld, J.P.; Foyer, C.H. ROS-related redox regulation and signaling in plants. In Proceedings of the Seminars in Cell & Developmental Biology; Academic Press: Cambridge, MA, USA, 2018; pp. 3–12.
73. Fang, H.; Liu, Z.; Jin, Z.; Zhang, L.; Liu, D.; Pei, Y. An emphasis of hydrogen sulfide-cysteine cycle on enhancing the tolerance to chromium stress in Arabidopsis. *Environ. Pollut.* **2016**, *213*, 870–877. [CrossRef]
74. Wang, H.R.; Che, Y.H.; Wang, Z.H.; Zhang, B.N.; Huang, D.; Feng, F.; Ao, H. The multiple effects of hydrogen sulfide on cadmium toxicity in tobacco may be interacted with CaM signal transduction. *J. Hazard. Mater.* **2021**, *403*, 123651. [CrossRef]
75. Chen, J.; Wang, W.H.; Wu, F.H.; You, C.Y.; Liu, T.W.; Dong, X.J.; He, J.X.; Zheng, H.L. Hydrogen sulfide alleviates aluminum toxicity in barley seedlings. *Plant Soil.* **2013**, *362*, 301–318. [CrossRef]
76. Shivaraj, S.M.; Vats, S.; Bhat, J.A.; Dhakte, P.; Goyal, V.; Khatri, P.; Kumawat, S.; Singh, A.; Prasad, M.; Sonah, H.; et al. Nitric oxide and hydrogen sulfide crosstalk during heavy metal stress in plants. *Physiol. Plant.* **2020**, *168*, 437–455. [CrossRef] [PubMed]
77. Zhu, C.Q.; Zhang, J.H.; Sun, L.M.; Zhu, L.F.; Abliz, B.; Hu, W.J.; Zhong, C.; Bai, Z.G.; Sajid, H.; Cao, X.C. Hydrogen sulfide alleviates aluminum toxicity via decreasing apoplast and symplast Al contents in rice. *Front. Plant Sci.* **2018**, *9*, 294. [CrossRef] [PubMed]
78. Fang, H.; Jing, T.; Liu, Z.; Zhang, L.; Jin, Z.; Pei, Y. Hydrogen sulfide interacts with calcium signaling to enhance the chromium tolerance in Setaria italica. *Cell Calcium* **2014**, *56*, 472–481. [CrossRef] [PubMed]
79. Yu, Y.; Zhou, X.; Zhu, Z.; Zhou, K. Sodium hydrosulfide mitigates cadmium toxicity by promoting cadmium retention and inhibiting its translocation from roots to shoots in Brassica napus. *J. Agric. Food Chem.* **2018**, *67*, 433–440. [CrossRef] [PubMed]

80. Kabil, O.; Banerjee, R. Redox biochemistry of hydrogen sulfide. *J. Biol. Chem.* **2010**, *285*, 21903–21907. [CrossRef]
81. Jia, H.; Wang, X.; Shi, C.; Guo, J.; Ma, P.; Ren, X.; Wei, T.; Liu, H.; Li, J. Hydrogen sulfide decreases Cd translocation from root to shoot through increasing Cd accumulation in cell wall and decreasing Cd^{2+} influx in Isatis indigotica. *Plant Physiol. Biochem.* **2020**, *155*, 605–612. [CrossRef]
82. He, H.; Li, Y.; He, L.F. The central role of hydrogen sulfide in plant responses to toxic metal stress. *Ecotoxicol. Environ. Saf.* **2018**, *157*, 403–408. [CrossRef]
83. Islam, F.; Xie, Y.; Farooq, M.A.; Wang, J.; Yang, C.; Gill, R.A.; Zhu, J.; Zhou, W. Salinity reduces 2, 4-D efficacy in Echinochloa crusgalli by affecting redox balance, nutrient acquisition, and hormonal regulation. *Protoplasma* **2018**, *255*, 785–802. [CrossRef]
84. Long, M.; Shou, J.; Wang, J.; Hu, W.; Hannan, F.; Mwamba, T.M.; Farooq, M.A.; Zhou, W.; Islam, F. Ursolic acid limits salt-induced oxidative damage by interfering with nitric oxide production and oxidative defense machinery in rice. *Front. Plant Sci.* **2020**, *11*, 697. [CrossRef]
85. Huang, D.; Huo, J.; Liao, W. Hydrogen sulfide: Roles in plant abiotic stress response and crosstalk with other signals. *Plant Sci.* **2021**, *302*, 110733. [CrossRef] [PubMed]
86. Huang, Q.; Farooq, M.A.; Hannan, F.; Chen, W.; Ayyaz, A.; Zhang, K.; Zhou, W.; Islam, F. Endogenous nitric oxide contributes to chloride and sulphate salinity tolerance by modulation of ion transporter expression and reestablishment of redox balance in *Brassica napus* cultivars. *Environ. Exp. Bot.* **2022**, *194*, 104734. [CrossRef]
87. Cui, P.; Liu, H.; Islam, F.; Li, L.; Farooq, M.A.; Ruan, S.; Zhou, W. OsPEX11, a peroxisomal biogenesis factor 11, contributes to salt stress tolerance in *Oryza sativa*. *Front. Plant Sci.* **2016**, *7*, 1357. [CrossRef] [PubMed]
88. Islam, F.; Ali, B.; Wang, J.; Farooq, M.A.; Gill, R.A.; Ali, S.; Wang, D.; Zhou, W. Combined herbicide and saline stress differentially modulates hormonal regulation and antioxidant defense system in *Oryza sativa* cultivars. *Plant Physiol. Biochem.* **2016**, *107*, 82–95. [CrossRef] [PubMed]
89. Islam, F.; Wang, J.; Farooq, M.A.; Yang, C.; Jan, M.; Mwamba, T.M.; Hannan, F.; Xu, L.; Zhou, W. Rice responses and tolerance to salt stress: Deciphering the physiological and molecular mechanisms of salinity adaptation. In *Advances in Rice Research for Abiotic Stress Tolerance*; Elsevier: Amsterdam, The Netherlands, 2019; pp. 791–819. [CrossRef]
90. Palmgren, M.G. Plant plasma membrane H^+-ATPases: Powerhouses for nutrient uptake. *Annu. Rev. Plant Biol.* **2001**, *52*, 817–845. [CrossRef]
91. Khan, M.N.; Mukherjee, S.; Al-Huqail, A.A.; Basahi, R.A.; Ali, H.M.; Al-Munqedhi, B.; Siddiqui, M.H.; Kalaji, H.M. Exogenous Potassium (K^+) Positively regulates Na^+/H^+ antiport system, carbohydrate metabolism, and ascorbate-glutathione cycle in H_2S-dependent manner in NaCl-stressed tomato seedling roots. *Plants* **2021**, *10*, 948. [CrossRef]
92. Khan, M.N.; Siddiqui, M.H.; Mukherjee, S.; Alamri, S.; Al-Amri, A.A.; Alsubaie, Q.D.; Al-Munqedhi, B.M.; Ali, H.M. Calcium-hydrogen sulfide crosstalk during K^+-deficient NaCl stress operates through regulation of Na^+/H^+ antiport and antioxidative defense system in mung bean roots. *Plant Physiol. Biochem.* **2021**, *159*, 211–225. [CrossRef]
93. Li, J.; Yu, Z.; Choo, S.; Zhao, J.; Wang, Z.; Xie, R. Chemico-proteomics reveal the enhancement of salt tolerance in an invasive plant species via H_2S signaling. *ACS Omega* **2020**, *5*, 14575–14585. [CrossRef]
94. Li, L.; Jia, Y.; Li, P.; Yin, S.; Zhang, G.; Wang, X.; Wang, Y.; Zang, X.; Ding, Y. Expression and activity of V-H^+-ATPase in gill and kidney of marbled eel *Anguilla marmorata* in response to salinity challenge. *J. Fish Biol.* **2015**, *87*, 28–42. [CrossRef]
95. Deng, Y.Q.; Bao, J.; Yuan, F.; Liang, X.; Feng, Z.T.; Wang, B.S. Exogenous hydrogen sulfide alleviates salt stress in wheat seedlings by decreasing Na^+ content. *Plant Growth Regul.* **2016**, *79*, 391–399. [CrossRef]
96. Shabala, S.; Cuin, T.A. Potassium transport and plant salt tolerance. *Physiol. Plant.* **2008**, *133*, 651–669. [CrossRef] [PubMed]
97. Cheng, P.; Zhang, Y.; Wang, J.; Guan, R.; Pu, H.; Shen, W. Importance of hydrogen sulfide as the molecular basis of heterosis in hybrid *Brassica napus*: A case study in salinity response. *Environ. Exp. Bot.* **2022**, *193*, 104693. [CrossRef]
98. Mostofa, M.G.; Saegusa, D.; Fujita, M.; Tran, L.S.P. Hydrogen sulfide regulates salt tolerance in rice by maintaining Na^+/K^+ balance, mineral homeostasis and oxidative metabolism under excessive salt stress. *Front. Plant Sci.* **2015**, *6*, 1055. [CrossRef] [PubMed]
99. Siddiqui, M.H.; Khan, M.N.; Mukherjee, S.; Alamri, S.; Basahi, R.A.; Al-Amri, A.A.; Alsubaie, Q.D.; Al-Munqedhi, B.M.; Ali, H.M.; Almohisen, I.A. Hydrogen sulfide (H_2S) and potassium (K^+) synergistically induce drought stress tolerance through regulation of H+-ATPase activity, sugar metabolism, and antioxidative defense in tomato seedlings. *Plant Cell Rep.* **2021**, *40*, 1543–1564. [CrossRef] [PubMed]
100. Jiang, J.L.; Tian, Y.; Li, L.; Yu, M.; Hou, R.P.; Ren, X.M. H_2S alleviates salinity stress in cucumber by maintaining the Na^+/K^+ balance and regulating H_2S metabolism and oxidative stress response. *Front. Plant Sci.* **2019**, *10*, 678. [CrossRef] [PubMed]
101. Li, H.; Shi, J.; Wang, Z.; Zhang, W.; Yang, H. H_2S pretreatment mitigates the alkaline salt stress on Malus hupehensis roots by regulating Na+/K+ homeostasis and oxidative stress. *Plant Physiol. Biochem.* **2020**, *156*, 233–241. [CrossRef]
102. Li, C.; Huang, D.; Wang, C.; Wang, N.; Yao, Y.; Li, W.; Liao, W. NO is involved in H_2-induced adventitious rooting in cucumber by regulating the expression and interaction of plasma membrane H^+-ATPase and 14-3-3. *Planta* **2020**, *252*, 9. [CrossRef] [PubMed]
103. Flowers, T.; Troke, P.; Yeo, A. The mechanism of salt tolerance in halophytes. *Annu. Rev. Plant Physiol.* **1977**, *28*, 89–121. [CrossRef]
104. Amooaghaie, R.; Enteshari, S. Role of two-sided crosstalk between NO and H_2S on improvement of mineral homeostasis and antioxidative defense in *Sesamum indicum* under lead stress. *Ecotoxicol. Environ. Saf.* **2017**, *139*, 210–218. [CrossRef]
105. Levine, A.; Tenhaken, R.; Dixon, R.; Lamb, C. H_2O_2 from the oxidative burst orchestrates the plant hypersensitive disease resistance response. *Cell* **1994**, *79*, 583–593. [CrossRef]

106. Janicka, M.; Reda, M.; Czyżewska, K.; Kabała, K. Involvement of signalling molecules NO, H_2O_2 and H_2S in modification of plasma membrane proton pump in cucumber roots subjected to salt or low temperature stress. *Funct. Plant Biol.* **2017**, *45*, 428–439. [CrossRef]
107. Asif, M.; Jamil, H.M.A.; Hayat, M.T.; Mahmood, Q.; Ali, S. Use of Phytohormones to Improve Abiotic Stress Tolerance in Wheat. In *Wheat Production in Changing Environments*; Springer: Berlin/Heidelberg, Germany, 2019; pp. 465–479. [CrossRef]
108. Javid, M.G.; Sorooshzadeh, A.; Moradi, F.; Modarres Sanavy, S.A.M.; Allahdadi, I. The role of phytohormones in alleviating salt stress in crop plants. *Aust. J. Crop Sci.* **2011**, *5*, 726–734.
109. Singh, V.P.; Singh, S.; Kumar, J.; Prasad, S.M. Hydrogen sulfide alleviates toxic effects of arsenate in pea seedlings through up-regulation of the ascorbate–glutathione cycle: Possible involvement of nitric oxide. *J. Plant Physiol.* **2015**, *181*, 20–29. [CrossRef] [PubMed]
110. Hancock, J.T.; Whiteman, M. Hydrogen sulfide and cell signaling: Team player or referee? *Plant Physiol. Biochem.* **2014**, *78*, 37–42. [CrossRef] [PubMed]
111. Aroca, A.; Gotor, C.; Bassham, D.C.; Romero, L.C. Hydrogen sulfide: From a toxic molecule to a key molecule of cell life. *Antioxidants* **2020**, *9*, 621. [CrossRef] [PubMed]
112. Lin, Y.T.; Li, M.Y.; Cui, W.T.; Lu, W.; Shen, W.B. Haem oxygenase-1 is involved in hydrogen sulfide-induced cucumber adventitious root formation. *J. Plant Growth Regul.* **2012**, *31*, 519–528. [CrossRef]
113. Scuffi, D.; Lamattina, L.; García-Mata, C. Gasotransmitters and stomatal closure: Is there redundancy, concerted action, or both? *Front. Plant Sci.* **2016**, *7*, 277. [CrossRef]
114. Hou, Z.; Liu, J.; Hou, L.; Li, X.; Liu, X. H_2S may function downstream of H_2O_2 in jasmonic acid-induced stomatal closure in Vicia faba. *Chin. Bull. Bot.* **2011**, *46*, 396. [CrossRef]
115. Raya-González, J.; López-Bucio, J.S.; Prado-Rodríguez, J.C.; Ruiz-Herrera, L.F.; Guevara-García, Á.A.; López-Bucio, J. The MEDIATOR genes MED12 and MED13 control Arabidopsis root system configuration influencing sugar and auxin responses. *Plant Mol. Biol.* **2017**, *95*, 141–156. [CrossRef]
116. Mei, Y.; Chen, H.; Shen, W.; Shen, W.; Huang, L. Hydrogen peroxide is involved in hydrogen sulfide-induced lateral root formation in tomato seedlings. *BMC Plant Biol.* **2017**, *17*, 162. [CrossRef]
117. Khan, M.S.S.; Basnet, R.; Islam, S.A.; Shu, Q. Mutational analysis of OsPLDα1 reveals its involvement in phytic acid biosynthesis in rice grains. *J. Agric. Food Chem.* **2019**, *67*, 11436–11443. [CrossRef] [PubMed]
118. Khan, M.S.S.; Basnet, R.; Ahmed, S.; Bao, J.; Shu, Q. Mutations of OsPLDa1 increase lysophospholipid content and enhance cooking and eating quality in rice. *Plants* **2020**, *9*, 390. [CrossRef] [PubMed]
119. Du, J.; Jin, H.; Yang, L. Role of hydrogen sulfide in retinal diseases. *Front. Pharmacol.* **2017**, *8*, 588. [CrossRef] [PubMed]
120. Wojtyla, Ł.; Lechowska, K.; Kubala, S.; Garnczarska, M. Different modes of hydrogen peroxide action during seed germination. *Front. Plant Sci.* **2016**, *7*, 66. [CrossRef]
121. Guo, M.; Liu, J.H.; Ma, X.; Luo, D.X.; Gong, Z.H.; Lu, M.H. The plant heat stress transcription factors (HSFs): Structure, regulation, and function in response to abiotic stresses. *Front. Plant Sci.* **2016**, *7*, 114. [CrossRef]
122. Jin, Z.; Wang, Z.; Ma, Q.; Sun, L.; Zhang, L.; Liu, Z.; Liu, D.; Hao, X.; Pei, Y. Hydrogen sulfide mediates ion fluxes inducing stomatal closure in response to drought stress in Arabidopsis thaliana. *Plant Soil* **2017**, *419*, 141–152. [CrossRef]
123. Li, Z.G.; Jin, J.Z. Hydrogen sulfide partly mediates abscisic acid-induced heat tolerance in tobacco (*Nicotiana tabacum* L.) suspension cultured cells. *Plant Cell Tissue Organ Cult.* **2016**, *125*, 207–214. [CrossRef]
124. Jin, Z.; Xue, S.; Luo, Y.; Tian, B.; Fang, H.; Li, H.; Pei, Y. Hydrogen sulfide interacting with abscisic acid in stomatal regulation responses to drought stress in Arabidopsis. *Plant Physiol. Biochem.* **2013**, *62*, 41–46. [CrossRef]
125. García-Mata, C.; Lamattina, L. Hydrogen sulphide, a novel gasotransmitter involved in guard cell signalling. *New Phytol.* **2010**, *188*, 977–984. [CrossRef]
126. Lisjak, M.; Srivastava, N.; Teklic, T.; Civale, L.; Lewandowski, K.; Wilson, I.; Wood, M.; Whiteman, M.; Hancock, J.T. A novel hydrogen sulfide donor causes stomatal opening and reduces nitric oxide accumulation. *Plant Physiol. Biochem.* **2010**, *48*, 931–935. [CrossRef]
127. Honda, K.; Yamada, N.; Yoshida, R.; Ihara, H.; Sawa, T.; Akaike, T.; Iwai, S. 8-Mercapto-cyclic GMP mediates hydrogen sulfide-induced stomatal closure in Arabidopsis. *Plant Cell Physiol.* **2015**, *56*, 1481–1489. [CrossRef] [PubMed]
128. Papanatsiou, M.; Scuffi, D.; Blatt, M.R.; García-Mata, C. Hydrogen sulfide regulates inward-rectifying K^+ channels in conjunction with stomatal closure. *Plant Physiol.* **2015**, *168*, 29–35. [CrossRef] [PubMed]
129. Wang, L.; Ma, X.; Che, Y.; Hou, L.; Liu, X.; Zhang, W. Extracellular ATP mediates H_2S-regulated stomatal movements and guard cell K^+ current in a H_2O_2-dependent manner in Arabidopsis. *Sci. Bull.* **2015**, *60*, 419–427. [CrossRef]
130. Jeon, B.W.; Acharya, B.R.; Assmann, S.M. The Arabidopsis heterotrimeric G-protein β subunit, AGB 1, is required for guard cell calcium sensing and calcium-induced calcium release. *Plant J.* **2019**, *99*, 231–244. [CrossRef]
131. Brault, M.; Amiar, Z.; Pennarun, A.-M.; Monestiez, M.; Zhang, Z.; Cornel, D.; Dellis, O.; Knight, H.; Bouteau, F.; Rona, J.P. Plasma membrane depolarization induced by abscisic acid in Arabidopsis suspension cells involves reduction of proton pumping in addition to anion channel activation, which are both Ca^{2+} dependent. *Plant Physiol.* **2004**, *135*, 231–243. [CrossRef]
132. Siegel, R.S.; Xue, S.; Murata, Y.; Yang, Y.; Nishimura, N.; Wang, A.; Schroeder, J.I. Calcium elevation-dependent and attenuated resting calcium-dependent abscisic acid induction of stomatal closure and abscisic acid-induced enhancement of calcium sensitivities of S-type anion and inward-rectifying K^+ channels in Arabidopsis guard cells. *Plant J.* **2009**, *59*, 207–220. [CrossRef]

133. Belin, C.; de Franco, P.O.; Bourbousse, C.; Chaignepain, S.; Schmitter, J.M.; Vavasseur, A.; Giraudat, J.; Barbier-Brygoo, H.; Thomine, S. Identification of features regulating OST1 kinase activity and OST1 function in guard cells. *Plant Physiol.* **2006**, *141*, 1316–1327. [CrossRef]
134. Chen, S.; Jia, H.; Wang, X.; Shi, C.; Wang, X.; Ma, P.; Wang, J.; Ren, M.; Li, J. Hydrogen sulfide positively regulates abscisic acid signaling through persulfidation of SnRK2.6 in guard cells. *Mol. Plant* **2020**, *13*, 732–744. [CrossRef]
135. Chen, J.; Zhou, H.; Xie, Y. SnRK2. 6 phosphorylation/persulfidation: Where ABA and H_2S signaling meet. *Trends Plant Sci.* **2021**, *26*, 1207–1209. [CrossRef]
136. Cavallari, N.; Artner, C.; Benkova, E. Auxin-regulated lateral root organogenesis. *Cold Spring Harb. Perspect. Biol.* **2021**, *13*, a039941. [CrossRef]
137. Zhang, J.; Zhou, M.; Zhou, H.; Zhao, D.; Gotor, C.; Romero, L.C.; Shen, J.; Ge, Z.; Zhang, Z.; Shen, W.; et al. Hydrogen sulfide, a signaling molecule in plant stress responses. *J. Integr. Plant Biol.* **2021**, *63*, 146–160. [CrossRef] [PubMed]
138. Shen, J.; Zhang, J.; Zhou, M.; Zhou, H.; Cui, B.; Gotor, C.; Romero, L.C.; Fu, L.; Yang, J.; Foyer, C.H. Persulfidation-based modification of cysteine desulfhydrase and the NADPH oxidase RBOHD controls guard cell abscisic acid signaling. *Plant Cell* **2020**, *32*, 1000–1017. [CrossRef] [PubMed]
139. Liu, H.; Xue, S. Interplay between hydrogen sulfide and other signaling molecules in the regulation of guard cell signaling and abiotic/biotic stress response. *Plant Commun.* **2021**, *2*, 100179. [CrossRef] [PubMed]
140. Batool, S.; Uslu, V.V.; Rajab, H.; Ahmad, N.; Waadt, R.; Geiger, D.; Malagoli, M.; Xiang, C.-B.; Hedrich, R.; Rennenberg, H. Sulfate is incorporated into cysteine to trigger ABA production and stomatal closure. *Plant Cell* **2018**, *30*, 2973–2987. [CrossRef] [PubMed]
141. Bittner, F.; Oreb, M.; Mendel, R.R. ABA3 is a molybdenum cofactor sulfurase required for activation of aldehyde oxidase and xanthine dehydrogenase in *Arabidopsis thaliana*. *J. Biol. Chem.* **2001**, *276*, 40381–40384. [CrossRef]
142. Rajab, H.; Khan, M.S.; Malagoli, M.; Hell, R.; Wirtz, M. Sulfate-induced stomata closure requires the canonical ABA signal transduction machinery. *Plants* **2019**, *8*, 21. [CrossRef]
143. Fancy, N.N.; Bahlmann, A.K.; Loake, G.J. Nitric oxide function in plant abiotic stress. *Plant Cell Environ.* **2017**, *40*, 462–472. [CrossRef]
144. Mishra, V.; Singh, P.; Tripathi, D.K.; Corpas, F.J.; Singh, V.P. Nitric oxide and hydrogen sulfide: An indispensable combination for plant functioning. *Trends Plant Sci.* **2021**, *26*, 1270–1285. [CrossRef]
145. Christou, A.; Fotopoulos, V.; Manganaris, G.A. Hydrogen sulfide confers systemic tolerance to salt and polyethylene glycol stress in strawberry plants. *Mol. Approaches Plant Abiotic Stress* **2011**. Available online: http://ktisis.cut.ac.cy/handle/10488/5071 (accessed on 17 February 2022).
146. Wang, Y.; Li, L.; Cui, W.; Xu, S.; Shen, W.; Wang, R. Hydrogen sulfide enhances alfalfa (Medicago sativa) tolerance against salinity during seed germination by nitric oxide pathway. *Plant Soil* **2012**, *351*, 107–119. [CrossRef]
147. Lisjak, M.; Teklic, T.; Wilson, I.D.; Whiteman, M.; Hancock, J.T. Hydrogen sulfide: Environmental factor or signalling molecule? *Plant Cell Environ.* **2013**, *36*, 1607–1616. [CrossRef] [PubMed]
148. Gong, T.; Li, C.; Bian, B.; Wu, Y.; Dawuda, M.M.; Liao, W. Advances in application of small molecule compounds for extending the shelf life of perishable horticultural products: A review. *Sci. Hortic.* **2018**, *230*, 25–34. [CrossRef]
149. Li, D.; Limwachiranon, J.; Li, L.; Du, R.; Luo, Z. Involvement of energy metabolism to chilling tolerance induced by hydrogen sulfide in cold-stored banana fruit. *Food Chem.* **2016**, *208*, 272–278. [CrossRef] [PubMed]
150. Peng, R.; Bian, Z.; Zhou, L.; Cheng, W.; Hai, N.; Yang, C.; Yang, T.; Wang, X.; Wang, C. Hydrogen sulfide enhances nitric oxide-induced tolerance of hypoxia in maize (*Zea mays* L.). *Plant Cell Rep.* **2016**, *35*, 2325–2340. [CrossRef] [PubMed]
151. Mukherjee, S. Recent advancements in the mechanism of nitric oxide signaling associated with hydrogen sulfide and melatonin crosstalk during ethylene-induced fruit ripening in plants. *Nitric Oxide* **2019**, *82*, 25–34. [CrossRef] [PubMed]
152. Whiteman, M.; Li, L.; Kostetski, I.; Chu, S.H.; Siau, J.L.; Bhatia, M.; Moore, P.K. Evidence for the formation of a novel nitrosothiol from the gaseous mediators nitric oxide and hydrogen sulphide. *Biochem. Biophys. Res. Commun.* **2006**, *343*, 303–310. [CrossRef] [PubMed]
153. Zhang, H.; Tang, J.; Liu, X.P.; Wang, Y.; Yu, W.; Peng, W.Y.; Fang, F.; Ma, D.F.; Wei, Z.J.; Hu, L.Y. Hydrogen sulfide promotes root organogenesis in *Ipomoea batatas*, *Salix matsudana* and *Glycine max*. *J. Integr. Plant Biol.* **2009**, *51*, 1086–1094. [CrossRef]
154. Ma, Y.; Wang, L.; Zhang, W. The role of hydrogen sulfide and its relationship with hydrogen peroxide and nitric oxide in brassinosteroid-induced stomatal closure of *Vicia faba* L. *S. Afr. J. Bot.* **2022**, *146*, 426–436. [CrossRef]
155. Ma, Y.; Shao, L.; Zhang, W.; Zheng, F. Hydrogen sulfide induced by hydrogen peroxide mediates brassinosteroid-induced stomatal closure of *Arabidopsis thaliana*. *Funct. Plant Biol.* **2020**, *48*, 195–205. [CrossRef]
156. Jing, L.; Hou, Z.; Liu, G.H.; Hou, L.X.; Xin, L. Hydrogen sulfide may function downstream of nitric oxide in ethylene-induced stomatal closure in *Vicia faba* L. *J. Integr. Agric.* **2012**, *11*, 1644–1653. [CrossRef]
157. Shi, C.; Qi, C.; Ren, H.; Huang, A.; Hei, S.; She, X. Ethylene mediates brassinosteroid-induced stomatal closure via $G\alpha$ protein-activated hydrogen peroxide and nitric oxide production in Arabidopsis. *Plant J.* **2015**, *82*, 280–301. [CrossRef] [PubMed]
158. Liu, J.; Hou, L.; Liu, G.; Liu, X.; Wang, X. Hydrogen sulfide induced by nitric oxide mediates ethylene-induced stomatal closure of Arabidopsis thaliana. *Chin. Sci. Bul.* **2011**, *56*, 3547–3553. [CrossRef]
159. Rather, B.A.; Mir, I.R.; Sehar, Z.; Anjum, N.A.; Masood, A.; Khan, N.A. The outcomes of the functional interplay of nitric oxide and hydrogen sulfide in metal stress tolerance in plants. *Plant Physiol. Biochem.* **2020**, *155*, 523–534. [CrossRef] [PubMed]

160. Shi, H.; Ye, T.; Chan, Z. Nitric oxide-activated hydrogen sulfide is essential for cadmium stress response in bermudagrass (*Cynodon dactylon* (L). Pers.). *Plant Physiol. Biochem.* **2014**, *74*, 99–107. [CrossRef]
161. Palma, J.M.; Mateos, R.M.; López-Jaramillo, J.; Rodríguez-Ruiz, M.; González-Gordo, S.; Lechuga-Sancho, A.M.; Corpas, F.J. Plant catalases as NO and H_2S targets. *Redox Biol.* **2020**, *34*, 101525. [CrossRef]
162. Casimiro, I.; Marchant, A.; Bhalerao, R.P.; Beeckman, T.; Dhooge, S.; Swarup, R.; Graham, N.; Inzé, D.; Sandberg, G.; Casero, P.J. Auxin transport promotes Arabidopsis lateral root initiation. *Plant Cell* **2001**, *13*, 843–852. [CrossRef]
163. Overvoorde, P.; Fukaki, H.; Beeckman, T. Auxin control of root development. *Cold Spring Harb. Perspect. Biol.* **2010**, *2*, a001537. [CrossRef]
164. De Smet, I.; Lau, S.; Voß, U.; Vanneste, S.; Benjamins, R.; Rademacher, E.H.; Schlereth, A.; De Rybel, B.; Vassileva, V.; Grunewald, W. Bimodular auxin response controls organogenesis in Arabidopsis. *Proc. Natl. Acad. Sci. USA* **2010**, *107*, 2705–2710. [CrossRef]
165. Fang, T.; Cao, Z.; Li, J.; Shen, W.; Huang, L. Auxin-induced hydrogen sulfide generation is involved in lateral root formation in tomato. *Plant Physiol Biochem.* **2014**, *76*, 44–51. [CrossRef]
166. Wu, X.; Du, A.; Zhang, S.; Wang, W.; Liang, J.; Peng, F.; Xiao, Y. Regulation of growth in peach roots by exogenous hydrogen sulfide based on RNA-Seq. *Plant Physiol. Biochem.* **2021**, *159*, 179–192. [CrossRef]
167. Li, J.; Chen, S.; Wang, X.; Shi, C.; Liu, H.; Yang, J.; Shi, W.; Guo, J.; Jia, H. Hydrogen sulfide disturbs actin polymerization via S-sulfhydration resulting in stunted root hair growth. *Plant Physiol.* **2018**, *178*, 936–949. [CrossRef] [PubMed]
168. Jia, H.; Hu, Y.; Fan, T.; Li, J. Hydrogen sulfide modulates actin-dependent auxin transport via regulating ABPs results in changing of root development in Arabidopsis. *Sci. Rep.* **2015**, *5*, 8251. [CrossRef] [PubMed]
169. Lanza, M.; Garcia-Ponce, B.; Castrillo, G.; Catarecha, P.; Sauer, M.; Rodriguez-Serrano, M.; Páez-García, A.; Sánchez-Bermejo, E.; Mohan, T.; del Puerto, Y.L. Role of actin cytoskeleton in brassinosteroid signaling and in its integration with the auxin response in plants. *Dev. Cell* **2012**, *22*, 1275–1285. [CrossRef]
170. Rahman, A.; Bannigan, A.; Sulaman, W.; Pechter, P.; Blancaflor, E.B.; Baskin, T.I. Auxin, actin and growth of the Arabidopsis thaliana primary root. *Plant J.* **2007**, *50*, 514–528. [CrossRef] [PubMed]
171. Muday, G.K.; Murphy, A.S. An emerging model of auxin transport regulation. *Plant Cell* **2002**, *14*, 293–299. [CrossRef] [PubMed]
172. Sun, H.; Basu, S.; Brady, S.R.; Luciano, R.L.; Muday, G.K. Interactions between auxin transport and the actin cytoskeleton in developmental polarity of Fucus distichus embryos in response to light and gravity. *Plant Physiol.* **2004**, *135*, 266–278. [CrossRef]
173. Kou, N.; Xiang, Z.; Cui, W.; Li, L.; Shen, W. Hydrogen sulfide acts downstream of methane to induce cucumber adventitious root development. *J. Plant Physiol.* **2018**, *228*, 113–120. [CrossRef]
174. Bai, X.; Todd, C.D.; Desikan, R.; Yang, Y.; Hu, X. N-3-oxo-decanoyl-L-homoserine-lactone activates auxin-induced adventitious root formation via hydrogen peroxide-and nitric oxide-dependent cyclic GMP signaling in mung bean. *Plant Physiol.* **2012**, *158*, 725–736. [CrossRef]
175. Qi, F.; Xiang, Z.; Kou, N.; Cui, W.; Xu, D.; Wang, R.; Zhu, D.; Shen, W. Nitric oxide is involved in methane-induced adventitious root formation in cucumber. *Physiol. Plant.* **2017**, *159*, 366–377. [CrossRef]
176. Mei, Y.; Zhao, Y.; Jin, X.; Wang, R.; Xu, N.; Hu, J.; Huang, L.; Guan, R.; Shen, W. L-Cysteine desulfhydrase-dependent hydrogen sulfide is required for methane-induced lateral root formation. *Plant Mol. Biol.* **2019**, *99*, 283–298. [CrossRef]
177. Liu, F.; Zhang, X.; Cai, B.; Pan, D.; Fu, X.; Bi, H.; Ai, X. Physiological response and transcription profiling analysis reveal the role of glutathione in H_2S-induced chilling stress tolerance of cucumber seedlings. *Plant Sci.* **2020**, *291*, 110363. [CrossRef] [PubMed]
178. Sun, Y.; Ma, C.; Kang, X.; Zhang, L.; Wang, J.; Zheng, S.; Zhang, T. Hydrogen sulfide and nitric oxide are involved in melatonin-induced salt tolerance in cucumber. *Plant Physiol. Biochem.* **2021**, *167*, 101–112. [CrossRef] [PubMed]
179. Zhang, X.W.; Liu, F.J.; Zhai, J.; Li, F.D.; Bi, H.G.; Ai, X.Z. Auxin acts as a downstream signaling molecule involved in hydrogen sulfide-induced chilling tolerance in cucumber. *Planta* **2020**, *251*, 69. [CrossRef] [PubMed]
180. Zhang, X.; Fu, X.; Liu, F.; Wang, Y.; Bi, H.; Ai, X. Hydrogen sulfide improves the cold stress resistance through the CsARF5-CsDREB3 module in cucumber. *Int. J. Mol. Sci.* **2021**, *22*, 13229. [CrossRef]
181. Marin, E.; Jouannet, V.; Herz, A.; Lokerse, A.S.; Weijers, D.; Vaucheret, H.; Nussaume, L.; Crespi, M.D.; Maizel, A. miR390, Arabidopsis TAS3 tasiRNAs, and their auxin response factor targets define an autoregulatory network quantitatively regulating lateral root growth. *Plant Cell* **2010**, *22*, 1104–1117. [CrossRef]
182. Zhang, L.; Pei, Y.; Wang, H.; Jin, Z.; Liu, Z.; Qiao, Z.; Fang, H.; Zhang, Y. Hydrogen sulfide alleviates cadmium-induced cell death through restraining ROS accumulation in roots of *Brassica rapa* L. ssp. pekinensis. *Oxid. Med. Cell. Longev.* **2015**, *2015*, 804603. [CrossRef]
183. Arnao, M.B.; Hernández-Ruiz, J. Is phytomelatonin a new plant hormone? *Agronomy* **2020**, *10*, 95. [CrossRef]
184. Jahan, M.S.; Shu, S.; Wang, Y.; Chen, Z.; He, M.; Tao, M.; Sun, J.; Guo, S. Melatonin alleviates heat-induced damage of tomato seedlings by balancing redox homeostasis and modulating polyamine and nitric oxide biosynthesis. *BMC Plant Biol.* **2019**, *19*, 414. [CrossRef]
185. Kaya, C.; Okant, M.; Ugurlar, F.; Alyemeni, M.N.; Ashraf, M.; Ahmad, P. Melatonin-mediated nitric oxide improves tolerance to cadmium toxicity by reducing oxidative stress in wheat plants. *Chemosphere* **2019**, *225*, 627–638. [CrossRef]
186. Mukherjee, S.; Bhatla, S.C. Exogenous melatonin modulates endogenous H_2S homeostasis and L-cysteine desulfhydrase activity in salt-stressed tomato (*Solanum lycopersicum* L. var. cherry) seedling cotyledons. *J. Plant Growth Regul.* **2021**, *40*, 2502–2514. [CrossRef]

187. Iqbal, N.; Fatma, M.; Gautam, H.; Umar, S.; Sofo, A.; D'Ippolito, I.; Khan, N.A. The Crosstalk of Melatonin and Hydrogen Sulfide Determines Photosynthetic Performance by Regulation of Carbohydrate Metabolism in Wheat under Heat Stress. *Plants* **2021**, *10*, 1778. [CrossRef] [PubMed]
188. Chen, J.; Zhang, J.; Kong, M.; Freeman, A.; Chen, H.; Liu, F. More stories to tell: Nonexpressor of pathogenesis-related genes1, a salicylic acid receptor. *Plant Cell Environ.* **2021**, *44*, 1716–1727. [CrossRef] [PubMed]
189. Khan, M.S.S.; Islam, F.; Chen, H.; Chang, M.; Wang, D.; Liu, F.; Fu, Z.Q.; Chen, J. Transcriptional Coactivators: Driving Force of Plant Immunity. *Front. Plant Sci.* **2022**, *13*, 823937. [CrossRef] [PubMed]
190. White, R. Acetylsalicylic acid (aspirin) induces resistance to tobacco mosaic virus in tobacco. *Virology* **1979**, *99*, 410–412. [CrossRef]
191. Dong, J.; Chen, C.; Chen, Z. Expression profiles of the Arabidopsis WRKY gene superfamily during plant defense response. *Plant Mol. Biol.* **2003**, *51*, 21–37. [CrossRef]
192. Liu, Z.; Fang, H.; Pei, Y.; Jin, Z.; Zhang, L.; Liu, D. WRKY transcription factors down-regulate the expression of H_2S-generating genes, LCD and DES in *Arabidopsis thaliana*. *Sci. Bullet.* **2015**, *60*, 995–1001. [CrossRef]
193. López-Martín, M.C.; Becana, M.; Romero, L.C.; Gotor, C. Knocking out cytosolic cysteine synthesis compromises the antioxidant capacity of the cytosol to maintain discrete concentrations of hydrogen peroxide in Arabidopsis. *Plant Physiol.* **2008**, *147*, 562–572. [CrossRef]
194. Álvarez, C.; Ángeles Bermúdez, M.; Romero, L.C.; Gotor, C.; García, I. Cysteine homeostasis plays an essential role in plant immunity. *New Phytol.* **2012**, *193*, 165–177. [CrossRef]
195. Tahir, J.; Watanabe, M.; Jing, H.C.; Hunter, D.A.; Tohge, T.; Nunes-Nesi, A.; Brotman, Y.; Fernie, A.R.; Hoefgen, R.; Dijkwel, P.P. Activation of R-mediated innate immunity and disease susceptibility is affected by mutations in a cytosolic O-acetylserine (thiol) lyase in Arabidopsis. *Plant J.* **2013**, *73*, 118–130. [CrossRef]
196. Glazebrook, J.; Zook, M.; Mert, F.; Kagan, I.; Rogers, E.E.; Crute, I.R.; Holub, E.B.; Hammerschmidt, R.; Ausubel, F.M. Phytoalexin-deficient mutants of Arabidopsis reveal that PAD4 encodes a regulatory factor and that four PAD genes contribute to downy mildew resistance. *Genetics* **1997**, *146*, 381–392. [CrossRef]
197. Feys, B.J.; Moisan, L.J.; Newman, M.A.; Parker, J.E. Direct interaction between the Arabidopsis disease resistance signaling proteins, EDS1 and PAD4. *EMBO J.* **2001**, *20*, 5400–5411. [CrossRef]
198. Pokotylo, I.; Kravets, V.; Ruelland, E. Salicylic acid binding proteins (SABPs): The hidden forefront of salicylic acid signalling. *Int. J. Mol. Sci.* **2019**, *20*, 4377. [CrossRef]
199. Chen, J.; Clinton, M.; Qi, G.; Wang, D.; Liu, F.; Fu, Z.Q. Reprogramming and remodeling: Transcriptional and epigenetic regulation of salicylic acid-mediated plant defense. *J. Exp. Bot.* **2020**, *71*, 5256–5268. [CrossRef]
200. Chen, J.; Mohan, R.; Zhang, Y.; Li, M.; Chen, H.; Palmer, I.A.; Chang, M.; Qi, G.; Spoel, S.H.; Mengiste, T. NPR1 promotes its own and target gene expression in plant defense by recruiting CDK8. *Plant Physiol.* **2019**, *181*, 289–304. [CrossRef]
201. Vlot, A.C.; Dempsey, D.M.A.; Klessig, D.F. Salicylic acid, a multifaceted hormone to combat disease. *Annu. Rev. Phytopathol.* **2009**, *47*, 177–206. [CrossRef]
202. Zhang, L.; Zhang, F.; Melotto, M.; Yao, J.; He, S.Y. Jasmonate signaling and manipulation by pathogens and insects. *J. Exp. Bot.* **2017**, *68*, 1371–1385. [CrossRef]
203. Kammerhofer, N.; Radakovic, Z.; Regis, J.M.; Dobrev, P.; Vankova, R.; Grundler, F.M.; Siddique, S.; Hofmann, J.; Wieczorek, K. Role of stress-related hormones in plant defence during early infection of the cyst nematode Heterodera schachtii in Arabidopsis. *New Phytol.* **2015**, *207*, 778–789. [CrossRef]
204. Martínez-Medina, A.; Fernandez, I.; Lok, G.B.; Pozo, M.J.; Pieterse, C.M.; Van Wees, S.C. Shifting from priming of salicylic acid to jasmonic acid-regulated defences by Trichoderma protects tomato against the root knot nematode *Meloidogyne incognita*. *New Phytol.* **2017**, *213*, 1363–1377. [CrossRef]
205. Criollo-Arteaga, S.; Moya-Jimenez, S.; Jimenez-Meza, M.; Gonzalez-Vera, V.; Gordon-Nunez, J.; Llerena-Llerena, S.; Ramirez-Villacis, D.X.; van't Hof, P.; Leon-Reyes, A. Sulfur Deprivation Modulates Salicylic Acid Responses via Nonexpressor of Pathogenesis-Related Gene 1 in *Arabidopsis thaliana*. *Plants* **2021**, *10*, 1065. [CrossRef]
206. Shan, C.; Sun, H.; Zhou, Y.; Wang, W. Jasmonic acid-induced hydrogen sulfide activates MEK1/2 in regulating the redox state of ascorbate in *Arabidopsis thaliana* leaves. *Plant Signal. Behav.* **2019**, *14*, 1629265. [CrossRef]
207. Foucher, J.; Ruh, M.; Preveaux, A.; Carrère, S.; Pelletier, S.; Briand, M.; Serre, R.-F.; Jacques, M.-A.; Chen, N.W. Common bean resistance to Xanthomonas is associated with upregulation of the salicylic acid pathway and downregulation of photosynthesis. *BMC Genom.* **2020**, *21*, 566.
208. Fu, L.H.; Hu, K.D.; Hu, L.Y.; Li, Y.H.; Hu, L.B.; Yan, H.; Liu, Y.S.; Zhang, H. An antifungal role of hydrogen sulfide on the postharvest pathogens *Aspergillus niger* and *Penicillium italicum*. *PLoS ONE* **2014**, *9*, e104206. [CrossRef]
209. Hu, K.D.; Wang, Q.; Hu, L.Y.; Gao, S.P.; Wu, J.; Li, Y.H.; Zheng, J.L.; Han, Y.; Liu, Y.S.; Zhang, H. Hydrogen sulfide prolongs postharvest storage of fresh-cut pears (*Pyrus pyrifolia*) by alleviation of oxidative damage and inhibition of fungal growth. *PLoS ONE* **2014**, *9*, e85524. [CrossRef]
210. Liu, D.; Li, J.; Li, Z.; Pei, Y. Hydrogen sulfide inhibits ethylene-induced petiole abscission in tomato (*Solanum lycopersicum* L.). *Hortic. Res.* **2020**, *7*, 14. [CrossRef]
211. Hou, Z.; Wang, L.; Liu, J.; Hou, L.; Liu, X. Hydrogen sulfide regulates ethylene-induced stomatal closure in *Arabidopsis thaliana*. *J. Integr. Plant Biol.* **2013**, *55*, 277–289. [CrossRef]

212. Al Ubeed, H.; Wills, R.; Bowyer, M.; Vuong, Q.; Golding, J. Interaction of exogenous hydrogen sulphide and ethylene on senescence of green leafy vegetables. *Postharvest Biol. Technol.* **2017**, *133*, 81–87. [CrossRef]
213. Du, X.; Jin, Z.; Liu, Z.; Liu, D.; Zhang, L.; Ma, X.; Yang, G.; Liu, S.; Guo, Y.; Pei, Y. H_2S Persulfidated and Increased Kinase Activity of MPK4 to Response Cold Stress in Arabidopsis. *Front. Mol. Biosci.* **2021**, *8*, 81. [CrossRef]
214. Carter, J.M.; Brown, E.M.; Irish, E.E.; Bowden, N.B. Characterization of dialkyldithiophosphates as slow hydrogen sulfide releasing chemicals and their effect on the growth of maize. *J. Agric. Food Chem.* **2019**, *67*, 11883–11892. [CrossRef]

International Journal of
Molecular Sciences

MDPI

Article

H_2S in Horticultural Plants: Endogenous Detection by an Electrochemical Sensor, Emission by a Gas Detector, and Its Correlation with L-Cysteine Desulfhydrase (LCD) Activity

María A. Muñoz-Vargas, Salvador González-Gordo, José M. Palma and Francisco J. Corpas *

Group of Antioxidants, Free Radicals and Nitric Oxide in Biotechnology, Food and Agriculture, Department of Biochemistry, Cell and Molecular Biology of Plants, Estación Experimental del Zaidín, Spanish National Research Council (CSIC), C/Profesor Albareda 1, E-18008 Granada, Spain; mangeles.munoz@eez.csic.es (M.A.M.-V.); salvador.gonzalez@eez.csic.es (S.G.-G.); josemanuel.palma@eez.csic.es (J.M.P.)

* Correspondence: javier.corpas@eez.csic.es

Abstract: H_2S has acquired great attention in plant research because it has signaling functions under physiological and stress conditions. However, the direct detection of endogenous H_2S and its potential emission is still a challenge in higher plants. In order to achieve a comparative analysis of the content of H_2S among different plants with agronomical and nutritional interest including pepper fruits, broccoli, ginger, and different members of the genus Allium such as garlic, leek, Welsh and purple onion, the endogenous H_2S and its emission was determined using an ion-selective microelectrode and a specific gas detector, respectively. The data show that endogenous H_2S content range from pmol to µmol $H_2S \cdot g^{-1}$ fresh weight whereas the H_2S emission of fresh-cut vegetables was only detected in the different species of the genus Allium with a maximum of 9 ppm in garlic cloves. Additionally, the activity and isozymes of the L-cysteine desulfhydrase (LCD) were analyzed, which is one of the main enzymatic sources of H_2S, where the different species of the genus Allium showed the highest activities. Using non-denaturing gel electrophoresis, the data indicated the presence of up to nine different LCD isozymes from one in ginger to four in onion, leek, and broccoli. In summary, the data indicate a correlation between higher LCD activity with the endogenous H_2S content and its emission in the analyzed horticultural species. Furthermore, the high content of endogenous H_2S in the Allium species supports the recognized benefits for human health, which are associated with its consumption.

Keywords: Allium; hydrogen sulfide; garlic; gas detector; ion-selective microelectrode; L-cysteine desulfhydrase; isozymes

Citation: Muñoz-Vargas, M.A.; González-Gordo, S.; Palma, J.M.; Corpas, F.J. H_2S in Horticultural Plants: Endogenous Detection by an Electrochemical Sensor, Emission by a Gas Detector, and Its Correlation with L-Cysteine Desulfhydrase (LCD) Activity. *Int. J. Mol. Sci.* **2022**, *23*, 5648. https://doi.org/10.3390/ijms23105648

Academic Editors: Yanjie Xie, Jisheng Li and Massimiliano Tattini

Received: 27 April 2022
Accepted: 17 May 2022
Published: 18 May 2022

Publisher's Note: MDPI stays neutral with regard to jurisdictional claims in published maps and institutional affiliations.

1. Introduction

H_2S is a key signaling molecule that plays multiple functions in many physiological and pathological processes in humans, regulating the basal metabolism, central nervous system, blood pressure, gastrointestinal motility, inflammation, the immune system, or cancer, among others [1–7]. In higher plants, H_2S also has relevant functions due to its direct or indirect implication in physiological functions including seed germination, root development, plant growth, stomatal closure, senescence, and fruit ripening as well as in the mechanism of response against adverse environmental conditions [8–13]. H_2S is part of the sulfur metabolism being enzymatically generated by different enzymes present in diverse subcellular compartments including cytosol, plastids, and mitochondria [14]. At the biochemical level, H_2S mediates the regulation of protein function by a posttranslational modification designated persufidation which involved the thiol group of cysteine residues from target proteins [15–17]. However, the detection of endogenous hydrogen sulfide and its possible emission continues to be a scientific challenge due to the complex biochemistry that is affected by its interaction with peptides and proteins, pH, cellular location, type of biological samples, etc. Different methodologies allow the detection of endogenous H_2S in different types of biological samples such as high-performance liquid chromatography (HPLC), gas chromatography (GC), colorimetric, specific fluorescence probes, ion-selective electrode (ISE), amperometric (polarographic) H_2S sensor, ozone-based chemiluminescence detection among others and all have advantages and disadvantages as well as different detection limits [18–21].

L-Cysteine desulfhydrase (LCD, EC 4.4.1.28) is considered one of the main cytosolic enzymatic sources of H_2S in Arabidopsis cells which is also generated by other enzymes such as the chloroplastic sulfite reductase (SiR, EC 1.8.7.1) or the mitochondrial bifunctional D-cysteine desulfhydrase/1-aminocyclopropane-1-carboxylate deaminase (DCDES1, EC 4.4.1.15) and D-cysteine desulfhydrase 2 (DCDES2, EC 4.4.1.15) [14,22,23]. The LCD catalyzes the following reaction: L-cysteine + $H_2O \rightarrow$ pyruvate + NH^{4+} + H_2S + H^+, and it requires pyridoxal 5′-phosphate (PLP) as cofactor [23,24]. This enzyme is involved in diverse processes such as root development [25], stomatal closure [26–28], leaf senescence [29], fruit ripening [30,31], and response to diverse stresses [32].

Aiming to correlate the potential relationship between endogenous H_2S, its potential emission, and the activity of the LCD, the present study provides a comparative analysis of these parameters in some horticultural plants such as pepper fruits, broccoli, ginger, fennel, eggplant, leek, garlic, Welsh and onion, which are relevant to human nutrition. The Allium species, particularly garlic cloves, are horticultural plants that present the highest values of H_2S that is well correlated with its very active metabolism of organosulfur compounds such as allicin, allyl sulfides, allyl thiosulfinate, ajoene, and S-allyl cysteine among others.

2. Results

Figure 1a–c illustrates the endogenous H_2S content in different horticultural plant species which was measured using an ion-selective microelectrode (Arrow H_2S^{TM}) H_2S measurement system. Accordingly, three well-differentiated data groups can be distinguished, while the species of the Allium genus give H_2S values in the range of μmol $H_2S \cdot g^{-1}$ FW, the leek with 1 μmol $H_2S \cdot g^{-1}$ FW being the species with the highest content, followed by garlic cloves with 0.54 μmol $H_2S \cdot g^{-1}$ FW. Other groups are in the range of nmol $H_2S \cdot g^{-1}$ FW such as broccoli but also in the range of pmol $H_2S \cdot g^{-1}$ FW such as pepper fruit at different stages of ripening green and red. On the other hand, in fennel and eggplant samples the H_2S detected was even lower, so they were not used for subsequent studies.

As part of the characterization of H_2S in these horticultural plants, the detection of H_2S gas emission was performed using 300 g of cut materials. Figure 2 indicates that the species of the genus Allium were the only ones that allowed it to be detected. It is noteworthy that the emission of H_2S is relatively quite fast and is maintained over time. Thus, leek reaches a maximum peak of 3.3 ppm of H_2S after 30 min, followed by spring onion with a maximum peak of 7.4 ppm after 75 min, purple onion with 8.1 ppm after 90 min, and finally garlic with 9.0 after 180 min. It is noteworthy that the garlic maintained the emission of H_2S up to 0.9 ppm after 21 h.

Assuming that the L-cysteine desulfhydrase (LCD) activity is considered the enzyme that most contributes to the generation of H_2S in the cell, its activity was measured spectrophotometrically and also in polyacrylamide gels under non-denaturing conditions. Figure 3a shows that species of the Allium genus have higher LCD activity, with garlic having the highest one, followed by broccoli, pepper, and ginger. On the other hand, Figure 3b shows the LCD isozymes profile in the different analyzed plant species. They were designated as I to IX according to their increasing electrophoretic mobility in the non-denaturing polyacrylamide gel. The number and relative abundance were quite different, while a single LCD isozyme is identified in ginger, three isozymes appear in garlic, broccoli, and green peppers, and up to four LCD isozymes in purple and Welsh onions. In pepper fruits, it is remarkable that the number of isozymes changes with ripening, having three in green pepper fruits and only one in red peppers.

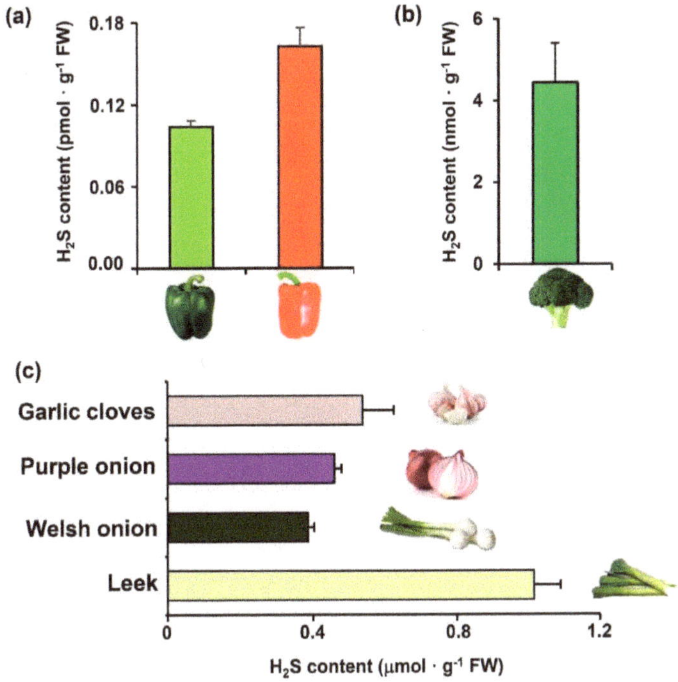

Figure 1. Endogenous H_2S detection in different plant species including (**a**) sweet pepper (*Capsicum annuum* L.) fruit at distinct ripening stages (fully green and fully red), (**b**) Broccoli (*Brassica oleracea* var. Itálica), (**c**) Allium species including garlic (*Allium sativum* L.) cloves, leek (*Allium ampeloprasum* var. porrum), welsh onion (*Allium fistulosum*), and purple onion (*Allium cepa*). H_2S was detected using a micro sulfide ion electrode.

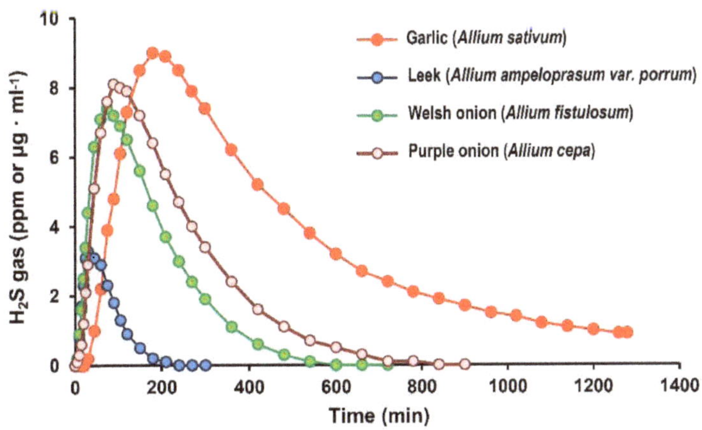

Figure 2. H_2S gas emission in different plant species including garlic (*Allium sativum* L.) cloves, leek (*Allium ampeloprasum* var. porrum), Welsh onion (*Allium fistulosum*), and purple onion (*Allium cepa*). H_2S emission was recorded using an H_2S sensor gas analyzer portable device. For each plant, 300 g of fresh material was used.

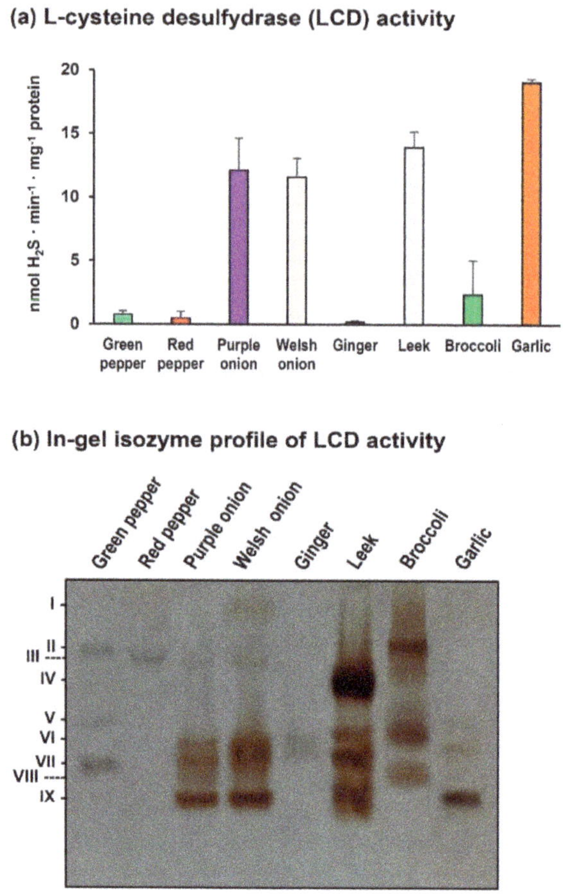

Figure 3. L-Cysteine desulfhydrase (LCD) activity in different plant samples. (**a**) Spectrophotometric assay. (**b**) In-gel isozyme profile of LCD activity. Protein samples (74 μg protein per lane) were separated by non-denaturing polyacrylamide gel electrophoresis (PAGE; 8% acrylamide) and the LCD activity was detected by lead acetate staining (see M&M for details).

3. Discussion

H_2S is recognized as a key molecule with a signaling function in animal and plant cells, with similar regulatory properties to those exerted by nitric oxide (NO) in higher plants under physiological and stress conditions [9,33–38]. With the aim of obtaining a better understanding of this molecule in different plants with agronomic interest, the endogenous content, its emission, as well as its possible correlation with LCD activity have been comparatively studied, since it is considered the most relevant enzyme in the H_2S production in plant cells.

One of the difficulties in determining the endogenous H_2S content in a specific sample is based on its chemistry because, being a weak acid, it can be dissociated to hydrosulfide (HS^-) and sulfide (S^{2-}) anions in an aqueous solution according to the following equations: $H_2S(gas) \leftrightarrow H_2S(aqueous\ solution) \leftrightarrow HS^- + H^+ \leftrightarrow S^{2-} + 2H^+$. Other considerations that should be taken into account are the pH of the medium, the nature of biological samples, that H_2S can interact with thiol groups present in peptides and proteins, the selectivity of the technical approach as well as the experimental conditions [2,15,18,39]. Otherwise, the development of new approaches such as the specific fluorescent probes has allowed the detection of H_2S at the subcellular level by bioimaging approaches [40–43]; however, its quantification is still a challenge. In an earlier study in cucumber (*Cucumis sativus* L.) plants, an H_2S concentration of 8 nmol $\cdot$ min^{-1} g^{-1} FW that was light-dependent [44] was reported and, after fumigation with SO_2, the H_2S content was 0.02 to 0.2 ng $\cdot$ g^{-1} dry weight [45]. In *Arabidopsis thaliana* and faba bean (*Vicia faba*) leaves, using a micro sulfide ion electrode, an H_2S content between 1 to 5 µmol $\cdot$ L^{-1} was reported [46,47]. More recently, Jin et al. [48], using both methylene blue and electrode methods, determined the H_2S content of 17 species in different developmental stages and different organs and they showed a wider range, from 0.177 to 0.708 µmol $\cdot$ g^{-1} FW, in which *Platycladus* sp. in the cypress family had the highest content whereas tobacco had the lowest content. Consequently, our data in the assayed horticultural plants are in good agreement with all these previous reports. Although the values of H_2S content determined with the micro sulfide ion electrode provide relative measurements accurately and reliably, the difficulty to measure the H_2S content when these values are low to 0.1 µmol $\cdot$ g^{-1} FW should be mentioned and, consequently, the obtained values should be interpreted with caution.

The emission of volatile H_2S is another aspect that could have great relevance in higher plants, although the available information is still very scarce. In an earlier study, Sekiya et al. [49], analyzed the H_2S emission by gas chromatography in leaf discs incubated in the presence of 10 mM L-Cys from nine species (*Cucumis sativus, Cucurbita pepo, Nicotiana tabacum, Coleus blumei, Beta vulgaris, Phaseolus vulgaris, Medicago sativa, Hordeum vulgare*, and *Gossypium hirsutum*) and they found an emission around 40 pmol H_2S $\cdot$ min^{-1} cm^{-2}; however, the H_2S emission was not observed in the presence of D-Cys. Rennenberg et al. [50], using a flame photometric sulfur analyzer, reported an H_2S emission between 38 to 91 pmol H_2S $\cdot$ min^{-1} cm^{-2} pumpkin leaf area. In our experimented conditions, we only observed H_2S emission in the species of the genus Allium, and this emission was maintained in the time, particularly in garlic which was up to 9 ppm. It is well known that these species of the genus Allium have a characteristic smell and taste which is due to the sulfur-containing volatile flavor compounds. These volatile constituents are generated by the action of the enzyme alliinase, with a molecular mass between 13 and 35 kDa depending on the Allium species, when plant tissue is disrupted, the alliinase catalyzes the conversion of odorless S-alk(en)yl-L-cysteine sulfoxides (SACs), known as alliin, into volatile smelling thiosulfinates [51,52]. These SACs are synthesized from glutathione in the cytosol and when the compartmentalization of the alliinase, present in the vacuole, is broken down, it allows the metabolization of these SACs and its emission in volatile thiosulfinates [53–55]. In fact, these groups of sulfur-containing natural products present in the different species of Allium are correlated with the benefits associated with human health when they are part of our diet [52,56]. Thus, it is well documented that garlic consumption reduces some risk factors related to cardiovascular diseases such as high blood pressure, high cholesterol, platelet aggregation, blood coagulation, and the increased content of reactive oxygen species (ROS) [52,57–60]. The detected H_2S emission in the different species of Allium is also well correlated with the beneficial effects exerted on human health [59]. In plants, these sulfur-containing molecules including H_2S have great relevance in the resistance of crops against diverse fungal diseases [61,62].

L-cysteine desulfhydrase (LCD) catalyzes the desulfuration of L-Cys to generate H_2S and is considered one of the main sources of this molecule in the cytosol of plant cells [63]. As was mentioned, biochemical and molecular approaches have revealed that LCD is implicated in diverse processes such as seed germination [64], root development [25,65], stomatal closure [26,27,66], drought tolerance [67], and fruit ripening [31,68].

In our experimental conditions, it is well correlated with the total LCD activity found in the Allium species with the highest level of endogenous H_2S content as well as its emission. However, the available information about the number of LCD isozymes present in a specific tissue or plant species and its corresponding functions is to our knowledge very scarcer. Considering all analyzed species and tissues, the present data indicate the existence of nine LCD isozymes indicating the great diversity in the number and relative abundance which seems to support the differential potential physiological regulatory functions that they could have. For example, in the pepper fruit samples, it was observed that total LCD activity was downregulated during ripening from green to red fruits, which was well correlated with the diminishment of the LCD isozymes since two of them were not detected in red fruits.

4. Materials and Methods

4.1. Plant Material

California-type sweet pepper (*Capsicum annuum* L., cv. Melchor) fruits were collected from plastic-covered experimental greenhouses (Syngenta Seeds, Ltd., El Ejido, Almería, Spain) whereas the other plants were acquired in the local market including broccoli (*Brassica oleracea* var. Itálica), ginger (*Zingiber officinale*) rhizome, fennel (*Foeniculum vulgare*), eggplant (*Solanum melongena*), leek (*Allium ampeloprasum* var. porrum), Welsh onion (*Allium fistulosum*), purple onion (*Allium cepa*) and garlic (*Allium sativum* L.) cloves.

4.2. Preparation of Plant Extracts

Plant samples were ground to a fine powder in liquid N_2 using an IKA® A11 basic analytical mill. The resulting powder was suspended in 0.1 M Tris-HCl buffer, pH 7.5, containing 1 mM EDTA, 0.1% (*v*/*v*) Triton X-100, 10% (*v*/*v*) glycerol to a final plant material/buffer (*w*/*v*) ratio of 1:1 for pepper, fennel, and leek fruit; 1:2 for eggplant, garlic, ginger, and broccoli; and, 2:1 for red onion and Welsh onion. Homogenates were then filtered through two layers of Miracloth and centrifuged at 27,000× *g* for 20 min. The supernatants were used for subsequent analyses. For H_2S endogenous quantification, the supernatants were mixed with antioxidant buffer (2 M NaOH, 170 mM sodium ascorbate, and 180 mM EDTA). In the case of enzymatic activity, the extraction buffer was similar but with a pH of 8.0. Protein content was determined by a standard Bradford assay using a reagent (Bio-Rad Laboratories, Hercules, CA, USA).

4.3. Endogenous H_2S Quantification

H_2S was measured in plant extracts for 5 min at 25 °C using a micro sulfide ion electrode (LIS-146AGSCM; Lazar Research Laboratories) attached to a voltage meter (Lazar Research Lab. Inc., Los Angeles, CA, USA, model ISM-146 H_2S-XS). H_2S concentrations were calculated from a standard curve made with sodium sulfide (Na_2S) according to the micro-electrode manufacturer's instructions.

4.4. H_2S Gas Emission Hydrogen Sulfide Gas Detector

H_2S gas emission was recorded using a high accuracy H_2S sensor gas analyzer portable device (NOBGP brand, model TK3Bcb4T78 model), which can measure H_2S gas in a range between 0~100 $\mu mol \cdot mL^{-1}$, a resolution of 0.1 with an accuracy less than or equal to ±5% full scale. In all cases, 300 g of fresh plant samples were cut into homogeneous pieces and placed in a methacrylate hermetic box (10 mm-thickness walls): 25 (large) × 25 (width) × 30 (height) cm = 15.34 L, furnished with a lid made on the same material. The H_2S gas detector was placed into the box and the H_2S emission was recorded for 18 h. In the case of garlic, the cloves were used but for leek and Welsh onion, the stringy roots and dark green leaves were chopped off.

4.5. Spectrophotometric Assay and In-Gel Isozyme Profile of L-Cysteine Desulfrydrase (LCD) Activity

L-cysteine desulfhydrase (LCD, E.C. 4.4.1.28) activity was spectrophotometrically determined by the release of H_2S from L-Cys as described previously [69,70]. Briefly, the enzyme assay contained 1 mM L-Cys, 100 mM Tris-HCl, pH 8.0, 1 mM dithiothreitol, and plant extract in a final volume of 1 mL. After 15 min incubation at 37 °C, the reaction was stopped by the addition of 100 µL of 30 mM $FeCl_3$ prepared in 1.2 N HCl and 100 µL of 20 mM N,N-dimethyl-p-phenylenediamine dihydrochloride prepared in 7.2 N HCl.

The formation of methylene blue was measured at 670 nm, and the enzyme activity was calculated using the extinction coefficient of 15×10^6 cm^2 mol^{-1}.

For in-gel isozyme profile analysis, protein samples were separated using non-denaturing polyacrylamide gel electrophoresis (PAGE) on 8% acrylamide gels. After the electrophoresis, the gels were incubated in the dark in a staining buffer containing Tris-HCl 100 mM, pH 7.5, L-cysteine 20 mM, lead acetate 0.4 mM, pyridoxal 5′-phosphate hydrate 50 μM and β-mercaptoethanol 20 mM until the appearance of brown bands [24,71].

5. Conclusions

H_2S is a signaling molecule in both animal and plant cells. The analysis of its endogenous content as to its possible emission in higher plants to determine its physiological functions and in response to environmental stresses has been a challenge for years [72]. However, the information concerning H_2S in horticultural species is still scarce; therefore, the present study provides new information on the content and emission of H_2S in horticultural plants, particularly in the species of the Allium genus. This may be of great importance in horticultural crops, considering that H_2S applied exogenously has been shown to exert multiple benefits for vegetables and fruits, since it has the capacity to preserve their quality during postharvest storage and prevents infections by pathogens because H_2S stimulates phytohormone, reactive oxygen, and nitrogen metabolism [73–78]. Furthermore, the identification of different LCD isozymes in the analyzed horticultural species indicates the relevance of this enzyme in H_2S metabolism and raises new questions about the specific function of each isozyme which could be modulated by environmental conditions or physiological processes such as was observed in the ripening of pepper fruits. Allium species have been recognized for a long time to have healthy properties [79], and among the sulfur compounds that it contains, H_2S seems to be of great relevance [58,80]. On the other hand, considering the high H_2S emission of Allium species, they should be regarded as a potential source of this gas for its possible biotechnological application in the horticulture industry since it can extend the quality of vegetables and fruits during postharvest storage.

Author Contributions: M.A.M.-V. and S.G.-G. performed biochemical analyses. F.J.C. and J.M.P. designed the work, drove and coordinated the tasks. F.J.C. wrote the first draft of the manuscript. All authors have read and agreed to the published version of the manuscript.

Funding: Our research is supported by a European Regional Development Fund cofinanced grants from the Spanish Ministry of Science and Innovation (PID2019-103924GB-I00) and Junta de Andalucía (P18-FR-1359), Spain.

Institutional Review Board Statement: Not applicable.

Informed Consent Statement: Not applicable.

Acknowledgments: M.A.M.-V. and S.G.-G. acknowledge a 'Formación de Personal Investigador' contract (PRE2020-093882 and BES-2016-078368, respectively) from the Ministry of Science and Innovation, Spain. The valuable technical assistance of Carmelo Ruiz-Torres is deeply acknowledged.

Conflicts of Interest: The authors declare no conflict of interest.

References

1. Banerjee, R.; Vitvitsky, V.; Garg, S.K. The undertow of sulfur metabolism on glutamatergic neurotransmission. *Trends Biochem. Sci.* **2008**, *33*, 413–419. [CrossRef] [PubMed]
2. Kabil, O.; Vitvitsky, V.; Banerjee, R. Sulfur as a signaling nutrient through hydrogen sulfide. *Annu Rev. Nutr.* **2014**, *34*, 171–205. [CrossRef]
3. Szabo, C. Hydrogen sulfide, an endogenous stimulator of mitochondrial function in cancer cells. *Cells* **2021**, *10*, 220. [CrossRef] [PubMed]
4. Parfenova, H.; Liu, J.; Hoover, D.T.; Fedinec, A.L. Vasodilator effects of sulforaphane in cerebral circulation: A critical role of endogenously produced hydrogen sulfide and arteriolar smooth muscle KATP and BK channels in the brain. *J. Cereb. Blood Flow Metab.* **2020**, *40*, 1987–1996. [CrossRef] [PubMed]

5. Kumar, R.; Banerjee, R. Regulation of the redox metabolome and thiol proteome by hydrogen sulfide. *Crit. Rev. Biochem. Mol. Biol.* **2021**, *56*, 221–235. [CrossRef]
6. Kimura, H. Hydrogen sulfide (H_2S) and polysulfide (H_2S_n) signaling: The First 25 Years. *Biomolecules* **2021**, *11*, 896. [CrossRef]
7. Deng, G.; Muqadas, M.; Adlat, S.; Zheng, H.; Li, G.; Zhu, P.; Nasser, M.I. Protective effect of hydrogen sulfide on cerebral ischemia-reperfusion injury. *Cell. Mol. Neurobiol.* 2022, in press. [CrossRef]
8. Corpas, F.J.; González-Gordo, S.; Cañas, A.; Palma, J.M. Nitric oxide and hydrogen sulfide in plants: Which comes first? *J. Exp. Bot.* **2019**, *70*, 4391–4404. [CrossRef]
9. Mishra, V.; Singh, P.; Tripathi, D.K.; Corpas, F.J.; Singh, V.P. Nitric oxide and hydrogen sulfide: An indispensable combination for plant functioning. *Trends Plant Sci.* **2021**, *26*, 1270–1285. [CrossRef]
10. Rai, P.; Singh, V.P.; Peralta-Videa, J.; Tripathi, D.K.; Sharma, S.; Corpas, F.J. Hydrogen sulfide (H_2S) underpins the beneficial silicon effects against the copper oxide nanoparticles (CuO NPs) phytotoxicity in *Oryza sativa* seedlings. *J. Hazard. Mater.* **2021**, *415*, 124907. [CrossRef]
11. Raza, A.; Tabassum, J.; Mubarik, M.S.; Anwar, S.; Zahra, N.; Sharif, Y.; Hafeez, M.B.; Zhang, C.; Corpas, F.J.; Chen, H. Hydrogen sulfide: An emerging component against abiotic stress in plants. *Plant Biol.* 2021, in press. [CrossRef] [PubMed]
12. Zhang, J.; Zhou, M.; Zhou, H.; Zhao, D.; Gotor, C.; Romero, L.C.; Shen, J.; Ge, Z.; Zhang, Z.; Shen, W.; et al. Hydrogen sulfide, a signaling molecule in plant stress responses. *J. Integr. Plant Biol.* **2021**, *63*, 146–160. [CrossRef] [PubMed]
13. Aroca, A.; Zhang, J.; Xie, Y.; Romero, L.C.; Gotor, C. Hydrogen sulfide signaling in plant adaptations to adverse conditions: Molecular mechanisms. *J. Exp. Bot.* **2021**, *72*, 5893–5904. [CrossRef] [PubMed]
14. González-Gordo, S.; Palma, J.M.; Corpas, F.J. Appraisal of H_2S metabolism in *Arabidopsis thaliana*: *In silico* analysis at the subcellular level. *Plant Physiol. Biochem.* **2020**, *155*, 579–588. [CrossRef]
15. Filipovic, M.R.; Zivanovic, J.; Alvarez, B.; Banerjee, R. Chemical biology of H_2S signaling through persulfidation. *Chem. Rev.* **2018**, *118*, 1253–1337. [CrossRef]
16. Corpas, F.J.; González-Gordo, S.; Muñoz-Vargas, M.A.; Rodríguez-Ruiz, M.; Palma, J.M. The modus operandi of hydrogen sulfide(H_2S)-dependent protein persulfidation in higher plants. *Antioxidants* **2021**, *10*, 1686. [CrossRef]
17. Corpas, F.J.; González-Gordo, S.; Palma, J.M. Nitric oxide and hydrogen sulfide modulate the NADPH-generating enzymatic system in higher plants. *J. Exp. Bot.* **2021**, *72*, 830–847. [CrossRef]
18. Zhao, D.; Zhang, J.; Zhou, M.; Zhou, H.; Gotor, C.; Romero, L.C.; Shen, J.; Yuan, X.; Xie, Y. Current approaches for detection of hydrogen sulfide and persulfidation in biological systems. *Plant Physiol. Biochem.* **2020**, *155*, 367–373. [CrossRef]
19. Luo, Y.; Zuo, Y.; Shi, G.; Xiang, H.; Gu, H. Progress on the reaction-based methods for detection of endogenous hydrogen sulfide. *Anal. Bioanal. Chem.* **2021**, *414*, 2809–2839. [CrossRef]
20. Quan, F.S.; Lee, G.J. Analytical methods for detection of gasotransmitter hydrogen sulfide released from live cells. *BioMed Res. Int.* **2021**, *2021*, 5473965. [CrossRef]
21. Li, B.; Kim, Y.L.; Lippert, A.R. Chemiluminescence measurement of reactive sulfur and nitrogen species. *Antioxid. Redox Signal.* **2022**, *36*, 337–353. [CrossRef] [PubMed]
22. Feldman-Salit, A.; Veith, N.; Wirtz, M.; Hell, R.; Kummer, U. Distribution of control in the sulfur assimilation in Arabidopsis thaliana depends on environmental conditions. *New Phytol.* **2019**, *222*, 1392–1404. [CrossRef] [PubMed]
23. Kurmanbayeva, A.; Bekturova, A.; Soltabayeva, A.; Oshanova, D.; Nurbekova, Z.; Srivastava, S.; Tiwari, P.; Dubey, A.K.; Sagi, M. Active O-acetylserine-(thiol) lyase A and B confer improved selenium resistance and degrade l-Cys and l-SeCys in Arabidopsis. *J. Exp. Bot.* **2022**, *73*, 2525–2539. [CrossRef]
24. Muñoz-Vargas, M.A.; Rodríguez-Ruiz, M.; González-Gordo, S.; Palma, J.M.; Corpas, F.J. Analysis of Plant L-Cysteine desulfhydrase (LCD) isozymes by non-denaturing polyacrylamide gel electrophoresis. In *Plant Abiotic Stress Signaling. Methods in Molecular Biology*; Couée, I., Ed.; Humana: New York, NY, USA, 2023; in press.
25. Mei, Y.; Zhao, Y.; Jin, X.; Wang, R.; Xu, N.; Hu, J.; Huang, L.; Guan, R.; Shen, W. L-Cysteine desulfhydrase-dependent hydrogen sulfide is required for methane-induced lateral root formation. *Plant Mol. Biol.* **2019**, *99*, 283–298. [CrossRef]
26. Scuffi, D.; Álvarez, C.; Laspina, N.; Gotor, C.; Lamattina, L.; García-Mata, C. Hydrogen sulfide generated by L-cysteine desulfhydrase acts upstream of nitric oxide to modulate abscisic acid-dependent stomatal closure. *Plant Physiol.* **2014**, *166*, 2065–2076. [CrossRef]
27. Shen, J.; Zhang, J.; Zhou, M.; Zhou, H.; Cui, B.; Gotor, C.; Romero, L.C.; Fu, L.; Yang, J.; Foyer, C.H.; et al. Persulfidation-based modification of cysteine desulfhydrase and the NADPH oxidase RBOHD controls guard cell abscisic acid signaling. *Plant Cell* **2020**, *32*, 1000–1017. [CrossRef] [PubMed]
28. Ma, Y.; Shao, L.; Zhang, W.; Zheng, F. Hydrogen sulfide induced by hydrogen peroxide mediates brassinosteroid-induced stomatal closure of *Arabidopsis thaliana*. *Funct. Plant Biol.* **2021**, *48*, 195–205. [CrossRef]
29. Hu, K.; Peng, X.; Yao, G.; Zhou, Z.; Yang, F.; Li, W.; Zhao, Y.; Li, Y.; Han, Z.; Chen, X.; et al. Roles of a cysteine desulfhydrase LCD1 in regulating leaf senescence in tomato. *Int. J. Mol. Sci.* **2021**, *22*, 13078. [CrossRef]
30. Muñoz-Vargas, M.A.; González-Gordo, S.; Cañas, A.; López-Jaramillo, J.; Palma, J.M.; Corpas, F.J. Endogenous hydrogen sulfide (H_2S) is up-regulated during sweet pepper (*Capsicum annuum* L.) fruit ripening. In vitro analysis shows that NADP-dependent isocitrate dehydrogenase (ICDH) activity is inhibited by H_2S and NO. *Nitric Oxide* **2018**, *81*, 36–45. [CrossRef]
31. Hu, K.D.; Zhang, X.Y.; Yao, G.F.; Rong, Y.L.; Ding, C.; Tang, J.; Yang, F.; Huang, Z.Q.; Xu, Z.M.; Chen, X.Y.; et al. A nuclear-localized cysteine desulfhydrase plays a role in fruit ripening in tomato. *Hortic. Res.* **2020**, *7*, 211. [CrossRef]

32. Kharbech, O.; Sakouhi, L.; Mahjoubi, Y.; Ben Massoud, M.; Debez, A.; Zribi, O.T.; Djebali, W.; Chaoui, A.; Mur, L.A.J. Nitric oxide donor, sodium nitroprusside modulates hydrogen sulfide metabolism and cysteine homeostasis to aid the alleviation of chromium toxicity in maize seedlings (*Zea mays* L.). *J. Hazard. Mater.* **2022**, *424 Pt A*, 127302. [CrossRef]
33. Gheibi, S.; Jeddi, S.; Kashfi, K.; Ghasemi, A. Regulation of vascular tone homeostasis by NO and H_2S: Implications in hypertension. *Biochem. Pharmacol.* **2018**, *149*, 42–59. [CrossRef] [PubMed]
34. Lv, B.; Chen, S.; Tang, C.; Jin, H.; Du, J.; Huang, Y. Hydrogen sulfide and vascular regulation—An update. *J. Adv. Res.* **2020**, *27*, 85–97. [CrossRef] [PubMed]
35. Kimura, H. Hydrogen polysulfide (H_2S_n) signaling along with hydrogen sulfide (H_2S) and nitric oxide (NO). *J. Neural Transm. (Vienna)* **2016**, *123*, 1235–1245. [CrossRef] [PubMed]
36. Kimura, H. Hydrogen sulfide signalling in the CNS—Comparison with NO. *Br. J. Pharmacol.* **2020**, *177*, 5031–5045. [CrossRef] [PubMed]
37. Corpas, F.J. Hydrogen sulfide: A new warrior against abiotic Stress. *Trends Plant Sci.* **2019**, *24*, 983–988. [CrossRef] [PubMed]
38. Kharbech, O.; Sakouhi, L.; Ben Massoud, M.; Jose Mur, L.A.; Corpas, F.J.; Djebali, W.; Chaoui, A. Nitric oxide and hydrogen sulfide protect plasma membrane integrity and mitigate chromium-induced methylglyoxal toxicity in maize seedlings. *Plant Physiol. Biochem.* **2020**, *157*, 244–255. [CrossRef]
39. Papenbrock, J.; Riemenschneider, A.; Kamp, A.; Schulz-Vogt, H.N.; Schmidt, A. Characterization of cysteine-degrading and H_2S-releasing enzymes of higher plants—From the field to the test tube and back. *Plant Biol.* **2007**, *9*, 582–588. [CrossRef]
40. Peng, B.; Chen, W.; Liu, C.; Rosser, E.W.; Pacheco, A.; Zhao, Y.; Aguilar, H.C.; Xian, M. Fluorescent probes based on nucleophilic substitution-cyclization for hydrogen sulfide detection and bioimaging. *Chemistry* **2014**, *20*, 1010–1016. [CrossRef]
41. Yu, F.; Han, X.; Chen, L. Fluorescent probes for hydrogen sulfide detection and bioimaging. *Chem. Commun.* **2014**, *50*, 12234–12249. [CrossRef]
42. Corpas, F.J.; Barroso, J.B.; González-Gordo, S.; Muñoz-Vargas, M.A.; Palma, J.M. Hydrogen sulfide: A novel component in Arabidopsis peroxisomes which triggers catalase inhibition. *J. Integr. Plant Biol.* **2019**, *61*, 871–883. [CrossRef] [PubMed]
43. Wang, J.; Xie, H.; Li, H.; Wang, R.; Zhang, B.; Ren, T.; Hua, J.; Chen, N. NIR Fluorescent probe for *in situ* bioimaging of endogenous H_2S in rice roots under Al^{3+} and flooding stresses. *J. Agric. Food Chem.* **2021**, *69*, 14330–14339. [CrossRef] [PubMed]
44. Wilson, L.G.; Bressan, R.A.; Filner, P. Light-dependent emission of hydrogen sulfide from plants. *Plant Physiol.* **1978**, *61*, 184–189. [CrossRef] [PubMed]
45. Winner, W.E.; Smith, C.L.; Koch, G.W.; Mooney, H.A.; Bewley, J.D.; Krouse, H.R. Rates of emission of H_2S from plants and patterns of stable sulphur isotope fractionation. *Nature* **1981**, *289*, 672–673. [CrossRef]
46. García-Mata, C.; Lamattina, L. Hydrogen sulphide, a novel gasotransmitter involved in guard cell signaling. *New Phytol.* **2010**, *188*, 977–984. [CrossRef]
47. Hou, Z.H.; Liu, J.; Hou, L.X.; Li, X.; Liu, X. H_2S may function downstream of H_2O_2 in jasmonic acid-induced stomatal closure in *Vicia faba*. *Chin. Bull. Bot.* **2011**, *4*, 396–406.
48. Jin, Z.; Wang, Z.; Yang, G.; Pei, Y. Diversity of hydrogen sulfide concentration in plant: A little spark to start a prairie fire. *Sci. Bull.* **2018**, *63*, 1314–1316. [CrossRef]
49. Sekiya, J.; Schmidt, A.; Wilson, L.G.; Filner, P. Emission of hydrogen sulfide by leaf tissue in response to L-cysteine. *Plant Physiol.* **1982**, *70*, 430–436. [CrossRef]
50. Rennenberg, H. Role of o-acetylserine in hydrogen sulfide emission from pumpkin leaves in response to sulfate. *Plant Physiol.* **1983**, *73*, 560–565. [CrossRef]
51. Krest, I.; Glodek, J.; Keusgen, M. Cysteine sulfoxides and alliinase activity of some Allium species. *J. Agric. Food Chem.* **2000**, *48*, 3753–3760. [CrossRef]
52. Yoshimoto, N.; Saito, K. S-Alk(en)ylcysteine sulfoxides in the genus Allium: Proposed biosynthesis, chemical conversion, and bioactivities. *J. Exp. Bot.* **2019**, *70*, 4123–4137. [CrossRef] [PubMed]
53. Lancaster, J.E.; Collin, H.A. 1981. Presence of alliinase in isolated vacuoles and of alkyl cysteine sulphoxides in the cytoplasm of bulbs of onion (*Allium cepa*). *Plant Sci. Lett.* **1981**, *22*, 169–176. [CrossRef]
54. Lancaster, J.E.; Dommisse, E.M.; Shaw, M.L. Production of flavor precursors [S-alk(en)yl-l-cysteine sulphoxides] in photomixotrophic callus of garlic. *Phytochemistry* **1988**, *27*, 2123–2124. [CrossRef]
55. Lancaster, J.E.; Reynolds, P.H.S.; Shaw, M.L.; Dommisse, E.M.; Munro, J. Intra-cellular localization of the biosynthetic pathway to flavour precursors in onion. *Phytochemistry* **1989**, *28*, 461–464. [CrossRef]
56. Nohara, T.; Fujiwara, Y.; El-Aasr, M.; Ikeda, T.; Ono, M.; Nakano, D.; Kinjo, J. Thiolane-type sulfides from garlic, onion, and Welsh onion. *J. Nat. Med.* **2021**, *75*, 741–751. [CrossRef]
57. Banerjee, S.K.; Maulik, S.K. Effect of garlic on cardiovascular disorders: A review. *Nutr. J.* **2002**, *1*, 4. [CrossRef]
58. Benavides, G.A.; Squadrito, G.L.; Mills, R.W.; Patel, H.D.; Isbell, T.S.; Patel, R.P.; Darley-Usmar, V.M.; Doeller, J.E.; Kraus, D.W. Hydrogen sulfide mediates the vasoactivity of garlic. *Proc. Natl. Acad. Sci. USA* **2007**, *104*, 17977–17982. [CrossRef]
59. Rose, P.; Moore, P.K.; Zhu, Y.Z. Garlic and Gaseous Mediators. *Trends Pharmacol. Sci.* **2018**, *39*, 624–634. [CrossRef]
60. Alali, F.Q.; El-Elimat, T.; Khalid, L.; Hudaib, R.; Al-Shehabi, T.S.; Eid, A.H. Garlic for cardiovascular disease: Prevention or treatment? *Curr. Pharm. Des.* **2017**, *23*, 1028–1041. [CrossRef]
61. Nwachukwu, I.D.; Slusarenko, A.J.; Gruhlke, M.C. Sulfur and sulfur compounds in plant defence. *Nat. Prod. Commun.* **2012**, *7*, 395–400. [CrossRef]

It also causes various physiological disorders, such as reduction in fruit size, low firmness and thin peel, tip burn, and oxidative stress [5,10–12]. It has also been reported that calcium deficiency causes bitter pit and decreases titratable acidity, total soluble solids, and vitamin C contents, resulting in poorer postharvest fruit quality [13]. Bitter pit is one of the major post-harvest disorders associated with apple production and can cause up to a 50% post-harvest loss [14].

Ca^{2+} is one of the most important second messengers in plant cell signaling [15]. Ca^{2+} signaling regulates numerous abiotic stress reactions [16]. Calmodulin (CaM) and calmodulin-like (CML) proteins [17] are conventional Ca^{2+}-binding proteins, which are generally comprised of one to six EF-hand motifs [18]. It has been reported that CMLs initiate cellular responses by binding to relevant transcription factors (TFs) (e.g., AP2/ERFs and bHLHs) and transmitting calcium signaling downstream [19]. It is reported that *SlCML44* has a critical effect on Ca^{2+} signaling during abiotic stress tolerance in tomato fruit [20]. In rice, *OsERF48*-OE induces modulation of *OsCML16*, thereby enhancing drought tolerance and root growth [21]. Nevertheless, whether and how ERFs/bHLHs regulate CMLs via calcium signaling in the postharvest fruit suffering from bitter pit due to calcium deficiency is still largely unknown. It has been reported that Ca^{2+} interacts with H_2S and plays a vital role in plant growth and development by regulating a range of physiological processes and imparting abiotic stress tolerance [22]. For example, the application of exogenous NaHS increases the intracellular Ca^{2+} content under both hypoxia and heat stress in tobacco [23,24]. However, there were few reports on the mode of interaction between Ca^{2+} and H_2S.

Hydrogen sulfide (H_2S) is an important signal molecule [25], participating in the regulation of plant development, and resistance to stress conditions [26]. In plants, the cysteine desulfhydrases (CDs) are responsible for the majority of endogenous H_2S production and L/D-cysteine desulfhydrase (LCD/DCD) catalyze the production of H_2S with L-cysteine and D-cysteine as the substrates, respectively [27]. Moreover, H_2S can also be produced by the O-acetyl-L-serine (thiol) lyase (OASTL) family proteins and sulfite reductase (SiR) [28]. A number of studies have reported that H_2S delayed postharvest senescence of various fruits and vegetables [29]. For example, postharvest treatment of H_2S was found to delay the softening of tomato [30] and Chilean strawberry (*Fragaria chiloensis*) fruit [31], thereby slowing postharvest senescence. In addition, H_2S can help to eliminate excessive ROS in harvested kiwifruit by activating antioxidative systems [32]. A previous report suggested a possible association between ROS and H_2S, and calcium homeostasis in wheat coleoptiles (*Triticum aestivum* L.) largely dictates the formation of ROS [33]. Thus, we propose a hypothesis that calcium deficiency may affect the production of endogenous H_2S and increase ROS, thereby reducing fruit quality during postharvest storage. However, the molecular mechanism of Ca^{2+} on the production of endogenous H_2S has rarely been reported, and the role of endogenous H_2S in postharvest quality deterioration are still unclear.

Apple (*Malus domestica*), a Rosaceae fruit, is of important economic value. Apple contains multiple beneficial and healthy components, which include vitamins and anthocyanins. Due to the antioxidant and anti-inflammatory properties, apples are often employed to maintain a well-balanced diet [34]. Mineral nutrients, particularly calcium, are critical to the growth of plants, fruit quality, and productivity [35]. Previous study has reported that calcium deficiency causes the formation of bitter pit on the surface of apple peels [14], which affects the storage quality and storage life of apples. Postharvest calcium treatments have been shown to be effective in reducing physiological diseases in apples, delaying aging, and greatly maintaining the quality of the fruit [36,37]. However, whether calcium deficiency affects the production of endogenous H_2S and the interaction between calcium, H_2S, and ROS are largely unknown in calcium deficient apples during postharvest storage.

In this study, the effects of calcium deficiency on ROS and H_2S production in apples were elucidated by the physiological parameters, principal component and correlation analysis. Then, differentially expressed calcium-regulated genes were screened from transcriptome data, and their expression was verified by RT-qPCR. Moreover, the expression of genes related to H_2S production was also analyzed to reveal the effect of calcium deficiency on H_2S production. Further, *ERF2-bHLH2* complex was found to activate the expression of *CML5* by apple transient expression assays and dual-luciferase reporter system experiments, thereby enhancing the understanding of the regulatory mechanism of calcium-deficient apple peels and the potential interaction between Ca^{2+} and endogenous H_2S, contributing to improving the appearance quality of apple fruit.

2. Results

2.1. Calcium Deficiency Significantly Affects the Phenotype of Apple Peels and Reduces H_2S Production

In this experiment, the phenotypes of calcium-deficient and calcium-sufficient apple fruits stored for different days after harvest were observed. As shown in Figure 1A, the apple fruit did not significantly differ in size, but the color and smoothness of apple peels were significantly different at 0, 7, 14, and 21 days after storage (DAS) between calcium-deficient and calcium-sufficient apples. The appearance of calcium-sufficient apple peels showed no obvious change with increasing days of storage. At 7 DAS, however, bitter pit appeared on calcium-deficient apple peels and increased continuously with increasing days of storage. At the same time, the peel color of calcium-deficient apples gradually deteriorated, and the surface became rough. Then, the calcium contents in calcium-deficient and calcium-sufficient apple peels at different storage times were measured. The calcium content of calcium-deficient apple peels was always lower than that of calcium-sufficient apple peels during storage, but significant difference between the two was observed at 7, 14, and 21 DAS ($p < 0.01$) (Figure 1B). It is evident that calcium deficiency in apples affects the phenotype of apple peels and leads to physiological disorders. Hence, it is important to investigate the molecular mechanism of calcium deficiency-induced disorders in apples.

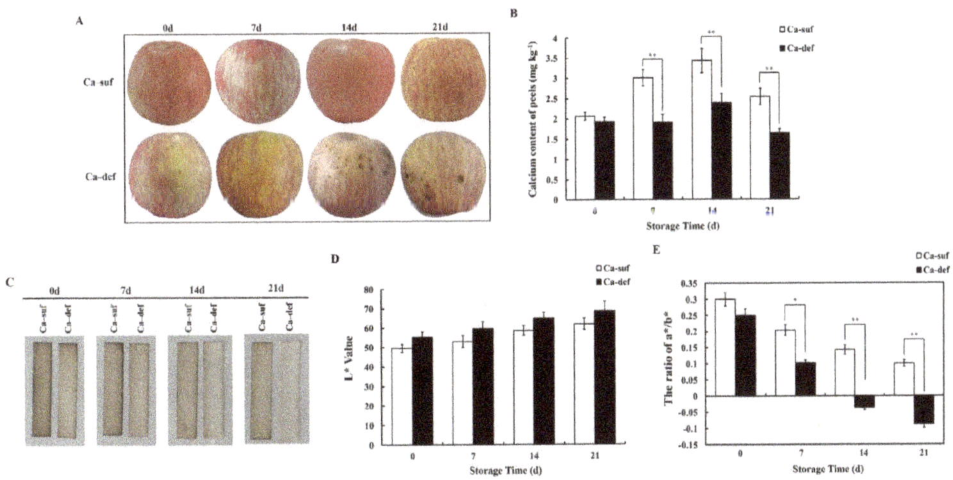

Figure 1. Comparison of phenotype and endogenous H_2S determination of apple peel at 0, 7, 14, and 21 DAS with calcium deficiency and calcium sufficiency. (**A**) Phenotypic change of apple peel at 0, 7, 14, and 21 DAS. Ca-suf, Ca sufficiency; Ca-def, Ca deficiency. Data are presented as means ± SD (n = 5). (**B**) Calcium contents of calcium-sufficient and calcium-deficient apple peels during storage. Data are presented as means ± SD (n = 3). (**C**) H_2S production capacity from calcium-sufficient and calcium-deficient apple peels during storage as detected by the brown precipitate, lead sulfide. Data are presented as means ± SD (n = 6). (**D**,**E**) represent the change of color-parameter L* value and the ratio of a*/b* corresponding to (**C**). L* represents luminance; a* represents a range from green to magenta; b* represents a range from yellow to blue. Data are presented as means ± SD (n = 6). Asterisks indicate statistical difference of the values at $p < 0.05$ (*); $p < 0.01$ (**).

In order to measure H_2S directly, lead sulfide method was applied to show the changes of H_2S production in calcium-deficient and sufficient apple peels. As shown in Figure 1C, more H_2S was produced by calcium-sufficient apple peels than by calcium-deficient apple peels at 0, 7, 14, 21 DAS, evidenced by the darkened precipitate due to lead sulfide formed on the strips. Luminance (L*) and color change (a*/b*) was measured by colorimeter, and they could show the effect of calcium deficiency on H_2S production. Higher L* values and lower ratio of a*/b* indicated that less H_2S was absorbed by the lead acetate filter paper. The L* values were higher for calcium-deficient apple peels than for calcium-sufficient apple peels (Figure 1D), but the ratio of a*/b* was reversed (Figure 1E). In summary, it is clear that calcium deficiency reduces H_2S production in apple peels.

2.2. Calcium Deficiency Increases the Contents of Flavonoids and Total Phenols but Decreases Anthocyanin Contents in Apple Peels

To further explore the mechanism by which calcium deficiency affects the phenotype of apple peels, the contents of flavonoids, total phenols, and anthocyanins were determined. As shown in Figure 2A, the flavonoid contents of calcium-deficient peels were almost indistinguishable from those of calcium-sufficient peels at 7 DAS, whereas at 0, 14, and 21 DAS, the flavonoid contents of calcium-deficient peels were significantly higher than those of calcium-sufficient peels ($p < 0.01$). The total phenol contents of calcium-deficient peels were higher than those of calcium-sufficient peels, and the difference was significant at 0, 14, and 21 DAS ($p < 0.01$ or $p < 0.05$) (Figure 2B). At the same time, there was no appreciable difference in anthocyanin contents between calcium-deficient and calcium-sufficient peels at 0 and 7 DAS, but the anthocyanin contents in calcium-deficient peels were significantly lower than those in calcium-sufficient peels at 14 and 21 DAS ($p < 0.01$) (Figure 2C).

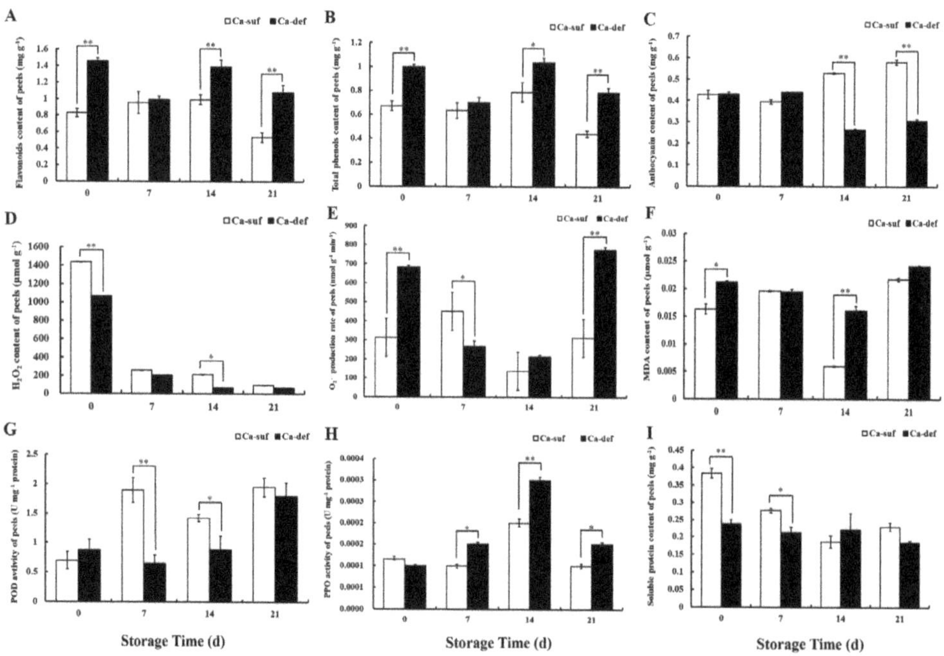

Figure 2. Effects of Ca-suf and Ca-def on the contents of (**A**) flavonoids, (**B**) total phenols, (**C**) anthocyanin, (**D**) hydrogen peroxide (H_2O_2), (**E**) superoxide anion ($O_2^{\bullet -}$), and (**F**) malonaldehyde (MDA) as well as on (**G**) peroxidase (POD) activity, (**H**) polyphenol oxidase (PPO) activity and (**I**) soluble protein in apple peels at 0, 7, 14, and 21 DAS. Ca-suf, Ca sufficiency; Ca-def, Ca deficiency. Data are presented as means ± SD (n = 3). Asterisks indicate statistical difference of the values at $p < 0.05$ (*); $p < 0.01$ (**).

2.3. Calcium Deficiency Induces the Production of ROS in Apple Peels

Excessive production of ROS and oxidative damage are usually observed during fruit storage. To investigate the effect of calcium deficiency on ROS production in apple peels, the accumulation of $O_2^{\bullet-}$, H_2O_2, and MDA was determined. Compared to calcium-sufficient peels, the H_2O_2 contents of calcium-deficient peels were lower during postharvest storage days; in particular, the most remarkable variation was at 0 DAS ($p < 0.01$) (Figure 2D). The $O_2^{\bullet-}$ production rate was significantly higher in calcium-deficient peels than in calcium-sufficient peels at 0 and 21 DAS ($p < 0.01$), but the difference was not significant at 14 DAS (Figure 2E). Similarly, the MDA contents in calcium-sufficient peels were lower than those in calcium-deficient peels, especially at 14 DAS ($p < 0.01$) (Figure 2F).

2.4. Changes in POD Activity, PPO Activity Soluble Protein Content between Calcium Deficiency and Sufficiency Conditions

To study the potential mechanism of calcium deficiency on the accumulation of ROS, the peroxidase (POD) and polyphenol oxidase (PPO) activities were determined in calcium deficiency and sufficiency apples. As shown in Figure 2G, the POD activity in calcium-sufficient peels was maintained at a higher level than that in calcium-deficient peels at 7 and 14 DAS ($p < 0.05$ or $p < 0.01$). The POD activity gradually increased with increasing days of storage and then decreased. The PPO activity in calcium-deficient peels was significantly higher than that in calcium-sufficient peels at 14 DAS ($p < 0.01$) (Figure 2H). To compare the effects of calcium deficiency on apple peel quality, soluble protein content was measured. The soluble protein content sustained at a lower level in calcium-deficient peels than that in the calcium-sufficient peels at 0, 7, 14, and 21 DAS (Figure 2I).

2.5. PCA Analysis of the Bioactive Substance Changes in Apple Peels

The principal component analysis (PCA), a statistical method of dimensionality reduction, was carried out to investigate the main factors of the bioactive substances in apple peels that were most affected by calcium deficiency, and $O_2^{\bullet-}$ production rate, ratio of a*/b*, L* value, POD and PPO activity, H_2O_2, calcium, soluble protein, flavonoid, total phenolics, anthocyanins, MDA contents were used in PCA (Figure 3). The contribution rates of PC1 and PC2 were 84.7% and 15.3%, respectively. There was a significant difference between calcium-deficient apple peels and calcium-sufficient apple peels. In PC1, H_2O_2, calcium content, $O_2^{\bullet-}$ production rate, and ratio of a*/b* were the main factors, while POD activity, L* value, and $O_2^{\bullet-}$ production rate were the main factors in PC2 (Table 1). The above results indicated that calcium deficiency caused physiological disorders and changes in postharvest storage quality in apple peels, which severely affected the ROS, H_2S production, and redox processes in apple peels.

Table 1. The factors score of all the metabolites by principal component analysis in apple peels.

Component Name	PC1 (84.7%)	PC2 (15.3%)
H_2O_2 content	9.97×10^{-1}	-8.25×10^{-2}
$O_2^{\bullet-}$ production rate	8.25×10^{-2}	9.97×10^{-1}
Calcium content	2.01×10^{-4}	-2.49×10^{-4}
Ratio of a*/b*	1.99×10^{-4}	-1.64×10^{-4}
Soluble protein content	9.53×10^{-5}	-7.17×10^{-5}
Flavonoid content	6.48×10^{-5}	3.87×10^{-4}
Total phenolics content	4.70×10^{-5}	1.37×10^{-4}
Anthocyanins content	1.46×10^{-5}	-1.74×10^{-4}
MDA content	5.56×10^{-8}	1.86×10^{-5}
PPO activity	-5.74×10^{-5}	-1.19×10^{-4}
POD activity	-5.61×10^{-4}	8.97×10^{-4}
L* value	-8.94×10^{-3}	9.04×10^{-3}

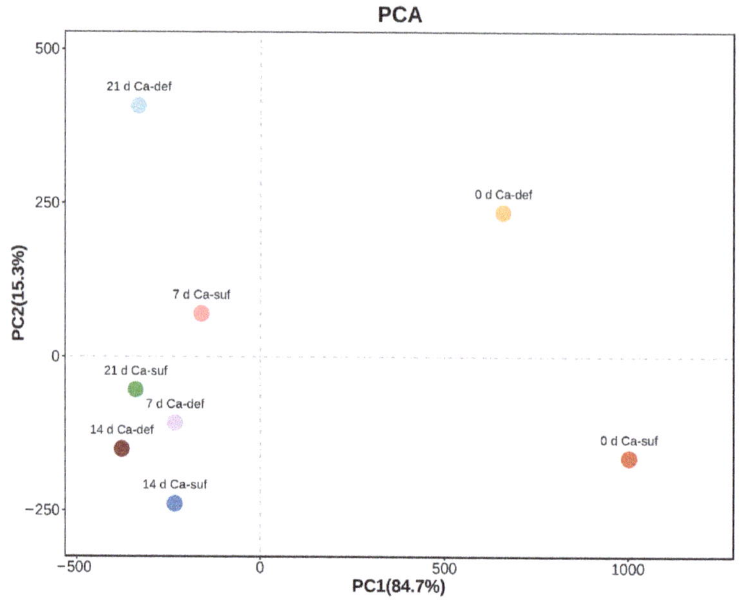

Figure 3. PCA of the main metabolites in apple peels during storage periods. PC1 and PC2, respectively, represented the contribution rate of principal components. Ca-suf, Ca sufficiency; Ca-def, Ca deficiency.

2.6. Identification and GO Classification and KEGG Enrichment Analysis of DEGs between Calcium-Deficient and Calcium-Sufficient Apple Peels

To investigate the molecular mechanisms of calcium regulation in apple peels, differentially expressed genes (DEGs) associated with calcium regulation between calcium-sufficient and calcium-deficient apple peels were obtained from published transcriptome data. Pairwise comparison of the three sequenced samples resulted in three sets of contrasts with T02/T01, T03/T01, and T03/T02 (T01, diseased peel of calcium-deficient apples; T02, healthy peel of calcium-deficient apples; and T03, peel of calcium-sufficient apples). In the T02/T01, T03/T01, and T03/T02 comparisons, 1323, 2880, and 2182 DEGs were upregulated, respectively, while 1573, 1606, and 269 DEGs were downregulated, respectively (Figure S1A–C). In total, 1031 DEGs were common to all three comparisons (Figure S1D).

GO classification analysis was used to investigate the gene expression profiles of calcium-deficient apple peels, and the T03/T01 comparison with the highest number of DEGs was selected for gene annotation analysis. The DEGs were classified into 51 functional groups based on their biological processes (Figure S1E). The major subcategories were as follows: 20 subcategories for biological process, 15 subcategories for cellular component, and 16 subcategories for molecular function. The DEGs in 'signaling' and 'antioxidant activity' played important roles during the calcium regulated processes. These results provided a comprehensive perspective for screening candidate genes involved in calcium regulation.

KEGG analysis provided information and further understanding of secondary metabolites induced by calcium signaling. As shown in Figure S1F, 20 pathways were enriched with over 20 DEGs. Plant hormone signal transduction was the most abundant DEG found in metabolism category followed by plant–pathogen interaction, biosynthesis of amino acids, and phenylpropanoid biosynthesis. These results suggested that the DEGs in the T03/T01 comparison were mainly enriched in metabolic processes related to plant signal transduction and plant pathogen infestation.

2.7. Analysis of Transcriptomics on Calcium-Deficient Peels and Gene Expression Validation

In order to explore the molecular mechanisms of calcium regulation in apple peels, calmodulin-like proteins (CMLs), pathogen-related proteins (PRs), and several costimulatory TFs, including AP2/ERFs, bHLHs, and MADS-box, were screened from the 1031 DEGs co-expressed in the three sets of sequenced samples. Heatmap analysis was performed on 5 CMLs, 4 PRs, 17 AP2/ERFs, 3 bHLHs, and 2 MADSs (Figure 4). These candidate genes were differentially expressed in apple peels due to calcium deficiency with two expression patterns. Among them, *CML1-3*, *PR2-4*, and *ERF8-16* were upregulated, and *CML4*, *CML5*, *PR1*, *ERF1-7*, *ERF17*, *bHLH1-3*, *MADS1*, and *MADS2* were downregulated. *CML1-3*, *CML5*, *PR1-4*, *ERF2*, *ERF5*, *ERF13-17*, *bHLH1-3*, *MADS1*, and *MADS2* had the most significantly different expression.

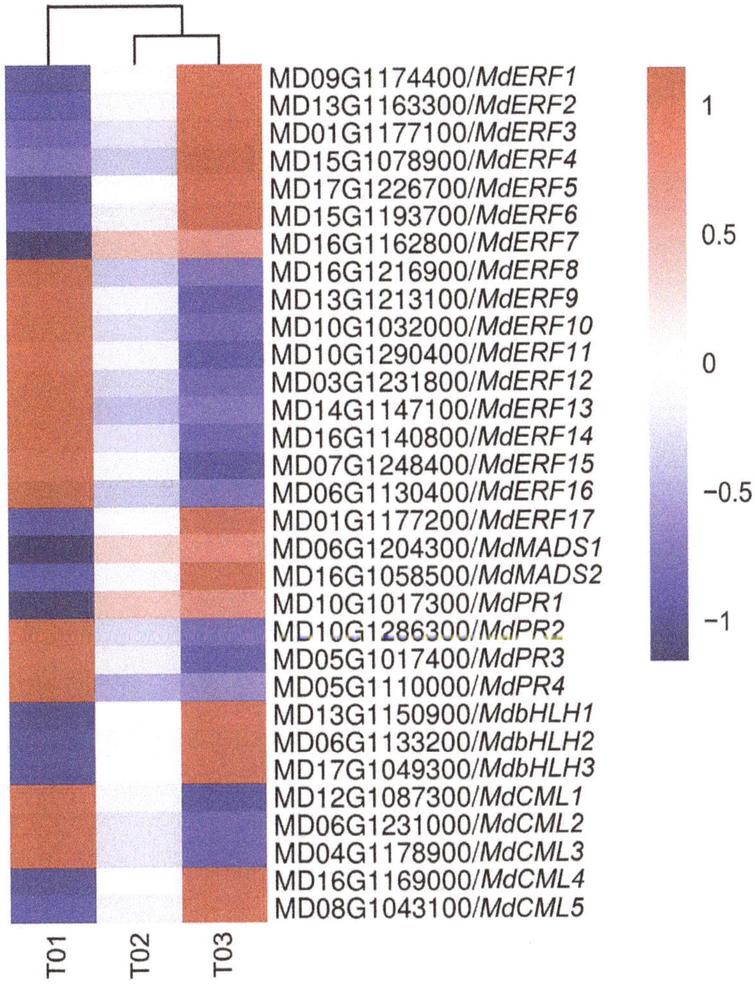

Figure 4. Heatmap of selected calcium related genes and co-expressed transcription factors. T01, diseased peels of calcium-deficient apples; T02, healthy peels of calcium-deficient apples; T03, peels of calcium-sufficient apples. The color of the scale label from red to blue represents the change in RPKM value from '1' to '−1' by Z-score normalization.

As heatmap analysis of DEGs yielded seven AP2/ERFs (*ERF2, ERF5, ERF13, ERF14, ERF15, ERF16,* and *ERF17*), three bHLHs (*bHLH1, bHLH2,* and *bHLH3*), two MADSs (*MADS1* and *MADS2*), four PRs (*PR1, PR2, PR3,* and *PR4*), and four CMLs (*CML1, CML2, CML3,* and *CML5*) with significant differential expression, real-time qPCR was utilized to further validate their expression patterns in calcium-sufficient and calcium-deficient apple peels (Figure 5 and Figure S2). The expression patterns of *ERF2, ERF5, ERF13, ERF14, ERF17, bHLH1, bHLH2, bHLH3, MADS1, MADS2, PR1, PR3, CML1, CML2, CML3,* and *CML5* were consistent with those obtained by transcriptome sequencing, and the correlation coefficient was 0.7722 (Figure S3). Among them, the expression of *ERF2* and *ERF17* was higher than that of *ERF5, ERF13,* and *ERF14,* and *bHLH2* expression level was also higher than that of *bHLH1* and *bHLH3*. *ERF2/17, bHLH2,* and *CML5* showed consistent expression patterns, and the differences in the expression of them were significant between calcium-deficient and calcium-sufficient peels.

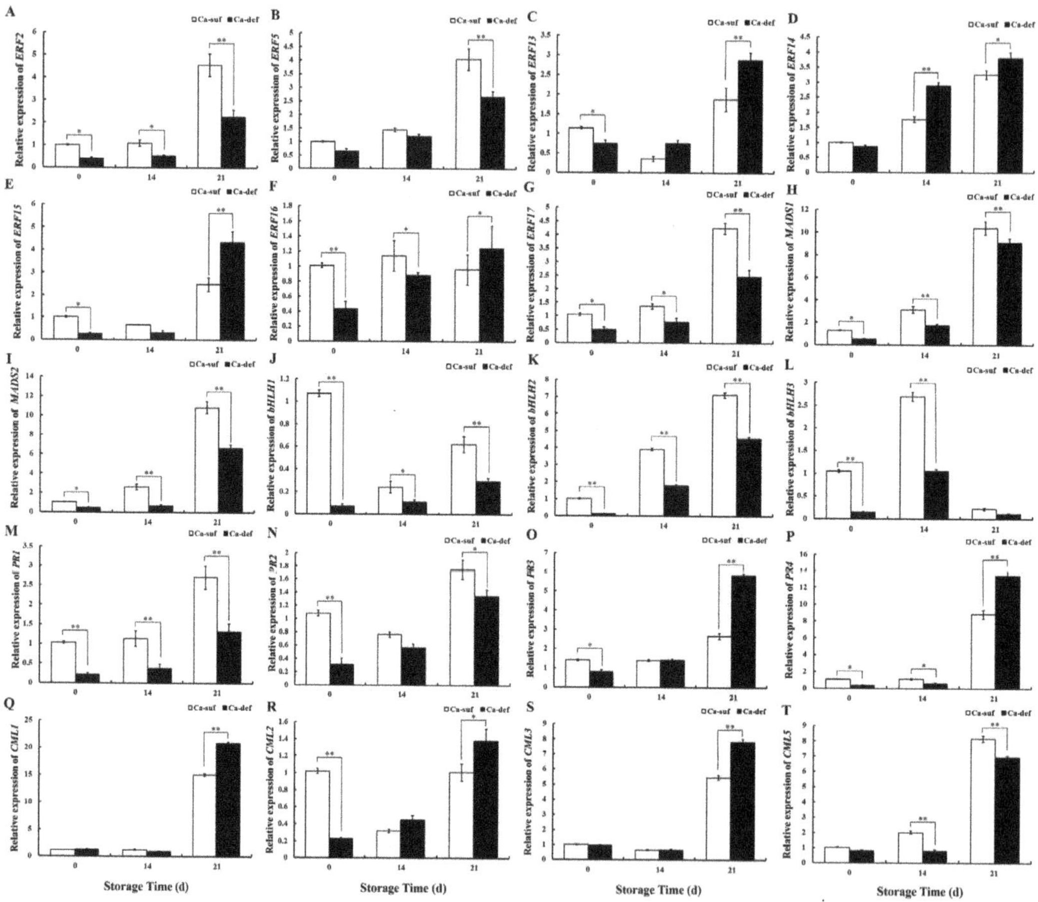

Figure 5. Expression pattern analysis of (**A**) *ERF2*, (**B**) *ERF5*, (**C**) *ERF13*, (**D**) *ERF14*, (**E**) *ERF15*, (**F**) *ERF16*, (**G**) *ERF17*, (**H**) *MADS1*, (**I**) *MADS2*, (**J**) *bHLH1*, (**K**) *bHLH2*, (**L**) *bHLH3*, (**M**) *PR1*, (**N**) *PR2*, (**O**) *PR3*, (**P**) *PR4*, (**Q**) *CML1*, (**R**) *CML2*, (**S**) *CML3*, and (**T**) *CML5* in apple peels under calcium sufficiency and deficiency with storage at 0, 14, and 21 DAS by RT-qPCR. Data are presented as means ± SD (n = 3). Asterisks indicate statistical difference of the values at $p < 0.05$ (*); $p < 0.01$ (**).

For the investigation of the expression of genes related to H_2S production in apple peels by calcium deficiency, four DCDs (*DCD1*, *DCD2*, *DCD3*, and *DCD4*), three LCDs (*LCD1*, *LCD2*, and *LCD3*), one SiR (*SiR1*), and one OASTL (*OASTL1*) genes were selected for real-time qPCR at 0 and 14 DAS. As shown in Figure 6, whether at 0 or 14 DAS, *DCD1*, *DCD2*, *DCD3*, *DCD4*, *LCD1*, *LCD2*, *LCD3*, *SiR1*, and *OASTL1* expressions in calcium-deficient apple peels were lower than in calcium-sufficient apple peels, while at 14 DAS, *DCD1*, *DCD2*, *DCD3*, *DCD4*, *LCD1*, *LCD2*, *LCD3*, *SiR1*, and *OASTL1* expressions in calcium-deficient peels were significantly lower than in calcium-sufficient peels ($p < 0.01$). These results were consistent with decreased H_2S production in calcium-deficient apple peels, suggesting that calcium deficiency caused a decrease in the production of H_2S from apple peels by reducing the expression of genes related to H_2S production.

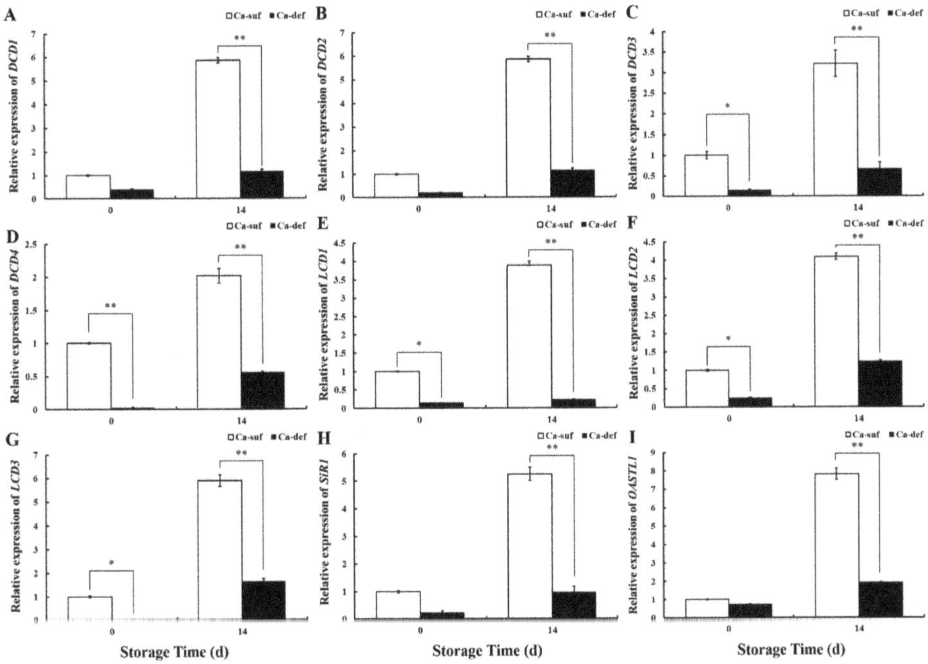

Figure 6. Expression pattern analysis of genes (**A**) *DCD1*, (**B**) *DCD2*, (**C**) *DCD3*, (**D**) *DCD4*, (**E**) *LCD1*, (**F**) *LCD2*, (**G**) *LCD3*, (**H**) *SiR1*, and (**I**) *OASTL1* related to H_2S production in apple peels under calcium sufficiency and deficiency with storage at 0 and 14 DAS by RT-qPCR. Data are presented as means $\pm$ SD (n = 3). Asterisks indicate statistical difference of the values at $p < 0.05$ (*); $p < 0.01$ (**).

2.8. Correlation Analysis between Genes Expression and Physiological Parameters in Apple Peels

The result of the correlation analysis between genes expression and physiological parameters in apple peels is shown in Figure 7. Calcium content was negatively correlated with MDA, $O_2^{\bullet -}$, H_2O_2 contents, and ratio of a*/b*, positively correlated with POD activity, H_2S production on filter papers, *DCD1*, *DCD2*, *DCD3*, *DCD4*, *LCD1*, *LCD2*, *LCD3*, *SiR1*, and *OASTL1* expression levels, indicating that calcium deficiency activated ROS production in apples but reduced H_2S production. The expression levels of *ERF2*, *ERF17*, *bHLH2*, and *CML5* were positively correlated with calcium content, PPO activity, POD activity, H_2S production on filter papers, *DCD1*, *DCD2*, *DCD3*, *DCD4*, *LCD1*, *LCD2*, *LCD3*, *SiR1*, and *OASTL1* expression levels, but negatively correlated with MDA, $O_2^{\bullet -}$, H_2O_2 contents, and ratio of a*/b*. Moreover, H_2S production showed positive correlation with MDA, $O_2^{\bullet -}$, H_2O_2 contents in apple peels. Based on Figure 7 and the above results, it could be concluded that there was a strong positive correlation between the expression of presumably *ERF2*, *ERF17*, *bHLH2*, and *CML5*, and they were highly expressed and differ significantly between calcium-deficient and calcium-sufficient apple peels at storage time. It was therefore hypothesized that *ERF2*, *ERF17*, and *bHLH2* may be involved in the regulation of *CML5* and H_2S production during postharvest apple fruit storage.

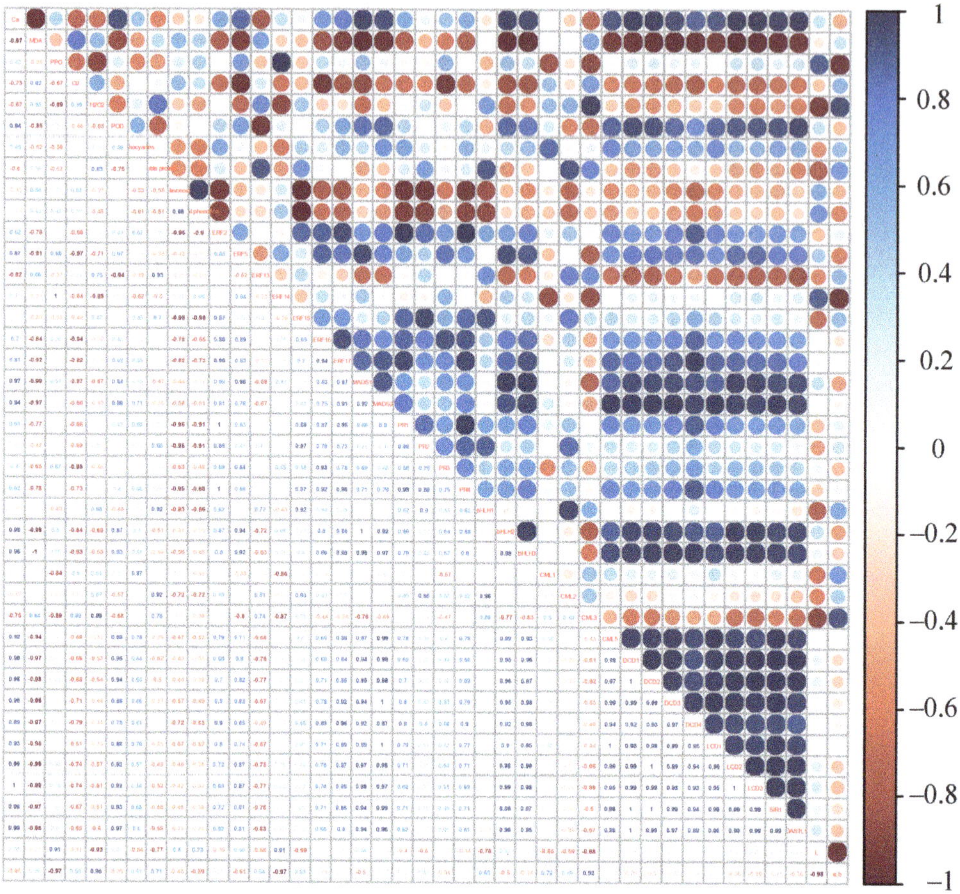

Figure 7. Correlation analysis between gene expression and identified indices in apple peels. R scripts were used to analyze Pearson's correlation coefficients. The color of the scale label from blue to red represents the change in Pearson's correlation coefficients from '1' to '−1' with the following indicators: '+' represents a positive correlation, '−' represents a negative correlation, 0.8–1 represents a highly strong correlation, 0.6–0.8 represents a strong correlation, 0.4–0.6 represents a moderate correlation, 0.2–0.4 represents a weak correlation, and 0–0.2 represents a very weak correlation or no correlation.

2.9. ERF2-bHLH2 Coregulation Promotes CML5 Expression in Apple Peels

To further investigate whether and how *ERF2/17* and *bHLH2* regulate *CML5* in apple peels, a transient transformation experiment was performed using 'Honeycrisp' apples at 60 days after flower blooming, and *ERF2/ERF17* and *bHLH2* were individually or cotransformed in apples via *Agroinfiltration*. Transient expression assays in apples showed that individual transformation of *ERF2* and *bHLH2* resulted in a minor deposition of anthocyanin and a slight loss of water in the peel at the injection site. In contrast, when *ERF17* was individually transformed, anthocyanin deposition was evident, and there was no water loss-induced wrinkling of the peel at the injection site. When *ERF2* and *bHLH2* were cotransformed, there was also minor anthocyanin deposition and significant water loss-induced wrinkling of the peel at the injection site, whereas when *ERF17* and *bHLH2* were cotransformed, there was significant anthocyanin deposition but no water loss-induced wrinkling (Figure 8A). The trends of the L* and a*/b* values in the injection regions shown in Figure 8B,C were consistent with the phenotypic results.

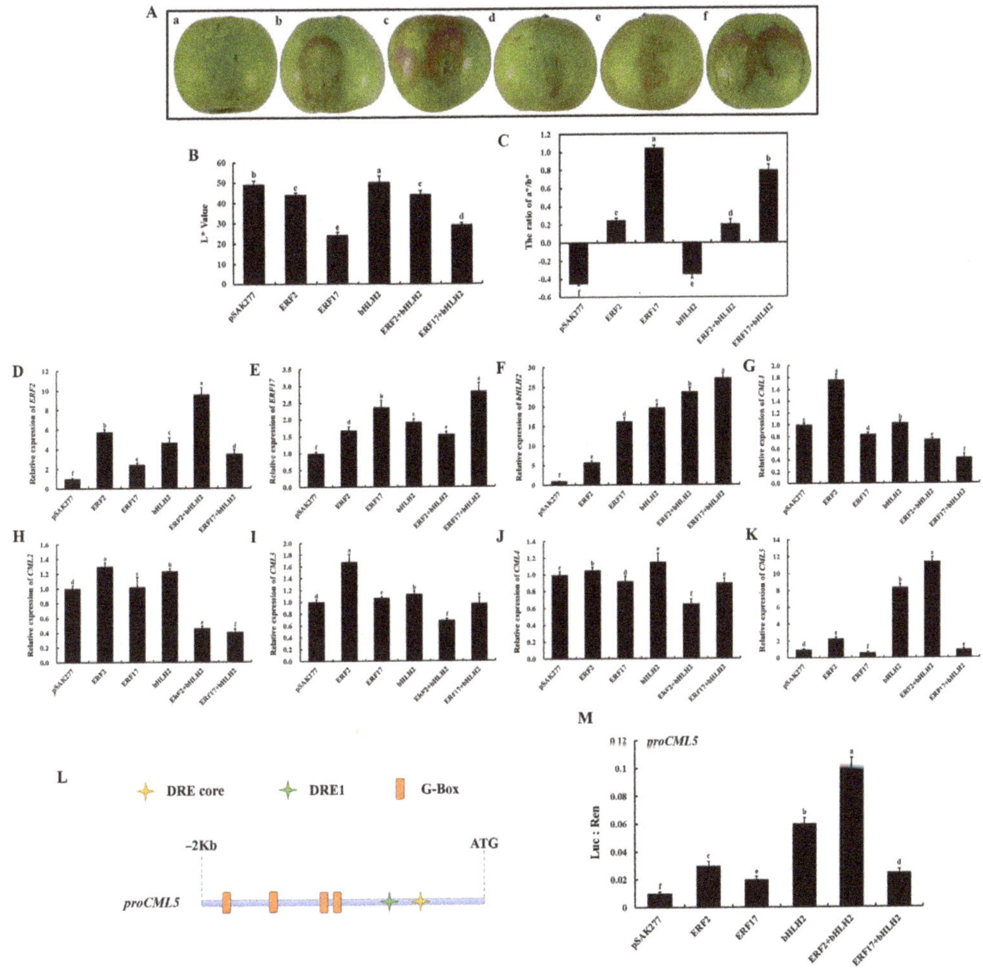

Figure 8. *ERF2-bHLH2* coregulation promotes *CML5* expression in apple peels. (**A**) Apple peel phenotypes are shown for the following transient transformations: (**a**), pSAK277; (**b**), *ERF2*; (**c**), *ERF17*; (**d**), *bHLH2*; (**e**), *ERF2* + *bHLH2*; (**f**), *ERF17* + *bHLH2*. Data are presented as means ± SD (n = 5). (**B**,**C**) The L* and a*/b* ratio color parameter values indicate the color changes. The values are presented as the means ± SD (n = 6). (**D**–**K**) Expression levels of *ERF2, ERF17, bHLH2, CML1, CML2, CML3, CML4,* and *CML5* in apple peels from (**A**(**a**–**f**)). The values are presented as the means ± SD (n = 3). (**L**) Predicted cis-acting elements in the upstream 2 kb promoter regions of *CML5, ERF2, ERF17,* and *bHLH2*. Pro, promoter. The DRE1 and DRE core are ERFs cis-acting elements, and the G-Box is bHLHs cis-acting element. (**M**) Validation of the activation effect by cotransformation of *ERF2*/*ERF17* and *bHLH2* on the *CML5* promoter using a dual-luciferase assay in tobacco leaves. The ratio of Luc to Ren indicates that TFs activate the promoter activity of *CML5*. The values are presented as the means ± SD (n = 3). Uppercase letters represent statistical difference at $p < 0.01$, and lowercase letters represent statistical difference at $p < 0.05$.

RT-qPCR was then used to analyze the expression of *ERF2, ERF17, bHLH2,* and calcium regulatory genes (*CML1, CML2, CML3, CML4,* and *CML5*). The expression of *ERF2, ERF17* and *bHLH2* was significantly increased when the apple peels were injected individually or cotransformed with *bHLH2* ($p < 0.01$, Figure 8D–F). Interestingly, *bHLH2* expression progressively increased with individual transformation of *ERF2, ERF17,* and *bHLH2* or when cotransformed with *bHLH2*. The expression of *CML1, CML2, CML3,* and *CML4* increased significantly ($p < 0.01$) when *ERF2* was individually transformed. There was a slight increase in *CML1, CML2, CML3,* and *CML4* expression when *bHLH2* was individually transformed. The expression of *CML1, CML2, CML3,* and *CML4* decreased when *ERF2, ERF17,* and *bHLH2* were cotransformed (Figure 8G–J). The expression of *CML5* increased significantly ($p < 0.01$) when *bHLH2* was individually transformed and with *ERF2* cotransformation, and the expression of *CML5* was slightly increased when *ERF2* was individually transformed (Figure 8K). In addition, the expression level of *CML5* were significantly higher than those of *CML1, CML2, CML3,* and *CML4* when *ERF2* and *bHLH2* were cotransformed Thus, these results demonstrated that individual transformation of *bHLH2* or cotransformation with *ERF2* significantly activates *CML5* transcription, suggesting that *ERF2* and *bHLH2* cooperate to activate *CML5* to regulate calcium signaling and postharvest apple storage quality.

Promoter cis-acting elements were first predicted for *CML5, ERF2/17,* and *bHLH2* and the results are shown in Figure 8L. The 2 kb region upstream of the *CML5* promoter contains one or more cis-acting elements of the DRE core/DRE1 of AP2/ERFs and the G-Box of bHLHs. Then, a dual-luciferase reporter system was used to verify the mode of regulation between *ERF2/17* and *bHLH2* and *CML5*. The results showed that *ERF2* and *bHLH2* co-injection significantly activated *CML5* transactivation (Figure 8M). From those it can be concluded that *ERF2-bHLH2* coregulation promotes *CML5* expression in apple peels, thereby regulating calcium signaling and postharvest storage quality.

3. Discussion

3.1. Calcium Deficiency Increases ROS Generation, Reduces H_2S Production and Postharvest Storage Quality in Apple Peels

Calcium is one of the essential elements for plant growth, and it has important functions in maintaining plant cell structure, resistance to stress and signal transduction [1–4]. In the present study, the peel of calcium-deficient apples became rough during postharvest storage due to water loss, and they developed distinct bitter pit compared to calcium-sufficient apples (Figure 1A,B). These phenotypic changes were consistent with those reported in previous studies [6,9,10]. In addition, compared to calcium-sufficient peels, calcium-deficient peels produced less H_2S (Figure 1C–E), significantly increased $O_2^{\bullet-}$, MDA, total phenol and flavonoid contents, and PPO activity, and decreased POD activity (Figure 2). The increase in ROS and antioxidants such as total phenol and flavonoid in calcium-deficient peels may be caused by bitter pit disorder due to calcium deficiency, which was also consistent with previous reports [5–7]. The result of PCA shows that calcium content, ROS, and H_2S production were main factors between calcium-deficient and calcium-sufficient apple peels. The above results indicated that calcium deficiency in apples reduced H_2S production and increased ROS production, thereby reducing postharvest fruit storage quality. These results are in agreement with those reported previously [38]. However, the specific regulatory network that exists between calcium signaling and the production of H_2S and ROS is currently unknown and rarely reported, and further research is needed.

3.2. The ERF2-bHLH2 Coactivate the CML5 Expression and Coregulate Downstream Responses in Calcium-Deficient Apple Peels

Calmodulin-like (CML) protein in plant tissues is an important calcium receptor protein in the process of signal transmission. In the present study, the expression of *CML5* was negatively correlated with ROS contents, but positively correlated with POD activity, flavonoids, total phenols, and calcium content (Figure 7), suggesting the potential importance of *CML5* in regulating ROS metabolism. There are numerous studies reporting that CMLs are involved in regulating a variety of abiotic stress processes. For example, *SlCML37* enhances cold tolerance in tomato fruit [39]; *CmCML13* improves drought resistance in *Arabidopsis* [40]; *CML21* is involved in abiotic stress response in grapevine [41]. Thus, it can be hypothesized that *CML5* may be involved in regulating postharvest storage quality and antioxidant-related processes in calcium-deficient apple peels.

Furthermore, the expression levels of *ERF2, ERF17,* and *bHLH2* were screened based on their significant positive correlation with *CML5* and they were significantly positively correlated with the H_2S production related genes (*DCD1, DCD2, DCD3, DCD4, LCD1, LCD2, LCD3, SiR1,* and *OASTL1*). The results of the apple transient transformation experiment and the dual-luciferase reporter system experiment showed that the co-transformation of *ERF2* and *bHLH2* significantly activated the transcription of *CML5* (Figure 8). Thus, the *ERF2-bHLH2-CML5* module may be not only involved in the regulation of calcium signaling, but also in the regulation of H_2S production in postharvest calcium-deficient apples. It was reported that H_2S could reduce ROS production and increase antioxidant capacity, thereby extending postharvest life of banana [42] and inhibiting enzymatic browning of fresh-cut Chinese water chestnuts [43]. Meanwhile, in *Arabidopsis thaliana* roots, H_2S induces ROS accumulation, thus inducing the appearance of Ca^{2+} signal [44,45]. In the present study, calcium deficiency caused an increase in ROS production, a decrease in H_2S production in apple peels and a reduction in post-harvest fruit quality, while further research was needed on how H_2S was involved in calcium regulation and affected post-harvest fruit quality. Recent studies have also shown that H_2S emission induced by chromium (Cr^{6+}) stress could be modulated by the Ca^{2+} level [46]. However, the molecular mechanism of Ca^{2+}-induced endogenous H_2S emission has rarely been reported. Therefore, the mechanism of how the *ERF2-bHLH2-CML5* module regulates calcium signaling and H_2S production in calcium-deficient apples, leading to reduced quality in postharvest storage apples, needs further investigation.

The anthocyanin content of calcium-deficient peels did not differ significantly from calcium-sufficient peels at 0 and 7 DAS, but was significantly lower at 14 and 21 DAS (Figure 2C), indicating that calcium deficiency reduced anthocyanin synthesis in postharvest stored apple peels. Moreover, in the apple transient expression assays experiment, anthocyanin deposition was observed in the peel at the injection site when *ERF2, ERF17,* and *bHLH2* were individually transformed or cotransformed (Figure 8A), suggesting that *ERF2, ERF17,* and *bHLH2* may not only be involved in calcium regulation in postharvest stored apples, but also in regulating the anthocyanin biosynthesis pathway in the peel of postharvest stored apples. The mechanism of *ERF2, ERF17,* and *bHLH2* in regulating anthocyanin biosynthesis still needs further investigation.

4. Conclusions

In summary, it can be concluded that the production of ROS, total phenols and flavonoids was increased in calcium-deficient peels compared to calcium-sufficient peels, but calcium content, H_2S production, anthocyanin content, and POD activity were reduced. Calcium content, ROS, and H_2S production were main factors between calcium-deficient and calcium-sufficient apple peels by PCA analysis. Four CMLs, seven AP2/ERFs, and three bHLHs were screened from transcriptome data analysis. In addition, *ERF2-bHLH2* co-activated *CML5* transcription and *ERF2, bHLH2,* and *CML5* expression levels were significantly positively correlated with H_2S production genes. Thus, the *ERF2-bHLH2-CML5* module is not only involved in the regulation of calcium signaling, but also in the regulation of postharvest quality of fruit and H_2S production under calcium deficiency in apples. These findings improved the understanding of the molecular basis of postharvest quality decline in calcium-deficient fruit, and the relationship between calcium and endogenous H_2S.

5. Materials and Methods

5.1. Plant Materials and Treatment

The calcium-sufficient and calcium-deficient 'Honeycrisp' apples used in this study were provided by the Shandong Academy of Agricultural Sciences (Tai'an, China). Postharvest calcium-sufficient and calcium-deficient apples of similar size and free from pathogen infection were selected for sealed storage in glass containers. In total, 20 calcium-sufficient apples and 20 calcium-deficient apples were placed in individual sealed containers, fumigated with distilled water to maintain a relative humidity of approximately 85% and stored at 25 °C for treatment. After 0, 7, 14, and 21 days after storage (DAS), five calcium-sufficient and calcium-deficient apples were randomly selected for peel sampling. Immediately after stripping, the peels were chilled in liquid nitrogen and stored at −80 °C. Some samples were utilized for the determination of physiological parameters, and some samples were utilized for RNA extraction. Three biological replicates were prepared for all samples. In addition, the 'Honeycrisp' apples at 60 days after full blooming used in this study were provided by Shandong Academy of Agricultural Sciences (Tai'an, Shandong Province, China) for transient transformation experiments, and were observed 5 days after injection. Apple peels from the injection sites were obtained to measure the related gene expression levels.

5.2. Determination of Calcium Content

A sample of freeze-dried peel (1 g) was dissolved in 1 mol·L^{-1} HNO_3. Three replicates of each sample were examined. The samples were then diluted with a 5% solution of $LaCl_3$. The calcium content of the apple peel samples was measured by atomic absorption spectrometer (Hitachi Z2000, Tokyo, Japan) with an air–acetylene flame [47]. The final calcium content in the peel was expressed as milligrams per kilogram dry weight.

5.3. Determination of H_2S Production in Apple Peels

Lead sulfide method [48] was used to measure H_2S production capacity. In the presence of excess substrate Cys and cofactor pyridoxal-5′-phosphate (PLP), a specific reaction between H_2S and lead acetate was used to form a black precipitate (lead sulfide) which can be captured and observed on filter paper containing lead acetate. A sample of fresh apple peel (1 g) was dissolved in PBS supplemented with 10 mM Cys and 10 mM PLP. Six lead acetate H_2S detection papers (Sigma, Darmstadt, Germany) were placed above the liquid phase in a closed container (covered 96-well plate) and incubated 2–5 h at 37 °C until lead sulfide darkening of the paper occurred.

The changes in H_2S production capacity were assayed according to Hu's method [49]. In brief, a lead acetate H_2S detection paper color was assayed by a colorimeter (model WSC-100, Konica Minolta, Tokyo, Japan) of each paper, and the values of L*, a*, and b* were obtained by averaging the data of three sites.

5.4. Determination of Total Phenol, Flavonoid, and Anthocyanin Contents

The method of Piri and Mullins [50] was used to determine the total phenol content of apple peels using a spectrophotometer at 280 nm. The calibration curve was generated using gallic acid as a reference standard.

By measuring the absorbance at 510 nm, the content of flavonoids was determined by aluminum chloride colorimetry [51]. Rutin was applied as a calibration standard.

The method proposed by Lee and Wicker [52] was used to obtain and measure the anthocyanin content. Apple peels (2 g) were ground with 10 mL of 1% hydrochloric acid methanol solution. The absorptions were measured at 530, 620, and 650 nm, and the anthocyanin content is presented as milligrams per gram fresh weight.

5.5. Assay of ROS (Superoxide Anion and Hydrogen Peroxide) and Malondialdehyde (MDA)

Superoxide anion ($O_2^{\bullet -}$), hydrogen peroxide (H_2O_2), and malondialdehyde (MDA) production were determined based on methods previously published by Hu [50]. The

MDA content of apple peels was measured using the thiobarbituric acid (TBA) reaction with values expressed as μmol g^{-1}. The content of H_2O_2 and the production of $O_2^{\bullet -}$ were presented as μmol g^{-1} and nmol g^{-1} min^{-1}, respectively.

5.6. Determination of POD, PPO Activity, and Soluble Protein Content

Guaiacol peroxidase (POD) activity was determined based on our previously published method [49] with some modifications. Apple peels (2 g) were homogenized in 2.5 mL of extraction buffer and centrifuged at 13,400× *g* for 30 min at 4 °C. The supernatant obtained was used to measure the activity of POD.

The methods of Benjamin and Montgomery [53] and Beaudoin-Eagan and Thorpe [54] were used to measure the activity of polyphenol oxidase (PPO). Apple peels (2 g) were used for the preparation of PPO enzyme by homogenization in 2 mL of sodium phosphate buffer (50 mM, pH 6.8). POD and PPO activities were expressed as active unit (U) per milligram of protein, and 1 U means 1 μmol product per min.

Bradford's [55] method was applied to measure the soluble protein content, and the absorbance value of soluble protein was recorded at 595 nm.

5.7. Transcriptome Data Analysis

Transcriptome data of calcium sufficient and calcium deficient apple peels were published in the NCBI database (https://www.ncbi.nlm.nih.gov/bioproject/PRJNA733599, accessed on 15 September 2021) by Qiu et al. [56]. The three sequenced conditions in this transcriptome were T01, diseased peel of calcium-deficient apples; T02, healthy peel of calcium-deficient apples; and T03, peel of calcium-sufficient apples, and each condition had two replicates. The sample peels used for transcriptome were taken from ripe apples. Differentially expressed genes (DEGs) were screened from this transcriptome data (the screening criteria for DEGs are generally: Fold Change $\geq$ 2 and FDR < 0.01.).

Gene Ontology (GO) is an officially standardized international classification system for the functional classification of genes, that describes the attributes of genes and their outputs in any organism. Kyoto Encyclopedia of Genes and Genomes (KEGG) analysis provided information and further understanding of secondary metabolites induced by calcium signaling. Then, DEGs were subjected to GO classification and KEGG enrichment analysis.

5.8. RNA Extraction and Real-Time Quantitative PCR Analysis

Total RNA from diseased peels of calcium-deficient apples and calcium-sufficient apple peels at 0, 14, and 21 DAS as well as transiently infected apple peel samples was extracted using the Plant RNA Isolation Kit (Foregene, Chengdu, China) and first strand ribonucleic acid was synthesized using a Primer Reverse Transcription Master Mix Kit (Takara, Tokyo, Japan). The RT-qPCR primers are shown in Table S1. The expression levels of *MdTUB* (*TUB*, accession number GO562615) and *MdUBQ* (*UBQ*, accession number MDU74358) were used as normalization genes, and the expression of relative genes was determined with the $2^{-\Delta\Delta CT}$ method. Three biological replicates were used for all analyses.

5.9. Prediction of Cis-Acting Elements in Gene Promoters

The upstream 2 kb promoter sequence of the target gene was taken from the NCBI (https://www.ncbi.nlm.nih.gov/, accessed on 24 September 2021), and the sequence was then input into the PlantPAN 3.0 (http://plantpan.itps.ncku.edu.tw/index.html, accessed on 24 September 2021) for cis-acting element prediction.

5.10. Transient Expression Assays in Apple Peels

In the transient transformation experiments, PCR amplification was performed using Phanta® Super-Fidelity DNA Polymerase (Vazyme, Nanjing, China), and the full-length coding sequences of *ERF2*, *ERF17*, and *bHLH2* were introduced into the pSAK277 vector using *EcoRI* and *XbaI* with the 35S promoter as a control. The primer sequences are listed in Table S2. The integration constructs were chemically transformed into the

Agrobacterium tumefaciens GV3101 strain, and the cells were cultivated at 28 °C for 2 d. The detailed approach for the infiltration experiment has been previously reported by Voinnet et al. [57]. Five apples of similar size and growth were selected for infiltration in each combination, with four sites infiltrated in each apple. At 5 days after infiltration, apple peels were collected for RNA extraction. The negative controls were empty vector infiltrations (pSAK277).

5.11. Dual-Luciferase Reporter Assay of Tobacco Leaves

To construct the dual-luciferase reporter vector, the 2 kb upstream promoter region of *CML5* (from the ATG start codon) was amplified from the genomic DNA of peels from 'Honeycrisp' apples and inserted into a pGreen II 0800-LUC binary vector. In addition, the injection solutions were prepared as in Section 5.10. *Agrobacterium* cells harboring the pGreen II 0800-LUC recombinant vector and pSAK277 vector, *ERF2/17*, *bHLH2* were mixed at a 1:9 ratio. The mixture of *Agrobacterium* cells was injected into young *N. tabacum* leaves that were 2 weeks old. At 48-72 h after infiltration, the LUC and Ren activity were measured with an E1910 Dual-Luciferase® Reporter Assay System (Promega, Madison, WI, USA).

5.12. Statistical Analysis

For the statistical analyses, Student's *t*-test and one-way ANOVA were performed. Significance was indicated by asterisks * ($p< 0.05$) or ** ($p< 0.01$) or different letters. The main factors were assessed by reducing dimensionality when conducting the PCA. R studio software was used for correlation analysis and heatmap analysis. All samples were assessed at least three times independently, and all data are represented as the mean ± SD.

Supplementary Materials: The following are available online at https://www.mdpi.com/article/10.3390/ijms222313013/s1.

Author Contributions: H.-Y.S., W.-W.Z., G.-F.Y., K.-D.H. and H.Z. conceived and designed the experiments; H.-Y.S., H.-Y.Q., S.-S.G., L.-X.L. and H.-H.S. performed the experiments; W.-W.Z. and H.-Y.Q. analyzed the data; H.-Y.S., G.-F.Y. and K.-D.H. wrote the paper; H.-Q.Y., G.-F.Y., K.-D.H., W.-J.L. and H.Z. interpreted the data and revised the manuscript. All authors have read and agreed to the published version of the manuscript.

Funding: This work was supported by the National Key R&D Program of China [grant number 2019YFD1000103], the National Natural Science Foundation of China [grant numbers 32170315, 31901993, 31970312, 31970200], the Natural Science Foundations of Anhui Province [grant number 1908085MC72], the Key Research and Development Program of Anhui Province [grant number 202003a06020011], the Fundamental Research Funds for the Central Universities [grant number JZ2021HGPA0063].

Institutional Review Board Statement: Not applicable.

Informed Consent Statement: Not applicable.

Acknowledgments: We thank Andrew C. Allan, Lin-Wang Kui and Richard Espley for the dual vector pGreen II 0800-LUC in The New Zealand Institute for Plant and Food Research Limited, Auckland, New Zealand.

Conflicts of Interest: The authors declare no conflict of interest.

Abbreviations

Calcium (Ca/Ca^{2+}); Reactive Oxygen Species (ROS); Hydrogen sulfide (H_2S); Cysteine desulfhydrases (CDs); L/D-cysteine desulfhydrase (LCD/DCD); O-acetyl-L-serine (thiol) lyase (OASTL); sulfite reductase (SiR); Superoxide Anion ($O_2^{\bullet -}$); Malondialdehyde (MDA); Hydrogen Peroxide (H_2O_2); Calmodulin (CaM); Calmodulin-Like Proteins (CML); APETALA2/ethylene-responsive factors (AP2/ERFs); Basic Helix-Loop-Helix (bHLH); Pathogen-Related Proteins (PRs); Days After Storage (DAS); Thiobarbituric Acid (TBA); Guaiacol Peroxidase (POD); Polyphenol Oxidase (PPO); Principal Component Analysis (PCA); Differentially Expressed Gene (DEG); Gene Ontology (GO); WEGO (Web Gene Ontology Annotation Plot); Kyoto Encyclopedia of Genes and Genomes (KEGG);

Reads Per Kilobase per Million mapped reads (RPKM); Real Time Quantitative PCR (RT-qPCR); Transcription factors (TFs); Pyridoxal-5′-phosphate (PLP).

References

1. Delian, E.; Chira, A.; Dulescu, L.B.; Chira, L. Calcium alleviates stress in plants: Insight into regulatory mechanisms. *AgroLife Sci. J.* **2014**, *3*, 19–28. [CrossRef]
2. Reddy, A.S.N.; Ali, G.S.; Celesnik, H.; Day, I.S. Coping with stresses: Roles of calcium-and calcium/calmodulin-regulated gene expression. *Plant Cell* **2011**, *23*, 2010–2032. [CrossRef] [PubMed]
3. Song, W.Y.; Zhang, Z.B.; Shao, H.B.; Guo, X.L.; Cao, H.X.; Zhao, H.B.; Hu, X.J. Relationship between calcium decoding elements and plant abiotic-stress resistance. *Int. J. Biol. Sci.* **2008**, *4*, 116–125. [CrossRef]
4. Jiang, T.; Zhan, X.; Xu, Y.; Zhou, L.; Zong, L. Roles of calcium in stress-tolerance of plants and its ecological significance. *J. Appl. Ecol.* **2005**, *16*, 971–976. [CrossRef]
5. Kowalik, P.; Lipa, T.; Michałojć, Z.; Chwil, M. Ultrastructure of cells and microanalysis in *Malus domestica* Borkh. 'Szampion' fruit in relation to varied calcium foliar feeding. *Molecules* **2020**, *25*, 4622. [CrossRef]
6. He, L.; Li, B.; Lu, X.; Yuan, L.; Yang, Y.; Yuan, Y.; Du, J.; Guo, S. The effect of exogenous calcium on mitochondria, respiratory metabolism enzymes and ion transport in cucumber roots under hypoxia. *Sci. Rep.* **2015**, *5*, 1–14. [CrossRef] [PubMed]
7. Tuteja, N.; Mahajan, S. Calcium signalling network in plants: An overview. *Plant Signal. Behav.* **2007**, *2*, 79–85. [CrossRef]
8. Silva, D.L.D.; Prado, R.D.M.; Tenesaca, L.F.L.; Silva, J.L.F.D.; Mattiuz, B.H. Silicon attenuates calcium deficiency by increasing ascorbic acid content, growth and quality of cabbage leaves. *Sci. Rep.* **2021**, *11*, 1770. [CrossRef] [PubMed]
9. Jain, V.; Chawla, S.; Choudhary, P.; Jain, S. Post-harvest calcium chloride treatments influence fruit firmness, cell wall components and cell wall hydrolyzing enzymes of Ber (*Ziziphus mauritiana* Lamk.) fruits during storage. *J. Food Sci. Technol.* **2019**, *56*, 4535–4542. [CrossRef] [PubMed]
10. Wang, W.; Wang, J.; Wei, Q.; Li, B.; Zhong, X.; Hu, T.; Hu, H.; Bao, C. Transcriptome-wide identification and characterization of circular RNAs in leaves of Chinese cabbage (*Brassica rapa* L. ssp. *pekinensis*) in response to calcium deficiency-induced tip-burn. *Sci. Rep.* **2019**, *9*, 14544. [CrossRef] [PubMed]
11. El Habbasha, S.F.; Faten, M.I. Calcium: Physiological function, deficiency and absorption. *Inter. J. Chem. Tech. Res.* **2015**, *8*, 196–202.
12. Hepler, P.K. Calcium: A central regulator of plant growth and development. *Plant Cell* **2005**, *17*, 2142–2155. [CrossRef] [PubMed]
13. Moradinezhad, F.; Dorostkar, M. Pre-harvest foliar application of calcium chloride and potassium nitrate influences growth and quality of apricot (*Prunus armeniaca* L.) fruit cv. 'Shahroudi'. *J. Soil Sci. Plant Nutr.* **2021**, *21*, 1642–1652. [CrossRef]
14. Rosenberger, D.A.; Schupp, J.R.; Hoying, S.A.; Cheng, L.; Watkins, C.B. Controlling bitter pit in 'Honeycrisp' apples. *Hort. Technol.* **2004**, *14*, 342–349. [CrossRef]
15. McAinsh, M.R.; Pittman, J.K. Shaping the calcium signature. *New Phytol.* **2009**, *181*, 275–294. [CrossRef] [PubMed]
16. Dodd, A.N.; Kudla, J.; Sanders, D. The language of calcium signaling. *Annu. Rev. Plant Biol.* **2010**, *61*, 593–620. [CrossRef] [PubMed]
17. Mohanta, T.K.; Yadav, D.; Khan, A.L.; Hashem, A.; Al-Harrasi, A. Molecular players of EF-hand containing calcium signaling event in plants. *Int. J. Mol. Sci.* **2019**, *20*, 1476. [CrossRef]
18. He, X.; Liu, W.; Li, W.; Liu, Y.; Wang, W.; Xie, P.; Kang, Y.; Liao, L.; Qian, L.; Liu, Z.; et al. Genome-wide identification and expression analysis of CaM/CML genes in *Brassica napus* under abiotic stress. *J. Plant Physiol.* **2020**, *255*, 153251. [CrossRef]
19. Shen, L.; Yang, S.; Guan, D.; He, S. *CaCML13* acts positively in pepper immunity against *Ralstonia solanacearum* infection forming feedback loop with *CabZIP63*. *Int. J. Mol. Sci.* **2020**, *21*, 4186. [CrossRef] [PubMed]
20. Munir, S.; Liu, H.; Xing, Y.; Hussain, S.; Ouyang, B.; Zhang, Y.; Li, H. Overexpression of calmodulin-like (*ShCML44*) stress-responsive gene from *Solanum habrochaites* enhances tolerance to multiple abiotic stresses. *Sci. Rep.* **2016**, *6*, 31772. [CrossRef]
21. Jung, H.; Chung, P.J.; Park, S.H.; Redillas, M.C.F.R.; Kim, Y.S.; Suh, J.W.; Kim, J.K. Overexpression of *OsERF48* causes regulation of *OsCML16*, a calmodulin-like protein gene that enhances root growth and drought tolerance. *Plant Biotech. J.* **2017**, *15*, 1295–1308. [CrossRef] [PubMed]
22. Xuan, L.; Li, J.; Wang, X.; Wang, C. Crosstalk between hydrogen sulfide and other signal molecules regulates plant growth and development. *Int. J. Mol. Sci.* **2020**, *21*, 4593. [CrossRef]
23. Peng, R.Y.; Bian, Z.Y.; Zhou, L.N.; Cheng, W.; Hai, N.; Yang, C.Q.; Yang, T.; Wang, X.Y.; Wang, C.Y. Hydrogen sulfide enhances nitric oxide-induced tolerance of hypoxia in maize (*Zea mays* L.). *Plant Cell Rep.* **2016**, *35*, 2325–2340. [CrossRef]
24. Li, Z.G.; Gong, M.; Xie, H.; Yang, L.; Li, J. Hydrogen sulfide donor sodium hydrosulfide-induced heat tolerance in tobacco (*Nicotiana tabacum* L.) suspension cultured cells and involvement of Ca^{2+} and calmodulin. *Plant Sci.* **2012**, *185*, 185–189. [CrossRef] [PubMed]
25. Hancock, J.T. Harnessing evolutionary toxins for signaling: Reactive oxygen species, nitric oxide and hydrogen sulfide in plant cell regulation. *Front. Plant Sci.* **2016**, *8*, 189. [CrossRef] [PubMed]
26. Yamasaki, H.; Cohen, M.F. Biological consilience of hydrogen sulfide and nitric oxide in plants: Gases of primordial earth linking plant, microbial and animal physiologies. *Nitric Oxide* **2016**, *55*, 91–100. [CrossRef] [PubMed]

27. Papenbrock, J.; Riemenschneider, A.; Kamp, A.; Schulz-Vogt, H.N.; Schmidt, A. Characterization of cysteine-degrading and H_2S-releasingenzymes of higher plants—From the field to the test tube and back. *Plant Biol.* **2007**, *9*, 582–588. [CrossRef] [PubMed]
28. Guo, H.; Xiao, T.; Zhou, H.; Xie, Y.; Shen, W. Hydrogen sulfide: A versatile regulator of environmental stress in plants. *Acta Physiol. Plant* **2016**, *38*, 16. [CrossRef]
29. Ali, S.; Nawaz, A.; Ejaz, S.; Haider, S.T.A.; Alam, M.W.; Javed, H.U. Effects of hydrogen sulfide on postharvest physiology of fruits and vegetables: An overview. *Sci. Hortic.* **2019**, *243*, 290–299. [CrossRef]
30. Yao, G.F.; Wei, Z.Z.; Li, T.T.; Tang, J.; Huang, Z.Q.; Yang, F.Y.; Li, Y.H.; Han, Z.; Hu, F.; Hu, L.Y.; et al. Enhanced antioxidant activity is modulated by hydrogen sulfide antagonizing ethylene in tomato fruit ripening. *J. Agric. Food Chem.* **2018**, *66*, 10380–10387. [CrossRef]
31. Molinett, S.A.; Alfaro, J.F.; Sáez, F.A.; Elgueta, S.; Moya-León, M.A.; Figueroa, C.R. Postharvest treatment of hydrogen sulfide delays the softening of Chilean strawberry fruit by downregulating the expression of key genes involved in pectin catabolism. *Int. J. Mol. Sci.* **2021**, *22*, 10008. [CrossRef] [PubMed]
32. Zhu, L.Q.; Wang, W.; Shi, J.Y.; Zhang, W.; Shen, Y.G.; Du, H.Y.; Wu, S.F. Hydrogen sulfide extends the postharvest life and enhances antioxidant activity of kiwifruit during storage. *J. Sci. Food Agric.* **2014**, *94*, 2699–2704. [CrossRef] [PubMed]
33. Kolupaev, Y.E.; Karpets, Y.V. The role of reactive oxygen species and calcium ions in the implementation of the stress-protective effect of brassinosteroids on plant cells. *Ukr. Biochem. J.* **2014**, *86*, 18–35. [CrossRef] [PubMed]
34. Kumari, A.; Parida, A.K.; Rangani, J.; Panda, A. Antioxidant activities, metabolic profiling, proximate analysis, mineral nutrient composition of *Salvadora persica* fruit unravel a potential functional food and a natural source of pharmaceuticals. *Front. Pharmacol.* **2017**, *8*, 61. [CrossRef]
35. Falchi, R.; Bonghi, C.; Drincovich, M.F.; Famiani, F.; Lara, M.V.; Walker, R.P. Sugar metabolism in stone fruit: Source-sink relationships and environmental and agronomical effects. *Front. Plant Sci.* **2020**, *11*, 573982. [CrossRef]
36. Gao, Q.; Xiong, T.; Li, X.; Chen, W.; Zhu, X. Calcium and calcium sensors in fruit development and ripening. *Sci. Hortic.* **2019**, *253*, 412–421. [CrossRef]
37. Zhang, L.; Wang, J.W.; Zhou, B.; Li, G.D.; Liu, Y.F.; Xia, X.L.; Xiao, Z.G.; Lu, F.; Ji, S.J. Calcium inhibited peel browning by regulating enzymes in membrane metabolism of 'Nanguo' pears during post-ripeness after refrigerated storage. *Sci. Hortic.* **2019**, *244*, 15–21. [CrossRef]
38. Fu, W.; Zhang, M.; Zhang, P.; Liu, Z.; Dong, T.; Zhang, S.; Ren, Y.; Jia, H.; Fang, J. Transcriptional and metabolite analysis reveal a shift in fruit quality in response to calcium chloride treatment on 'Kyoho' grapevine. *J. Food Sci. Technol.* **2021**, *58*, 2246–2257. [CrossRef]
39. Tang, M.; Xu, C.; Cao, H.; Shi, Y.; Chen, J.; Chai, Y.; Li, Z. Tomato calmodulin-like protein *SlCML37* is a calcium (Ca^{2+}) sensor that interacts with proteasome maturation factor *SlUMP1* and plays a role in tomato fruit chilling stress tolerance. *J. Plant Physiol.* **2021**, *153373*, 258–259. [CrossRef]
40. Yang, S.; Xiong, X.; Arif, S.; Gao, L.; Zhao, L.; Shah, I.H.; Zhang, Y. A calmodulin-like *CmCML13* from *Cucumis melo* improved transgenic *Arabidopsis* salt tolerance through reduced shoot's Na^+, and also improved drought resistance. *Plant Physiol. Biochem.* **2020**, *155*, 271–283. [CrossRef]
41. Aleynova, O.A.; Kiselev, K.V.; Ogneva, Z.V.; Dubrovina, A.S. The grapevine calmodulin-like protein gene *CML21* is regulated by alternative splicing and involved in abiotic stress response. *Int. J. Mol. Sci.* **2020**, *21*, 7939. [CrossRef]
42. Siddiqui, M.W.; Homa, F.; Lata, D.; Mir, H.; Aftab, T.; Mishra, P. Hydrogen sulphide infiltration downregulates oxidative metabolism and extends postharvest life of banana. *Plant Biol.* **2021**, *11*. [CrossRef]
43. Dou, Y.; Chang, C.M.; Wang, J.; Cai, Z.P.; Zhang, W.; Du, H.Y.; Gan, Z.Y.; Wan, C.P.; Chen, J.Y.; Zhu, L.Q. Hydrogen sulfide inhibits enzymatic browning of fresh-cut Chinese water chestnuts. *Front Nutr.* **2021**, *8*, 165. [CrossRef]
44. Li, J.; Jia, H.; Wang, J.; Cao, Q.; Wen, Z. Hydrogen sulfide is involved in maintaining ion homeostasis via regulating plasma membrane Na^+/H^+ antiporter system in the hydrogen peroxide-dependent manner in salt-stress *Arabidopsis thaliana* root. *Protoplasma* **2014**, *251*, 899–912. [CrossRef]
45. Kim, T.; Böhmer, M.; Hu, H.; Nishimura, N.; Schroeder, J. Guard cell signal transduction network: Advances in understanding abscisic acid, CO_2, and Ca^{2+} signaling. *Annu. Rev. Plant Biol.* **2010**, *61*, 561–591. [CrossRef]
46. Fang, H.H.; Liu, Z.Q.; Long, Y.P.; Liang, Y.L.; Jin, Z.P.; Zhang, L.P.; Liu, D.M.; Pei, Y.X. The Ca^{2+}/calmodulin2-binding transcription factor TGA3 elevates *LCD* expression and H_2S production to bolster Cr^{6+} tolerance in *Arabidopsis*. *Plant J.* **2017**, *91*, 1038–1050. [CrossRef] [PubMed]
47. Suliburska, J.; Krejpcio, Z. Evaluation of the content and bioaccessibility of iron, zinc, calcium and magnesium from groats, rice, leguminous grains and nuts. *J. Food Sci. Technol.* **2014**, *51*, 589–594. [CrossRef] [PubMed]
48. Hine, C.; Harputlugil, E.; Zhang, Y.; Ruckenstuhl, C.; Lee, B.C.; Brace, L.; Longchamp, A. Endogenous hydrogen sulfide production is essential for dietary restriction benefits. *Cell* **2015**, *160*, 132–144. [CrossRef] [PubMed]
49. Hu, L.Y.; Hu, S.L.; Wu, J.; Li, Y.H.; Zheng, J.L.; Wei, Z.J.; Liu, J.; Wang, H.L.; Liu, Y.S.; Zhang, H. Hydrogen sulfide prolongs postharvest shelf life of strawberry and plays an antioxidative role in fruits. *J. Agric. Food Chem.* **2012**, *60*, 8684–8693. [CrossRef]
50. Pirie, A.; Mullins, M.G. Changes in anthocyanin and phenolics content of grapevine leaf and fruit tissues treated with sucrose, nitrate and abscisic acid. *Plant Physiol.* **1976**, *58*, 468–472. [CrossRef] [PubMed]

51. Li, S.P.; Hu, K.D.; Hu, L.Y.; Li, Y.H.; Jiang, A.M.; Xiao, F. Hydrogen sulfide alleviates postharvest senescence of broccoli by modulating antioxidant defense and senescence-related gene expression. *J. Agric. Food Chem.* **2014**, *62*, 1119–1129. [CrossRef]
52. Lee, H.S.; Wicker, L. Anthocyanin pigments in the skin of lychee fruit. *J. Food Sci.* **2006**, *56*, 466–468. [CrossRef]
53. Benjamin, N.D.; Montgomery, M.W. Polyphenol oxidase of royal ann cherries: Purification and characterization. *J. Food Sci.* **1973**, *38*, 799–806. [CrossRef]
54. Beaudoin-Eagan, L.D.; Thorpe, T.A. Tyrosine and phenyl-alanine ammonia lyase activities during shoot initiation in tobacco callus cultures. *Plant Physiol.* **1985**, *78*, 438–441. [CrossRef]
55. Bradford, M.M. A rapid and sensitive method for the quantitation of microgram quantities of protein utilizing the principle of protein-dye binding. *Anal. Biochem.* **1976**, *72*, 248–254. [CrossRef]
56. Qiu, L.N.; Hu, S.S.; Wang, Y.Z.; Qu, H.Y. Accumulation of abnormal amyloplasts in pulp cells induces bitter pit in *Malus domestica*. *Front. Plant Sci.* **2021**, *12*, 2052. [CrossRef]
57. Voinnet, O.; Rivas, S.; Mestre, P.; Baulcombe, D. Retracted: An enhanced transient expression system in plants based on suppression of gene silencing by the p19 protein of tomato bushy stunt virus. *Plant J.* **2003**, *33*, 949–956. [CrossRef] [PubMed]

International Journal of *Molecular Sciences*

Article

Roles of a Cysteine Desulfhydrase LCD1 in Regulating Leaf Senescence in Tomato

Kangdi Hu [1,*,†], Xiangjun Peng [1,†], Gaifang Yao [1], Zhilin Zhou [2], Feng Yang [2], Wanjie Li [3], Yuqi Zhao [1], Yanhong Li [1], Zhuo Han [1], Xiaoyan Chen [1] and Hua Zhang [1,*]

1 School of Food and Biological Engineering, Hefei University of Technology, Hefei 230009, China; 2019111433@mail.hfut.edu.cn (X.P.); yaogaifang@hfut.edu.cn (G.Y.); 2019111429@mail.hfut.edu.cn (Y.Z.); yhl719@126.com (Y.L.); kanahan80@163.com (Z.H.); swspcxy@163.com (X.C.)
2 Xuzhou Institute of Agricultural Sciences of the Xuhuai District of Jiangsu Province, Xuzhou 221131, China; zhouzhilinting@163.com (Z.Z.); XZ-YANGFENG@163.com (F.Y.)
3 Key Laboratory of Cell Proliferation and Regulation Biology, Ministry of Education, College of Life Science, Beijing Normal University, Beijing 100875, China; lwj@bnu.edu.cn
* Correspondence: kangdihu@hfut.edu.cn (K.H.); hzhanglab@hfut.edu.cn (H.Z.)
† Co-Author: They contributed equally to this manuscript.

Abstract: Hydrogen sulfide (H_2S), a novel gasotransmitter in both mammals and plants, plays important roles in plant development and stress responses. Leaf senescence represents the final stage of leaf development. The role of H_2S-producing enzyme L-cysteine desulfhydrase in regulating tomato leaf senescence is still unknown. In the present study, the effect of an L-cysteine desulfhydrase LCD1 on leaf senescence in tomato was explored by physiological analysis. *LCD1* mutation caused earlier leaf senescence, whereas *LCD1* overexpression significantly delayed leaf senescence compared with the wild type in 10-week tomato seedlings. Moreover, *LCD1* overexpression was found to delay dark-induced senescence in detached tomato leaves, and the *lcd1* mutant showed accelerated senescence. An increasing trend of H_2S production was observed in leaves during storage in darkness, while *LCD1* deletion reduced H_2S production and *LCD1* overexpression produced more H_2S compared with the wild-type control. Further investigations showed that *LCD1* overexpression delayed dark-triggered chlorophyll degradation and reactive oxygen species (ROS) accumulation in detached tomato leaves, and the increase in the expression of chlorophyll degradation genes *NYC1*, *PAO*, *PPH*, *SGR1*, and *senescence-associated genes* (*SAGs*) during senescence was attenuated by *LCD1* overexpression, whereas *lcd1* mutants showed enhanced senescence-related parameters. Moreover, a correlation analysis indicated that chlorophyll content was negatively correlated with H_2O_2 and malondialdehyde (MDA) content, and also negatively correlated with the expression of chlorophyll degradation-related genes and *SAGs*. Therefore, these findings increase our understanding of the physiological functions of the H_2S-generating enzyme LCD1 in regulating leaf senescence in tomato.

Keywords: tomato; hydrogen sulfide; cysteine desulfhydrase; leaf senescence; reactive oxygen species

Citation: Hu, K.; Peng, X.; Yao, G.; Zhou, Z.; Yang, F.; Li, W.; Zhao, Y.; Li, Y.; Han, Z.; Chen, X.; et al. Roles of a Cysteine Desulfhydrase LCD1 in Regulating Leaf Senescence in Tomato. *Int. J. Mol. Sci.* **2021**, *22*, 13078. https://doi.org/10.3390/ijms222313078

Academic Editors: Yanjie Xie, Francisco J. Corpas and Jisheng Li

Received: 20 November 2021
Accepted: 30 November 2021
Published: 3 December 2021

Publisher's Note: MDPI stays neutral with regard to jurisdictional claims in published maps and institutional affiliations.

1. Introduction

Leaf senescence represents the final stage of leaf development, which is a genetically controlled process [1]. As leaves age, the decomposition of chloroplast is initiated, accompanied by the catabolism of macromolecules including nucleic acids, proteins, and lipids. The decomposed nutrients then transfer to other developing organs, such as young leaves and growing fruit [2]. Chloroplasts constitute approximately 70% of the total proteins in green leaves and chlorophyll degradation causes the first visible signs of leaf senescence [3]. Thus, the coordinated degradation of chlorophyll is crucial for the breakdown of chloroplasts. Terrestrial plants utilize two types of chlorophyll species (i.e., chlorophyll a and chlorophyll b) for photosynthesis [4]. Chlorophyll b has to be converted to chlorophyll a before it can be processed into the degradation pathway and NON-YELLOW COLORING 1 (NYC1) catalyzes the reduction of chlorophyll b to 7-hydroxymethyl chlorophyll a [5]. Chlorophyll a is further decomposed to pheophytin a by Mg-dechelatase NON-YELLOWINGs/STAY-GREENs (NYEs/SGRs) [6]. Pheophytin a is then hydrolyzed by the pheophytinase PPH to generate pheophorbide a which is further catalyzed by oxygenase PAO to produce a red chlorophyll catabolite (RCC) [7]. Moreover, hundreds of *senescence-associated genes* (*SAGs*), whose transcripts increase as leaves age [8,9], are also involved in the regulation of leaf senescence.

Plant hormones are major players influencing each stage of leaf senescence. For instance, ethylene, abscisic acid (ABA), and so on promote leaf senescence, while cytokinins (CKs), gibberellic acid (GA), and so on delay leaf senescence [10–12]. Furthermore, leaf senescence-linked events are often associated with the pronounced accumulation of reactive oxygen species (ROS) [13]. Among them, H_2O_2 is a well-defined inducer of leaf senescence. Recently, it was reported that transcription factor NAC075 delays leaf senescence by deterring ROS accumulation through directly binding the promoter of the antioxidant enzyme gene *catalase 2* (*CAT*) in *Arabidopsis* [14]. Hydrogen sulfide (H_2S) is an important gasotransmitter in both animals and plants [15]. H_2S has not only been implicated in seed germination and root development, but can also enhance plant tolerance to various stresses such as heavy metals, drought, salinity, and cold by enhancing the antioxidant system [16–18]. In addition, H_2S can extend the shelf life of bananas, grapes, strawberries, tomatoes, and so on [19–22]. The underlining mechanism of H_2S in alleviating postharvest senescence may involve the activation of the antioxidant system, the inhibition of ethylene synthesis and the signaling pathway, etc.

H_2S is endogenously produced in a precise and regulated manner. Cysteine degradation by cysteine desulfhydrases (CDes) to the formation of sulfide, ammonia, and pyruvate was believed to be an important source of H_2S [23]. Plant cells contain different CDes localized in the cytoplasm, plastids, and mitochondria [23]. DES1, an O-acetylserine(thiol)lyase homolog with L-cysteine desulfhydrase activity, regulates cysteine homeostasis in *Arabidopsis* [24]. Recently it was reported that the *des1* mutant was more sensitive to drought stress and displayed accelerated leaf senescence, while the leaves of *OE-DES1* contained adequate chlorophyll levels accompanied by significantly increased drought resistance, suggesting the role of DES1 in regulating leaf senescence [25]. In our previous work, a L-cysteine desulfhydrase named LCD1 was found to localize in the nucleus by a potential nuclear localization signal, and an *lcd1* mutation caused accelerated fruit ripening compared with the wild type [26]. However, whether and how LCD1 participates in leaf senescence or dark-induced senescence are still unknown. In the present study, we focused on the role of endogenous H_2S-producing enzyme LCD1 during leaf senescence. Moreover, a correlation analysis was applied to investigate the potential relations between H_2S, ROS, and chlorophyll breakdown.

2. Results

2.1. Role of LCD1 in Regulating Tomato Leaf Senescence

To elucidate the possible involvement of LCD1 in regulating leaf senescence, two previously reported tomato lines, *lcd1-7* and *lcd1-9*, with mutations near the PAM sequence were used as *lcd1* mutants, and two lines (hereafter called *LCD1-oe* and *LCD1-oe1*) with an increased expression of *LCD1* under the control of the CaMV *35S* promoter were also used. The overexpression efficacy of five *LCD1* overexpression lines was verified by RT-qPCR as shown in Figure S1 and the lines *LCD1-oe* and *LCD1-oe1*, which showed a higher *LCD1* expression, were used in the present study. To confirm the role of LCD1 in catalyzing H_2S production, the H_2S producing rates were determined in leaves of *lcd1* and *LCD1-oe* lines. The data in Figure 1B suggest that *lcd1* leaves had a lower H_2S producing rate compared with the wild type, while *LCD1* overexpression induced a significantly higher level of the H_2S producing rate. Besides, H_2S production was also evaluated by lead acetate H_2S detection strips, and the results showed that *lcd1* leaves produced less H_2S and *LCD1* overexpression produced more H_2S (Figure 1C). After 10 weeks of growth, the *LCD1* mutation caused earlier leaf senescence compared with the wild type. In contrast, *LCD1* overexpression significantly delayed leaf senescence (Figure 1D).

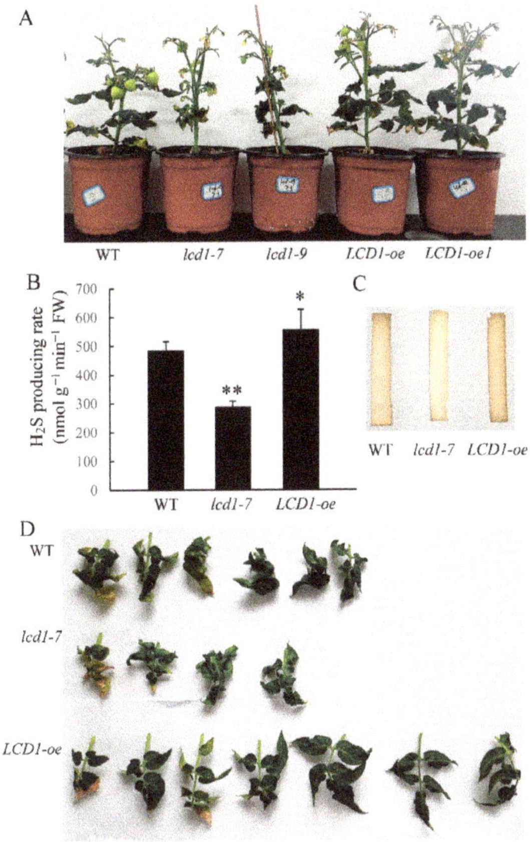

Figure 1. Phenotypic characterization of *lcd1* mutants and *LCD-oe* (over expression) tomatoes. (**A**) Phenotype of 10-week-old wild-type (WT), *lcd1-7*, *lcd1-9*, *LCD1-oe*, and *LCD1-oe1* plants. (**B**) H_2S producing rate in the mature leaves from wild type, *lcd1*, and *LCD1-oe* lines of 10 weeks growth. (**C**) H_2S production from the mature leaves of 10-week-old wild type, *lcd1*, and *LCD1-oe* lines detected by lead acetate H_2S detection strips (Sigma-Aldrich). (**D**) The leaves of different tomato lines in (**A**) were detached and photographed. Data are means of three biological replicates ± standard deviation (SD). The symbols * and ** stand for $p < 0.05$ and $p < 0.01$ as determined by the Student's *t*-test, respectively.

2.2. LCD1 Participates in Dark-Induced Senescence

Leaf senescence is an important phenomenon in the growth and development of plant leaves, and darkness is widely used as a tool to induce senescence in detached leaves. To study the role of LCD1 in dark-induced senescence, the mature leaves without visible senescence of 6-week-old wild type, *lcd1* mutant, and *LCD1-oe* were stored in darkness for 8 days. As shown in Figure 2A, *lcd1* showed the obvious syndrome of the leaf yellowing phenotype after 5 and 8 days in dark stress, whereas *LCD1* overexpression still maintained the green phenotype. To study the kinetics of tomato leaf H_2S production during senescence, H_2S production in the leaves at different developmental stages—young leaves (YL), mature leaves (ML), senescent leaves (SL), and late senescent leaves (LS)—was evaluated and the H_2S detection strips showed browning with senescence, suggesting

H_2S production increased during leaf senescence (Figure S2). Moreover, H_2S production in leaves of wild-type (WT), *lcd1*, and *LCD1-oe* tomatoes were also determined during dark-induced senescence (Figure 2B). Generally, an increasing trend of H_2S production was observed in all samples during storage, while *LCD1-oe* leaves showed a higher H_2S production compared with the wild-type control. In addition, the *lcd1* mutant produced a significantly lower level of H_2S compared with the wild type. Therefore, it can be concluded that *LCD1* deletion caused a lower H_2S release and the attenuated H_2S release may cause an accelerated senescence in the *lcd1* mutant. Overall, the present results indicate that *LCD1* plays a negative role in leaf senescence in both developmental and dark-induced senescence.

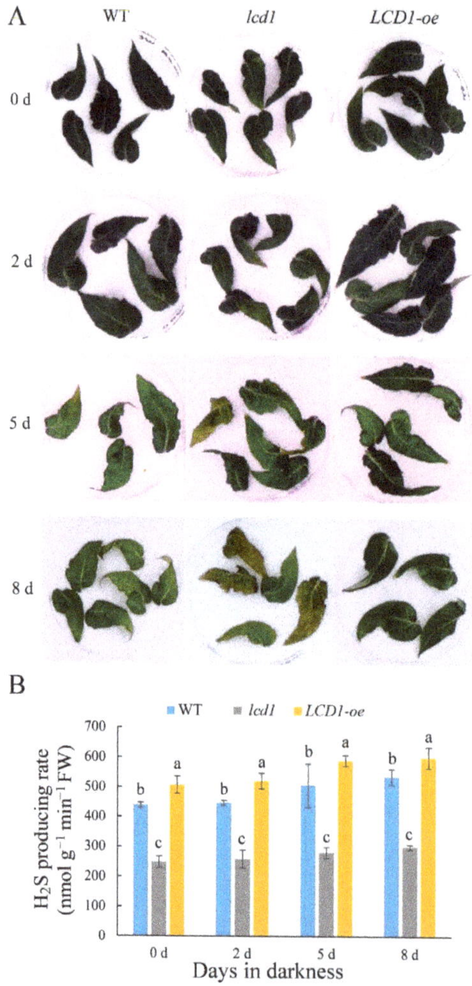

Figure 2. (**A**) Dark-induced senescence symptoms in detached leaves of 6-week-old wild-type (WT), *lcd1*, and *LCD1-oe* tomatoes for up to 8 days. (**B**) H_2S producing rate in detached leaves of 6-week-old wild-type (WT), *lcd1*, and *LCD1-oe* tomatoes stored in darkness for up to 8 days. Different letters above the columns stand for significant difference between two values ($p < 0.05$) at the same time point.

2.3. Effect of LCD1 on Dark-Triggered Chlorophyll Degradation and Reactive Oxygen Species Accumulation in Detached Tomato Leaves

Chlorophyll degradation is the one of the most significant changes during leaf senescence; thus, chlorophyll contents were determined in wild-type, *lcd1* mutant, and *LCD1-oe* leaves during dark-induced senescence. As shown in Figure 3A, the content of total chlorophyll in the wild type decreased gradually during storage in darkness for 8 days, whereas the content of chlorophyll in the *lcd1* mutant showed an obvious decrease on days 5 and 8 under darkness, and the value on day 8 was about 32.6% of the initial value on day 0. In contrast, *LCD1* overexpression maintained a relatively higher chlorophyll content compared with the wild type and the *lcd1* mutant on days 5 and 8 under darkness. After 8 days in darkness, the chlorophyll content in *LCD1* overexpression decreased to 84.6% of the initial value, suggesting the role of *LCD1* in delaying dark-induced senescence. As shown in Figure 3B, there were minor changes in the chlorophyll a content between different groups during storage. Moreover, only a slight decrease in chlorophyll a was observed during dark-induced senescence, except for a significant decline found in *lcd1* on day 8. Figure 3C shows the change pattern of chlorophyll b content in wild type, *lcd1* mutant, and *LCD1-oe* during dark-induced senescence. With the increase of storage days, the chlorophyll b content decreased in each group. At day 0, chlorophyll b content in *lcd1* leaves was about 53.4% of that in the *LCD1-oe* group, and decreased to 21.3% on day 8 compared with the value on day 0. Furthermore, the ratio of chlorophyll a/b was also evaluated in dark-stored detached leaves of wild-type, *lcd1*, and *LCD1-oe* tomatoes for 0, 2, 5, and 8 days. As shown in Figure 3D, the ratio of chlorophyll a/b in WT and *lcd1* mutant leaves increased during storage, while *LCD1* deletion caused the highest ratio compared with other groups. In contrast, the ratio of chlorophyll a/b in *LCD1* overexpression almost remained unchanged. The above results indicate that the *lcd1* mutation accelerated dark-induced leaf senescence and *LCD1* overexpression delayed leaf yellowing and chlorophyll degradation.

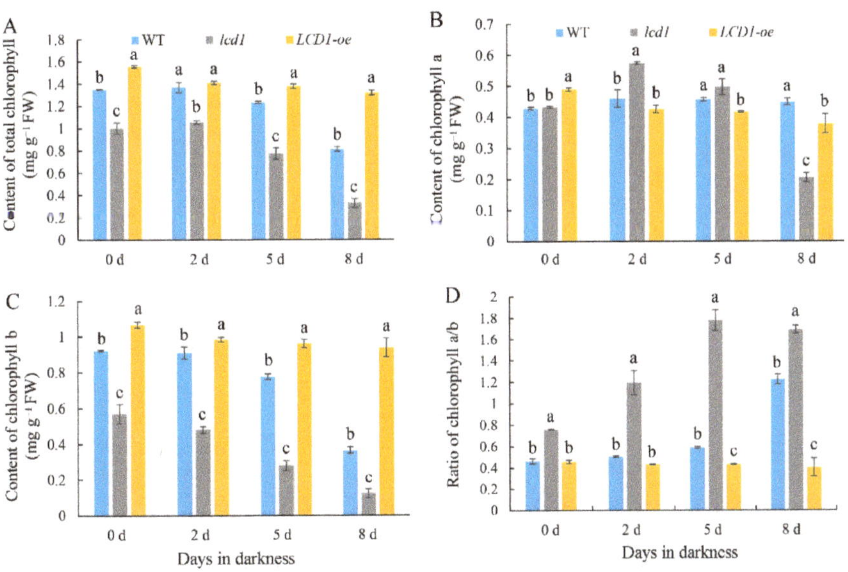

Figure 3. Changes in the contents of (**A**) total chlorophyll, (**B**) chlorophyll a, (**C**) chlorophyll b, and (**D**) the ratio of chlorophyll a/b in dark-stored detached leaves of 6-week-old wild-type (WT), *lcd1*, and *LCD1-oe* tomatoes for 0, 2, 5, and 8 days. Data are means of three biological replicates ± standard deviation (SD). Different letters above the columns stand for significant difference between two values ($p < 0.05$) at the same time point.

Leaf senescence is usually associated with the excessive accumulation of ROS; therefore, the levels of H_2O_2 and malondialdehyde (MDA) were monitored in wild-type, *lcd1* mutant, and *LCD1-oe* leaves during dark-induced senescence. As shown in Figure 4A, there was no significant difference in H_2O_2 content between the different groups on day 0. During the dark-induced senescence, the H_2O_2 content in each group showed an increasing trend, of which the *lcd1* group increased the fastest, followed by the wild-type and *LCD1-oe* group. However, H_2O_2 content in the *LCD1-oe* group increased slowly compared with other groups. As shown in Figure 4B, the change of MDA content among the groups also showed a similar trend to H_2O_2. The content of MDA in *lcd1* leaves during storage was the highest compared with other groups, and the lowest MDA content was observed in *LCD1-oe* leaves. Therefore, it can be concluded that the overexpression of *LCD1* could reduce the accumulation of ROS and MDA in leaves under dark-triggered senescence.

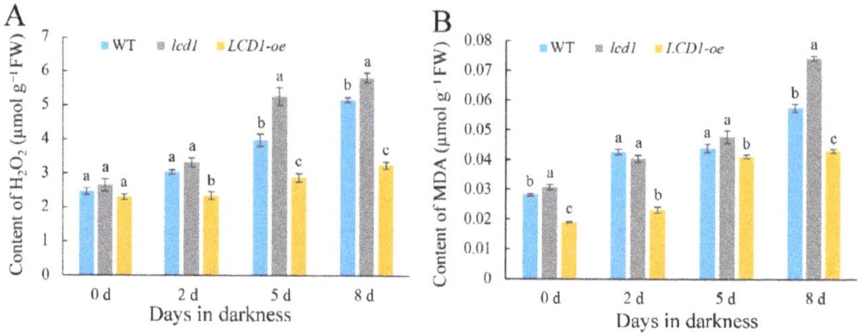

Figure 4. Changes in the contents of (**A**) H_2O_2 and (**B**) malondialdehyde (MDA) in dark-stored detached leaves of 6-week-old wild-type (WT), *lcd1*, and *LCD1-oe* tomatoes for 0, 2, 5, and 8 days. Data are means of three biological replicates ± standard deviation (SD). Different letters above the columns stand for significant difference between two values ($p < 0.05$) at the same time point.

2.4. Effect of LCD1 on the Expressions of Genes Related to Chlorophyll Degradation in Detached Tomato Leaves

Chlorophyll degradation marks the senescence stage of leaves. In order to explore the molecular mechanism of the differences in chlorophyll content of *lcd1*, *LCD1-oe*, and wild-type leaves during dark-induced senescence, the expression levels of key genes *NYC1*, *PAO*, *PPH*, and *SGR1* in the chlorophyll degradation pathway were analyzed by RT-qPCR. The present data showed *NYC1* was transcriptionally induced during dark-induced senescence in all groups (Figure 5A). In accordance with the early senescence phenotype of the *lcd1* mutant and late senescence in *LCD1-oe* leaves, the expression of *NYC1* was significantly higher in the *lcd1* mutant and was less expressed in *LCD1-oe* leaves during dark storage. Three other genes—*PAO* (Figure 5B), *PPH* (Figure 5C), and *SGR1* (Figure 5D)—were also analyzed at the transcriptional level in wild-type, *lcd1* mutant, and *LCD1-oe* leaves during dark-induced senescence and similar changes to that of the *NYC1* expression were observed. The higher expression of *PAO*, *PPH*, and *SGR1* in *lcd1* and lower expression in *LCD1-oe* again supported the role of *LCD1* in delaying leaf senescence. The results suggest that LCD1 may delay the chlorophyll degradation by down-regulating the transcription of key genes in the chlorophyll degradation pathway.

2.5. Effect of LCD1 on the Expressions of SAGs in Detached Tomato Leaves

To further analyze the senescence-alleviating role of LCD1, we conducted an RT-qPCR analysis to evaluate the expression patterns of senescence-associated genes (*SAGs*) in *lcd1*, *LCD1-oe*, and wild-type leaves during dark-induced senescence. As shown in Figure 6, *SAG12*, *SAG15*, and *SAG113* were transcriptionally induced during dark-induced senescence. Compared with *SAG15* and *SAG113*, *SAG12* showed more fold changes during leaf senescence, which was 109.6 times in the wild type on day 8 compared with day 0

(Figure 6A). In accordance with the early senescence phenotype of the *lcd1* mutant and late senescence in *LCD1-oe* leaves, the expression of the three *SAGs* was significantly higher in the *lcd1* mutant and was less expressed in *LCD1-oe* leaves during dark storage, especially on day 8.

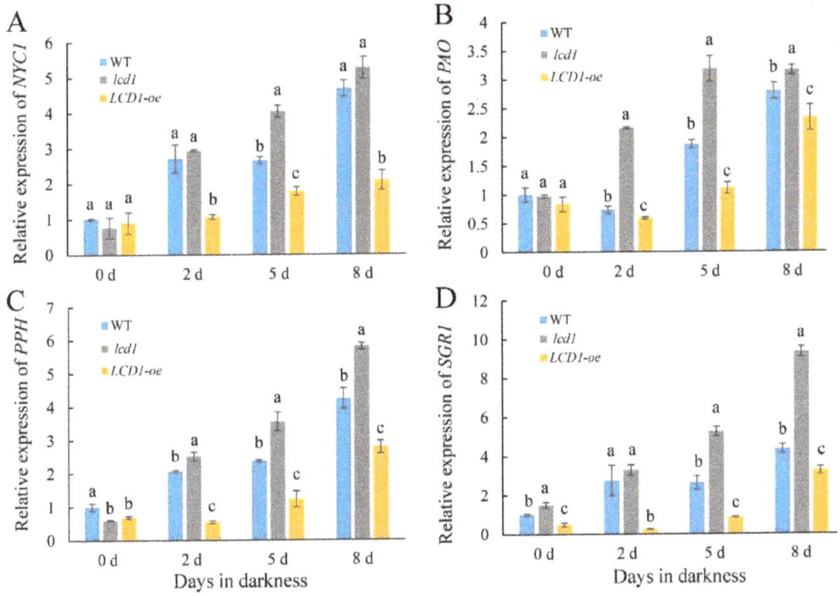

Figure 5. Changes in the gene expressions of chlorophyll degradation related genes: (**A**) *NYC1*, (**B**) *PAO*, (**C**) *PPH*, and (**D**) *SGR1* in detached leaves of 6-week-old wild-type (WT), *lcd1*, and *LCD1-oe* tomatoes stored in darkness for 0, 2, 5, and 8 days. Data are means of three biological replicates ± standard deviation (SD). Different letters above the columns stand for significant difference between two values ($p < 0.05$) at the same time point.

2.6. Correlation Analysis of Different Leaf Physiological Indexes and Senescence-Related Gene Expression

The correlation among total chlorophyll, chlorophyll a, chlorophyll b, chlorophyll a/b, H_2O_2, and MDA content and the gene expression of *NYC1, PAO, PPH, SGR1, SAG12, SAG15,* and *SAG113* was analyzed to investigate the potential relations among the indexes. As shown in Figure 7, chlorophyll content was negatively correlated with H_2O_2, MDA content, and with the expression of chlorophyll degradation-related genes *NYC1, PAO, PPH,* and *SGR1* and senescence-related genes *SAG12, SAG15,* and *SAG113*. Moreover, total chlorophyll and chlorophyll b showed a higher negative correlation to ROS content and senescence-related gene expression in comparison to chlorophyll a. The contents of H_2O_2 and MDA were positively correlated with the expression levels of *NYC1, PAO, PPH, SGR1, SAG12, SAG15,* and *SAG113*. Among them, the expressions of *PPH* and *NYC1* were highly positively correlated (r = 0.966), and the total chlorophyll content was highly negatively correlated with the H_2O_2 content (r = −0.882). Moreover, chlorophyll a/b also showed a significant positive correlation with H_2O_2 and MDA content. Through these analyses, the positive correlation between H_2O_2/MDA content and senescence-related gene expressions indicates that they may act synergistically to accelerate the senescence process of leaves.

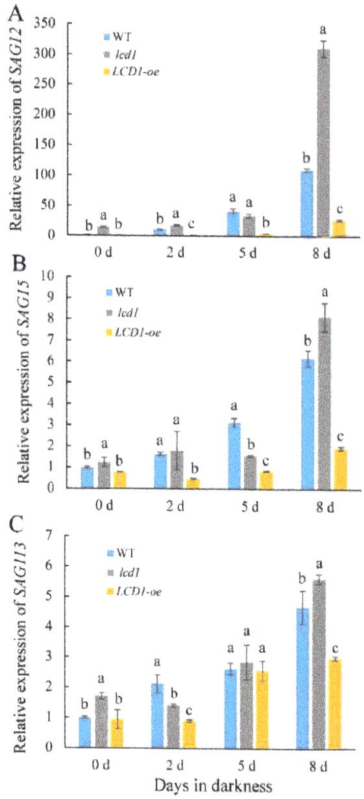

Figure 6. Changes in the gene expressions of senescence-related genes: (**A**) *SAG12*, (**B**) *SAG15*, and (**C**) *SAG113* in detached leaves of 6-week-old wild-type (WT), *lcd1*, and *LCD1-oe* tomatoes stored in darkness for 0, 2, 5, and 8 days. Data are means of three biological replicates ± standard deviation (SD). Different letters above the columns stand for significant difference between two values ($p < 0.05$) at the same time point.

2.7. Principal Component Analysis of Different Leaf Physiological Indexes and Senescence-Related Gene Expression

The principal component analysis (PCA) was performed based on the data of chlorophyll, H_2O_2, and MDA content and the expressions of chlorophyll degradation-related genes and *SAGs*. As shown in Figure 8, PC1 and 2 contributed to 81.1% and 10.4% of the variability of the data, respectively. It can be seen that *lcd1* 2 d and *lcd1* 5 d clustered together, and *lcd1* 8 d was distributed separately from other groups. The variety showing the highest positive load value in the direction of PC1 was *LCD1-oe* 0 d and the variety that showed the lowest load value in the direction of PC2 was *lcd1* 5 d. Therefore, it could be concluded that a decrease in the endogenous H_2S content in the *lcd1* mutant caused significant changes during dark-induced senescence compared with other groups.

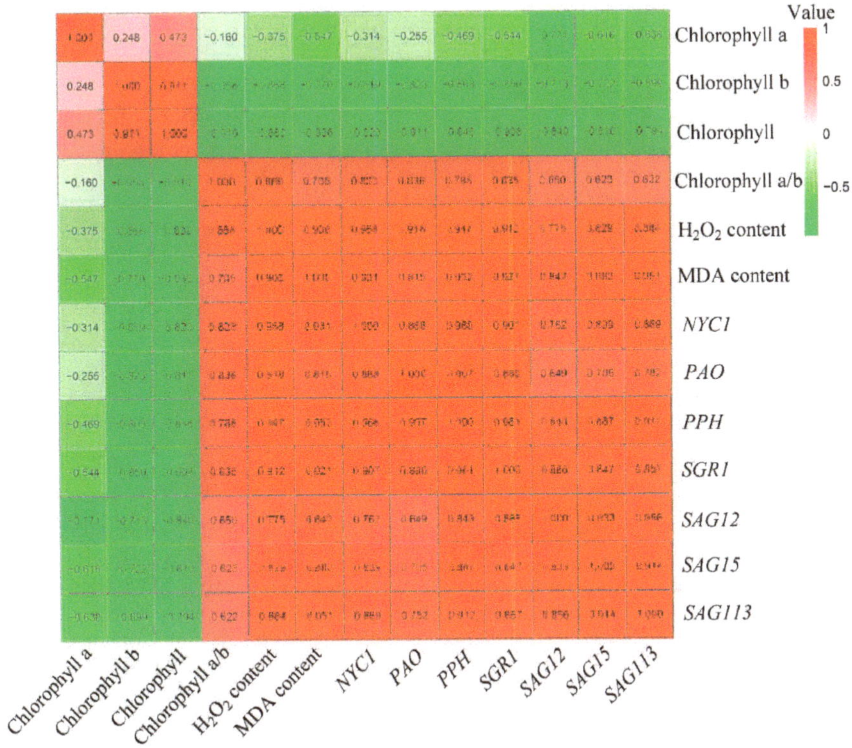

Figure 7. Correlation analysis among the parameters of chlorophyll, chlorophyll a, chlorophyll b, chlorophyll a/b, H_2O_2, malondialdehyde (MDA), and gene expressions of *NYC1*, *PAO*, *PPH*, *SGR1*, *SAG12*, *SAG15*, and *SAG113* in detached leaves of 6-week-old wild-type, *lcd1*, and *LCD1-oe* tomatoes stored in darkness for 0, 2, 5, and 8 days.

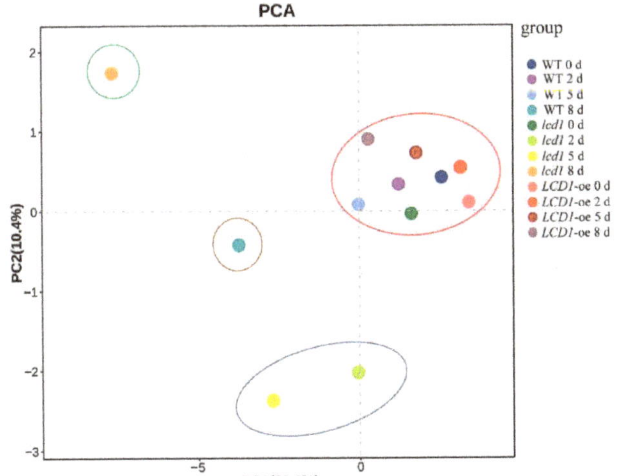

Figure 8. Principal component analysis based on the parameters of chlorophyll, chlorophyll a, chlorophyll b, chlorophyll a/b, H_2O_2, malondialdehyde (MDA), and gene expressions of *NYC1*, *PAO*, *PPH*, *SGR1*, *SAG12*, *SAG15*, and *SAG113* in detached leaves of 6-week-old wild-type, *lcd1*, and *LCD1-oe* tomatoes stored in darkness for 0, 2, 5, and 8 days.

3. Discussion

H_2S, a multifunctional signaling molecule in plants, was found to alleviate the postharvest senescence of broccoli, banana, grape, and tomato [19–22,27]. Furthermore, H_2S is implicated in suppressing the chlorophyll degradation of detached leaves of *Arabidopisis* by regulating a dark-dependent reaction [13]. DES1, an O-acetylserine(thiol)lyase homolog with L-cysteine desulfhydrase activity, regulates cysteine homeostasis in *Arabidopsis* [24]. Recently it was reported that a *des1* mutant displayed accelerated leaf senescence, while the leaves of *OE-DES1* contained adequate chlorophyll levels, suggesting the role of DES1 in regulating leaf senescence [25]. In our previous work, mutation of an L-Cys desulfhydrase named LCD1 caused accelerated fruit ripening compared with the wild type [26]. In the present study, to elucidate the possible involvement of LCD1 in regulating leaf senescence, two previously reported tomato lines, *lcd1-7* and *lcd1-9*, with mutations near the PAM sequence were used as *lcd1* mutants, and two lines (hereafter called *LCD1*-oe and *LCD1*-oe1) with an increased expression of *LCD1* under the control of the CaMV 35S promoter were also used. To confirm the role of *LCD1* in catalyzing H_2S production, the H_2S producing rates were determined in leaves of the *lcd1* and *LCD1*-oe lines; *lcd1* leaves showed a lower H_2S producing rate compared with the wild type, while *LCD1* overexpression induced a significantly higher level of the H_2S producing rate, suggesting the efficiency of LCD1 in producing H_2S.

Leaf senescence is a highly programmed degeneration process during the final stage of leaf development. To study the role of H_2S-producing enzyme LCD1 in regulating natural leaf senescence, we compared 10-week old tomatoes of *lcd1* mutants, and the two lines of *LCD1-oe* with the enhanced expression of *LCD1*, and found that *lcd1* developed more senescence syndrome while *LCD1* overexpression maintained more functional leaves.

Prolonged darkness is often used to initiate rapid and synchronous senescence in detached leaves [13]. The roles of LCD1 on dark-induced senescence were evaluated in tomato leaves. The *lcd1* mutant leaves showed an obvious syndrome of the leaf yellowing phenotype after 5 and 8 days in dark stress, whereas *LCD1* overexpression still maintained the green phenotype (Figure 2A). All this evidence suggests the role of H_2S in alleviating the dark-induced senescence of detached leaves. In accordance with the phenotype of accelerated senescence in the *lcd1* mutant and delayed senescence in *LCD1-oe* leaves, chlorophyll decreased significantly in *lcd1*, but the decrease was attenuated in *LCD1-oe* leaves. Interestingly, we found that chlorophyll b may contribute more to the decrease in total chlorophyll compared with chlorophyll a (Figure 3). By analyzing the ratio of chlorophyll a/b during leaf storage, it was found that the ratio in *LCD1* deletion increased significantly, implying that more chlorophyll b was decomposed to chlorophyll a in the *lcd1* mutant. Moreover, the ratio in *LCD1* overexpression almost remained unchanged, suggesting the significant impact of H_2S content on the ratio of chlorophyll a/b. Then, the expression levels of key genes *NYC1, PAO, PPH,* and *SGR1* in the chlorophyll degradation pathway were analyzed by RT-qPCR. It was found that *NYC1, PAO, PPH,* and *SGR1* transcript abundance increased during darkness in all groups, especially in *lcd1* mutant leaves, whereas this response was significantly inhibited by *LCD1* overexpression (Figure 5). Senescence-associated gene (*SAG*) *12, 15,* and *113* are widely used as molecular markers for leaf senescence [28] and their transcriptions were also analyzed in detached tomato leaves. In agreement with the phenotype, significant increases in *SAGs* expression were observed in all groups, especially in *lcd1* mutant leaves, and the increase was greatly attenuated by *LCD1* overexpression (Figure 6). The results suggest that LCD1 may delay the chlorophyll degradation by down-regulating the transcription of key genes in the chlorophyll degradation pathway and *SAGs*.

Leaf senescence is often associated with the pronounced accumulation of ROS [13]. Recently, it was reported that transcription factor NAC075 delays leaf senescence by deterring ROS accumulation through directly activating the expression of the antioxidant enzyme gene *catalase 2* (*CAT*) in *Arabidopsis* [14]. To unveil the relations between H_2S and ROS metabolism in leaf senescence, H_2O_2 and MDA contents were determined during leaf senescence in darkness. During the dark-induced senescence, the H_2O_2 and MDA content in each group showed an increasing trend, but the overexpression of *LCD1* could reduce the accumulation of ROS and MDA in leaves under dark-triggered senescence. Furthermore, the correlation analysis indicated that the ROS and MDA content showed a higher negative correlation to total chlorophyll and chlorophyll b in comparison to chlorophyll a. The contents of H_2O_2 and MDA were positively correlated with the expression levels of *NYC1*, *PAO*, *PPH*, *SGR1*, *SAG12*, *SAG15*, and *SAG113*. The positive correlation between H_2O_2 and MDA content and senescence-related genes indicates that they may act synergistically to accelerate the senescence process of leaves, whereas *LCD1* overexpression delayed leaf senescence by inhibiting ROS accumulation and senescence-related gene expressions. Interestingly, increasing H_2S production was observed in natural senescence leaves (Figure S2). We hypothesize that endogenous H_2S production was activated to counteract the effect of increasing ROS in senescence leaves. In our previous reports, exogenous H_2S fumigation delayed the postharvest senescence of broccoli in a dose-dependent manner and H_2S maintained higher levels of chlorophyll, carotenoids, anthocyanin, and ascorbate, suggesting the role of H_2S in delaying the postharvest senescence of broccoli [21]. Moreover, H_2S treatment effectively alleviates ethylene-induced banana peel yellowing and fruit softening [22]. The above results suggest that H_2S is an effective signal in delaying the postharvest senescence of fruits and vegetables. In the present research, an increasing trend of H_2S production was observed during leaf senescence, suggesting that H_2S generation may be activated in response to leaf senescence as ROS metabolites (H_2O_2 and MDA) accumulate during dark-induced senescence. Leaf senescence, once initiated, cannot be stopped. Though more H_2S is produced during leaf senescence, leaves still undergo senescence during storage in darkness. Moreover, compared with the early senescence phenotype of the *lcd1* mutant, *LCD1* overexpression induced more H_2S production and showed a delayed leaf senescence, clearly suggesting the role of H_2S in delaying leaf senescence. In all, the data suggest that senescent leaves produced more H_2S, but reduction in H_2S production of the *lcd1* mutant caused early senescence in both natural and dark-induced senescence. Moreover, the principal component analysis (PCA) in Figure 8 shows that *lcd1* 2 d and *lcd1* 5 d clustered together, and *lcd1* 8 d was distributed separately from other groups, suggesting that the decreased endogenous H_2S content in *lcd1* caused significant changes during dark-induced senescence compared with other groups.

4. Conclusions

In the present study, the role of a cysteine desulfhydrase LCD1 in regulating leaf senescence in tomato was explored. The *LCD1* mutation caused an earlier leaf senescence, whereas *LCD1* overexpression significantly delayed leaf senescence compared with the wild type in 10-week tomato seedlings. Furthermore, LCD1 was found to play a negative role in dark-induced senescence in detached tomato leaves. Further investigations showed that LCD1 delayed dark-triggered chlorophyll degradation and ROS accumulation in detached tomato leaves, and the increase in chlorophyll degradation and *SAGs* related gene expression was attenuated by *LCD1* overexpression. Moreover, a correlation analysis indicated that chlorophyll content was negatively correlated with H_2O_2 and MDA content, and also negatively correlated with the expression of chlorophyll degradation-related genes *NYC1*, *PAO*, *PPH*, and *SGR1* and senescence-related genes *SAG12*, *SAG15*, and *SAG113*. Therefore, these findings increase our understanding of the physiological functions of the H_2S-generating enzyme LCD1 in regulating leaf senescence in tomato.

5. Materials and Methods

5.1. Plant Material and Growth Conditions

Solanum lycopersicum cv. "Micro-Tom" was used as the control in this study. The mutants *lcd1–7*, which contained a T residue inserted near the PAM sequence, and *lcd1–9*, which had a deletion of G near the PAM as previously reported were used as *lcd1* mutants [26]. The coding sequence of tomato cysteine desulfhydrylase LCD1 (LOC101258894) was obtained from NCBI (http://www.ncbi.nlm.nih.gov/, accessed on 11 September 2018) and the primers including the restriction enzyme sites (LCD1-F: CGCGGATCCTAATCCTAAATGGAACCGGC; LCD1-R: CCGCTCGAGTTCTGAGTGAAGCATCTTAC, the underlines stand for BamHI and XhoI sites, respectively) were used to amplified the coding sequence of *LCD1*. Then, the coding sequence of *LCD1* was inserted into the pBI121 vector at the sites of BamHI and XhoI and transformed into wild-type tomato by *Agrobacterium tumefaciens*, which contained the recombinant *LCD1*-pBI121. The efficiency of *LCD1* over-expression was verified by RT-qPCR. The seeds of tomatoes were grown in a nutrient soil:vermiculite (3:1, v/v) in growth pots 10 cm in diameter in an environment-controlled growth room (23 ± 2 °C; 50–70% relative humidity, RH) under 16 h light/8 h dark and 250 mol/m^2/s light.

5.2. Determination of H_2S Producing Rate and H_2S Detection in Tomato Leaves

The H_2S producing rate was measured as described previously [29]. Tomato leaves at 1 g were ground to a fine powder in liquid nitrogen and homogenized in 9 mL of 20 mM Tris-HCl, pH 8.0. After centrifugation at 12,000× *g* for 15 min, the protein content of the supernatant was sampled and the H_2S producing rate was detected by monitoring the release of H_2S from L-cysteine in the presence of dithiothreitol (DTT). The assay was performed in a total volume of 1 mL comprising 0.8 mM L-cysteine, 2.5 mM DTT, 100 mM Tris-HCl, pH 8.0, and 100 µL of protein solution. The reaction was incubated for 15 min at 37 °C, and terminated by adding 100 µL of 30 mM $FeCl_3$ dissolved in 1.2 N HCl and 100 µL of 20 mM N,N-dimethyl-phenylenediamine dihydrochloride dissolved in 7.2 N HCl. The formation of methylene blue was determined at 670 nm.

The end-point detection of H_2S production from tomato leaves by lead acetate strips (cat. number WHA2602501A, Sigma, Darmstadt, Germany) were performed as previously described [30]. One gram of fresh tomato leaves was ground to a fine powder in liquid nitrogen and then homogenized in 10 mL of Phosphate Buffered Saline (pH 6.8) supplemented with 10 mM L-cysteine and 10 µM pyridoxal-5′-phosphate (PLP), and then the mixture was placed in petri dishes. The H_2S detection strips were attached to the inner part of the upper lid of the petri dishes and incubated at 37 °C for 2–5 h until lead sulfide darkening of the strip occurred.

5.3. Dark Treatment of Tomato Leaves

For dark-induced leaf senescence experiments, detached mature leaves from 6-week-old wild-type, *lcd1*, and *LCD1-oe* transgenic plants were placed on filter papers which were moistened by 2 mL of sterile water in petri dishes with a 9 cm diameter and the adaxial side facing upwards. The petri dishes which contained 5–6 detached leaves were kept in darkness at 23 °C for 8 days. The leaves were sampled and rapidly frozen in liquid nitrogen and stored at −80 °C until analysis.

5.4. Determination of Chlorophyll Content

Tomato leaves at 2 ± 0.01 g were homogenized in liquid nitrogen and subsequently extracted ethanol and 80% (v/v) acetone solution in a ratio of 1:1 (v/v) according to the method in [31]. The absorbance of the supernatant was read at 663 and 645 nm. The experiments were repeated three biological times, and the chlorophyll levels were expressed as mg/g fresh weight (FW).

5.5. Determination of H_2O_2 and Malondialdehyde (MDA) Content in Tomato Leaves

The contents of H_2O_2 and malondialdehyde (MDA) were assayed as described by Ge et al. [22] and Hu et al. [27]. For the determination of H_2O_2 content, 2.0 ± 0.01 g of tomato leaves were homogenized in 3 mL of precooled acetone, and centrifuged at 12,000× *g* for 30 min. The content of H_2O_2 was measured by determining the absorbance value at 508 nm. For the determination of MDA content, 2.0 ± 0.01 g of tomato leaf powder was homogenized with 5% trichloroacetic acid, and the supernatant was obtained by centrifugation at 12,000× *g* for 30 min. The absorbance values were measured at 600 nm, 532 nm, and 450 nm, respectively. The experiments were repeated three times, and the contents of H_2O_2 and MDA were expressed as μmol/g fresh weight (FW).

5.6. Gene Expression Analysis

Total RNA from 0.1 g of tomato leaves was extracted using an Extraction Kit (Tiangen, Beijing, China) and cDNA was synthesized by a reverse transcription kit (PrimeScript RT Master Mix; Takara, Kyoto, Japan). Then the cDNA products were used for gene expression analysis by quantitative polymerase chain reaction (qPCR) performed using a Bio-Rad IQ5 (Hercules, CA, USA). The specific primers used for qPCR were designed based on the coding sequence of the genes as shown in the SGN database (https://solgenomics.net/, accessed on 12 April 2021). *Tubulin* gene expression in control tomato plants was used for the normalization of the data. The experiments were repeated in three technical replicates.

5.7. Data Analysis

The statistical analysis of data was based on Student's *t*-tests. Significant differences were evaluated using multiple pair wise *t*-test comparisons at $p < 0.05$. The correlation among the contents of chlorophyll, H_2O_2, MDA, and the expression of chlorophyll degradation related genes and *SAGs* in tomato leaves and the principal component analysis (PCA) of the data above were analyzed by the tools on the OmicShare platform (https://www.omicshare.com/, accessed on 20 November 2021).

Supplementary Materials: The following are available online at https://www.mdpi.com/article/10.3390/ijms222313078/s1.

Author Contributions: Conceptualization, K.H., X.P. and H.Z.; data curation, K.H.; formal analysis, K.H., X.P., F.Y. and X.C.; funding acquisition, K.H., G.Y., Z.Z. and H.Z.; investigation, K.H., X.P. and G.Y.; methodology, X.P., Z.Z., F.Y. and W.L.; project administration, H.Z.; software, K.H. and X.P.; supervision, K.H. and H.Z.; validation, K.H., G.Y., F.Y. and W.L.; visualization, K.H.; writing—original draft, K.H. and X.P.; writing—review and editing, K.H., X.P., Z.Z., Y.Z., Y.L., Z.H., X.C. and H.Z. All authors have read and agreed to the published version of the manuscript.

Funding: This research was supported by the National Natural Science Foundation of China (32170315, 31970200, 31970312, 31901993, 31670278), the Fundamental Research Funds for the Central Universities (JZ2021HGPA0063), the National Key R&D Program of China (2019YFD1000700, 2019YFD1000701), the National Key R&D Program of China (2019YFD1001300, 2019YFD1001303), and the Natural Science Foundations of Anhui Province (1908085MC72).

Data Availability Statement: Data are contained within the article or Supplementary Material.

Conflicts of Interest: The authors declare no conflict of interest.

References

1. Guo, Y.; Ren, G.; Zhang, K.; Li, Z.; Miao, Y.; Guo, H. Leaf senescence: Progression, regulation, and application. *Mol. Horti.* **2021**, *1*, 5. [CrossRef]
2. Lim, P.O.; Kim, H.J.; Nam, H.G. Leaf senescence. *Annu. Rev. Plant Biol.* **2007**, *58*, 115–136. [CrossRef]
3. Kusaba, M.; Ito, H.; Morita, R.; Iida, S.; Sato, Y.; Fujimoto, M.; Kawasaki, S.; Tanaka, R.; Hirochika, H.; Nishimura, M.; et al. Rice NONYELLOW COLORING1 is involved in light-harvesting complex II and grana degradation during leaf senescence. *Plant Cell* **2007**, *19*, 1362–1375. [CrossRef] [PubMed]
4. Chen, M. Chlorophyll modifications and their spectral extension in oxygenic photosynthesis. *Annu. Rev. Biochem.* **2014**, *83*, 317–340. [CrossRef] [PubMed]
5. Sato, Y.; Morita, R.; Katsuma, S.; Nishimura, M.; Tanaka, A.; Kusaba, M. Two short-chain dehydrogenase/reductases, NON-YELLOW COLORING 1 and NYC1-LIKE, are required for chlorophyll b and light-harvesting complex II degradation during senescence in rice. *Plant J.* **2009**, *57*, 120–131. [CrossRef]
6. Shimoda, Y.; Ito, H.; Tanaka, A. Arabidopsis STAY-GREEN, Mendel's green cotyledon gene, encodes magnesium-dechelatase. *Plant Cell* **2016**, *28*, 2147–2160. [CrossRef]
7. Pruzinska, A.; Tanner, G.; Anders, I.; Roca, M.; Hortensteiner, S. Chlorophyll breakdown: Pheophorbide a oxygenase is a Rieske-type iron-sulfur protein, encoded by the accelerated cell death 1 gene. *Proc. Natl. Acad. Sci. USA* **2003**, *100*, 15259–15264. [CrossRef]

8. Schippers, J.H. Transcriptional networks in leaf senescence. *Curr. Opin. Plant Biol.* **2015**, *27*, 77–83. [CrossRef]
9. Guo, Y.; Gan, S. Leaf senescence: Signals, execution, and regulation. *Curr. Top. Dev. Biol.* **2005**, *71*, 83–112.
10. Gan, S.; Amasino, R.M. Making sense of senescence (molecular genetic regulation and manipulation of leaf senescence). *Plant Physiol.* **1997**, *113*, 313–319. [CrossRef]
11. Li, Z.; Peng, J.; Wen, X.; Guo, H. Ethylene-insensitive3 is a senescence-associated gene that accelerates age-dependent leaf senescence by directly repressing miR164 transcription in Arabidopsis. *Plant Cell* **2013**, *25*, 3311–3328. [CrossRef] [PubMed]
12. Zhang, K.; Halitschke, R.; Yin, C.; Liu, C.J.; Gan, S.S. Salicylic acid 3-hydroxylase regulates Arabidopsis leaf longevity by mediating salicylic acid catabolism. *Proc. Natl. Acad. Sci. USA* **2013**, *110*, 14807–14812. [CrossRef]
13. Wei, B.; Zhang, W.; Chao, J.; Zhang, T.; Zhao, T.; Noctor, G.; Liu, Y.; Han, Y. Functional analysis of the role of hydrogen sulfide in the regulation of dark-induced leaf senescence in Arabidopsis. *Sci. Rep.* **2017**, *7*, 2615. [CrossRef]
14. Kan, C.; Zhang, Y.; Wang, H.L.; Shen, Y.; Xia, X.; Guo, H.; Li, Z. Transcription factor NAC075 delays leaf senescence by deterring reactive oxygen species accumulation in Arabidopsis. *Front. Plant Sci.* **2021**, *12*, 634040. [CrossRef]
15. Wang, R. Physiological implications of hydrogen sulfide: A whiff exploration that blossomed. *Physiol. Rev.* **2012**, *92*, 791–896. [CrossRef]
16. Zhang, H.; Hu, L.Y.; Hu, K.D.; He, Y.D.; Wang, S.H.; Luo, J.P. Hydrogen sulfide promotes wheat seed germination and alleviates oxidative damage against copper stress. *J. Integr. Plant Biol.* **2008**, *50*, 1518–1529. [CrossRef] [PubMed]
17. Zhang, H.; Tang, J.; Liu, X.P.; Wang, Y.; Yu, W.; Peng, W.-Y.; Fang, F.; Ma, D.-F.; Wei, Z.-J.; Hu, L.Y. Hydrogen sulfide promotes root organogenesis in *Ipomoea batatas*, *Salix matsudana* and *Glycine max*. *J. Integr. Plant Biol.* **2009**, *51*, 1086–1094. [CrossRef]
18. Jin, Z.P.; Wang, Z.Q.; Ma, Q.X.; Sun, L.M.; Zhang, L.P.; Liu, Z.Q.; Liu, D.; Hao, X.; Pei, Y. Hydrogen sulfide mediates ion fluxes inducing stomatal closure in response to drought stress in *Arabidopsis thaliana*. *Plant Soil* **2017**, *419*, 141–152. [CrossRef]
19. Hu, K.D.; Zhang, X.Y.; Wang, S.S.; Tang, J.; Yang, F.; Huang, Z.Q.; Deng, J.Y.; Liu, S.Y.; Zhao, S.J.; Hu, L.Y.; et al. Hydrogen sulfide inhibits fruit softening by regulating ethylene synthesis and signaling pathway in tomato (*Solanum lycopersicum*). *HortScience* **2019**, *54*, 1824–1830. [CrossRef]
20. Yao, G.F.; Wei, Z.Z.; Li, T.T.; Tang, J.; Huang, Z.Q.; Yang, F.; Li, Y.H.; Han, Z.; Hu, F.; Hu, L.Y.; et al. Modulation of enhanced antioxidant activity by hydrogen sulfide antagonization of ethylene in tomato fruit ripening. *J. Agric. Food Chem.* **2018**, *66*, 10380–10387. [CrossRef]
21. Ni, Z.J.; Hu, K.D.; Song, C.B.; Ma, R.-H.; Li, Z.-R.; Zheng, J.-L.; Fu, L.-H.; Wei, Z.-J.; Zhang, H. Hydrogen sulfide alleviates postharvest senescence of grape by modulating the antioxidant defenses. *Oxid. Med. Cell. Longev.* **2016**, *2016*, 4715651. [CrossRef]
22. Ge, Y.; Hu, K.D.; Wang, S.S.; Hu, L.Y.; Chen, X.Y.; Li, Y.H.; Yang, Y.; Yang, F.; Zhang, H. Hydrogen sulfide alleviates postharvest ripening and senescence of banana by antagonizing the effect of ethylene. *PLoS ONE* **2017**, *12*, e0180113. [CrossRef]
23. Jin, Z.P.; Pei, Y.X. Physiological implications of hydrogen sulfide in plants: Pleasant exploration behind its unpleasant odour. *Oxid. Med. Cell Longev.* **2015**, *2015*, 397502. [CrossRef]
24. Álvarez, C.; Calo, L.; Romero, L.C.; Garcıá, I.; Gotor, C. An O-acetylserine(thiol)lyase homolog with L-cysteine desulfhydrase activity regulates cysteine homeostasis in Arabidopsis. *Plant Physiol.* **2010**, *152*, 656–669. [CrossRef] [PubMed]
25. Jin, Z.; Sun, L.; Yang, G.; Pei, Y. Hydrogen sulfide regulates energy production to delay leaf senescence induced by drought stress in Arabidopsis. *Front. Plant Sci.* **2018**, *9*, 1722. [CrossRef]
26. Hu, K.D.; Zhang, X.Y.; Yao, G.F.; Rong, Y.L.; Ding, C.; Tang, J.; Yang, F.; Huang, Z.Q.; Xu, Z.M.; Chen, X.Y.; et al. A nuclear-localized cysteine desulfhydrase plays a role in fruit ripening in tomato. *Hortic. Res.* **2020**, *7*, 211. [CrossRef] [PubMed]
27. Li, S.P.; Hu, K.D.; Hu, L.Y.; Li, Y.H.; Jiang, A.M.; Xiao, F.; Han, Y.; Liu, Y.S.; Zhang, H. Hydrogen sulfide alleviates postharvest senescence of broccoli by modulating antioxidant defense and senescence-related gene expression. *J. Agric. Food Chem.* **2014**, *62*, 1119–1129. [CrossRef]
28. James, M.; Masclaux-Daubresse, C.; Marmagne, A.; Azzopardi, M.; Laîné, P.; Goux, D.; Etienne, P.; Trouverie, J. A new role for SAG12 cysteine protease in roots of *Arabidopsis thaliana*. *Front. Plant Sci.* **2019**, *9*, 1998. [CrossRef]
29. Riemenschneider, A.; Wegele, R.; Schmidt, A.; Papenbrock, J. Isolation and characterization of a D-cysteine desulfhydrase protein from *Arabidopsis thaliana*. *FEBS J.* **2005**, *272*, 1291–1304. [CrossRef] [PubMed]
30. Hine, C.; Harputlugil, E.; Zhang, Y.; Ruckenstuhl, C.; Lee, B.C.; Brace, L.; Longchamp, A.; Treviño-Villarreal, J.H.; Mejia, P.; Ozaki, C.K.; et al. Endogenous hydrogen sulfide production is essential for dietary restriction benefits. *Cell* **2015**, *160*, 132–144. [CrossRef]
31. Lichtenthaler, H.K.; Wellburn, A.R. Determinations of total carotenoids and chlorophylls a and b of leaf extracts in different solvents. *Biochem. Soc. Trans.* **1983**, *11*, 591–592. [CrossRef]

International Journal of
Molecular Sciences

MDPI

Review

The Role of Hydrogen Sulfide in Plant Roots during Development and in Response to Abiotic Stress

Hua Li [1,2,*], Hongyu Chen [1], Lulu Chen [1] and Chenyang Wang [3,4,*]

[1] College of Life Science, Henan Agricultural University, Zhengzhou 450002, China; CHY19970927@163.com (H.C.); Nancy061798@163.com (L.C.)
[2] State Key Laboratory of Crop Biology, Shandong Agricultural University, Taian 271018, China
[3] College of Agronomy, Henan Agricultural University, Zhengzhou 450002, China
[4] State Key Laboratory of Wheat and Maize Crop Science, Henan Agricultural University, Zhengzhou 450002, China
* Correspondence: lihua@henau.edu.cn (H.L.); xmzxwang@henau.edu.cn (C.W.)

Abstract: Hydrogen sulfide (H_2S) is regarded as a "New Warrior" for managing plant stress. It also plays an important role in plant growth and development. The regulation of root system architecture (RSA) by H_2S has been widely recognized. Plants are dependent on the RSA to meet their water and nutritional requirements. They are also partially dependent on the RSA for adapting to environment change. Therefore, a good understanding of how H_2S affects the RSA could lead to improvements in both crop function and resistance to environmental change. In this review, we summarized the regulating effects of H_2S on the RSA in terms of primary root growth, lateral and adventitious root formation, root hair development, and the formation of nodules. We also discussed the genes involved in the regulation of the RSA by H_2S, and the relationships with other signal pathways. In addition, we discussed how H_2S regulates root growth in response to abiotic stress. This review could provide a comprehensive understanding of the role of H_2S in roots during development and under abiotic stress.

Keywords: hydrogen sulfide; root growth; nitric oxide; auxin; heavy metal; salt

Citation: Li, H.; Chen, H.; Chen, L.; Wang, C. The Role of Hydrogen Sulfide in Plant Roots during Development and in Response to Abiotic Stress. *Int. J. Mol. Sci.* **2022**, 23, 1024. https://doi.org/10.3390/ijms23031024

Academic Editors: Yanjie Xie, Francisco J. Corpas and Jisheng Li

Received: 27 December 2021
Accepted: 17 January 2022
Published: 18 January 2022

1. Introduction

The root system is an important vegetative organ of plants. In terrestrial environments, the root system provides structural support, uptakes water and nutrition from soil, and is where some amino acids, endogenous hormones, and other substances are synthesized. The growth and development of the root system largely determines the water and nutrient absorption efficiency of plants. With the improvement of genomics, genetics, molecular biology, and other research methods, as well as the generation of a mutant library related to root development, more and more functional genes and regulatory genes that affect root growth have been identified [1,2]. In addition to being regulated by internal genes, the physical environment can also have a regulatory effect on the growth and development of plant roots, such as the soil temperature, moisture, nutrients, and pH [3–7]. Plant signal molecules act as a bridge between the physical environment and root-growth-related genes, and hence determine how plants respond to environmental stress and changes. The plant signal molecules regulate the expression of root-growth-related genes through the transmission and transduction of environmental signals, which can lead to changes in the root system architecture (RSA) (including primary growth, the formation of lateral or adventitious roots, and the distribution and length of root hairs). These signal molecules include plant hormones, nitric oxide (NO), carbon monoxide (CO), reactive oxygen species (ROS), and hydrogen sulfide (H_2S). The importance of plant hormones, NO, and ROS to root growth and development has been reviewed in many articles [8–14]. Hydrogen sulfide plays an important role in plant growth and development, and responds to various environmental stresses. In the past ten years, more than 2000 literatures have reported and discussed the impact of H_2S on plant physiology. The importance of H_2S for regulating plant responses to abiotic stresses (such as drought, salt, heat, and heavy metals) [15–17] and the effects on stomatal movement, seed germination, leaf senescence, and fruit ripening [18,19] have been extensively studied and discussed. In this review, we focused on the role of H_2S in root growth and development. The effects of H_2S on primary root growth, lateral and adventitious root formation, root hair development, and nodules were summarized here. As H_2S can alleviate the negative impacts of abiotic stress on plant root growth, we also reviewed how H_2S regulates root growth in response to abiotic stress.

2. The Role of Hydrogen Sulfide during Root Development

The literature on the regulation of H_2S on root growth and development were shown in Table 1. According to these studies, we summarized a model for the regulatory mechanism of H_2S on root growth and development (Figure 1).

Table 1. The role of hydrogen sulfide in root during development and its interaction with other signals.

Plant Species	Signal Involved	Root Response	Reference
Ipomoea batatas, Salix matsudana, Glycine max	Auxin and NO	Adventitious root formation	[20]
Cucumis sativus	HO-1/CO	Adventitious root formation	[21]
Cucumis sativus	Methane	Adventitious root development	[22]
Solanum lycopersicum	Auxin	Lateral root formation	[23]
Capsicum annuum	Cinnamaldehyde	Lateral root formation	[24]
Solanum lycopersicum	H_2O_2	Lateral root formation	[25]
Solanum lycopersicum	Methane	Lateral root formation	[26]
Kandelia obovata	Brassinosteroid, carbohydrate metabolism, cellular redox homeostasis, protein metabolism, secondary metabolism, and amino acid metabolism	Lateral root development	[27]
Arabidopsis	ROS, NO, MPK6	Primary root growth	[28]
Arabidopsis	Actin dynamics	Root hair growth	[29]
Glycine max	Nitrogen-fixation ability	Nodulation	[30]
Glycine max	Nitrogen-fixation ability	Nodulation	[31]
Arabidopsis	Actin-dependent auxin transport	Root development and growth	[32]
Prunus persica	Auxin biosynthesis, transport, and signal transduction.	Root development and growth	[33]
Fragaria × ananassa	H_2O_2 and soluble sugar accumulation	Root development and growth	[34]

2.1. *Hydrogen Sulfide Regulates the Formation and Growth of Lateral and Adventitious Roots*

Hydrogen sulfide has a concentration-dependent effect on the regulation of root growth. In tomato, 0.01–1.0 mM sodium hydrosulfide (NaHS) (a H_2S donor) can significantly promote the initiation and length of lateral roots (LRs) and can increase the number and density of LRs. However, a high concentration of NaHS (10 mM) inhibits the LRs' growth [23]. A similar phenomenon was found in the mangrove plant *Kandelia obovata*, where a concentration of 0.01–1.0 mM NaHS led to a notable increase in the length and total surface area of LRs [27]. H_2S was essential for the formation of pepper LRs, where a concentration of 0.5–8.0 mM NaHS significantly increased the number of LRs. In contrast, different concentrations of the H_2S scavenger hypotaurine (HT) markedly inhibited the formation of LRs [24]. In peach, H_2S had a notable effect on the formation of LRs, with a concentration of 0.2 mM NaHS leading to a significant increase in the number of LRs [33]. Our previous research also found that H_2S promoted the growth and development of lateral roots in wheat, with a concentration of 0.4 mM NaHS resulting in an increase in the number, density, and length of LRs [35]. However, in *Arabidopsis*, the effects of H_2S on lateral roots were slightly different. H_2S can promote the occurrence of LRs, but inhibits the LRs' length [32], which may be related to the concentration of NaHS used in the treatment.

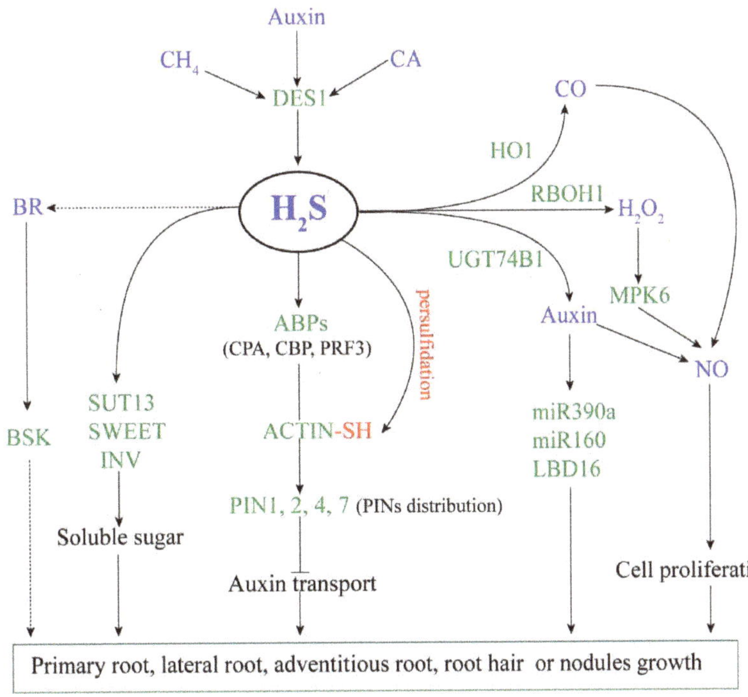

Figure 1. The proposed model of H_2S regulating plant root growth. Arrow and bar ends indicate activation and inhibitory effects, respectively. Green fonts represent genes or proteins, blue fonts represent signal molecules. H_2S: hydrogen sulfide; CH_4: methane; CA: cinnamaldehyde; BR: brassinosteroid; CO: carbon monoxide; NO: nitric oxide; DES1: L-cysteine desulfhydrase 1; HO1: haem oxygenase-1; RBOH1: respiratory burst oxidase 1; UGT74B1: UDP-glycosyltransferase 74B1; MPK6: mitogen-activated protein kinase 6; ABPs: actin-binding proteins; PINs: pin-formed family; SUT13: sugar transport protein 13; SWEET: bidirectional sugar transporter; INV: invertase; BSK: BR-signaling kinase; LBD16: LOB domain-containing protein 16.

The effect of H_2S on the formation and growth of adventitious roots is the same as that for lateral roots. The application of the appropriate concentration of H_2S promoted the number and length of adventitious roots in sweet potato [20]. In addition, the same result was obtained for excised willow, soybeans [20], and cucumber [21,22].

2.2. *Hydrogen Sulfide Inhibits the Growth of Primary Roots and Root Hair*

Several studies have shown that H_2S has a toxic effect on the growth of primary roots. The concentration of exogenous NaHS used in treatments was negatively correlated with the rate of growth (length) of primary roots [28,29,32]. This inhibitory effect of H_2S on primary roots may be related to the reduction in the meristem cell division potential, as the length of the root meristematic zones were reduced when treated with NaHS [28]. Although H_2S inhibited the length of root meristematic zones, Li et al. [29] found that the length from the root apex to a root hair for the seedlings was promoted by H_2S. This may be due to the inhibitory effect of H_2S on root hairs. H_2S is known to inhibit the initiation of root hair; that is, the starting site of the root hair zones may be further away from the root apex [29], resulting in a longer distance from the root hair to root apex, even when H_2S

inhibits the meristem zones. In addition, H_2S not only inhibited the initiation of root hair growth, but also significantly decreased the root hair length and density [29].

2.3. *Hydrogen Sulfide Promotes the Formation of Root Nodules and Nitrogen Fixation*

Root nodules are a special organ formed by symbiosis between leguminous plants (Fabaceae) and rhizobia. The formation and growth of the nodules are strictly controlled by plant hormones [36]. As a recognized signal molecule that interacts with plant hormones to regulate plant growth and development, H_2S is known to influence the growth of root nodules [30,31]. Endogenous H_2S production in both young soybean nodules (14 days post-inoculation [DPI] with the *Sinorhizobium fredii Q8* strain) and mature nodules (28 DPI) can be detected by fluorescent probes SF7-A, whereas no significant fluorescence was observed in the nascent soybean nodules (7 DPI). This suggested that H_2S may mediate the growth of root nodules [31]. Indeed, the application of NaHS significantly increased the number of soybean nodules and enhanced nitrogenase (Nase) activity after 7 DPI and 24 DPI, respectively. In addition, H_2S was found to affect rhizobial infection, where a greater abundance of developing infection threads and cortex infection threads was found in NaHS-treated soybean roots than those in untreated controls at 5 DPI and 7 DPI, respectively [30]. On the contrary, an endogenous H_2S production deficit rhizobia mutant ΔCSE (cystathionine γ-lyase) symbiosis with soybean roots significantly reduced the nitrogenase activity and H_2S content in nodule cells. Moreover, higher contents of H_2O_2 (hydrogen peroxide), MDA (malondialdehyde), and protein carbonyl were observed in ΔCSE root nodules; that is, the H_2S-induced nitrogen-fixation ability of root nodules may be related to its regulation of the antioxidant system that protects nodule cells from oxidative damage [31]. These studies suggested that H_2S might have a positive effect on the soybean–rhizobium symbiosis system and may enhance nitrogen fixation.

2.4. *Hydrogen Sulfide Interacts with Other Signaling Molecules to Regulate Root Development*

2.4.1. Auxin

The inhibition of primary root growth by H_2S, and the promotion of lateral and adventitious root formation was consistent with the known effects of auxin on root development. It is not difficult to associate H_2S and auxin signaling to RSA. The change in the endogenous IAA (indole acetic acid) content was similar to that reported for H_2S, but with different time-courses in sweet potato explants. The increase in the H_2S content during the formation of sweet potato adventitious roots preceded changes to the IAA content [20]. The research of Wu et al. (2021) [33] on peach roots also obtained similar results: NaHS induced a significant increase in the endogenous H_2S content in roots at 1 DAT (days after treatment), while it increased the concentration of endogenous auxin in roots by 44.50% at 5 DAT. Moreover, it was found that treatment with NaHS significantly increased the production of IAA, and that N-1-naphthylphthalamic acid (an IAA transport inhibitor, NPA) weakened the effect of H_2S on the number of adventitious roots in sweet potato, soybean, and willow [20]. These results showed that IAA may be located downstream of H_2S in order to mediate root development. However, the results in tomato indicated H_2S might partially act as a downstream component of the auxin signaling to trigger lateral root formation [23]. The depletion of auxin down-regulated the transcription of *SlDES1* (L-cysteine desulfhydrase 1, a H_2S synthesis gene), DES activity, and endogenous H_2S contents in tomato roots, and the inhibitory effect of NPA on lateral root formation was offset by NaHS, whereas the inhibition of lateral root formation by HT was not reversed by naphthalene acetic acid (NAA) [23]. In addition, H_2S not only induced auxin synthesis, but also affected the auxin response and transport. After the application of NaHS, the expression of the indicator of the auxin response DR5::GUS (synthetic auxin-responsive promoter::β-glucuronidase) was attenuated in the quiescent center (QC), columella initial cells, and mature columella cells of the root apex, and was concentrated to the QC [32]. The movement of auxin in the root acropetal and basipetal was reduced by an increase in the NaHS concentration, which implied that an increase in H_2S levels reduces the IAA transport capacity. Further research showed that the inhibition of IAA transport by H_2S was related to the polar subcellular localization of PIN proteins (PIN1, PIN2, PIN4, and PIN7) [32].

2.4.2. Reactive Oxygen Species

High concentrations of ROS (reactive oxygen species) often cause oxidative damage to plants, but low concentrations of ROS are necessary for signaling to maintain plant growth and development. The ROS-related regulation of root development has been reported for *Arabidopsis* [37], tomato [38], maize [39,40], and sweet potato [41]. The relationship between ROS and H_2S for the regulation of root growth was also discussed in several studies [25,28,34]. These studies found that ROS signaling might be downstream of H_2S to mediate RSA. For example, H_2S could induce the expression of *RBOH1* (respiratory burst oxidase 1) in tomato roots and could enhance the accumulation of H_2O_2, thereby promoting lateral root formation. These H_2S-related effects on lateral roots were destroyed by DMTU (dimethylthiourea, a H_2O_2 scavenger) and DPI (diphenylene idonium, an inhibitor of NADPH oxidase) [25]. The inhibitory effect of H_2S on primary root growth depended on the ROS pathway, as the relative root growth in *rbohF* and *rbohD/F* was higher than that in WT for the NaHS treatment, which meant that respiratory burst oxidase homolog mutants (*rboh*) were less sensitive to treatment with NaHS [28]. The promoting effect of H_2S on strawberry roots during plug transplant production could also be attributed (in part) to the elevated H_2O_2 [34].

2.4.3. Nitric Oxide and Carbon Monoxide

Nitric oxide (NO), carbon monoxide (CO), and H_2S are the three gas signal molecules in organisms. NO and CO also participate in root growth and development [42–47]. Therefore, the relationship between H_2S and NO or CO has attracted attention in the regulation of RSA. The H_2S-mediated adventitious root formation was alleviated by 2-(4-carboxyphenyl)-4,4,5,5-tetramethylimidazoline-1-oxyl-3-oxide (cPTIO, an NO scavenger) in sweet potato, willow, and soybean [20]. The toxic effect of H_2S on the primary root of *Arabidopsis* was reduced in NO synthase mutants (*nia1/2* and *noa1*), or when treated with cPTIO and NG-nitro-L-Arg-methyl ester (L-NAME, NO synthesis inhibitor) [28]. These results indicated that H_2S acts upstream of NO signal transduction pathways when regulating adventitious root formation and primary root growth. From the results reported by Lin et al. (2012) [21], it is known that haem oxygenase-1/carbon monoxide (HO-1/CO) also acts as a downstream signal system during H_2S-induced adventitious root formation. NaHS up-regulated *HO1* gene expression and promoted HO1 protein accumulation, thereby increasing the number of cucumber adventitious roots. These phenomena were inhibited by ZnPPIX (zinc protoporphyrin IX, an inhibitor of HO-1), whereas the removal of H_2S by HT did not affect the CO-induced adventitious rooting.

2.4.4. Brassinosteroid, Methane, and Cinnamaldehyde

Brassinosteroid (BR) contributes to the maintenance of root meristems, root cell elongation, lateral root development, root hair formation, and rhizosphere symbiosis [48–51]. At present, there is no direct evidence that H_2S interacts with BR to regulate root development, but a recent proteomic analysis in *Kandelia obovata* has shown that H_2S induced the accumulation of the BR-positive regulator protein BSK [27]. An RNA-seq analysis also showed that differentially expressed genes (DEGs) in peach roots, regulated by H_2S, were significantly enriched in the "Brassinosteroid biosynthesis" pathway [33]. These results implied that H_2S-mediated RSA might depend on the BR signal pathway.

Methane (CH_4) plays an important role in the response to abiotic stress (such as heavy metal, salinity, and osmotic stress) [52]. In recent years, the role of CH_4 in the formation of lateral and adventitious roots has been elucidated [22,26,53–56]. Both NO and CO signaling pathways were involved in CH_4-induced adventitious root formation in cucumber [53,54]. Hydrogen peroxide (H_2O_2) signaling is also known to mediate the effects of CH_4 on tomato lateral root formation [56]. As expected, H_2S was confirmed to be located downstream of CH_4 in order to regulate adventitious and lateral root formation in both cucumber and tomato. Methane induced the DES enzyme activity and promoted the production of endogenous H_2S. These methane-related effects on the adventitious roots of cucumber were blocked by HT [22]. The same results were reported for the relationship between CH_4 and H_2S on the formation of lateral roots in tomato [26].

Cinnamaldehyde (CA) is a natural plant essential oil with antibacterial properties. It is widely used as a food additive and in medicines [57]. Recently, CA has also been used as a biological agent for plant disease resistance. For example, CA showed significant antibacterial activity against *Pseudomonas syringae pv. actinidiae*, which causes bacterial canker disease in kiwifruit [58]. Cinnamaldehyde reduced the number of *Meloidogyne incognita* galls and eggs on the roots of soybean plants to approximately 14% and 7%, respectively [59]. In addition, CA was found to play an important role in root development, as it markedly induced the formation of lateral roots in pepper, but without any inhibitory effect on primary root growth. Further study showed that H_2S participated in this regulation process. Cinnamaldehyde increased the DES activity and promoted endogenous H_2S production, thereby increasing the number of lateral roots. However, treatment with HT counteracted the effect of CA on endogenous H_2S and lateral roots [24].

2.5. *The Genes Involved in Hydrogen Sulfide-Mediated Root Development*

Root system architecture is continuously adjusted in response to changes in various endogenous and exogenous factors (such as plant hormones, light, nutrition, and water). The regulation of these factors on root development involves many genes, including genes related to auxin synthesis, transport, and response, and genes related to cytokinin, abscisic acid (ABA), nitrate sensing and transport, and photoreceptors. The roles of these genes in plant growth and development were reviewed by Satbhai et al. (2015) [60]. In addition, many miRNAs are also involved in root development and architecture [2]. It is therefore important for researchers to have a clear understanding of which genes are involved in H_2S signaling, and hence the regulation of root development. We have carried out a detailed discussion and summary of gene regulation below.

2.5.1. Genes Associated with the Auxin Signaling Pathway

The RNA-seq results for peach roots showed that 963 and 1113 DEGs were detected after H_2S treatments for 1 day and 5 days, respectively [33]. These DEGs were significantly enriched in the "Glutathione metabolism", "Plant-pathogen interaction", "Plant hormone signal transduction", "Brassinosteroid biosynthesis", and "Cyanoamino acid metabolism" pathways. In particular, the pathway for "Plant hormone signal transduction" was significantly enriched when treated with H_2S for 1 day and 5 days. A significant proportion (73.68%) of the genes associated with this pathway were related to auxin. More specifically, there were 2, 7, and 17 genes involved in auxin biosynthesis, transport, and signal transduction, respectively. These auxin-related genes included *UGT74B1*, *TAA1*, *PINs*, *ABCBs*, *ARFs*, *Aux/IAAs*, *GH3*, and *SAUR*. The auxin-synthesis-related gene *UGT74B1* was up-regulated 1.95-fold when subjected to the H_2S treatment. This might explain the H_2S-induced increase in the root auxin content [20,33]. *PINs* exhibited different expression patterns over time under the NaHS treatment. After treatment with NaHS, *PIN1* was up-regulated during 3 to 6 h and recovered to the control levels by 6 h, and the expression of *PIN2* and *PIN7* increased during 3 to 6 h, whereas it decreased in 12 or 24 h. On the contrary, the expression of *PIN4* decreased after being treated with NaHS for 3 to 12 h, but recovered by 24 h. Although H_2S had different effects on the expression of the *PIN* genes, its effect on the subcellular distribution of the PIN proteins was consistent. H_2S disrupted the polar distribution of the PIN proteins (PIN1, PIN2, PIN4, and PIN7) on the plasma membrane in the root epidermal cells, and a large amount of PIN::GFP signals were found to dissociate from the plasma membrane upon cytoplasmic entry. Therefore, H_2S inhibited auxin transport through its effect on the polarity distribution of PIN proteins, thus promoting the initiation of lateral roots [32]. It has been noted that the location of

PIN proteins on the membrane was affected by F-actin [61,62], while H_2S significantly reduced the occupancy rate of F-actin bundles in each cell. This led to the disappearance of thick actin cables [32]. This implied that the influence of H_2S on the distribution of PIN proteins depended on the actin cytoskeleton, which is directly controlled by different ABPs (actin-binding proteins) [63]. Therefore, the expression of *ABPs* (*CPA*, *CBP*, and *PRF3*) was found to be up-regulated by H_2S, whereas the effects of H_2S on the percentage occupancy of the F-actin bundles was partially removed in the *cpa*, *cbp*, and *prf3* mutants [32]. In addition, some auxin signal transduction genes were found to be regulated by H_2S during root development. *CsAux22D-like* and *CsAux22B-like* were up-regulated by H_2S during the formation of cucumber adventitious roots [22]. Hydrogen sulfide induced *miR390a* and *miR160*, and thus inhibited the expression of their target genes *ARF4* and *ARF16* in both tomato and *Arabidopsis* roots [25,26]. *AtGATA23* and *AtLBD16* were down-regulated in the *Atdes1* mutant compared to WT, whereas *AtGH3.1* and *AtIAA28* were up-regulated in the *Atdes1* mutant [26].

2.5.2. Genes Associated with Cell Proliferation

Cell proliferation is the basis for root growth and development, so the expression of cell-proliferation-related genes is very important during root growth. In the tomato root, H_2S up-regulated *SlCDKA;1*, *SlCYCA2;1*, and *AtCYCA2;3*, but down-regulated *SlKRP2* and *AtKRP2* [25,26]. These genes are involved in the cell cycle. *DNAJ-1*, a gene phase that specifically regulates the G2/M cell cycle, was significantly induced by H_2S in cucumber roots [21,22]. In addition, the expression of *CsCDC6* (a cell-division-related gene) also increased in response to the NaHS treatment [22]. Interestingly, these cell proliferation-related genes also responded to auxin, CO, and CH_4, which are closely related to the H_2S signaling pathway. From the results of the RNA-seq work on peach roots, researchers identified that three cyclin genes and thirteen cell wall formation and remodeling-related genes were regulated by H_2S [33]. All three cyclin genes (*LOC109950471*, *LOC18790988*, and *LOC18784990*) were up-regulated by H_2S. In contrast, the cell wall formation and remodeling-related genes showed different patterns of expression in response to the H_2S treatment [33].

2.5.3. Transcription Factors and Protein Kinases

Both transcription factors (TFs) and protein kinases are regulatory genes that mediate plant growth and development. Wu et al. (2021) [33] found that 36 transcription factors in peach roots were regulated by H_2S, including LBD, MYB, and the AP2/ERF family. The overexpression of the peach *PpLBD16*, which was induced by H_2S, significantly increased the number of lateral roots in *Arabidopsis*, whereas the *Arabidopsis* mutant *ldb16* and *ldb18* showed a decrease in the number of lateral roots [64]. These results strengthened our understanding of LBD-mediated lateral root growth. Interestingly, LBDs (such as *AtLBD16*, *AtLBD18*, and *AtLBD29*) have been shown to be directly regulated by *ARFs* when regulating the formation of lateral roots [65,66], which implies that H_2S may interact with the auxin signaling pathway to regulate the growth of lateral roots, partly dependent on *LBD* genes. In *Kandelia obovata* roots, other TFs were also found to respond to H_2S, such as trihelix transcription factor GT-3b (*GT-3B*), the zinc finger CCCH domain-containing protein 14 (*ZC3H14*), and the MADS-box transcription factor [27].

Previous studies have shown that several protein kinases respond to H_2S during root development. The calmodulin kinases *CsCDPK1* and *CsCDPK5* were up-regulated by H_2S in cucumber roots [21]. MPK6 was involved in H_2S-inhibited primary root growth. When subjected to the NaHS treatment, the root length of the mutant *mpk6* was significantly longer than that for WT. Moreover, MPK6 was shown to function downstream of H_2S-induced ROS and upstream of NO [35]. In addition, in peach roots, the DEGs in the H_2S treatment for five days were significantly enriched in the mitogen-activated protein kinase (MAPK) signaling pathway, relative to the control group [33]. These results suggested that *CDPK* and *MAPK* may play an important role in H_2S-regulated root development.

2.5.4. Genes Associated with Carbohydrate Metabolism

Hu et al. (2020) [34] reported that H_2S induced the accumulation of soluble sugar in strawberry roots during plug production. Subsequently, the transcriptome and proteome data showed that the H_2S-regulated genes in roots were significantly enriched in "Starch and sucrose metabolism" [27,33]. These data indicated that soluble sugar was either directly or indirectly involved with H_2S-regulated root development. The sucrose transport protein SUT13, bidirectional sugar transporter SWEET, and invertase (INV) were found to be up-regulated by H_2S in *Kandelia obovata* roots, which led researchers to speculate that H_2S may facilitate sucrose transport and promote the hydrolysis of sucrose to provide metabolites and energy for root growth.

3. The Role of Hydrogen Sulfide in Roots Exposed to Abiotic Stress

Abiotic stress often stimulates oxidative damage by generating ROS, which leads to the inhibition of plant growth and even death. Plant root growth is sensitive to abiotic stress factors in the soil, such as heavy metals (HMs), aluminum, salinity, and hypoxia. It has been reported that H_2S could alleviate the inhibitory effect of abiotic stress on root growth in many plants (Table 2). Here, we discussed and reviewed the role and mechanism of H_2S on root growth when exposed to an abiotic stress (Figure 2).

Table 2. Hydrogen sulfide promotes root growth and its regulation mechanism under abiotic stress.

Abiotic Stress	H_2S Action	Plant Species	Reference
Cadmium	H_2S improved oxidation resistance, and NO was involved in the NaHS-induced alleviation of Cd toxicity	*Medicago sativa*	[67]
	H_2S removed excessive ROS and reduced cell oxidative damage	*Brassica rapa*	[68]
	H_2S inhibited the ROS burst, and H_2S-Cys cycle system plays an important role in it	*Arabidopsis*	[69]
	H_2S mediated the phytotoxicity of Cd by regulating UPB1s-modulated balance between H_2O_2 and O_2^-	*Brassica rapa*	[70]
	H_2S relieved-Cd stress was involved in MeJA signal	*Setaria italica*	[71]
	H_2O_2 raised H_2S content in root tissues independently from the desulfhydrase activity, and protected V-ATPase	*Cucumis sativus*	[72]
	H_2S reduced Cd uptake/translocation and decreased MDA, H_2O_2, and O_2^- accumulation	*Hordeum vulgare*	[73]
	H_2S activated glutathione biosynthetic and AsA-GSH cycle enzymes, and maintained redox status of ascorbate and glutathione	*Solanum lycopersicum*	[74]
	H_2S inhibited Cd-induced cell death by reducing ROS accumulation, activating the antioxidant system, inhibiting mitochondrial Cyt c release, and reducing the opening of the MPTP	*Cucumis sativus*	[75]
	H_2S improved Cd tolerance by modulating growth biomarkers and antioxidative system	*Brassica rapa*	[76]
	H_2S operates downstream of CH_4, enhancing tolerance against Cd stress	*Medicago sativa*	[77]
Chromium	H_2S increased Cys accumulation by up-regulating the Cys generation-related genes, enhanced glutathione generation, and activated phytochelatins (PCs) synthesis	*Arabidopsis*	[78]
	H_2S improved the physiological and biochemical attributes of Cr-stressed plants, and decreased Cr content in different parts of Cr-stressed plants	*Brassica oleracea botrytis*	[79]
Aluminum	H_2S protected plants against Al toxicity by inducing the activities of antioxidant enzymes, increasing citrate secretion and citrate transporter gene expression, and enhancing the expression of PM H^+-ATPase.	*Hordeum vulgare*	[80]

Table 2. *Cont.*

Abiotic Stress	H_2S Action	Plant Species	Reference
	H_2S alleviated Al toxicity by decreasing the Al content in the apoplast and symplast	*Oryza sativa*	[81]
Lead	H_2S lowered the Pb concentration in roots, improved the cell structure, and presented the well-developed nucleus with continuous cell membrane	*Brassica napus*	[82]
	H_2S alleviated Pb toxicity by improvement of nitrate reductase activity and glutathione content and regulation of amino acids metabolism	*Zea mays*	[83]
Nickel	H_2S induced Ni tolerance that required the entry of extracellular Ca^{2+} into cells across the plasma membrane and the mediation of intracellular CaM	*Cucurbita pepo*	[84]
Salt	H_2S enhanced plant responses against salinity stress by reducing oxidative damage, which might have a possible interaction with NO	*Medicago sativa*	[85]
	H_2S increased salt tolerance by maintaining Na^+ and K^+ ion homeostasis, which was mediated by NO signal	*Hordeum vulgare*	[86]
	H_2S alleviated growth inhibition by maintaining a lower Na^+ concentration under NaCl stress via the regulation of NSCCs and SOS1 pathways	*Triticum aestivuml*	[87]
	H_2S up-regulated the Na^+/H^+ antiport system, which promoted exchange of Na^+ with H^+ across the PM and simultaneously restricted the channel-mediated K^+ loss	*Populus euphratica* and *Populus popularis*	[88]
	H_2S acts downstream of NO in the mitigation of NaCl-induced oxidative stress	*Solanum lycopersicum*	[89]
Hypoxia	H_2S protected root tip cell membranes from ROS damage induced by hypoxia, and stimulated a quiescence strategy through inhibiting ethylene production	*Vigna radiata*	[90]
	H_2S enhanced endogenous Ca^{2+} levels, as well as the Ca^{2+}-dependent activity of alcohol dehydrogenase (ADH), improved the capacity for antioxidant defense, and thus increased the NO-induced hypoxia tolerance in maize	*Zea mays*	[91]

3.1. Heavy Metals

Cadmium (Cd) is regarded as the most toxic of the heavy metals (HMs) for plants. The function of H_2S in plants subject to Cd-related stress has been extensively studied. Hydrogen sulfide could alleviate the Cd-induced inhibition of root growth in *Arabidopsis* [69], *Medicago sativa* [67,77], *Brassica rapa* [68,70,76], *Setaria italica* [71], *Cucumis sativus* [72,75], *Solanum lycopersicum* [74], and *Hordeum vulgare* [73]. Furthermore, the suppression of plant root growth caused by other heavy metals, such as chromium (Cr), lead (Pb), mercury (Hg), and nickel (Ni), could also be relieved when exposed to H_2S [73,78,79,82–84,92,93].

When plants are exposed to HM-related stress, they first reduce the absorption of the HMs, or translocate the HMs to vacuoles to reduce oxidative damage to cells. As expected, the alleviation of the HM-related stress by H_2S is partly dependent on these pathways. The Cd content in NaHS-pretreated root tissues was 33–37% lower than that for untreated Cd-stressed plant samples [67]. Pretreatment with NaHS markedly reduced the Pb content in maize roots [83]. The Ni content declined in the NaHS+Ni treatment of zucchini roots in comparison to the alone treatment of Ni-stressed [84]. Hydrogen sulfide also had a significant inhibitory effect on the absorption of Cr in cauliflower roots, stems, leaves, and flowers [79].

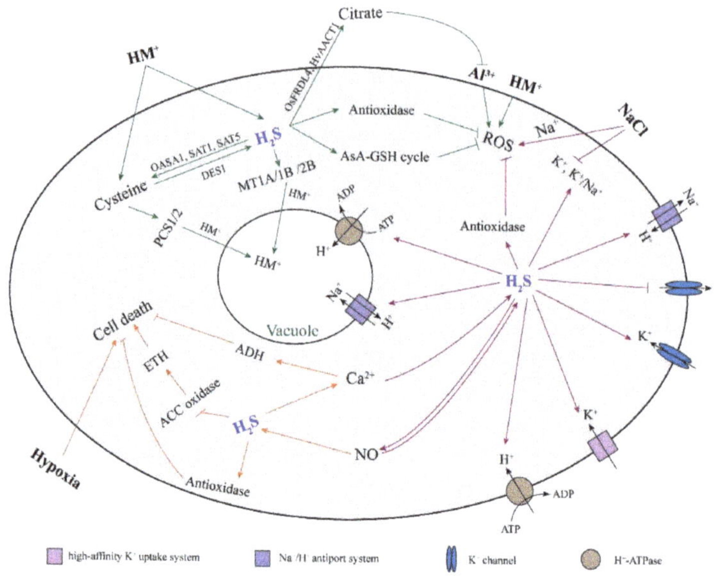

Figure 2. The proposed model of H_2S alleviating plant root cell damage under abiotic stress. Arrow and bar ends indicate activation and inhibitory effects, respectively. The green, orange, and purple arrows refer to the response of H_2S in plant roots to heavy metals, hypoxia, and salt stress, respectively. H_2S: hydrogen sulfide; HM: heavy metal; ROS: reactive oxygen species; NO: nitric oxide; AsA-GSH cycle: ascorbate–glutathione cycle; ETH: ethylene; ADH: alcohol dehydrogenase; DES1: L-cysteine desulfhydrase 1; OASA1: o-acetylserine lyase isoform A1; SAT1: serine acetyltransferase 1; SAT5: serine acetyltransferase 5; PCS1/5: phytochelatin synthase 1/5; MT1A/1B/2B: metallothionein 1A/1B/2B; FRDL4: citrate efflux transporter; AACT1: citrate transporter.

One of the most significant effects of HMs on plants is the production of large amounts of ROS, which, in turn, leads to oxidative damage to cells. The role of H_2S in oxidative stress has been a major research focus for many years. Not surprisingly, the antioxidant function of H_2S plays an important role in the response of plants to exposure to HM stress. When subjected to Cd-stress, treatment with NaHS reduced the accumulation of ROS and lipid peroxidation in *Brassica rapa* and barley [68,73]. The antioxidant function of H_2S also had an effective response to Cr, Pb, and Hg-induced stress [78,79,82,83,93]. Generally, H_2S reduced oxidative damage to cells, mainly by inducing antioxidant-related enzyme activity. The NaHS pretreatment significantly increased the activity of SOD (superoxide dismutase), APX (ascorbate peroxidase), CAT (catalase), and POD (peroxidase) in Cr-stressed cauliflower [79]. In addition, the positive effect of H_2S on anti-oxidation was attributed to glutathione (GSH) homeostasis. The treatment with Cd significantly decreased the content of GSH and homoglutathione (hGSH) and increased the content of GSSG (oxidized GSH) and hGSSGh (oxidized hGSH), whereas the decreased ratio of hGSH/hGSSGh and GSH/GSSG in alfalfa seedlings was obviously inhibited by H_2S [77]. In the presence of HT, the activities of the AsA-GSH-cycle-related enzymes (ascorbate peroxidase APX, monodehydroascorbate reductase MDHAR, dehydroascorbate reductase DHAR, and glutathione reductase GR) were reduced compared to the activities in the untreated Cd-stressed group, and thus the ratio of AsA/DHA (ascorbate/ dehydroascorbate) and GSH/GSSG declined [74].

In plants, H_2S is mainly produced from cysteine desulfhydrylase (CDes) catalyzing the degradation of cysteine (Cys). Like H_2S, when exposed to HM stress, the endogenous Cys content increased in *Arabidopsis* roots. The addition of exogenous Cys can significantly alleviate the inhibitory effect of HMs on primary growth. Cys synthesis could be induced by H_2S by up-regulating the expression of the Cys-generation-related genes *OASTLa*, *SAT1*, and *SAT5*. Subsequently, Cys promoted GSH accumulation and induced the expression of phytochelatin (PC) genes (*PCS1* and *PCS2*) counteracting Cd^{2+} and Cr^{6+} toxicity, and H_2S could up-regulate the metallothionein gene (*MT1A*, *MT1B*, *MT2B*) to alleviate the Cd^{2+} and Cr^{6+} toxicity [69,78]. These results indicated that the H_2S-Cys cycle system played a key role in plant responses to HM-related stress.

Hydrogen sulfide interacted with other signaling molecules to regulate heavy-metal-induced oxidative damage to cells. For example, the accumulation of NO was enhanced by NaHS treatment during exposure to Cd-induced stress. Hydrogen sulfide reduced Cd-induced oxidative damage in alfalfa roots, which was reversed by cPTIO [67]. MeJA enhanced Cd tolerance and alleviated growth inhibition in foxtail millet, whereas these effects were weakened by HT [71]. Similarly, the effects of CH_4 on redox imbalance and cell death in alfalfa roots subjected to Cd stress was dependent on the induction of H_2S metabolism. Treatment with either HT or PAG (propargylglycine, a H_2S biosynthesis inhibitor) reduced the alleviating effects of CH_4 on Cd-stressed plants [77].

3.2. Aluminum

Like heavy metal stress, excessive aluminum (Al) can also cause a large amount of ROS production in cells, resulting in oxidative damage and even cell death. Several studies have reported that H_2S can weaken the inhibitory effect of aluminum toxicity on plant root growth by inhibiting Al^{3+} absorption and enhancing the antioxidant system. For example, both in barley and rice, the Al content in the leaves and roots of Al-stressed plants treated with NaHS was much lower than for untreated plants [80,81,94]. Further, the inhibitory effect of H_2S on the absorption of Al^{3+} may be related to an increase in the secretion of citrate. In rice, the expression of *OsFRDL4*, a gene that regulates the efflux of citrate, was significantly up-regulated by NaHS treatments in Al-stressed conditions, and a simultaneous increase in citrate secretion from roots was found in the NaHS-pretreated group compared with the untreated Al-stressed plants [81]. Similarly, research on barley has shown that H_2S could promote citrate secretion in roots and could up-regulate the expression of the citrate transporter gene (*HvAACT1*) when the plants were subjected to Al stress [80]. The increase in the rate of citrate secretion reduced the deposition of Al on the surface of roots. Therefore, the promotion of citrate secretion by H_2S could lead to a reduction in the Al content of roots. Moreover, H_2S-induced antioxidant-related enzyme activity also contributed to the mitigation of aluminum toxicity. The NaHS pretreatment significantly increased SOD, APX, CAT, and POD activity in Al-stressed rice [81]. Similarly, H_2S was found to enhance SOD, CAT, POD, and GR activity in roots under Al stress in barley [94].

3.3. Salinity

According to previous studies, the inhibition of root growth by salinity (NaCl) stress could be attributed (at least in part) to a decrease in K^+ concentrations and the K^+/Na^+ ratio in the cytoplasm. This would have disrupted ion homeostasis and hence caused cell death [95]. It displayed a net K^+ efflux after exposure to NaCl, whereas NaHS could restrict the NaCl-induced K^+ efflux in both salt-tolerant or salt-sensitive grape roots. Furthermore, H_2S promoted Na^+ efflux and the influx of H^+ by up-regulating the Na^+/H^+ antiport system to maintain the plasma membrane (PM) polarity, thereby reducing the K^+ loss by inhibiting PM depolarization-activated K^+ channels [88]. K^+ and Na^+ homeostasis was an important adaptation by plants to salt stress. Researchers found that H_2S significantly reduced the Na^+ content and Na^+/K^+ ratio in wheat roots. The H_2S facilitated the exclusion of Na^+ and absorption of K^+ by regulating selective absorption and transport of K^+ over Na^+ [87]. The content of H_2S in a *Brassica napus* hybrid was more than that of the two parents. When exposed to salt stress, the expression of *NHX1* (Na^+/H^+ antiporter), *AKT1* (inward-rectifying potassium channel), and *HAK5* (potassium transporter) was significantly higher in the hybrid, in which, the Na^+ content and Na^+/K^+ ratio was reduced, and the K^+ content increased. The hybrid, therefore, had a higher salt tolerance than the parents.

However, these beneficial effects in the hybrid were eliminated by HT and PAG [96]. These results indicated that H_2S improved the salt tolerance of plants by maintaining Na^+ and K^+ homeostasis. Other studies have shown that the regulation of Na^+ and K^+ homeostasis in salt-stressed plants by H_2S involved the Ca^{2+} and NO signal pathways. Ca^{2+} and H_2S had a synergistic effect on the induction of the Na^+/H^+ antiport system in mung bean roots. In contrast, the HT treatment negated the beneficial effects of Ca^{2+} on salt stress. Furthermore, a supplementation of Ca^{2+} could enhance H_2S biosynthesis through promoting a cysteine pool. This implied the downstream functioning of H_2S during the Ca^{2+}-mediated regulation of plant adaptive responses to NaCl stress [97]. Both NO and H_2S could increase the K^+/Na^+ ratio in alfalfa roots, whereas the treatment with cPTIO reduced the H_2S-induced K^+/Na^+ ratio and antioxidant capacity of H_2S [85]. When barley roots were exposed to salt stress, H_2S could decrease the net K^+ efflux by increasing the transcriptional expression of *HvAKT1* (inward-rectifying potassium channel) and *HvHAK4* (a high-affinity K^+ uptake system), promote Na^+ export by increasing the expression of PM H^+-ATPase (*HvHA1*) and Na^+/H^+ antiporter (*HvSOS1*), and transfer excess Na^+ into vacuoles by increasing the gene expression of vacuolar Na^+/H^+ antiporter (*HvVNHX2*), H^+-ATPase subunit β (*HvVHA-β*), and the accumulating vacuolar Na^+/H^+ antiporter (NHE1) protein. However, these effects induced by H_2S were quenched by the addition of cPTIO [86]. These results mean that the H_2S is upstream of NO in order to maintain ion homeostasis and improve salt tolerance. However, Da Silva et al. (2018) [89] proposed that H_2S may act downstream of NO in the mitigation of salt-induced oxidative stress. Researchers found that, after treatment with NaCl, the accumulation of H_2S in tomato roots occurred later than the accumulation of NO, and that NO could increase the expression of the H_2S synthesis gene (*L-DES*) and H_2S production, whereas H_2S could not induce the accumulation of NO. H_2S and NO have shown complex interactions when regulating other physiological processes [98]. Therefore, the relationship between H_2S and NO in plant root growth regulation under salt stress needs more research and discussion.

3.4. Hypoxia

Hypoxia leads to root cell death. However, H_2S could reduce the rate of root tip cell death by inducing antioxidant enzyme activity and by inhibiting ACC oxidase (ACO) activity and ethylene production [90]. H_2S also promoted endogenous Ca^{2+} accumulation and the Ca^{2+}-dependent activity of alcohol dehydrogenase (ADH). It therefore improved the antioxidant defensive capabilities of the plants, and thus increased the rate of maize root tip cell survival in hypoxic conditions [91]. Subsequent studies have shown that the regulation of root tip cell death by H_2S mediated the NO signal pathway. The NO-induced tolerance of hypoxia was enhanced by the application of NaHS, but was eliminated by HT [91].

4. Conclusions and Future Prospects

The effects of H_2S on plant root growth and development have been widely recognized. In this review, we summarized the regulatory effects of H_2S on lateral roots, adventitious roots, primary roots, root hairs, and root nodules. The mitigation effect of H_2S on root growth under abiotic stress was also discussed here. Hydrogen sulfide interacts with a variety of other signals to regulate root growth. These signals mainly included auxin, NO, CO, ROS, and CH_4. In addition, there are many genes involved in H_2S-regulated root growth. However, there are still many issues that need to be clarified to explain how H_2S regulates root growth. For example, H_2S interacts with other signal molecules to regulate root growth, so finding the key genes that connect H_2S and other signal molecules is crucial for understanding the complex interactions between H_2S and signal molecules. Previous studies have found that many genes contribute to the regulation of H_2S during root growth and development. A great number of genes involved in the regulation of H_2S on root systems were identified through transcriptome and proteome, but the involvement of these genes was based on the effects of H_2S on their expression. The importance of these genes to the H_2S-regulated root growth pathway requires further functional verification. Finally, in recent years, studies have found that H_2S could directly regulate the S-sulfhydration of proteins by converting Cys-SH to Cys-SSH. This affected the activity of proteins, and, thus, mediated plant growth and development and responses to stresses [99–103]. The ACTIN2 protein, associated with the development of root hair, has been found to be S-sulfhydrated at Cys-287 by H_2S, thereby mediating H_2S-regulated root hair growth [29]. This implied that there might be more proteins involved in root development that are S-sulfhydrated by H_2S that still need to be identified.

Author Contributions: H.L. and C.W. conceived and planned this review paper. H.L. and H.C. prepared and drafted the manuscript. L.C. and C.W. revised the manuscript. All authors have read and agreed to the published version of the manuscript.

Funding: This work was supported by grants from the State Key Laboratory of Crop Biology Open Fund (2021KF02) and Natural Science Foundation of Henan Province (212300410352).

Institutional Review Board Statement: Not applicable.

Informed Consent Statement: Not applicable.

Data Availability Statement: Not applicable.

Conflicts of Interest: The authors declare no conflict of interest.

References

1. Motte, H.; Vanneste, S.; Beeckman, T. Molecular and environmental regulation of root development. *Annu. Rev. Plant Biol.* **2019**, *70*, 465–488. [CrossRef] [PubMed]
2. Khan, G.A.; Declerck, M.; Sorin, C.; Hartmann, C.; Crespi, M.; Lelandais-Brière, C. MicroRNAs as regulators of root development and architecture. *Plant Mol. Biol.* **2011**, *77*, 47–58. [CrossRef]
3. Karlova, R.; Boer, D.; Hayes, S.; Testerink, C. Root plasticity under abiotic stress. *Plant Physiol.* **2021**, *187*, 1057–1070. [CrossRef]
4. Balliu, A.; Zheng, Y.; Sallaku, G.; Fernández, J.A.; Gruda, N.S.; Tuzel, Y. Environmental and cultivation factors affect the morphology, architecture and performance of root systems in soilless grown plants. *Horticulturae* **2021**, *7*, 243. [CrossRef]
5. Bouain, N.; Krouk, G.; Lacombe, B.; Rouached, H. Getting to the root of plant mineral nutrition: Combinatorial nutrient stresses reveal emergent properties. *Trends Plant Sci.* **2019**, *24*, 542–552. [CrossRef] [PubMed]
6. Bao, Y.; Aggarwal, P.; Robbins, N.E.; Sturrock, C.J.; Thompson, M.C.; Tan, H.Q.; Tham, C.; Duan, L.; Rodriguez, P.L.; Vernoux, T.; et al. Plant roots use a patterning mechanism to position lateral root branches toward available water. *PNAS* **2014**, *111*, 9319–9324. [CrossRef]
7. Robbins, N.E.; Dinneny, J.R. The divining root: Moisture-driven responses of roots at the micro-and macro-scale. *J. Exp. Bot.* **2015**, *66*, 2145–2154. [CrossRef] [PubMed]
8. Demecsová, L.; Tamás, L. Reactive oxygen species, auxin and nitric oxide in metal-stressed roots: Toxicity or defence. *Biometals* **2019**, *32*, 717–744. [CrossRef]
9. Xu, P.; Zhao, P.X.; Cai, X.T.; Mao, J.L.; Miao, Z.Q.; Xiang, C.B. Integration of jasmonic acid and ethylene into auxin signaling in root development. *Front. Plant Sci.* **2020**, *11*, 271. [CrossRef]
10. Hebelstrup, K.H.; Shah, J.K.; Igamberdiev, A.U. The role of nitric oxide and hemoglobin in plant development and morphogenesis. *Physiol. Plant.* **2013**, *148*, 457–469. [CrossRef]
11. Lindsey, K.; Rowe, J.; Liu, J. Hormonal crosstalk for root development: A combined experimental and modeling perspective. *Front. Plant Sci.* **2014**, *5*, 116.
12. Corpas, F.J. Barroso, J.B. Functions of nitric oxide (NO) in roots during development and under adverse stress conditions. *Plants* **2015**, *4*, 240–252. [CrossRef] [PubMed]
13. Pacifici, E.; Polverari, L.; Sabatini, S. Plant hormone cross-talk: The pivot of root growth. *J. Exp. Bot.* **2015**, *66*, 1113–1121. [CrossRef] [PubMed]
14. Tsukagoshi, H. Control of root growth and development by reactive oxygen species. *Curr. Opin Plant Biol.* **2016**, *29*, 57–63. [CrossRef]
15. Zhang, J.; Zhou, M.; Zhou, H.; Zhao, D.; Gotor, C.; Romero, L.C.; Shen, J.; Ge, Z.; Zhang, Z.; Shen, W.; et al. Hydrogen sulfide, a signaling molecule in plant stress responses. *J. Integr. Plant Biol.* **2021**, *63*, 146–160. [CrossRef]
16. Pandey, A.K.; Gautam, A. Stress responsive gene regulation in relation to hydrogen sulfide in plants under abiotic stress. *Physiol. Plant.* **2020**, *168*, 511–525. [CrossRef]
17. Huang, D.; Huo, J.; Liao, W. Hydrogen sulfide: Roles in plant abiotic stress response and crosstalk with other signals. *Plant Sci.* **2021**, *302*, 110733. [CrossRef] [PubMed]
18. Arif, Y.; Hayat, S.; Yusuf, M.; Bajguz, A. Hydrogen sulfide: A versatile gaseous molecule in plants. *Plant Physiol. Bioch.* **2021**, *158*, 372–384. [CrossRef]
19. Xuan, L.; Li, J.; Wang, X.; Wang, C. Crosstalk between hydrogen sulfide and other signal molecules regulates plant growth and development. *Int. J. Mol. Sci.* **2020**, *21*, 4593. [CrossRef]

20. Zhang, H.; Tang, J.; Liu, X.P.; Wang, Y.; Yu, W.; Peng, W.Y.; Fang, F.; Ma, D.F.; Wei, Z.J.; Hu, L.Y. Hydrogen sulfide promotes root organogenesis in *Ipomoea batatas, Salix matsudana* and *Glycine max*. *J. Integr. Plant Biol.* **2009**, *51*, 1086–1094. [CrossRef] [PubMed]
21. Lin, Y.T.; Li, M.Y.; Cui, W.T.; Lu, W.; Shen, W.B. Haem oxygenase-1 is involved in hydrogen sulfide-induced cucumber adventitious root formation. *J. Plant Growth Regul.* **2012**, *31*, 519–528. [CrossRef]
22. Kou, N.; Xiang, Z.; Cui, W.; Li, L.; Shen, W. Hydrogen sulfide acts downstream of methane to induce cucumber adventitious root development. *J. Plant Physiol.* **2018**, *228*, 113–120. [CrossRef]
23. Fang, T.; Cao, Z.; Li, J.; Shen, W.; Huang, L. Auxin-induced hydrogen sulfide generation is involved in lateral root formation in tomato. *Plant Physiol. Bioch.* **2014**, *76*, 44–51. [CrossRef]
24. Xue, Y.F.; Zhang, M.; Qi, Z.Q.; Li, Y.Q.; Shi, Z.; Chen, J. Cinnamaldehyde promotes root branching by regulating endogenous hydrogen sulfide. *J. Sci. Food Agric.* **2016**, *96*, 909–914. [CrossRef]
25. Mei, Y.; Chen, H.; Shen, W.; Shen, W.; Huang, L. Hydrogen peroxide is involved in hydrogen sulfide-induced lateral root formation in tomato seedlings. *BMC Plant Biol.* **2017**, *17*, 1–12. [CrossRef] [PubMed]
26. Mei, Y.; Zhao, Y.; Jin, X.; Wang, R.; Xu, N.; Hu, J.; Huang, L.Q.; Guan, R.Z.; Shen, W. L-Cysteine desulfhydrase-dependent hydrogen sulfide is required for methane-induced lateral root formation. *Plant Mol. Biol.* **2019**, *99*, 283–298. [CrossRef]
27. Li, H.; Ghoto, K.; Wei, M.Y.; Gao, C.H.; Liu, Y.L.; Ma, D.N.; Zheng, H.L. Unraveling hydrogen sulfide-promoted lateral root development and growth in mangrove plant Kandelia obovata: Insight into regulatory mechanism by TMT-based quantitative proteomic approaches. *Tree Physiol.* **2021**, *41*, 1749–1766. [CrossRef]
28. Zhang, P.; Luo, Q.; Wang, R.; Xu, J. Hydrogen sulfide toxicity inhibits primary root growth through the ROS-NO pathway. *Sci Rep.* **2017**, *7*, 868. [CrossRef] [PubMed]
29. Li, J.; Chen, S.; Wang, X.; Shi, C.; Liu, H.; Yang, J.; Shi, W.; Guo, J.; Jia, H. Hydrogen sulfide disturbs actin polymerization via S-sulfhydration resulting in stunted root hair growth. *Plant Physiol.* **2018**, *178*, 936–949. [CrossRef]
30. Zou, H.; Zhang, N.N.; Pan, Q.; Zhang, J.H.; Chen, J.; Wei, G.H. Hydrogen sulfide promotes nodulation and nitrogen fixation in soybean–rhizobia symbiotic system. *Mol. Plant Microbe. Interact.* **2019**, *32*, 972–985. [CrossRef] [PubMed]
31. Zou, H.; Zhang, N.N.; Lin, X.Y.; Zhang, W.Q.; Zhang, J.H.; Chen, J.; Wei, G.H. Hydrogen sulfide is a crucial element of the antioxidant defense system in *Glycine max–Sinorhizobium fredii* symbiotic root nodules. *Plant Soil.* **2020**, *449*. [CrossRef]
32. Jia, H.; Hu, Y.; Fan, T.; Li, J. Hydrogen sulfide modulates actin-dependent auxin transport via regulating ABPs results in changing of root development in *Arabidopsis*. *Sci. Rep.* **2015**, *5*, 1–13. [CrossRef]
33. Wu, X.; Du, A.; Zhang, S.; Wang, W.; Liang, J.; Peng, F.; Xiao, Y. Regulation of growth in peach roots by exogenous hydrogen sulfide based on RNA-Seq. *Plant Physiol. Bioch.* **2021**, *159*, 179–192. [CrossRef]
34. Hu, J.; Li, Y.; Liu, Y.; Kang, D.I.; Wei, H.; Jeong, B.R. Hydrogen sulfide affects the root development of strawberry during plug transplant production. *Agriculture.* **2020**, *10*, 12. [CrossRef]
35. Zhang, X.; Zhang, Y.; Zhang, L.; Zhao, H.; Li, H. Hydrogen sulphide improves iron homeostasis in wheat under iron-deficiency. *Plant Sci.* **2017**, *5*, 170.
36. Bensmihen, S. Hormonal control of lateral root and nodule development in legumes. *Plants* **2015**, *4*, 523–547. [CrossRef] [PubMed]
37. Dunand, C.; Crèvecoeur, M.; Penel, C. Distribution of superoxide and hydrogen peroxide in Arabidopsis root and their influence on root development: Possible interaction with peroxidases. *New Phytol.* **2007**, *174*, 332–341. [CrossRef] [PubMed]
38. Ivanchenko, M.G.; Den Os, D.; Monshausen, G.B.; Dubrovsky, J.G.; Bednářová, A.; Krishnan, N. Auxin increases the hydrogen peroxide (H_2O_2) concentration in tomato (*Solanum lycopersicum*) root tips while inhibiting root growth. *Ann. Bot.* **2013**, *112*, 1107–1116. [CrossRef] [PubMed]
39. Liszkay, A.; van der Zalm, E.; Schopfer, P. Production of reactive oxygen intermediates (O^{2-}, H_2O_2, and OH) by maize roots and their role in wall loosening and elongation growth. *Plant physiol.* **2004**, *136*, 3114–3123. [CrossRef]
40. Voothuluru, P.; Mäkelä, P.; Zhu, J.; Yamaguchi, M.; Cho, I.J.; Oliver, M.J.; Simmonds, J.; Sharp, R.E. Apoplastic hydrogen peroxide in the growth zone of the maize primary root. Increased levels differentially modulate root elongation under well-watered and water-stressed conditions. *Front. Plant Sci.* **2020**, *11*, 392. [CrossRef] [PubMed]
41. Deng, X.P.; Cheng, Y.J.; Wu, X.B.; Kwak, S.S.; Chen, W.; Eneji, A.E. Exogenous hydrogen peroxide positively influences root growth and exogenous hydrogen peroxide positively influences root growth and metabolism in leaves of sweet potato seedlings. *Aust. J. Crop. Sci.* **2012**, *6*, 1572.
42. Böhm, F.M.L.Z.; Ferrarese, M.D.L.L.; Zanardo, D.I.L.; Magalhaes, J.R.; Ferrarese-Filho, O. Nitric oxide affecting root growth, lignification and related enzymes in soybean seedlings. *Acta Physiol. Plant.* **2010**, *32*, 1039–1046. [CrossRef]
43. Chen, Y.H.; Chao, Y.Y.; Hsu, Y.Y.; Hong, C.Y.; Kao, C.H. Heme oxygenase is involved in nitric oxide-and auxin-induced lateral root formation in rice. *Plant Cell Rep.* **2012**, *31*, 1085–1091. [CrossRef]
44. Fernández-Marcos, M.; Sanz, L.; Lorenzo, O. Nitric oxide: An emerging regulator of cell elongation during primary root growth. *Plant Signal. Behav.* **2012**, *7*, 196–200. [CrossRef] [PubMed]
45. Lin, Y.; Zhang, W.; Qi, F.; Cui, W.; Xie, Y.; Shen, W. Hydrogen-rich water regulates cucumber adventitious root development in a heme oxygenase-1/carbon monoxide-dependent manner. *J. Plant Physiol.* **2014**, *171*, 1–8. [CrossRef] [PubMed]
46. Sanz, L.; Albertos, P.; Mateos, I.; Sánchez-Vicente, I.; Lechón, T.; Fernández-Marcos, M.; Lorenzo, O. Nitric oxide (NO) and phytohormones crosstalk during early plant development. *J. Exp. Bot.* **2015**, *66*, 2857–2868. [CrossRef]
47. Chen, Y.; Wang, M.; Hu, L.; Liao, W.; Dawuda, M.M.; Li, C. Carbon monoxide is involved in hydrogen gas-induced adventitious root development in cucumber under simulated drought stress. *Front. Plant Sci.* **2017**, *8*, 128. [CrossRef]

48. Hacham, Y.; Holland, N.; Butterfield, C.; Ubeda-Tomas, S.; Bennett, M.J.; Chory, J.; Savaldi-Goldstein, S. Brassinosteroid perception in the epidermis controls root meristem size. *Development* **2011**, *138*, 839–848. [CrossRef] [PubMed]
49. Gupta, A.; Singh, M.; Laxmi, A. Interaction between glucose and brassinosteroid during the regulation of lateral root development in *Arabidopsis*. *Plant Physiol.* **2015**, *168*, 307–320. [CrossRef]
50. Wei, Z.; Li, J. Brassinosteroids regulate root growth, development, and symbiosis. *Mol. Plant* **2016**, *9*, 86–100. [CrossRef]
51. Kang, Y.H.; Breda, A.; Hardtke, C.S. Brassinosteroid signaling directs formative cell divisions and protophloem differentiation in *Arabidopsis* root meristems. *Development* **2017**, *144*, 272–280. [CrossRef] [PubMed]
52. Li, L.; Wei, S.; Shen, W. The role of methane in plant physiology: A review. *Plant Cell Rep.* **2020**, *39*, 171–179. [CrossRef]
53. Cui, W.; Qi, F.; Zhang, Y.; Cao, H.; Zhang, J.; Wang, R.; Shen, W. Methane-rich water induces cucumber adventitious rooting through heme oxygenase1/carbon monoxide and Ca^{2+} pathways. *Plant Cell Rep.* **2015**, *34*, 435–445. [CrossRef]
54. Qi, F.; Xiang, Z.; Kou, N.; Cui, W.; Xu, D.; Wang, R.; Zhu, D.; Shen, W. Nitric oxide is involved in methane-induced adventitious root formation in cucumber. *Physiol. Plant.* **2017**, *159*, 366–377. [CrossRef] [PubMed]
55. Jiang, X.; He, J.; Cheng, P.; Xiang, Z.; Zhou, H.; Wang, R.; Shen, W. Methane control of adventitious rooting requires γ-glutamyl cysteine synthetase-mediated glutathione homeostasis. *Plant Cell Physiol.* **2019**, *60*, 802–815. [CrossRef]
56. Zhao, Y.; Zhang, Y.; Liu, F.; Wang, R.; Huang, L.; Shen, W. Hydrogen peroxide is involved in methane-induced tomato lateral root formation. *Plant Cell Rep.* **2019**, *38*, 377–389. [CrossRef]
57. Vergis, J.; Gokulakrishnan, P.; Agarwal, R.K.; Kumar, A. Essential oils as natural food antimicrobial agents: A review. *Crit Rev. Food Sci. Nutr.* **2015**, *55*, 1320–1323. [CrossRef] [PubMed]
58. Song, Y.R.; Choi, M.S.; Choi, G.W.; Park, I.K.; Oh, C.S. Antibacterial activity of cinnamaldehyde and estragole extracted from plant essential oils against Pseudomonas syringae pv. actinidiae causing bacterial canker disease in kiwifruit. *Plant Pathol. J.* **2016**, *32*, 363. [CrossRef]
59. Jardim, I.N.; Oliveira, D.F.; Silva, G.H.; Campos, V.P.; de Souza, P.E. (E)-cinnamaldehyde from the essential oil of Cinnamomum cassia controls Meloidogyne incognita in soybean plants. *J. Pest. Sci.* **2018**, *91*, 479–487. [CrossRef]
60. Satbhai, S.B.; Ristova, D.; Busch, W. Underground tuning: Quantitative regulation of root growth. *J. Exp. Bot.* **2015**, *66*, 1099–1112. [CrossRef]
61. Muday, G.K.; Murphy, A.S. An emerging model of auxin transport regulation. *Plant Cell.* **2002**, *14*, 293–299. [CrossRef]
62. Dhonukshe, P.; Grigoriev, I.; Fischer, R.; Tominaga, M.; Robinson, D.G.; Hašek, J.; Paciorek, T.; Petrásek, J.; Seifertová, D.; Tejos, R.; et al. Auxin transport inhibitors impair vesicle motility and actin cytoskeleton dynamics in diverse eukaryotes. *Proc. Natl Acad Sci. USA* **2008**, *105*, 4489–4494. [CrossRef]
63. Pollard, T.D.; Cooper, J.A. Actin, a central player in cell shape and movement. *Science* **2009**, *326*, 1208–1212. [CrossRef]
64. Lee, H.W.; Kim, N.Y.; Lee, D.J.; Kim, J. LBD18/ASL20 regulates lateral root formation in combination with LBD16/ASL18 downstream of ARF7 and ARF19 in *Arabidopsis*. *Plant Physiol.* **2009**, *151*, 1377–1389. [CrossRef] [PubMed]
65. Okushima, Y.; Fukaki, H.; Onoda, M.; Theologis, A.; Tasaka, M. ARF7 and ARF19 regulate lateral root formation via direct activation of LBD/ASL genes in *Arabidopsis*. *Plant Cell.* **2007**, *19*, 118–130. [CrossRef]
66. Porco, S.; Larrieu, A.; Du, Y.; Gaudinier, A.; Goh, T.; Swarup, K.; Swarup, R.; Kuempers, B.; Bishopp, A.; Lavenus, J.; et al. Lateral root emergence in *Arabidopsis* is dependent on transcription factor LBD29 regulation of auxin influx carrier LAX3. *Development* **2016**, *143*, 3340–3349.
67. Li, L.; Wang, Y.; Shen, W. Roles of hydrogen sulfide and nitric oxide in the alleviation of cadmium-induced oxidative damage in alfalfa seedling roots. *Biometals* **2012**, *25*, 617–631. [CrossRef]
68. Zhang, L.; Pei, Y.; Wang, H.; Jin, Z.; Liu, Z.; Qiao, Z.; Fang, H.; Zhang, Y. Hydrogen sulfide alleviates cadmium-induced cell death through restraining ROS accumulation in roots of *Brassica rapa* L. ssp. *pekinensis*. *Oxid Med. Cell Longev.* **2015**, *2015*, 804603.
69. Jia, H.; Wang, X.; Dou, Y.; Liu, D.; Si, W.; Fang, H.; Zhao, C.; Chen, S.; Xi, J.; Li, J. Hydrogen sulfide-cysteine cycle system enhances cadmium tolerance through alleviating cadmium-induced oxidative stress and ion toxicity in *Arabidopsis* roots. *Sci Rep.* **2016**, *6*, 39702. [CrossRef]
70. Lv, W.; Yang, L.; Xu, C.; Shi, Z.; Shao, J.; Xian, M.; Chen, J. Cadmium disrupts the balance between hydrogen peroxide and superoxide radical by regulating endogenous hydrogen sulfide in the root tip of *Brassica rapa*. *Front. Plant Sci.* **2017**, *8*, 232. [CrossRef]
71. Tian, B.; Zhang, Y.; Jin, Z.; Liu, Z.; Pei, Y. Role of hydrogen sulfide in the methyl jasmonate response to cadmium stress in foxtail millet. *Front. Biosci.* **2017**, *22*, 530–538.
72. Kabała, K.; Zboińska, M.; Głowiak, D.; Reda, M.; Jakubowska, D.; Janicka, M. Interaction between the signaling molecules hydrogen sulfide and hydrogen peroxide and their role in vacuolar H^+-ATPase regulation in cadmium-stressed cucumber roots. *Physiol. Plant.* **2019**, *166*, 688–704. [CrossRef]
73. Fu, M.M.; Dawood, M.; Wang, N.H.; Wu, F. Exogenous hydrogen sulfide reduces cadmium uptake and alleviates cadmium toxicity in barley. *Plant Growth Regul.* **2019**, *89*, 227–237. [CrossRef]
74. Alamri, S.; Kushwaha, B.K.; Singh, V.P.; Siddiqui, M.H. Dose dependent differential effects of toxic metal cadmium in tomato roots: Role of endogenous hydrogen sulfide. *Ecotoxicol. Environ. Saf.* **2020**, *203*, 110978. [CrossRef]
75. Luo, S.; Tang, Z.; Yu, J.; Liao, W.; Xie, J.; Lv, J.; Feng, Z.; Dawuda, M.M. Hydrogen sulfide negatively regulates cd-induced cell death in cucumber (*Cucumis sativus* L) root tip cells. *BMC Plant Biol.* **2020**, *20*, 480. [CrossRef]
76. Li, G.; Shah, A.A.; Khan, W.U.; Yasin, N.A.; Ahmad, A.; Abbas, M.; Ali, A.; Safdar, N. Hydrogen sulfide mitigates cadmium induced toxicity in *Brassica rapa* by modulating physiochemical attributes, osmolyte metabolism and antioxidative machinery. *Chemosphere* **2021**, *263*, 127999. [CrossRef]

77. Yang, X.; Kong, L.; Wang, Y.; Su, J.; Shen, W. Methane control of cadmium tolerance in alfalfa roots requires hydrogen sulfide. *Environ. Pollut.* **2021**, *284*, 117123. [CrossRef]
78. Fang, H.; Liu, Z.; Jin, Z.; Zhang, L.; Liu, D.; Pei, Y. An emphasis of hydrogen sulfide-cysteine cycle on enhancing the tolerance to chromium stress in *Arabidopsis*. *Environ. Pollut.* **2016**, *213*, 870–877. [CrossRef]
79. Ahmad, R.; Ali, S.; Rizwan, M.; Dawood, M.; Farid, M.; Hussain, A.; Wijaya, L.; Alyemeni, M.; Ahmad, P. Hydrogen sulfide alleviates chromium stress on cauliflower by restricting its uptake and enhancing antioxidative system. *Physiol. Plant.* **2020**, *168*, 289–300. [CrossRef]
80. Chen, J.; Wang, W.H.; Wu, F.H.; You, C.Y.; Liu, T.W.; Dong, X.J.; He, J.X.; Zheng, H.L. Hydrogen sulfide alleviates aluminum toxicity in barley seedlings. *Plant Soil.* **2013**, *362*, 301–318. [CrossRef]
81. Zhu, C.Q.; Zhang, J.H.; Sun, L.M.; Zhu, L.F.; Abliz, B.; Hu, W.J.; Zhong, C.; Bai, Z.G.; Sajid, H.; Cao, X.C.; et al. Hydrogen sulfide alleviates aluminum toxicity via decreasing apoplast and symplast Al contents in rice. *Front. Plant Sci.* **2018**, *9*, 294. [CrossRef]
82. Ali, B.; Mwamba, T.M.; Gill, R.A.; Yang, C.; Ali, S.; Daud, M.K.; Wu, Y.; Zhou, W. Improvement of element uptake and antioxidative defense in *Brassica napus* under lead stress by application of hydrogen sulfide. *Plant Growth Regul.* **2014**, *74*, 261–273. [CrossRef]
83. Zanganeh, R.; Jamei, R.; Rahmani, F. Role of salicylic acid and hydrogen sulfide in promoting lead stress tolerance and regulating free amino acid composition in *Zea mays* L. *Acta Physiol. Plant.* **2019**, *41*, 94. [CrossRef]
84. Valivand, M.; Amooaghaie, R.; Ahadi, A. Interplay between hydrogen sulfide and calcium/calmodulin enhances systemic acquired acclimation and antioxidative defense against nickel toxicity in zucchini. *Environ. Exp. Bot.* **2019**, *158*, 40–50. [CrossRef]
85. Wang, Y.; Li, L.; Cui, W.; Xu, S.; Shen, W.; Wang, R. Hydrogen sulfide enhances alfalfa (*Medicago sativa*) tolerance against salinity during seed germination by nitric oxide pathway. *Plant Soil* **2012**, *351*, 107–119. [CrossRef]
86. Chen, J.; Wang, W.H.; Wu, F.H.; He, E.M.; Liu, X.; Shangguan, Z.P.; Zheng, H.L. Hydrogen sulfide enhances salt tolerance through nitric oxide-mediated maintenance of ion homeostasis in barley seedling roots. *Sci. Rep.* **2015**, *5*, 12516. [CrossRef] [PubMed]
87. Deng, Y.Q.; Bao, J.; Yuan, F.; Liang, X.; Feng, Z.T.; Wang, B.S. Exogenous hydrogen sulfide alleviates salt stress in wheat seedlings by decreasing Na^+ content. *Plant Growth Regul.* **2016**, *79*, 391–399. [CrossRef]
88. Zhao, N.; Zhu, H.; Zhang, H.; Sun, J.; Zhou, J.; Deng, C.; Zhang, Y. Chen, S. Hydrogen sulfide mediates K^+ and Na^+ homeostasis in the roots of salt-resistant and salt-sensitive poplar species subjected to NaCl stress. *Front. Plant Sci.* **2018**, *9*, 1366. [CrossRef]
89. da-Silva, C.J.; Mollica, D.C.; Vicente, M.H.; Peres, L.E.; Modolo, L.V. NO, hydrogen sulfide does not come first during tomato response to high salinity. *Nitric Oxide* **2018**, *76*, 164–173. [CrossRef]
90. Cheng, W.; Zhang, L.; Jiao, C.; Su, M.; Yang, T.; Zhou, L.; Peng, R.; Wang, R.; Wang, C. Hydrogen sulfide alleviates hypoxia-induced root tip death in *Pisum sativum*. *Plant Physiol. Biochem.* **2013**, *70*, 278–286. [CrossRef]
91. Peng, R.; Bian, Z.; Zhou, L.; Cheng, W.; Hai, N.; Yang, C.; Yang, T.; Wang, X.; Wang, C. Hydrogen sulfide enhances nitric oxide-induced tolerance of hypoxia in maize (*Zea mays* L.). *Plant Cell Rep.* **2016**, *35*, 2325–2340. [CrossRef]
92. Ali, S.; Farooq, M.A.; Hussain, S.; Yasmeen, T.; Abbasi, G.H.; Zhang, G. Alleviation of chromium toxicity by hydrogen sulfide in barley. *Environ. Toxicol. Chem.* **2013**, *32*, 2234–2239. [CrossRef]
93. Chen, Z.; Chen, M.; Jiang, M. Hydrogen sulfide alleviates mercury toxicity by sequestering it in roots or regulating reactive oxygen species productions in rice seedlings. *Plant Physiol. Biochem.* **2017**, *111*, 179–192. [CrossRef]
94. Dawood, M.; Cao, F.; Jahangir, M.M.; Zhang, G.; Wu, F. Alleviation of aluminum toxicity by hydrogen sulfide is related to elevated ATPase, and suppressed aluminum uptake and oxidative stress in barley. *J. Hazard. Mater.* **2012**, *209*, 121–128. [CrossRef]
95. Basu, S.; Kumar, A.; Benazir, I.; Kumar, G. Reassessing the role of ion homeostasis for improving salinity tolerance in crop plants. *Physiol. Plant.* **2021**, *171*, 502–519. [CrossRef]
96. Cheng, P.; Zhang, Y.; Wang, J.; Guan, R.; Pu, H.; Shen, W. Importance of hydrogen sulfide as the molecular basis of heterosis in hybrid *Brassica napus*: A case study in salinity response. *Environ. Exp. Bot.* **2021**, *193*, 104693. [CrossRef]
97. Khan, M.N.; Siddiqui, M.H.; Mukherjee, S.; Alamri, S.; Al-Amri, A.A.; Alsubaie, Q.D.; Ai-Munqedhi, B.M.A.; Ali, H.M. Calcium-hydrogen sulfide crosstalk during K+-deficient NaCl stress operates through regulation of Na^+/H^+ antiport and antioxidative defense system in mung bean roots. *Plant Physiol. Biochem.* **2021**, *159*, 211–225. [CrossRef]
98. Mishra, V.; Singh, P.; Tripathi, D.K.; Corpas, F.J.; Singh, V.P. Nitric oxide and hydrogen sulfide: An indispensable combination for plant functioning. *Trends Plant Sci.* **2021**, *26*, 1270–1285. [CrossRef]
99. Jia, H.; Chen, S.; Liu, D.; Liesche, J.; Shi, C.; Wang, J.; Ren, M.; Wang, X.; Yang, J.; Shi, W.; et al. Ethylene-induced hydrogen sulfide negatively regulates ethylene biosynthesis by persulfidation of ACO in tomato under osmotic stress. *Front. Plant Sci.* **2018**, *9*, 1517. [CrossRef]
100. Shen, J.; Zhang, J.; Zhou, M.; Zhou, H.; Cui, B.; Gotor, C.; Romero, L.C.; Fu, L.; Yang, J.; Foyer, C.H.; et al. Persulfidation-based modification of cysteine desulfhydrase and the NADPH oxidase RBOHD controls guard cell abscisic acid signaling. *Plant Cell.* **2020**, *32*, 1000–1017. [CrossRef]
101. Chen, S.; Jia, H.; Wang, X.; Shi, C.; Wang, X.; Ma, P.; Wang, J.; Ren, M.; Li, J. Hydrogen sulfide positively regulates abscisic acid signaling through persulfidation of SnRK2.6 in guard cells. *Mol. Plant* **2020**, *13*, 732–744. [CrossRef]
102. Li, J.; Shi, C.; Wang, X.; Liu, C.; Ding, X.; Ma, P. Hydrogen sulfide regulates the activity of antioxidant enzymes through persulfidation and improves the resistance of tomato seedling to Copper Oxide nanoparticles (CuO NPs)-induced oxidative stress. *Plant Physiol. Biochem.* **2020**, *156*, 257–266. [CrossRef]
103. Aroca, A.; Zhang, J.; Xie, Y.; Romero, L.C.; Gotor, C. Hydrogen sulfide signaling in plant adaptations to adverse conditions: Molecular mechanisms. *J. Exp. Bot.* **2021**, *72*, 5893–5904. [CrossRef]

MDPI
St. Alban-Anlage 66
4052 Basel
Switzerland
Tel. +41 61 683 77 34
Fax +41 61 302 89 18
www.mdpi.com

International Journal of Molecular Sciences Editorial Office
E-mail: ijms@mdpi.com
www.mdpi.com/journal/ijms